Lecture Notes in Physics

Edited by J. Ehlers, München, K. Hepp, Zürich, and
H. A. Weidenmüller, Heidelberg
Managing Editor: W. Beiglböck, Heidelberg

39

International Symposium on Mathematical Problems in Theoretical Physics

January 23–29, 1975, Kyoto University, Kyoto/Japan

Edited by H. Araki

Springer-Verlag
Berlin · Heidelberg · New York 1975

Editor
Prof. Huzihiro Araki
Research Institute
for Mathematical Sciences
Kyoto University
Kyoto 606/Japan

Library of Congress Cataloging in Publication Data

International Symposium on Mathematical Problems in
    Theoretical Physics, Kyoto University, 1975.
    Proceedings of the International Symposium on
Mathematical Problems in Theoretical Physics (M∧Σ)
January 23-29, 1975, Kyoto University, Kyoto, Japan.

    (Lecture notes in physics ; 39)
    Bibliography: p.
    Includes index.
    1. Mathematical physics--Congresses. I. Araki,
Huzihiro, 1932-   II. Series.
QC19.2.I57 1975   530.1'5   75-17580

ISBN 3-540-07174-1 Springer-Verlag Berlin Heidelberg New York
ISBN 0-387-07174-1 Springer-Verlag New York Heidelberg Berlin

<u>FOREWORD</u>

This volume is the Proceedings of the International Symposium on Mathematical Problems in Theoretical Physics which was held at Research Institute for Mathematical Sciences (RIMS) and Research Institute for Fundamental Physics (RIFP) of Kyoto University, Kyoto, Japan on January 23-29, 1975.

The subject of the Symposium is mainly mathematical problems in quantum field theory and statistical mechanics along with related mathematics. The Symposium was attended by both mathematicians and theoretical physicists working in different fields such as theory of elementary particles, quantum field theory, statistical mechanics, theory of probability, theory of operator algebras, theory of hyperfunctions, mathematical scattering theory, linear and nonlinear differential equations. The aim of the Symposium is to provide an opportunity to find out recent results and outstanding problems in boundary areas of mathematics and physics and to promote an interaction between mathematicians and physicists. On the side of physics, attention is focussed on methods in quantum field theory and statistical mechanics and as a consequence phenomenology is not included. On the side of mathematics, attention is focussed on those areas which have lively interactions with quantum field theory and statistical mechanics. Obviously there are other areas of physics, such as fluid mechanics and general relativity, which have lively interaction with mathematics but which could not be included.

This kind of international meetings of mathematical physicists was started by the International Conference on Mathematical Problems in Quantum Field Theory and Quantum Statistics held at Steklov Mathematical Institute in Moscow, USSR, on December 12-18, 1972. This Symposium is the second in the series. The symbol mark $M \cap \Phi$ adopted in both meetings denotes the intersection of mathematics and physics and well indicates the interdisciplinary character of these meetings. It is hoped that this series of international meetings of mathematical physicists continue to be held biannually or triannually.

In writing contributions in this Proceedings, each speaker of the Symposium has been asked to make it suitable for Lecture Notes in Physics. A full paper with technical details is supposed to be published elsewhere. Review of a subject, report on recent new results, description of outstanding problems and well-defined proposal or conjecture, along with appropriate references, are supposed to be included. Thus this volume is hoped to be useful for graduate students in finding out an appropriate subject for his research and for mathematicians and physicists in finding out what is happening in their neighbouring fields as well as in his own field.

Discussions on each contribution in the Symposium are written mostly on the spot by relevant persons and are given to the speaker. Each speaker is then given a complete freedom in deciding whether he includes a certain question or comment into "Discussion" part of his manuscripts for this Proceedings.

Contributions are grouped into 30 Sessions, and arranged in the order of presentation in each Session. Each morning Session (numbered 1 to 6) includes two one-hour review talks. Afternoon Sessions, which were done in parallel 2 at a time and 4 altogether in each afternoon, are arranged with odd numbered Sessions (held at RIMS) first and even numbered Sessions (held at RIFP) afterwards in each afternoon. Otherwise Sessions are arranged in chronological order. Authors of the last two articles were unable to attend the Symposium but kindly sent in their manuscripts which they had already prepared for presentation at the Symposium.

Grouping of the contributions into Sessions were originally planned according to the subject of contributions. However, due to various reasons, last minutes rearrangement of the schedule caused some mixing. In addition, some contributions are related with more than one subjects and some others do not fit well into the original classification. Therefore we delete the subject title of each Session in this Proceedings. Instead we provide a Subject Index in which a contribution may appear more than once under different subjects.

The members of the Organizing Committee were as follows:

Huzihiro ARAKI, Hiroshi FUJITA, Masakuni IDA, Teruo IKEBE, Kyozi KAWASAKI, Ziro MAKI, Sigeru MIZOHATA, Noboru NAKANISHI, Kazuhiko NISHIJIMA, David RUELLE, Mikio SATO, Vasilie Sergeevich VLADIMIROV, Hisaaki YOSHIZAWA(Chairman).

The Symposium was sponsored by International Mathematical Union, Japan Society for the Promotion of Science and Science Council of Japan. We wish to thank these organizations for their financial and moral

support of the Symposium. We wish to express our appreciations and
thanks to all authors of contributions in this Proceedings for their
splendid works and for their co-operation with assigned page limitation
and early deadline, to all participants of the Symposium and to all
secretarial and administrative staffs of host Institutes for their effort
in bringing this Symposium to success.

March, 1975

Huzihiro ARAKI
Executive Secretary of the Symposium

Table of Contents

Session 1 (Chairman: H. J. Borchers)

D. Iagolnitzer: Analyticity properties of scattering
amplitudes: a review of some recent developments. 1

M. Sato: Recent development in hyperfunction theory
and its application to physics. (Microlocal
analysis of S-matrices and related quantities.) 13

Session 11 (Chairman: H. Komatsu)

M. Kashiwara: Micro-local calculus. 30

T. Kawai and H. P. Stapp: Microlocal study of
S-matrix singularity structure. 38

M. Morimoto: Convolutors for ultrahyperfunctions. 49

Session 13 (Chairman: M. Sato)

T. Kinoshita and A. Ukawa: Mass singularities of
Feynman amplitudes. 55

K. Pohlmeyer: Large momentum behaviour of Feynman
amplitudes. 59

G. Sommer: Spectral dependence of the analyticity domain
of local vertex functions. 66

Session 12 (Chairman: K. Nishijima)

E. Brüning and P. Stichel: Asymptotics and light-cone
singularities in quantum field theory. 72

S. Schlieder: Structure of singularities of Wightman-
distributions and the Wilson-Zimmermann-expansion
respectively lightcone-expansion. 85

B. Schroer: Conformal invariance in Minkowskian quantum
field theory and global operator expansions. 92

K. Symanzik: On some massless superrenormalizable
and non-renormalizable theories. 102

Session 14 (Chairman: Y. Ohnuki)

Y. Ohnuki and S. Kamefuchi: The locality condition in
parafermi field theory. 107

M. Ninomiya and K. Watanabe: Bound state nature and deep
inelastic structure functions. 111

V. P. Shelest: The physical content of the statistical
bootstrap and high energy hadron interaction. 114

K. Kinoshita: Field theoretical approach to composite
particle reactions. 116

Session 2 (Chairman: J. R. Klauder)

J. Glimm and A. Jaffe: Particles and bound states and
progress toward unitarity and scaling. 118

K. Kikkawa: Field theory of relativistic strings. 128

Session 21 (Chairman: N. Mugibayashi)

K. Hepp: Results and problems in irreversible statistical
mechanics of open systems. 138

J. S. Feldman and K. Osterwalder: The Wightman axioms
and the mass gap for weakly coupled $(\Phi^4)_3$ quantum
field theories. 151

Session 23 (Chairman: H. Ezawa)

J. R. Klauder: Soluble models and the meaning of
nonrenormalizability. 160

Y. Kato: Bound states and asymptotic fields in model
field theories. 170

Session 22 (Chairman: M. Ida)

M. Minami: Feynman propagators associated with the
Veneziano model. 175

T. Yoneya: Dual string models and quantum gravity. 180

G. Konisi and T. Saito: Physical states satisfying
supergauge conditions in dual resonance models. 184

Y. Chikashige and K. Kamimura: The canonical quantization
of a relativistic string. 187

Session 24 (Chairman: K. Kawarabayashi)

F. Gürsey: Algebraic methods and quark structure. 189

K. Sekine: Unitarity and asymptotic condition in a model
with dipole ghost. 196

K. Yokoyama and Reijiro Kubo: Invariant gauge families
inherent in Abelian-gauge field theory. 199

N. Nakazawa and M. Yamada: Space-time approach to anomalies
in the Ward-Takahashi identities and implications in
physical processes. .202

Session 3 (Chairman: V. S. Vladimirov)

K. Nishijima: Dispersion approach to Ward-Takahashi
identities. 205

A. A. Slavnov: Renormalization of supersymmetric gauge
theories. 214

Session 31 (Chairman: T. Hida)

K. Itô: Stochastic calculus. 218

Masatoshi Fukushima: Asymptotic properties of the spectral
distributions of disordered systems. 224

M. Miyamoto: A remark to Harris's theorem on percolation. . . .228

Session 33 (Chairman: K. Itô)

I. Kubo: The ergodicity of the motion of a particle in
a potential field. 230

T. Niwa: Ergodicity of some simple model systems of
infinitely many particles. 236

Session 32 (Chairman: Z. Maki)

A. Hosoya and J. Ishida: New exact solutions of the
classical Yang-Mills field equation. 238

J. Arafune, P. G. O. Freund, and C. J. Goebel: Topology of
Higgs fields. 240

T. Maskawa and H. Nakajima: Spontaneous breaking of
chiral symmetry in a vector-gluon model. 242

N. Nakanishi: Quantum field-theoretical approach to
spontaneously broken gauge invariance. 245

H. Reeh: Reviews on axiomatic study of symmetry breaking. . . .249

Session 34 (Chairman: N. Nakanishi)

O. I. Zavialov: Structure of renormalized Feynman amplitudes. .256

Y. Fujii and Y. Takahashi: On the regularization in the
Callan-Symanzik equation. 261

Session 4 (Chairman: N. M. Hugenholtz)

S. Doplicher: The statistics of particles in local quantum
theories. 264

Ryogo Kubo: Relaxation and fluctuation of macrovariables. . . .274

Session 41 (Chairman: K. Sekine)

V. S. Vladimirov: Holomorphic functions with a
nonnegative imaginary part in the future tube. 281

H. J. Borchers: Algebraic aspects of Wightman quantum
field theory. 283

G. Reents: The infrared problem and non-Fock representations
of the canonical commutation relations. 293

G. Lassner: Continuous representations of the test
function algebra and the existence problem for
quantum fields. 297

Session 43 (Chairman: T. Nakano)

D. W. Robinson: Unbounded derivations of C^* algebras. 303

R. Arens: A method for making relativistic non-quantum
systems by constructing the Hamiltonian and the
other nine generating functions. 312

G. G. Emch: An algebraic approach to the theory of
K-flows and K-entropy. 315

Session 42 (Chairman: R. Abe)

R. Jackiw: Symmetry restoration at finite temperature. 319

G. Jona-Lasinio: Critical behaviour of stationary
random fields. 326

J. Zittartz: Phase transitions of continuous order. 330

Session 44 (Chairman: S. Katsura)

H. Nakano: Order of the phase transition related to the
degeneracy and interaction with proposal for an exact
problem. 336

T. Ogawa: The Heisenberg model on the Cayley tree. 339

C. Di Castro: Generalized Gell-Mann and Low group transformations: scaling variables and cross-over effects. . . 342

Y. Yamazaki: Generalizations and applications of Callan-
Symanzik equations to statistical mechanics in critical
phenomena. 349

T. Izuyama: On the rigorous definition of superfluidity
and superconductivity. 353

Session 5 (Chairman: D. Ruelle)

J.-L. Lions: Variational problems and free boundary
problems. 356

J. L. Lebowitz: Uniqueness, analyticity and decay properties
of correlations in equilibrium systems. 370

Session 51 (Chairman: H. Fujita)

S. Matsuura: On the propagation of support of solutions
to general systems of partial differential equations. . . .380

M. Toda: Wave propagation in a non-linear lattice. 387

R. Jackiw: Quantization of non-linear waves. 394

Session 53 (Chairman: M. Yamaguchi)

Y. Kametaka: On the non-linear diffusion equation of
Kolmogorov-Petrovskii-Piskunov type. 401

T. Nishida and M. Mimura: Global solutions to the
Broadwell's model of Boltzmann equation for a simple
discrete velocity gas. 408

Session 52 (Chairman: N. Saito)

H. Mori: Scaling method for asymptotic evaluation in
nonequilibrium statistical mechanics. 413

Y. Kuramoto: Self-entrainment of a population of coupled
non-linear oscillators. 420

M. Suzuki: Thermodynamic limit of non-equilibrium systems
——extensive property, fluctuation and nonlinear
relaxation. 423

Session 54 (Chairman: H. Nakano)

J. L. Lebowitz: Equilibrium states and ergodic properties
of infinite systems. 432

H. Hasegawa: Variational principles for Markov processes
and Onsager principle. 433

B. Mühlschlegel: Functional averages in statistical
physics. 437

M. Takahashi: On the validity of collective variable
description of Bose systems. 446

Session 6 (Chairman: G. G. Emch)

D. Ruelle: Equilibrium statistical mechanics of one-
dimensional classical lattice systems. 449

T. Ikebe: A look at the development of spectral and
scattering theory in Japan. 458

Session 61 (Chairman: T. Ikebe)

J. M. Combes: On the Born-Oppenheimer approximation. 467

S. T. Kuroda: Abstract approaches to spectral and
scattering theory, contributions from Japan. 472

Y. Saitō: Eigenfunction expansions for differential
operators with operator-valued coefficients and their
applications to the Schrödinger operators with long-
range potentials. 476

Session 63 (Chairman: S. T. Kuroda)

L. E. Thomas: Asymptotic completeness in three-particle
quantum mechanical scattering. 483

K. Mochizuki: Decay and asymptotics for wave equations
with dissipative term. 486

T. Sasakawa: Treatment of three-body problems in
coordinate space. 491

Session 62 (Chairman: H. Matsuda)

B. Souillard: Links between decay properties of correlations and analyticity of the pressure and correlation
functions. 497

A. Verbeure: Linear response, stability, cluster properties. . .504

G. L. Sewell: Equilibrium and metastable states of classical
systems. 510

J. T. Lewis and J. V. Pulè: Dynamical theories of Brownian
motion. 516

Session 64 (Chairman: K. Kawasaki)

K. Ikeda: Distribution of zeros and the equation of
state. 520

Y. Karaki: New theorems on algebraic equation and its
application to statistical physics. 524

Masahisa Fukushima: Operator-valued-measure approach
to spectra of two-dimensional classical harmonic
lattices. 528

M. E. Mayer: Models for relativistic statistical
mechanics. 532

Contributions by mail

C. DeWitt-Morette: Path integrals in Riemannian
manifolds. 535

B. Simon: Bose quantum field theory as an Ising
ferromagnet: recent developments. 543

Author Index. 554
Subject Index. 555

ANALYTICITY PROPERTIES OF SCATTERING AMPLITUDES :

A REVIEW OF SOME RECENT DEVELOPMENTS

D. Iagolnitzer
Service de Physique Théorique
Centre d'Etudes Nucléaires de Saclay
BP n°2 - 91190 Gif-sur-Yvette
France

Abstract

We present some recent developments on momentum-space analyticity properties
of scattering amplitudes for multiparticle processes in relativistic quantum theory.
We describe some results obtained in axiomatic quantum field theory and S-matrix
theory, and some useful related mathematical results.

INTRODUCTION

We consider systems of massive particles with short-range interactions in
relativistic quantum physics. The basic quantities of interest are then the scattering
functionals $S_{II'}$ between sets I and I' of initial and final (free) particles, the
set of which represents the S-matrix[1]. The S-matrix being a bounded operator, each
$S_{II'}$, or its connected part $S_{II'}^c$ is a tempered distribution defined on the space
of all real on mass shell initial and final energy-momentum 4-vectors p_j
$(p_j^2 = (p_j)_o^2 - \vec{p}_j^2 = m_j^2 , (p_j)_o > 0)$. Moreover, (apart from the exceptional points where
all 4-momenta are colinear), it can be written in the factorized form :

$$S_{II'}^c = T_{II'} \times \delta^4 \left(\sum_{i \in I} p_i - \sum_{i' \in I'} p_{i'} \right) \tag{1}$$

which exhibits energy-momentum conservation. The scattering function $T_{II'}$ is a distribution now defined on the physical-region $\mathcal{M}_{II'}$ of the process $I \rightarrow I'$ (i.e. the
set of all real 4-vectors p_j such that $p_j^2 = m_j^2 , (p_j)_o > 0$ and $\sum_{i \in I} p_i = \sum_{i' \in I'} p_{i'}$).

For simplicity, we consider a theory with one spinless particle of mass μ ;

I and I' are then characterized by the numbers m, m' of initial and final particles and we shall put m + m' = n.

To discuss analyticity away from the physical region, it is also useful to change, by convention, the signs of the initial 4-vectors, in which case energy-momentum conservation reads $\sum_{j=1}^{n} p_j = 0$.

The physical problem discussed below is the derivation of momentum space analyticity properties on the <u>complexified mass-shell manifold</u> $\mathcal{M}_n^c$ (which is the set of all complex 4-vectors $k_j = p_j + iq_j$ such that $k_j^2 = \mu^2$, $\sum_{j=1}^{n} k_j = 0$) ; $\mathcal{M}_n^c$ admits different real connected parts associated with the physical regions of various "crossed" processes.

Parts I and III will present some developments in the respective axiomatic frameworks of axiomatic field theory and S-matrix theory. Part II is an outlook at mathematical results, useful in parts I and III, on the local analytic structure of distributions and the theory of essential supports.

I – <u>AXIOMATIC QUANTUM FIELD THEORY</u>.

General references are given in ref. 2.3, those of ref. 3. being more specifically concerned with momentum-space analyticity properties.

In that theory, a detailed space-time description of interactions is introduced on the <u>microscopic</u> level in terms of underlying fields. The Haag-Ruelle theory allows one to reobtain <u>free-particle states</u> asymptotically and hence to define an S-matrix, which is unitary if the axiom of asymptotic completeness is included.

The link between the scattering functions $T_{m,m'}$ and the fields is provided by the well known LSZ reduction formulae

$$T_{m,m'}(p_1 \ldots p_n) \;=\; \tau(p_1 \ldots p_n)\big|_{\mathcal{M}_{m,m'}} \tag{2}$$

where τ is the distribution obtained, after factorizing out energy-momentum conservation, from the Fourier transform of the amputated, connected vacuum expectation value of the chronological product of the fields $A(x_1) \ldots A(x_n)$, at space-time points $x_1 \ldots x_n$. The distribution τ is defined on $\mathbb{R}^{4(n-1)}$ and the right-hand side of (2) is its restriction to the manifold $\mathcal{M}_{m,m'}$. Formulae (2), including the existence of this restriction have been put on a rigorous basis by Hepp (see ref.2c).

The program of axiomatic field theory is then to derive analyticity properties on the complex mass-shell $\mathcal{M}^c$.

The <u>linear program</u> uses the axioms of <u>microscopic</u> causality of the fields and the spectral condition. The <u>non linear programs</u> make use moreover of positivity properties and (on-shell or "off-shell") unitarity.

Among the old results of the linear program is the existence for any given n of a unique n-point function $H_n(k)$ defined and analytic in a certain primitive domain D_n of $\mathbb{C}^{4(n-1)}$; the distributions $\tau_n(p)$ in various real regions are boundary values of $H_n(k)$ from various directions.

The intersection of D_n with the complex mass-shell $\mathcal{M}^c$ is however empty, and the problem is then to find the holomorphy envelope $\mathcal{H}(D_n)$ of D_n .

Powerful results have been obtained long ago for $n = 4$, on $\mathcal{M}^c$: crossing, dispersion relations,.... Further results have also been obtained in that case in the non linear program[4].

One is unfortunately still far from analogous results for arbitrary n. However, a number of interesting results, outlined below, have been obtained in the last years.

a) <u>Local study</u>.

By local study is meant the study of the local analytic structure of the scattering functions $T_{m,m'}$ in the neighborhood of their physical regions $\mathcal{M}_{m,m'}$.

This study by Bros, Epstein, Glaser (1972)[5] is an elaboration, concerned more specifically with analyticity, of the work previously carried out by Hepp[2c]. A new version of the BEG work which is more direct and simple at some point and which we use here is also been given in ref. 6.

In view of proving the existence of the restriction of the distribution $\tau(p)$ to $\mathcal{M}_{m,m'}$ and of determining its local analyticity properties, one is led, according to part II, to study the <u>essential support</u> $S_p(\tau_n)$ of τ_n at each real point $P = (P_1 \ldots P_n)$ of $\mathcal{M}^c_{m,m'}$.

Microcausality provides <u>support properties</u> in x-space for the vacuum expectation values of chronological products. The spectral condition provides on the other hand various relations in p-space, which depend on the point P considered. In view of a simple lemma, support properties become <u>essential support</u> properties for the generalized Fourier transform <u>at P</u>. The point is that one may now take advantage of the relations in p-space which hold at P to obtain a region of exponential fall-off possibly much better than the previous region of vanishing of the usual Fourier transform.

As a matter of fact, $S_p(\tau_n)$ is contained in the union of a finite number of well specified convex cones C_β , (which depend on P) whose intersection with the conormal space $N(P)$ at P to $\mathcal{M}$ is empty.

This condition ensures (see end of part II) the existence of the restriction of τ_n to $\mathcal{M}$, and the results of part II yield corresponding decompositions of $T_{m,m'}$ as a sum of boundary values of functions $(F_{m,m'})_\beta$ analytic in domains of $\mathcal{M}_n^c$.

It has moreover been proved in 5) that there always exists a part of $\mathcal{M}_{m,m'}$ where $T_{m,m'}$ is the boundary value of a unique analytic function $F_{m,m'}$, the best results being obtained for $2 \to 3$ processes.

For physical reasons (see part III) one actually wants $T_{m,m'}$ to be the boundary value of a unique analytic function $F_{m,m'}$ in the <u>whole physical region</u> $\mathcal{M}_{m,m'}$, apart from some exceptional points ($F_{m,m'}$ will eventually coincide with an analytic continuation of H_n). Moreover $F_{m,m'}$ should be analytic at real points outside $+\alpha$Landau surfaces.

Such results cannot probably be achieved by the linear program alone, but it is hoped that they will eventually be obtained in the non linear program.

<u>Relativistic covariance</u>.

It allows one to show that :

$$S_p(\tau_n) \ \subset \ \mathcal{L}_p^\perp , \tag{3}$$

where $\mathcal{L}_p^\perp$ is the manifold conormal at P to the Lorentz orbit of P. This result is useful in improving the previous results, in particular for $2 \to 3$ processes (BEG). Recent more refined results are due to Berceam, Gheorghe, Mihul .

<u>Positivity</u>

The positivity (of absoptive parts) also allows one to improve the essential support in both local and global ways. In particular, Bros has shown that Martin's ellipses for $2 \to 2$ processes can be viewed as a localized form of the cut-tube theorem described below in subsection b , in the neighborhood of points at infinity $(1/s = 0)$.

There is some hope of extending in that way Martin's results to more general processes.

<u>Unitarity</u>

We briefly recall the situation for $2 \to 2$ processes. Unitarity is written in the form :

$$T - T^* \ = \ T\,T^* \tag{4}$$

which can be viewed as a <u>Fredholm</u> equation. If an appropriate technical assumption

(on zeroes in the denominator of the solution) is made, in order to exclude pathologies such as those described by Martin, it follows that T is indeed <u>analytic</u> in the
elastic region $4m^2 < s < 9m^2$ ($s = (p_1 + p_2)^2$).

Glaser has obtained recently substantial results of this type for $2 \to 3$
processes : in particular, he shows that $T_{2,3}$ is again the boundary value of just <u>one</u>
analytic function in the region $9m^2 < s < 16m^2$ with the cut $\text{Im} s < 0$ being again
removed.

This technique, together possibly with the general theorems on products and
integrals of part II$^+$ should also provide more general results.

b) <u>Some results in the global study.</u>

<u>Linear program</u>

An (implicit) computation of the holomorphy envelope of the 3-point function is due to Sommer, as described in his lecture.

On the other hand, the method of the generalized Fourier transform allows
one to compute holomorphy envelope of a certain type. Investigations in this domain
concerning the n-point functions are due to Bros, Epstein, Glaser.

<u>Positivity</u>

The following cut-tube theorem which is an extension of the usual edge-of-
the wedge theorem when $\Delta f = f^+ - f^-$ is <u>not</u> zero but is of a positive type, is due to
Bros and Glaser :

"Let $F^+(s,t)$, $F^-(s,t)$ be analytic in respective tubes T^+, T^- whose bases
in $(\text{Im} s, \text{Im} t)$ space are open convex cones C^+, C^- with apex at the origin, lying in
the region $\text{Im} s > 0$, resp $\text{Im} s < 0$. Let f^+, f^- denote their boundary value (assumed to
be distributions) at real points. If $\Delta f = f^+ - f^-$ is, for all real s, of positive
type with respect to t (i.e. $\int \Delta f(s, t-t')\, \varphi(t)\, \varphi^+(t')\, dt\, dt' \geq 0$ for any Schwartz
test function φ), then F^+, resp. F^- can be analytically continued in the tube
$(T^+ \cup T^-)^c \cap \{\text{Im} s < 0\}$, resp $(T^+ \cup T^-)^c \cap \{\text{Im} s < 0\}$, where $(T^+ \cup T^-)^c$ is the convex
envelope of T^+ and T^- ."

Adaptations of this result, which allows one in particular to reobtain
Martin's ellipses : see subsection a), might provide new results for general processes.

$^+$These theorems are useful for computing essential supports of unitarity integrals,
or "bubble diagram functions", as in the structure theorem of part III. The difference is that the essential supports of the bubbles are much worse here than in part
III.

<u>Non linear program in the sense of Symanzik.</u>

This is a program first proposed by Symanzik in 1960[7] and recently developed by Bros-Lassalle. Its aim is :

1) to define rigorously, in the axiomatic framework, n-point functions $H_n^{(p)}$ which will be <u>p-particle irreducible</u> with respect to a given, or several, channels, where p is here a positive integer. (In perturbation theory, such functions are defined formally as sums of "p-particle irreducible" graphs).

2) to show that the p-particle irreducible functions have <u>nice analyticity properties</u> : for instance in the simple case n = 4, $H_4^{(2)}$ should be analytic (on $\mathcal{M}^c$) up to $s = 9m^2$.

3) to derive corresponding analyticity properties for the complete n-point function H_n : for instance analyticity (on $\mathcal{M}^c$) in the elastic region $4m^2 < s < 9m^2$ for $2 \to 2$ processes.

This program makes use of the GLZ (generalized unitarity) equations and again of technical assumptions, in order to avoid pathologies. It has been carried out by Bros[8] (1968) for n=4 ; $H_4^{(2)}$ is then defined by a generalized Bethe-Salpeter equation.

In view of treating more general cases, the following preliminary result has been achieved by Lassalle[9] (1973) : "The n-point primitive analytic and algebraic structure in p-space is preserved by generalized Feynman convolution".

Primitive structure means that derived from the linear program, the algebraic part refering to the Ruelle-Araki relations.

A "generalized Feynman convolution" is an integral of a product of n-point functions associated with each vertex of a given graph. The integration is made over the 4-momenta of the internal lines with a suitable prescription depending on the external momenta (Integration over the euclidean region when the external momenta are in the euclidean region. To explore general domains, suitable distortions are used).

A number of results on the 1-particle and 2-particle irreducible functions (for arbitrary n) have then been obtained more recently by Bross-Lassalle[10] (including a neat study of the technical assumptions). The form of their results should lead to progress in more general situations.

II - <u>LOCAL ANALYTIC STRUCTURE OF DISTRIBUTIONS AND ESSENTIAL SUPPORTS</u>[11]

Consider the n-dimension real vector space $R^n_{(p)}$ of a variable $p = (p_1,\ldots,p_n)$, its dual (real) space $R^n_{(v)}$ of the variable $v = (v_1,\ldots,v_n)$ and the complexified space $C^n_{(k)}$ of the variable $k = p + iq$. $(k = k_1, \ldots, k_n)$.

The following result is easily proved for any tempered distribution f on $R^n_{(p)}$: there always exist $n + 1$ distributions f_i $(i = 1, \ldots n+1)$, each of which is the boundary value (in R^n) of an analytic function F_i from the directions q of an open convex cone Γ_i, such that :

$$f = \sum f_i \qquad \text{in} \quad R^n. \tag{5}$$

In the one-dimensional case $(n = 1)$, this is the well known result :

$$f = f^+ - f^- \tag{6}$$

where f^+ and f^- are boundary values of functions analytic in the respective half-spaces $q > 0$ and $q < 0$.

The cone Γ_i in q-space is here independent of the real point P considered in $R^n_{(p)}$. The above result therefore cannot take into account the more refined <u>local</u> analytic structure that f may have in various real regions.

<u>Problem</u> :   find all possible decompositions of f as a sum of boundary values f_i of analytic functions F_i, from directions q of open convex cones $(\Gamma_i)_p$, which <u>may depend</u> on the real point p. (Such a function F_i will be analytic in a domain of C^n called a "tuboid" of "profile" $\Gamma_i = \underset{p}{U} \; (p, (\Gamma_i)_p.)$

As in (5), such decompositions are in general not unique (if there is more than one function F involved). The basic intrinsic notion which allows one to characterize them is that of <u>essential support</u>.

Let P be a given point in p-space. The generalized Fourier transform of f at P is defined (for $v_o > 0$) by :

$$\mathcal{F}(v, v_o, P) = \int f(p) \; e^{-iv \cdot p - (p - P)^2 v_o} \; dp. \tag{7}$$

The <u>essential support</u> $S_P(f)$ of f at P is a closed cone in v-space with apex at the origin defined as follows : a direction V is outside $S_P(f)$ if there exists a neighboring cone v of V in v-space with apex at the origin), $\alpha > 0$ and $\gamma_o > 0$ such that :

$$\left| \mathcal{F}(v, \gamma |v| ; P) \right| < \mathcal{P}(|v|) \; e^{-\alpha \gamma |v|} \tag{8}$$

in the region $v \in v$, for all positive γ less than γ_o, $\mathcal{P}$ being a polynomial. (If f does not decrease sufficiently fast at infinity, (8) is to be slightly modified.)

In others words $S_P(f)$ is the set of directions in v-space along which the generalized Fourier transform of f at P does <u>not</u> decrease exponentially (in the sense of (8).

The essential support $S_\Omega(f)$, of f over a real domain Ω , is defined as :

$$S_\Omega(f) = \bigcup_{p \in \Omega} (p, S_p(f)). \qquad\qquad (9)$$

More general notions of essential support in which the rate α of exponential fall-off and the region in (v, v_o) space where it holds are specified, and in which the exponent $(p-P)^2$ of (7) is possibly replaced by more general functions Φ , are also useful.

Finally, the notion of essential support can be extended to distributions defined on __manifolds__ rather than on R^n.

Decomposition theorems

Various decompositions of f then follow (for distributions defined on R^n or on manifolds) from corresponding coverings of the essential support.

For instance, if $S_p(f) \subset \bigcup_j C_j$ where each C_j is a (closed, salient) convex cone (with apex at the origin), then one has equivalently, in the neighborhood of P, $f = \Sigma_j f_j$, where each f_j is the boundary value at P of an analytic function F_j from the directions of the open dual cone Γ_j of C_j .

If $S_\Omega(f) \subset \bigcup_j \Sigma_j$ where each Σ_j is a closed subset of the cotangent bundle $T^+\Omega$ to Ω such that $(\Sigma_j)_p$ is a (closed, salient) convex cone for every p in Ω , then $f = \Sigma_j f_j$ in Ω, where each f_j is the boundary value in Ω of an analytic function F_j, obtained at each point $p \in \Omega$ from the directions of the dual cone $(\Gamma_j)_p$ of $(\Sigma_j)_p$.

Finally, results in which the complex domains of analyticity are precisely specified are also associated with the more general notions of essential support in (v, v_o)-space mentioned earlier.

Products, integrals, restrictions.

Simple and direct proofs of results previously proved in the mathematical framework of ref.12 have also been obtained in the framework of the essential support theory. These results are well adapted to the physical context of parts I and III where causality is precisely expressed in terms of exponential fall-off properties.

1) Products

Let f_1, f_2 be two tempered distributions on a real domain Ω . If $v_1 + v_2 \neq 0$ for any (non zero) v_1 in $S_p(f_1)$, v_2 in $S_p(f_2)$ and any p in Ω , then $f_1 \times f_2$ is well defined in Ω and

$$S_p(f_1 \times f_2) \subset S_p(f_1) + S_p(f_2). \qquad (\forall p \in \Omega)$$

where $S_p(f_1) + S_p(f_2)$ is the cone of v of the form $v_1 + v_2$ with $v_i \in S_p(f_i)$. (i = 1,2).

2) Integrals

Let p,p' be now two sets of independent variables, let f be a distribution on (p,p')-space with compact support in the variables p' and let

$$g(p) \; = \; \int \; f(p,p') \; dp'$$

(in the sense of distributions).

If, for any p' in the support of f, (V,0) is outside $S_{P,p'}(f)$ then $V \notin S_P(g)$.

3) Restrictions

Let $\mathcal{M}$ be a submanifold of R^n, let f be a distribution defined on R^n, and let N(P) be the conormal space to $\mathcal{M}$ at a point P. The restriction $f/\mathcal{M}$ of f to $\mathcal{M}$ is well defined in the neighborhood of P if $N(P) \cap S_P(f)$ is empty, and

$$S_P(f/\mathcal{M}) \; = \; S_P(f)/N(P).$$

III - S-MATRIX THEORY.

General references on S-matrix theory are given in ref. 13.

In contrast to field theory, no space-time description on the microscopic level is introduced in S-matrix theory and one starts only from assumptions expressed in terms of the initial and final free particle states and of a macroscopic notion of space-time. Correspondingly, on-shell quantities are considered from the outset.

Decisive advances, described below, have been made at the end of the sixties and developed since that time, in the general derivation and understanding of the analytic structure of the scattering functions $T_{m,m'}$ in the neighborhood of their physical regions. In particular macroscopic causality (see below) entails the following fundamental result:for each process $m \rightarrow m'$ there exists a unique analytic function $F_{m,m'}$ (defined in a domain of $\mathcal{M}^c$) to which $T_{m,m'}$ is equal at all points which do not lie on (codimension 1) + α-Landau surfaces of connected graphs, and from which it is a "plus iε" boundary value at almost all + α- Landau points (The +α-Landau surfaces for $m = m' = 2$ are $s = (\ell\mu)^2$, $\ell = 2,3,4,\ldots$).

As described in part I, these results, or equivalently macroscopic causality, have not yet been derived in general from the axioms of field theory.

On the other hand, in contrast to field theory, there is no link a priori between the scattering functions of various processes, and one has here to <u>assume</u> the existence of analytic continuations (on $\mathcal{M}^c$) away from the physical region. We recall that properties such as crossing (i.e. the fact that such continuations for crossed process coincide) have been derived long ago (Olive, Stapp...) in S-matrix theory from arguments of internal consistency. This derivation makes use however of assumptions on the analytic continuations involved which are not easy to control.

There has been some conceptual progress on these questions more recently, but no decisive advance and we shall not discuss them here.

a) Macrocausality[14)]

Macrocausality is an appropriate mathematical expression of a certain classical limit of quantum theory in terms of particles, namely of the principle that any energy-momentum transfer over large distances which cannot be attributed to real stable physical particles according to classical ideas, gives effects that are damped exponentially with distance ("Short-range" effects).

Quantum (free) particles are in fact _asymptotically_ localized in space-time along "classical trajectories of (free) point particles", up to exponential fall-off, for appropriate momentum-space wave functions . Macrocausality is then stated as an exponential fall-off property of transition probabilities in "non causal" situations, i.e. when these initial and final trajectories cannot be linked via intermediate particles according to the laws of classical kinematics. (It is therefore more than a strict causality requirement.)

This property turns out mathematically to be precisely a condition on the essential support $S_P(T_{m,m'})$ of $T_{m,m'}$ at each point P of $\mathcal{M}_{m,m'}$, and the analyticity properties of $T_{m,m'}$ then follow from the decomposition theorems recalled in part II. As a matter of fact, $S_P(T)$ is empty if P is not a $+\alpha$-Landau point ; it is composed of a unique direction, (or of those of a convex (salient) cone) at almost all $+\alpha$-Landau points.

The fundamental result announced earlier on $F_{m,m'}$ follows. If the exceptional points which lie in the intersection of several $+\alpha$-Landau surfaces are included, $T_{m,m'}$ can generally be written as a sum of boundary values of analytic functions, each of which being associated with a given $+\alpha$-Landau surface.

b) Unitarity

From unitarity, discontinuity formulae have moreover been derived for the functions $F_{m,m'}$ around the $+\alpha$-Landau singularities[15)16)] . These formulae are an adaptation of Cutkosky formulae in perturbation theory. If the $+\alpha$-Landau surface is associated with a graph G with no multiple lines, the discontinuity is the integral, over the on-shell internal momenta of the product of the scattering functions associated with each vertex of G . A slightly different form is also obtained by Coster and Stapp for graphs with multiple lines.

These results play an important role not only for determining the nature of the $+\alpha$-Landau singularities, but also for the study of analyticity properties away from the physical region (crossing, ...) and for measurement theory.

More recently, Coster and Stapp[17] have moreover obtained important results on multiparticle optical theorems (Related results in field theory are also due to Cahill and Stapp[18]).

Besides an algebraic work, these results make use of a basic preliminary result, first proved by Stapp and called structure theorem, which proves analyticity properties for "bubble diagram functions". (The latter, which arise in unitarity equations, are integrals over on-shell 4-momenta of products of scattering functions T, or $T^- = T^+$).

A proof of a general form of this theorem in terms of essential support, in which an unnecessary technical assumption of the original proof is removed, has been recently given in ref.19), following a method proposed by F. Pham[20]. This proof is a consequence of the mathematical results on products and integrals mentioned at the end of part II.

A technical assumption (on " $u = 0$ points") is still used : some progress on this is presented by Kawaï and Stapp at this Conference.[*]

---

[*] Although it is presented in the very different mathematical framework of ref.12), their proof for $u \neq 0$ points is essentially the same as that given previously in ref.19). It is our opinion that their new results for $u = 0$ points could also be obtained in the framework of essential support theory.

References

1. For details, see for instance D. Iagolnitzer, "Introduction to S-matrix theory", A.D.T., 21 rue Olivier Noyer, 75015 Paris (France) (1973).
2. General references on axiomatic field theory :
   a) R.F. Streater and A.S. Wightman, "PCT, Spin-Statistics and all that", Benjamin, New York (1964).
   b) R. Jost, "The general theory of quantized fields", American Mathematical Society, Providence, Rhode Island (1965).
   c) K. Hepp in "Lectures in particle symmetries and axiomatic field theory", Vol.1, ed. by M. Chretien and S. Deser, Gordon and Breach, New York (1966).
3. a) H. Epstein in the book already quoted in ref.3c)
   b) J. Bros, H. Epstein, V. Glaser, Nuovo Cimento $\underline{31}$, 1265 (1964).
   c) J. Bros in "De l'axiomatique à l'expérience", Institut National de Physique Nucléaire et des Particules, Paris (1972).
4. A. Martin, "Scattering theory : unitarity, analyticity and crossing", Springer Verlag, Heidelberg (1970).
5. J. Bros, H. Epstein, V. Glaser, Helv. Phys. Acta $\underline{45}$, 149 (1972).
6. J. Bros, D. Iagolnitzer in "Proceedings of the 1972 Moscow International Conference on Mathematical Methods in Quantum Field Theory and Quantum Statistics".
7. K. Symanzik, J. Math. Phys. $\underline{1}$, 249 (1960) and in Symposium on Theoretical Physics, Vol.3, New York, Plenum Press (1967).
8. J. Bros in "Analytic methods in mathematical physics", Gordon and Breach, New York (1970).
9. M. Lassalle, Comm. Math. Phys. $\underline{36}$, 185 (1974).
10. J. Bros, M. Lassalle, to be published in Comm. Math. Phys.
11. This theory is due to a collaboration of J. Bros and D. Iagolnitzer. See :
    - J. Bros, D. Iagolnitzer in "Proceedings of the 1971 Marseille Meeting on Renormalization Theory" and Ann. Inst. H. Poincaré, vol.$\underline{18}$, n°2, p.147 (1973).
    - J. Bros in "Proceedings of the November 1971 Strasbourg RCP Meeting.
    - D. Iagolnitzer in ref.1 and "Essential support of a distribution - An introduction", Proceedings of the 1973 Nice Conference on Hyperfunction Theory and Applications, and part I of ref.19 .

    It has very close links with results obtained independently in :

12. M. Sato, T. Kawaï, M. Kashiwara in "Hyperfunctions and pseudo-differential equations", 1971 Katata Conference Lecture Notes in Mathematics n°287, Springer Verlag, Heidelberg (1973).
13. Basic historical references :
    J.A. Wheeler, Phys. Rev. $\underline{52}$, 1107 (1937) ; W. Heisenberg, Z. Physik $\underline{120}$, 513 and 673 (1943).
    Recent general references :
    G.F. Chew, "The analytic S-matrix", Benjamin, New York (1966) ;
    R.J. Eden, P.V. Landshoff, D.I. Olive, "The analytic S-matrix", Cambridge University Press (1966) ;
    Ref.1 and other references quoted there.
14. D. Iagolnitzer and H.P. Stapp, Comm. Math. Phys. $\underline{14}$, 15 (1969), this work has been developed in a more elaborate form in ref.1 and in D. Iagolnitzer in the Proceedings of the 1973 Nice Conference on Hyperfunction Theory and its Applications.
    Earlier works on macrocausality are due in particular to C. Chandler and H.P. Stapp, R. Omnès, F. Pham, G. Wanders.
15. J. Coster and H.P. Stapp, J. Math. Phys. $\underline{10}$, 371 (1969) and $\underline{11}$, 2743 (1969).
16. M.I.W. Bloxmam, P.I. Olive, P.C. Polkinghorne, J. Math. Phys. $\underline{10}$, 494 (1969) ; $\underline{10}$, 545 (1969) ; $\underline{10}$, 553 (1969).
17. J. Coster, H.P. Stapp, Optical theorems for three to three processes, Berkeley report, to be published and references quoted therein.
18. K. Cahill, H.P. Stapp (1973) and K. Cahill (1974).
19. D. Iagolnitzer, "The structure theorem in S-matrix theory", Saclay-report (1974), to be published in Comm. Math. Phys.
20. F. Pham in "Proceedings of the 1973 Nice Conference on Hyperfunction Theory and Applications", and private communication.

RECENT DEVELOPMENT IN HYPERFUNCTION THEORY
AND ITS APPLICATION TO PHYSICS
(MICROLOCAL ANALYSIS OF S-MATRICES AND RELATED QUANTITIES)

Mikio Sato
RIMS, Kyoto University, Kyoto

1. - Through recent development of the theory of microfunctions
and microdifferential (=pseudodifferential) equations ([1]-[5*];[6],[7]) a
systematic method of handling (hyper-)functions has been evolved. The
principle is this: Each (hyper-)function of 'natural origin' would
microlocally satisfy a system of sufficiently many microdifferential
equations, i.e. a *holonomic system* (= a maximally overdetermined system)
of microdifferential equations. This being so, the characteristic
manifold of that system must be the (locally finite) union of a number
of *holonomic manifolds* (= Lagrangean manifolds) Λ_1, $\Lambda_2, \cdots$, and now the
hyperfunction in question will be most effectively analyzed by carefully
observing the configuration of these holonomic manifolds.

The '*holonomy diagram*' is a convenient way of visualizing this
configuration; it is a diagram consisting of vertices 1,2,... which
represent the constituent holonomic manifold $\Lambda_1, \Lambda_2, \cdots$ and of segments
joining them which represent 1-*codimensional* intersection manifolds
where two or more holonomic manifolds meet together not only in the
geometrical sense but also in the analytical sense, i.e. where the given
holonomic system does not split into summands supported on each of
relevant   holonomic manifolds.

The holonomy diagrams of holonomic systems satisfied by
$(x_1^2+\cdots+x_n^2)^s$ (n>2) and by (det $X\cdot {}^tX)^s$ (where X denotes an m×n matrix
with $n\geq2m$ whose entries x_{ij} are all independent variables, and ${}^tX$
denotes the transposed of X) are, for example, as Fig.1 and Fig.2,
respectively.

This method, the 'microlocal calculus' as we call it, appeared
very fruitful (and especially so when the multiplicity of the holonomic
system is 1) in studying various problems of mathematics related to
(hyper-)functions of 'natural background' ([8]-[10]).

An interesting example is the beautiful algorithm due to M.

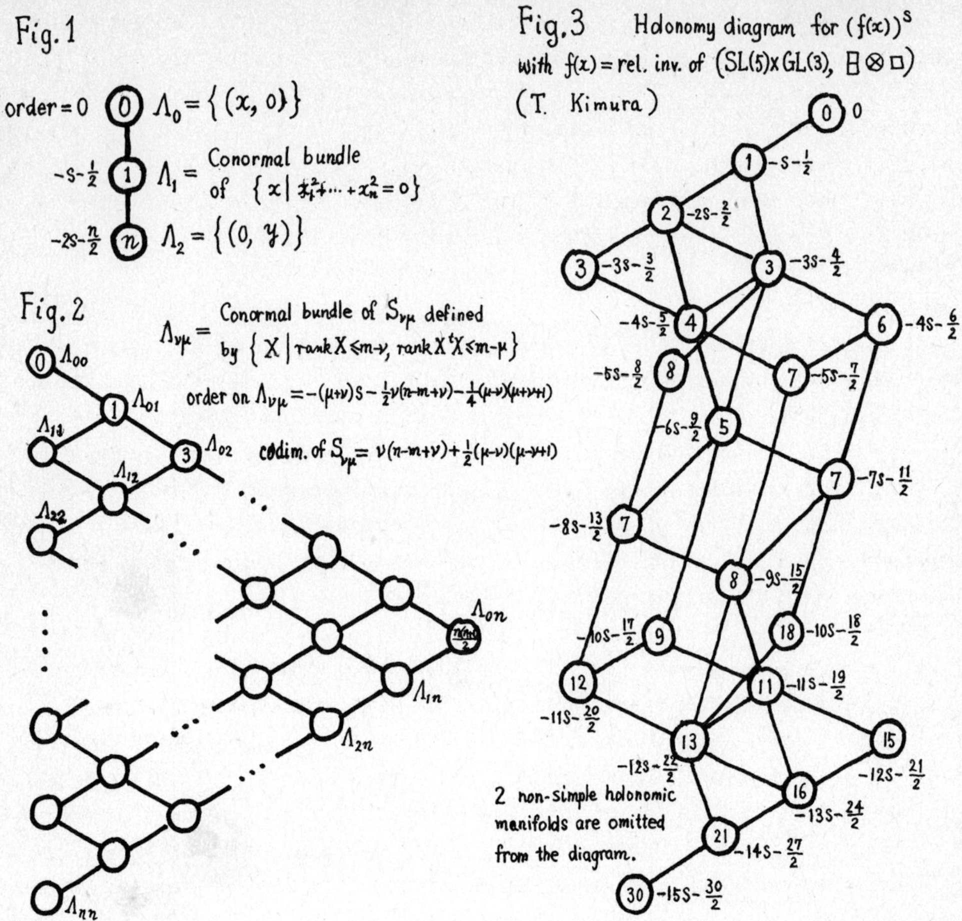

Kashiwara of computing the Fourier transform of relative invariant
hyperfunctions on a prehomogeneous vector space. Thus, the Fourier
transforms of hyperfunctions cited above, for example, are easily obtained
by his formula simply by superposing elementary transformations corresponding to each segments appearing in Fig.1 and Fig.2, respectively.

2. - Our present objective is to introduce the microlocal calculus
into the investigation of S-matrix and related quantities (time ordered
Green's functions and correlation functions). In this connection,
F. Pham [11] observed that *macroscopic causality* in S-matrix theory is
mathematically interpreted as *microanalyticity* at non-causal codirections.
Otherwise stated, his postulate of microanalyticity says that the
singular spectrum of the S-matrix viewed as a microfunction is contained
in the union of various positive α Landau singularities (in the (p,iu)-
space).

Another observation due to him is that the Cutkosky-type discontinuity formula simply says that the S-matrix, when viewed microlocally at the Landau singularity corresponding to a Feynman graph G, reduces to (the integral of) a product of S-matrices corresponding to constituent elementary subgraphs of G.   D.I.Olive [12], on the other side, emphasized that, once the microanalyticity is postulated, the discontinuity relations are the equivalent of unitarity relations for S-matrix ([13], [14]).

Now it seems to me that the above observations strongly suggest to introduce an even stronger postulate into the S-matrix theory without violating the unitarity and discontinuity relations.   Namely let us propose: The S-matrix should satisfy a holonomic system of micro-differential equations whose characteristic manifold is described by various Landau equations. This system is supposed to be simple, i.e. of multiplicity 1, and the 'order' of the system at each constituent holonomic manifold is easy to predict.

Further it seems that, to achieve dynamical completeness, one should take account of the 'off-the-mass-shell' properties, that is, one should study the microlocal properties of the (time-ordered) *Green's functions* or τ-*functions* as well.

It should be noticed that, in either case, the control of a holonomic system over the solution quantity inevitably extends beyond the real region or the physical region.

We will give a precise description of our suppositions and results after the following summary of our previous result on how holonomic systems really control the *Feynman integrals*.

3. — <u>Feynman Integrals</u> [15] — After the pioneer work of L. D. Landau and N. Nakanishi in 1959 on the singularity structure of general Feynman integrals several important results on it were found by R. E. Cutkosky, by F. Pham and others, by the Cambridge school and by many others. Here we brief our way of describing Feynman integrals in terms of microlocal analysis [16].

Consider an oriented (resp. non-oriented) linear graph G consisting of n vertices and N oriented (resp. non-oriented) internal lines joining 2 distinct vertices. Hence 2^N different oriented linear graphs correspond to one and the same non-oriented graph. To each of N lines be assigned a positive constant (the mass) $m_\ell > 0$ $(\ell = 1, \cdots, N)$. If G is oriented, the incidence number $[j:\ell]$ between j-th vertex and ℓ-th line is defined as in [15].

Let ν be the dimensionality of the space-time and $M = \mathbb{R}^\nu$ denote the corresponding energy-momentum space equipped with the Minkowskian metric. For brevity we set $\xi_j(k) = \xi_j(k_1, \ldots, k_N) = \sum_{\ell=1}^{N} [j:\ell] k_\ell$ and $\eta_\ell(x) = \eta_\ell(x_1, \ldots, x_n) = \sum_{j=1}^{n} [j:\ell] x_j$, k_ℓ and x_j denoting vectors in M and its dual vector space, respectively. k_ℓ and $\xi_j(k)$ physically represent the energy-momentum of the ℓ-th internal line and the sum of energy-momenta of internal lines flowing into the vertex j, respectively, while x_j and $\eta_\ell(x)$ represent the space-time coordinates of the vertex j and the space-time displacement between the initial and final end points of the ℓ-th internal line, respectively.

The Feynman integral $F_G(p) = F_G(p_1, \ldots, p_n)$ corresponding to the (oriented) linear graph G is formally defined as follows as a function on νn-dimensional manifold $M^n = M \times \cdots \times M$ (n-copies of M):

$$F_G(p) = \int \cdots \int_{M^N} \Phi_G(p,k) \, d^\nu k_1 \ldots d^\nu k_N$$

where the integrand function $\Phi_G(p,k) = \Phi_G(p_1, \ldots, p_n, k_1, \ldots, k_N)$ is given by

$$\Phi_G(p,k) = \prod_{j=1}^{n} \delta^\nu(p_j - \xi_j(k)) \cdot \prod_{\ell=1}^{N} \frac{1}{k_\ell^2 - m_\ell^2 + i0} .$$

We note that the definition of F_G does not depend on the orientation on G one employs.

Let further σ denote a map: $\{1, \ldots, n'\} \to \{1, \ldots, n\}$ and let our graph G be supplemented by 'external energy-momenta' $q_i \in M$, $i = 1, \ldots, n'$, issuing from the vertex $\sigma(i)$. Then, setting $p_j(q) = \sum_{i \in \sigma^{-1}(j)} q_i$ (which represents 0 if $\sigma^{-1}(j)$ is empty), the Feynman integral corresponding to this supplemented graph or 'Feynman graph' (G, σ) will by definition be given by $F_{G,\sigma}(q) = F_G(p(q))$.

It will be shown below that the integral $F_G(p)$ is *microlocally* well-defined even in some cases when it is not well-defined in a customary sense (even as a hyperfunction).

The general rule on multiplication and integration([3]-[5]) immediately tells that Φ_G is always well-defined and its singular spectrum is given by

$$SS \ \Phi_G = \{(p,k;i\infty(x,v))\,|\,p_j-\xi_j(k)=0 \quad (j=1,\ldots,n), \ \alpha_\ell\cdot(k_\ell^2-m^2)=0,$$

$$\eta_\ell(x)+v_\ell = \alpha_\ell\cdot k_\ell \quad \text{for some } \alpha_\ell\geq 0, \ (\ell=1,\ldots,N)\},$$

while F_G will be well-defined microlocally at $(p, i\infty x)$ providing that the projection map to the $(p, i\infty x)$-space from the $v=0$ portion of SS Φ_G is a proper map (meaning that its fibers or the inverse images are locally uniformly compact) in the neighborhood of this point, and that, assuming that this is the case, its singular spectrum does not extend beyond the image of this projection:

$$SS \ F_G \subset \{(p, \ i\infty x)\,|\,\exists\,k_\ell(\ell=1,\ldots,N), \ s.t. \ (p,k; \ i\infty(x,0))\in SS\Phi_G\}$$

$$= \{(p, \ i\infty x)\,|\,\exists\,k_\ell\in M, \alpha_\ell\geq 0 \ (\ell=1,\ldots,N), \ s.t. \ p_j-\xi_j(k)=0 \ (j=1,\ldots,n),$$

$$\alpha_\ell\cdot(k_\ell^2-m_\ell^2)=0, \ \eta_\ell(x)=\alpha_\ell\cdot k_\ell \quad (\ell=1,\ldots,N)\}.$$

Let τ denote a map: $\{1,\ldots,\ n\} \to \{1,\ldots,\ n_1\}$ such that the inverse image $\tau^{-1}(j_1)$, regarded as a subset of vertices in G, is connected and non-empty for $j_1=1,\ldots,n_1$. The reduced or contracted linear graph $G_1=\tau(G)$ with n_1 vertices is then defined in an obvious manner; internal lines of G_1 are those of G whose both ends go to different points in G_1 by the map τ. This G_1, or the 'contraction' τ, will be of the 1st kind if $\tau^{-1}(j_1)$ are all tree graphs (i.e. contain no loop); otherwise it will be of the 2nd kind.

Now define Λ_G^+ (resp. $\Lambda_G^{\mathbb{C}}$) as the totality of $(p, i\infty x)$, with p_j and x_j being real (resp. complex) vectors, satisfying the following *Landau equations*:

$$p_j-\xi_j(k)=0 \ (j=1,\ldots,\ n), \ k_\ell^2-m_\ell^2=0, \ \eta_\ell(x)=\alpha_\ell\cdot k_\ell \quad (\ell=1,\ldots,\ N) \quad \text{for}$$

some real (resp. complex) vectors k_ℓ and real non-negative (resp. complex) numbers α_ℓ.

More generally, define $\Lambda_{G,\sigma}^+$ (resp. $\Lambda_{G,\sigma}^{\mathbb{C}}$), which reduces to Λ_G^+ (resp. $\Lambda_G^{\mathbb{C}}$) when $\sigma=$id=identity map, to be the totality of those $(q, i\infty y)=(q_1,\ldots,$ $q_{n'}; \ i\infty(y_1,\ldots,\ y_{n'}))$, with q_i and y_i being again real (resp. complex), for which one can find $(p, i\infty x)\in \Lambda_G^+$ (resp. $\Lambda_G^{\mathbb{C}}$) satisfying the equalities $p_j(q) = p_j$ $(j=1,\ldots, n)$ and $y_i = x_{\sigma(i)}$ $(i=1,\ldots, n')$.

Employing the notations just introduced, the above result on SS F_G is readily expressed as follows: F_G is microlocally well-defined everywhere except on the union of $\Lambda_{\tau(G),\tau}^+$ corresponding to the 2nd kind τ , and outside there SS F_G is contained in the union of $\Lambda_{\tau(G),\tau}^+$ corresponding to the 1st kind τ — including of course the case $\tau =\mathrm{id}$ (the *leading* Landau singularity).

The passage to $F_{G,\sigma}(q)$ from $F_G(p)$ is to be checked again by the general rule on *restriction* of microfunctions([3]-[5]), namely $F_{G,\sigma}(q)$ will be microlocally well-defined at $(q_0, i\infty y_0)$ if $\{(p,i\infty x)$ $\in$SS $F_G|$ $p_j=p_j(q)$ $(j=1,\ldots,n)$, $y_i=x_{\sigma(i)}$ $(i=1,\ldots, n')\}$ is compact in a uniform manner with respect to $(q, i\infty y)$ sufficiently close to $(q_0, i\infty y_0)$. (What matter here are those vertices j for which $\sigma^{-1}(j)$ is empty.) Thus we have SS $F_{G,\sigma} \subset \bigcup_\tau \Lambda_{\tau(G),\tau\sigma}^+$, admitting that for some (well-specified) τ, $F_{G,\sigma}$ may not be well-defined on $\Lambda_{\tau(G),\tau\sigma}^+$. ($\tau\sigma$ denotes the composition of maps). See [17].

4. - The above is a verification that microanalyticity postulate holds for Feynman integrals so long as they are well-defined. As was stated with S-matrices and Green's functions, however, this claim is essentially a consequence of the assertion that they are solutions of holonomic systems of microdifferential equations having complex Landau singularities $\Lambda_{\tau(G),\tau}^{\mathbb{C}}$ as their characteristic manifolds. (It is an elementary fact that the singular spectrum of a solution function of a system of microdifferential equation is contained in its characteristic manifold([1]-[3]).)

In the case of the leading singularity $\Lambda_G^{\mathbb{C}}$ this latter assertion is confirmed as follows.

First, we note that, setting $H(x) = \sum_{\ell=1}^N m_\ell \sqrt{(\eta_\ell(x))^2}$, we have

$$\Lambda_G^+ = \text{closure of } \{(p, i\infty x)| \; p_j = \frac{\partial}{\partial x_j} H(x) \; (j=1,\ldots,n), \; (\eta_\ell(x))^2 > 0$$
$$(\ell=1,\ldots, N)\},$$

This is not by accident because, generally speaking, a holonomic manifold in $(p, i\infty x)$- space whose projection to x-space is generically surjective should necessarily be of the form $\{(p,i\infty x)\mid p=\mathrm{grad}_x H(x)\}$ with suitable $H(x)$ locally defined, holomorphic and homogeneous of degree 1 in x (and vice versa). Furthermore, a simple holonomic system having this holonomic manifold as its characteristic manifold can always be written in the following canonical form (in which u signifies the unknown function)

$$(p_j - \Phi_j(-i\tfrac{\partial}{\partial p}))u = 0 \qquad (j=1,2,\dots)$$

with Φ_j constructed as follows: Let $A(x)=1+A_{-1}(x)+A_{-2}(x)+\cdots$ be a formal series of locally defined holomorphic functions $A_{-\mu}(x)$, homogeneous of degree $-\mu$ in x, such that $\frac{1}{\mu}\sqrt[\mu]{|A_{-\mu}(x)|}$ is bounded as μ increases, locally uniformly in x. (This is equivalent to saying that $\log A(x) = C_{-1}(x)+C_{-2}(x)+\cdots$ be a formal series of the same nature.) Let $\Psi_0(x)$ be a locally defined invertible holomorphic function, homogeneous of an arbitrary degree, say d, in x. Set $\Psi(x) = e^{-iH(x)}\cdot\Psi_0(x)\cdot A(x)$ formally, and then Φ_j is defined formally by $\Phi_j(x) = i\frac{\partial}{\partial x_j}\log\Psi(x)$. Hence we have

$$\Phi_j(x) = \frac{\partial H(x)}{\partial x_j} + i\left(\frac{1}{\Psi_0(x)}\frac{\partial\Psi_0(x)}{\partial x_j} + \sum_{\mu=1}^{\infty}\frac{\partial C_{-\mu}(x)}{\partial x_j}\right)$$

and imposed conditions on $\Psi_0(x)$ and $A(x)$ guarantee $\Phi_j(-i\frac{\partial}{\partial p})$ to be a well-defined microdifferential operator.

The order of this system is given by $d + \frac{1}{2}\times$dimension of the underlying space. In the above, $A(x)$ is uniquely determined by the given system while $\Psi_0(x)$ and hence $\Psi(x)$ are unique up to an arbitrary constant factor. Indeed, $\Psi(x)$ is characterized by the formal equation $(i\frac{\partial}{\partial x_j} - \Phi_j(x))\Psi(x) = 0 \ (j=1,2,\dots)$ or equivalently, by the formal equation obtained from the given holonomic system for $u(p)$ by replacing p_j and $\partial/\partial p_j$ by $i\partial/\partial x_j$ and ix_j respectively. Suppose now that the solution $u(p)$ of the given system is the Fourier transform of some $\tilde{u}(x)$:

$$u(p) = \mathrm{const.}\int e^{ip\cdot x}\,\tilde{u}(x)dx$$

and that $\tilde{u}(tx)$ as $t\to\infty$ has an asymptotic expression of the form $\Psi(tx) = e^{-it\cdot H(x)}\cdot\Psi_0(x)(1+t^{-1}\cdot A_{-1}(x)+t^{-2}\cdot A_{-2}(x)+\cdots)$. Then this formal expression $\Psi(x)$ is exactly the one which describes the above canonical form.

The holonomic system satisfied by F_G on the leading $\Lambda_G^{\mathbb{C}}$ is derived in [16]; the result is as follows.

$$\Psi(x) = \text{const.}\ \Pi_{\ell=1}^{N}\ F_{\nu/2-1}(m_\ell\sqrt{(\eta_\ell(x))^2-i0})$$

with $F_{\nu/2-1}(t) =_{\text{def}} e^{-it}\cdot t^{-(\nu-1)/2}\cdot\ {}_2F_0(\tfrac{\nu-1}{2},\ \tfrac{-\nu+3}{2};\ \tfrac{1}{-2it})$

$${}_2F_0(\alpha,\beta;z) =_{\text{def}} 1+\tfrac{\alpha\cdot\beta}{1!}z + \tfrac{\alpha(\alpha+1)\cdot\beta(\beta+1)}{2!}z^2+\cdots \quad (\text{divergent}),$$

$$\Phi_j(x) = i\frac{\partial}{\partial x_j}\log\Psi(x) = \sum_{\ell=1}^{N}[j:\ell]\frac{m_\ell\eta_\ell(x)}{\sqrt{(\eta_\ell(x))^2-i0}}\ L_{\nu/2-1}(m_\ell\sqrt{(\eta_\ell(x))^2-i0})$$

i.e. $\Phi_j(x)=\xi_j(\Phi^*(x))=\sum_{\ell=1}^{N}[j:\ell]\Phi_\ell^*(x)$ with

$$\Phi_\ell^*(x) = \frac{m_\ell\eta_\ell(x)}{\sqrt{(\eta_\ell(x))^2-i0}}\ L_{\nu/2-1}(m_\ell\sqrt{(\eta_\ell(x))^2 - i0}\ ),$$

with $L_{\nu/2-1}(t) =_{\text{def}} i\frac{d}{dt}\log F_{\nu/2-1}(t)= 1 - \frac{\nu-1}{-2it}$

$$- \frac{(\nu+1)(-\nu+3)}{2(-2it)^2}\cdot {}_2F_0(\tfrac{\nu+1}{2},\ \tfrac{-\nu+5}{2};\ \tfrac{1}{-2it})\Big/{}_2F_0(\tfrac{\nu-1}{2},\ \tfrac{-\nu+3}{2};\tfrac{1}{-2it})$$

$$= {}_2F_0(\tfrac{\nu+1}{2},\ \tfrac{-\nu+1}{2};\tfrac{1}{-2it}\ )\ \Big/\ {}_2F_0(\tfrac{\nu-1}{2},\ \tfrac{-\nu+3}{2};\ \tfrac{1}{-2it})$$

$$= 1+2(\gamma+\tfrac{1}{2})\cdot\tfrac{1}{2it}+2(\gamma^2-\tfrac{1}{4})\cdot\frac{1}{(2it)^2} - 4(\gamma^2-\tfrac{1}{4})\cdot\frac{1}{(2it)^3}$$

$$- 2(\gamma^2-\tfrac{1}{4})(\gamma^2-\tfrac{25}{4})\cdot\frac{1}{(2it)^4} + 16(\gamma^2-\tfrac{1}{4})(\gamma^2-\tfrac{13}{4})\cdot\frac{1}{(2it)^5}$$

$$+ 4(\gamma^2-\tfrac{1}{4})(\gamma^4-\tfrac{114}{4}\gamma^2 + \tfrac{1073}{16})\cdot\frac{1}{(2it)^6} + \cdots \quad (\gamma= \tfrac{\nu}{2}-1),$$

The equation: $(p_j - \Phi_j(-i\frac{\partial}{\partial p}))F_G(p) = 0, \quad j=1,\ldots,n,$

The order: $-\frac{\nu-1}{2}N + \tfrac{1}{2}\nu n.$

The following observation will show that this result is quite natural in view of what we mentioned above:

$F_G(p)$ is the Fourier transform of $\tilde{F}_G(x) =_{\text{def}}$

const. $\Pi^N_{\ell=1} \Delta_F(\eta_\ell(x), m^2_\ell)$ where Δ_F denotes the Feynman's fundamental solution for the Klein-Gordon equation. It is given by $\Delta_F(x,m^2) = (t^{-(\nu/2-1)} H^{(2)}_{\nu/2-1}(t))_{t\mapsto m\sqrt{x^2-io}}$ by means of Hankel function $H^{(2)}_{\nu/2-1}(t)$. The asymptoic behavior of this function $t^{-(\nu/2-1)} H^{(2)}_{\nu/2-1}(t)$ as $t \to +\infty$ is, however, given exactly by const. $F_{\nu/2-1}(t)$. Hence the asymptotic behavior of $\tilde{F}_G(x)$ as $(\eta_\ell(x))^2$, $\ell=1,\ldots,$ N, go to $+\infty$ simultaneously is given by $\Psi(x)$, as one will expect.

5. - The simple holonomic system which F_G satisfies on $\Lambda^{\mathbb{C}}_{\tau(G),\tau}$ for the 1st kind contraction τ is also easy to give. It is

$$\begin{cases} (\sum_{j\in\tau^{-1}(j_1)} p_j - \sum^{N_1}_{\ell=1}[j_1:\ell]_{\tau(G)} \Phi^*_\ell(-i\frac{\partial}{\partial p}))F_G(p) = 0 \quad (j_1=1,\ldots,n_1) \\[2ex] (\eta_\ell(-i\frac{\partial}{\partial p}) - 2i\Theta_\ell(p,-i\frac{\partial}{\partial p})\cdot((\Theta_\ell(p,-i\frac{\partial}{\partial p}))^2-m^2)^{-1})F_G(p)=0 \quad (\ell=N_1+1,\ldots N) \end{cases}$$

where it is supposed that the N internal lines of G are so labeled that the survivors in $\tau(G)$ correspond to $\ell=1,\ldots,N_1$. $[j_1:\ell]_{\tau(G)}$ signifies the incidence number in $\tau(G)$. Finally $\Theta_\ell(p,-i\frac{\partial}{\partial p})$, $\ell=N_1+1,\ldots,N$ is a (vector valued) microdifferential operator of the 0th order constructed as follows:

First, one solves the linear equations $p_j - \xi_j(k) = 0$ $(j=1,\ldots,n)$ with respect to $k_{N_1+1},\ldots, k_N$ and obtains expressions of the form

$$k_\ell = L_\ell(p,k) = \sum^n_{j=1} a_{\ell j}p_j + \sum^{N_1}_{\ell'=1}b_{\ell\ell'} k_{\ell'}, \quad (\ell=N_1+1,\ldots,N).$$

(The coefficients $a_{\ell j}$, $b_{\ell\ell'}$ are +1, -1 or 0.) This is possible because τ is of the 1st kind. L_ℓ are uniquely determined modulo linear combination of $\sum_{j\in\tau^{-1}(j_1)} p_j - \sum^{N_1}_{\ell=1}[j_1:\ell]_{\tau(G)}k_\ell$, $j_1=1,\ldots,n_1$. Θ_ℓ is then given by

$$\Theta_\ell(p,x) = L_\ell(p,\Phi^*(x)) = \sum^n_{j=1}a_{\ell j}p_j + \sum^{N_1}_{\ell'=1} b_{\ell\ell'}\Phi^*_{\ell'}(x).$$

The holonomy diagram built by these holonomy manifolds $\Lambda^{\mathbb{C}}_{\tau(G),\tau}$ for the 1st kind τ's has the following shape: Calling the vertex representing $\Lambda^{\mathbb{C}}_{\tau(G),\tau}$ 'the vertex τ' for brevity, vertices τ and τ' are joined by a segment if and only if one of the linear graphs corresponding to them, say $\tau'(G)$, is obtained from the other, $\tau(G)$, by

contracting two vertices which are joined by a *simple* internal line.
Further, this segment represents a *simple* intersection of holonomy
manifolds, and the order of the system on the vertex τ' or the manifold
$\Lambda^{\mathbb{C}}_{\tau'(G),\tau'}$ is less by 1/2 than that on the vertex τ.

The full holonomy diagram including vertices corresponding to the
2nd kind τ's is not so simple as this. There again will be those
segments each of which corresponds to contraction of a simple internal
line in some $\tau(G)$, giving rise to a *simple* intersection of holonomy
manifolds; there appear however segments of more complicated nature
which correspond to contraction of one or more loops in $\tau(G)$ whose
size is 'small' in relation to the dimensionality ν . (See [17].)

On the other hand, if every loop present in the graph $\tau(G)$ is
'big' in relation to ν in the sense as is defined in Kashiwara's note
[17], i.e. if the inequality $-(\nu-1)N+\nu n \geq (\nu+1) \times$ (the number of connected components in the graph) holds, not only with the graph $\tau(G)$ itself
but also with every subgraph of $\tau(G)$ (such a graph we will refer to
as an *'ample'* graph), then $\Lambda^{+}_{\tau(G),\tau}$ (resp. $\Lambda^{\mathbb{C}}_{\tau(G),\tau}$) is the conormal
bundle of a real (resp. complex) submanifold of codimension $-(\nu-1)N_{\tau}+\nu n_{\tau}$
in the νn-dimensional p-space (N_{τ} and n_{τ} denote the numbers of internal
lines and vertices in $\tau(G)$), and $F_G(p)$ is, microlocally on $\Lambda^{+}_{\tau(G),\tau}$,
equal to the δ-function of this submanifold multiplied by an analytic
function which is given by a beautiful *explicit formula*. This remarkable
important result was obtained by Kashiwara by applying his powerful
theory of *principal symbols for holonomic systems and microfunction
solutions* [9].

6. - As to $\tau(G)$ which is not necessarily ample, one derives
the following results, which we owe to T. Kawai and M. Kashiwara:

(1) Let τ_1 denote the contraction of two vertices in G joined
by a simple line ℓ_1. Then in a neighborhood of any 'interior' point
of $\Lambda_G \cap \Lambda_{\tau_1(G),\tau_1}$ (meaning a point $(p, i\infty x)$ in it where $\eta_{\ell}(x) \neq 0$
hold for $\ell \neq \ell_1$) we can find a microdifferential operator $Q(p, \frac{\partial}{\partial p})$ such
that

$$F_{\tau_1(G)}(p) = Q(p, \frac{\partial}{\partial p})F_G(p)$$

holds there.

(2) Let $\tau(G)$ be a contraction of G of the 1st kind. Then in a neighborhood of the 'interior' of $\Lambda_{\tau(G),\tau}$ (meaning that $\eta_\ell(x) \neq 0$ hold for those lines ℓ which survive in $\tau(G)$)

$$F_G(p) = a(p) \cdot F_{\tau(G)}(p)$$

holds with an analytic function $a(p) \neq 0$.

This generalizes a result of L. M. Brown [19] where n-vertex loop graph is discussed.

It should be remembered that these theorems as well as the following theorem (3) are logical consequences of the above cited micro-local property of Feynman integral — the very simple nature of its holonomy diagram as long as the 1st kind τ's are concerned.

Microlocal calculus also gives us the explicit form of local singularity structure of a solution hyperfunction of a holonomic system as are exemplified in Kashiwara [17]. One can accordingly find explicit forms of $F_G(p)$ at various singular points in the p-space, providing of course that this Feynman integral is well-defined. Here we will content ourselves with one example taken from [17]:

Let τ_1 be as in (1). As holonomy manifolds in the $(p, ix\infty)$-space Λ_G and $\Lambda_{\tau_1(G),\tau_1}$ are the conormal bundles of their respective projections to the p-space, which will be called L_G and $L_{\tau_1(G),\tau_1}$ (the so-called Landau 'surfaces' — actually they are submanifolds in the p-space in general). Assume now that L_G, whence $L_{\tau_1(G),\tau_1}$ too, is of codimension $\nu+1$ in the p-space, i.e. of codimension 1 within the ν-codimensional linear subspace defined by the energy-momentum conservation $\sum_{j=1}^n p_j=0$, so that they are defined by single equations $f(p')=0$ and $g(p')=0$, respectively, with p' denoting $(p_j; j=1,\ldots, n-1)$ with p_n missing. Assume further that they are both non-singular along the 'interior' of their (2-codimensional) intersection manifold $L_G \cap L_{\tau_1(G),\tau_1}$. Then, along there they osculate together exactly to the 2nd order, (Fig.4), so that one can write $f = g + h^2$ with an analytic h, where one chooses the sign of h so that Λ_G^+ corresponds to $h \geq 0$. Now we state:

(3) Assumptions being as above, the Feynman integral $F_G(p)$ is

Fig. 4

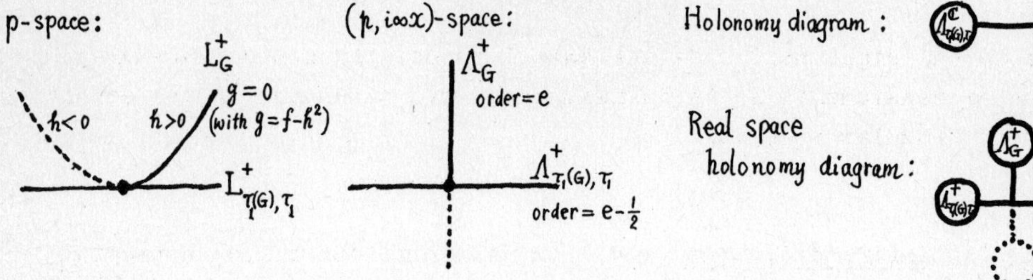

expressed as follows in a neighborhood of the interior of $L_G \cap L_{\tau_1(G),\tau_1}$
(= $\{p \mid \sum_{j=1}^{n}p_j = 0,\ f(p') = 0,\ g(p') = 0\}$) by suitably choosing analytic
functions $\varphi_0(p')\ (\neq 0)$, $\varphi_1(p')$ and $\psi(p')$:

$$F_G(p) = (\varphi_0(p')v_0(f(p'),h(p')) + \varphi_1(p')v_1(f(p'),h(p'))$$

$$+ \psi(p'))\cdot\delta^\nu(\sum_{j=1}^{n}p_j)$$

where $v_0(t_1,t_2)$ denotes the hyperfunction of (t_1,t_2) characterized
up to a constant factor by the holonomic system of differential equations

$$(t_1\frac{\partial}{\partial t_1} + \frac{1}{2}t_2\frac{\partial}{\partial t_2} + e + \frac{1}{2})v_0 = 0,\quad \frac{\partial}{\partial t_2}(\frac{\partial}{\partial t_2} + 2t_2\frac{\partial}{\partial t_1})v_0 = 0$$

whose characteristic manifold is the union of Λ and Λ_1 given by

$$\Lambda = \{(t_1,t_2;i\infty(\eta_1,\eta_2))\mid t_1-t_2^2 = 0,\ 2t_2\eta_1 + \eta_2 = 0\},$$

$$\Lambda_1 = \{(t_1,t_2;i\infty(\eta_1,\eta_2))\mid t_1 = 0,\ \eta_2 = 0\},$$

and by the additional condition $SS\ v_0 \subset \{\text{points in }\Lambda \mid \eta_1 \geq 0,\ t_2 \geq 0\} \cup$
$\{\text{points in }\Lambda_1 \mid \eta_1 \geq 0\}$, whereas $v_1(t_1, t_2)$ is defined to be
$(\frac{\partial}{\partial t_2}\Big/\frac{\partial}{\partial t_1})v_0(t_1, t_2)$.

Explicitly they are given by

$$v_0(t_1,t_2) = t_1^{-e/2-1/4}(t_1-t_2^2)^{-e/2-1/4}\ Q_{e-1/2}^{e+1/2}(t_2/\sqrt{t_1}),$$

$$v_1(t_1,t_2) = t_1^{-e/2+1/4}(t_1-t_2^2)^{-e/2-1/4}\ Q_{e-3/2}^{e+1/2}(t_2/\sqrt{t_1}),$$

with

$$Q_\nu^\mu(z) = \text{associated Legendre function}$$

$$= \text{const. } z^{-\nu-1}\left(1-\frac{1}{z^2}\right)^{\mu/2} \cdot {}_2F_1\left(\frac{\mu+\nu+1}{2} \ \frac{\mu+\nu+2}{2}; \ \nu+\frac{3}{2} \ ; \frac{1}{z^2}\right), \quad |z| > 1,$$

where on the right hand side one should choose the appropriate branch
of the expressions so as to satisfy the above-cited additional condition
on their singular spectrum.

7. <u>Infinite Systems</u> – We now return to our intended program
and try to formulate a general theory of an *infinite system* of functions
(for definiteness, say that of τ-functions for a neutral scalar field
of mass $m > 0$) $F_n(p_1,\ldots, p_n)$, $n=0,1,2,\ldots$, on the dual basis of
microlocal calculus and (generalized) *unitarity-discontinuity relations*.

Le M and $M^{\mathbb{C}}$ denote the ν-dimensional vector space (of energy-
momenta) and its complexification, and let $\mathcal{E}_n$ and $\mathcal{C}_n$ denote the
sheaf of microdifferential operators on the cotangent bundle $T^*(M^{\mathbb{C}})^n$
(i.e. on the complexified n-particle $(p,i\infty x)$-space) and the sheaf of
microfunctions on iS^*M^n , respectively. By an 'infinite system' we
will mean the system of hyperfunctions $F_n = F_n(p_1,\ldots, p_n)$ on M^n,
$n=0,1,2,\ldots$, symmetric in their n sets of variables and satisfying
the equation $(p_1+\ldots+p_n)\cdot F_n = 0$ (the total energy-momentum conservation)
and the following postulates I (properties imposed on each F_n) and
II (relations imposed between various F_n).

<u>Postulate I</u>. – The singular spectrum of F_n is contained in
the union of positive α Landau holonomy manifolds $\Lambda^+_{G,\sigma}$ for various
pairs of linear graph G and a map $\sigma:\{1,\ldots, n\} \to \{\text{vertices of } G\}$,
and there it satisfies the following conditions:

(1) In the interior of each $\Lambda^+_{G,\sigma}$, F_n satisfies a *simple* holonomic
system of microdifferential equations whose characteristic manifold is
$\Lambda^{\mathbb{C}}_{G,\sigma}$ in the microlocal sense, and F_n has the order $-\frac{\nu-1}{2}N_G + \frac{\nu}{2}n_G$
there; (2) In the interior of $\Lambda^+_{G,\sigma} \cap \Lambda^+_{\tau_1(G),\tau_1\sigma}$ corresponding to
a simple contraction τ_1, F_n satisfies a *simple* holonomic system whose
characteristic manifold is $\Lambda^{\mathbb{C}}_{G,\sigma} \cup \Lambda^{\mathbb{C}}_{\tau_1(G),\tau_1\sigma}$ in the microlocal sense.

Note that, as we mentioned in §6, these two conditions are
sufficient to yield theorems (1)-(3) of §6 (with Feynman integrals now
replaced by F_n). Our conditions are exactly those which the Feynman
integrals F_G are verified to satisfy. Therefore it will be natural

to further introduce the following condition : (3) At each point
$(p_0, i\infty x_0)$ of the singular spectrum $\cup_{G,\sigma} \Lambda^+_{G,\sigma}$, excepting those
points to be specified below, F_n satisfies a holonomic system of
microdifferential equations which has $\cup_{G,\sigma} \Lambda^{\mathbb{C}}_{G,\sigma}$ as its characteristic
manifold in the microlocal sense, and furthermore this holonomic system
is of the same nature as the one satisfied by the corresponding Feynman
integral. (The last part of the statement is not satisfactory and
needs further clarification — this is a future problem.)

In (3) one has to exclude those points where the family of $\Lambda^{\mathbb{C}}_{G,\sigma}$
is not locally finite because if an infinite number of $\Lambda^{\mathbb{C}}_{G,\sigma}$ clusters
at $(p_0, i\infty x_0)$ their union can not be the characteristic manifold of
a holonomic system at this point. (Note that the finiteness theorem
of Stapp [20] on the positive α Landau surfaces does not apply to
our case of *complex* holonomic manifold.)

<u>Postulate II</u> - The system $F_n(p_1, \ldots, p_n)$, n=0,1,2,..., satisfies
the *generalized unitarity relations* as are formulated in Nishijima [21]
and consequently, satisfies the Cutkosky type *discontinuity formulas*
which are to be derived from the generalized unitarity relations on
the hypothesis of microanalyticity.

Such derivation of discontinuity from unitarity has been extensively studied by the Cambridge school and by H. P. Stapp, especially in the
case of S-matrix. (See [13], [14] and the book in preparation by Stapp.)

One the other hand, existence of holonomic system (Postulate I)
is verified for the S-matrix at 'very simple' point of Landau surfaces
by Kawai-Stapp [18] on the basis of discontinuity formula.

It is an interesting problem to determine the structure of the
relevant holonomic system at points which is not covered by (1) or (2).
Defining $\mathcal{J}_n$ to be the left ideal of $\mathcal{E}_n$ consisting of $P \in \mathcal{E}_n$ such
that $PF_n = 0$ holds and setting $\mathcal{M}_n = \mathcal{E}_n / \mathcal{J}_n$ we obviously have $F_n \in$
$\mathrm{Hom}_{\mathcal{E}_n}(\mathcal{M}_n, \mathcal{C}_n)$ and support of $\mathcal{M}_n = \mathrm{SS}\, F_n$, and the problem now is to
determine the structure of the stalks of $\mathcal{M}_n$ at various points of
$\mathrm{SS}\, F_n$. (A stalk of a sheaf is the set of its germs at a point.)
However, little is known about this. In some cases we can confirm that
$\mathcal{M}_n$ is the proper direct image of a coherent $\mathcal{E}$-module.

8. <u>Models</u> - The n-site correlation function $\overset{\gamma}{\phi}_n(\nu_1,\ldots,\nu_n)$, $\nu_i \in \mathbb{Z}$, for the Ising model of 1-dimensional crystal lattice without external field is easily computed to be $= \exp(-\alpha \cdot M(\nu_1,\ldots,\nu_n))$ for even n, and $= 0$ for odd n, where $\alpha = - \log \tanh J > 0$ (i.e. $J>0$) denotes a given constant and $M(\nu_1,\ldots,\nu_n)$ for even n stands for $\nu'_1-\nu'_2+\nu'_3-\nu'_4+\cdots-\nu'_n$ with $\nu'_1 \geq \nu'_2 \geq \cdots \geq \nu'_n$ denoting the rearranged set of $\nu_1,\nu_2,\cdots,\nu_n$. We now admit the parameter α to be a complex number with positive real part (i.e. Im J to be in between $\pm\pi/4$) and further admit the limit case of pure imaginary α. The Fourier transform $\Phi_n(\theta_1,\ldots,\theta_n) = \sum_\nu \overset{\gamma}{\phi}_n(\nu_1,\ldots,\nu_n) \cdot \exp$ $i(\nu_1\theta_1+\cdots+\nu_n\theta_n)$ is then a hyperfunction on $(\mathbb{R}/2\pi\mathbb{Z})^n$. By taking the continuum limit $\lim_{\varepsilon\to 0} \varepsilon^n \Phi_n(\varepsilon p_1,\ldots,\varepsilon p_n)$ with $\alpha = i\varepsilon m$ ($m>0$), we obtain a system of hyperfunctions $F_n(p_1,\ldots,p_n)$ on $\mathbb{R}^n$ given by

$$\sum \frac{i}{p_1-m+i0} \frac{i}{p_1+p_2+i0} \frac{i}{p_1+p_2+p_3-m+i0} \cdots \frac{i}{p_1+\cdots+p_{n-1}-m+i0} \cdot 2\pi\delta(p_1+\cdots+p_n)$$

for even n and by 0 for odd n; here the sum is to be taken over n! permutations of suffixes.

On considering the present situation of '1-dimensional space-time' we define graphs and holonomy manifolds associated to them as follows: Let I denote a partition of $\{1,\ldots,n\}$ for even n into subsets A,B,... where, further, those subsets with odd number of suffixes (the number of such subsets is obviously even) are to be divided into equal numbers of '+' class and '-' class. Set $\varepsilon(A)=+1$, -1, or 0 according as the subset A is of '+' class, '-' class, or even class. The complex holonomy manifold $\Lambda_I^{\mathbb{C}}$ is then defined by these set of equations: $\sum_{j\in A} p_j - \varepsilon(A) m = 0$ and $x_{j_1} = x_{j_2} = \ldots$ for each subset $A=\{j_1,j_2,\ldots\}$ in I. Let λ denote a linear ordering A_1, A_2,... of subsets in I such that $\sum_{i=1}^r \varepsilon(A_i)$ is 0 or $+1$ for $r=1,2,\ldots$. A 'positive α' holonomy manifold $\Lambda_{I,\lambda}^+$ is then defined to be the real points in $\Lambda_I^{\mathbb{C}}$ satisfying the inequality $x_{j_1} \geq x_{j_2} \geq \ldots$ for $j_1 \in A_1$, $j_2 \in A_2,\ldots$. Now we have the following results <u>A</u> and <u>B</u> [22]:

<u>Theorem A</u> - The functions F_n, n=0,1,2,... (where F_0 represents the constant 1) form an infinite system satisfying Postulates I and II. More specifically we have: (I) F_n satisfies a holonomic system of microdifferential equations $\mathcal{M}_n$ defined everywhere in $T^*\mathbb{C}^n$ (the complex (p,i∞x)space) whose characteristic manifold is the union of $\Lambda_I^{\mathbb{C}}$ for verious partitions I of $\{1,\ldots,n\}$. SS F_n coincides with the union of $\Lambda_{I,\lambda}^+$ for various (I, λ). $\mathcal{M}_n$ is simple and obeys the situation (1) and (2) of Postulate I (and even the situation for 'ample graphs'

in §5) everywhere.

(II$_1$) The *generalized unitarity relations* (the Fourier transformed form of (2.15) in K. Nishijima [21]) hold for n=1,2,...:

$$\sum_{\ell=0}^{\infty} \frac{(-1)^{\ell}}{\ell!} \; \sum_{k=0}^{n} \; \sum{}^{*} \; F_k^{(\ell)}(p_1,\ldots,p_k) \cdot \overline{F}_{n-k}^{(\ell)}(p_{k+1},\ldots,p_n) = 0$$

where $F_k^{(\ell)}(p_1,\ldots,p_k)$ represents $((q_1-m)\ldots(q_\ell-m)F_{k+\ell}(p_1,\ldots,p_k,$
$-q_1,\ldots,q_\ell))_{q_1\mapsto m,\ldots,q_\ell\mapsto m}$, $\overline{F}_k^{(\ell)}$ denotes its complex conjugate, and Σ^*
signifies the sum over $\binom{n}{k}$ combinations of k suffixes from $\{1,\ldots,n\}$.
—— Indeed, we have $F_k^{(\ell)}=0$ except for $\ell=0$ and 1 and these surviving
terms (finite in number) are shown to add up together to 0.

(II$_2$) The *discontinuity relations* present themselves as follows,
owing to the 'ample graph' situation mentioned in (I). Let (I,λ) be
a linearly ordered partition $(A_1,A_2,\ldots)$ as introduced above and suppose
that A_1 consists of $\{1,\ldots,k\}$. Then the principal symbol (of Kashiwara)
of $F_n(p_1,\ldots,p_n)$ on the interior of $\Lambda_{I,\lambda}^{+}$ is factorized as: $\sigma_{I,\lambda}^{+}(F_n)$
$=\sigma_{I_1,\lambda_1}^{+}(F_k)\cdot\sigma_{I',\lambda'}^{+}(F_{n-k})$ if k is even, $=\sigma_{I_1,\lambda_1}^{+}(F_{k+1})\cdot\sigma_{I',\lambda'}^{+}(F_{n-k+1})$
if k is odd   (with constant factors being suitably adjusted). Here,
in the case of even k (I_1, λ_1) and (I',λ') represent the ordered
partitions (A_1) and $(A_2,\ldots)$ while in the opposite case they mean
$(A_1, \overline{B})$ and $(B, A_2,\ldots)$ where $\overline{B}$ (resp. B) stands for a newly introduced set of − (resp. +) class consisting of a single letter; F_{k+1} and
F_{n-k+1} accordingly stand for expressions of the form $F_{k+1}(p_1,\ldots,p_k,-q)$
and $F_k(q, p_{k+1},\ldots, p_n)$. By further applying the same process on the
second factor one arrives at the desired discontinuity formula: $\sigma_{I,\lambda}^{+}=$
$\sigma_{I_1,\lambda_1}^{+} \; \sigma_{I_2,\lambda_2}^{+} \cdots$.

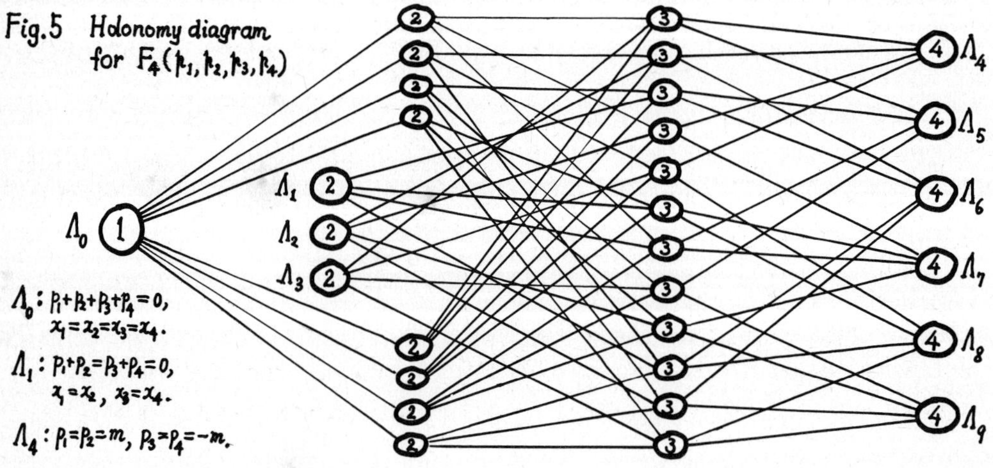

Theorem B - If an infinite system of hyperfunctions $F_n^*(p_1,\ldots,p_n)$ $n=0,1,2,\ldots$ satisfying $(p_1+\ldots+p_n)F_n^*=0$ possesses the properties (I) and (II$_2$) of the preceding theorem and further satisfies the following condition (III) on the growth order then $F_n^* = F_n$ results.

(III) The principal symbol of the holonomic system $\mathcal{M}_n$ on (the interior of) $\Lambda_{I_0}^{\mathbb{C}} = \{(p,\ i\infty x)\,|\,\sum p_j=0,\ x_1=\ldots=x_n=0\}$ (which corresponds to 'the partition into the single set') tends to 0 at infinity. — This means that the (extended) scattering amplitude tends to 0 for large values of complex energy-momenta (T. Miwa).

Similar results as above are obtained also for $\Phi_n(\theta_1,\ldots,\theta_n)$ considered as an infinite system of hyperfunctions on the *torus* manifold. A simple gas model gives another example of F_n. In the dilute limit we obtain $F_n(p_1,\ldots,p_n) = \sum_G^{(n)} F_G^*(p_1,\ldots,\,p_n)$, where the sum is to be taken over all linear graphs with n vertices and F_G^* denotes the Feynman integral of §3 with the 'propagator function' $1/(k_\ell^2-m_\ell^2+i0)$ replaced by one and the same (given) function $\Delta(k_\ell)$ (which describes the interaction between two molecules) [22].

References

1. M.Sato, in Proc. Intern. Conf. on Functional Analysis and Related
    Topics, Univ. of Tokyo Press(1969).
2. M.Sato-T.Kawai, in Reports of Katata Symp., Shinkokai(1969)(Japanese).
3. M.Sato-M.Kashiwara, Sugaku no Ayumi 15, 9(1970)(Japanese).
4. M.Sato, in Proc. Nice Congress 2, Gauthier-Villars, Paris(1970).
5. M.Sato-T.Kawai-M.Kashiwara, in Proc. Katata Conference 1971, Lecture
    Notes in Math. 287, Springer Verlag(1973).
6. L.Hörmander, Comm. Pure Appl. Math. 24, 671(1971);Acta Math.127,79(1971).
7. D.Iagolnitzer, in Proc. Nice Conf. 1973; and in this Proceedings.
8. M.Sato, Lectures at Nagoya Univ. 1974, Kokyuroku 226, RIMS. Kyoto
    Univ., 114(1975) (Notes by M.Jimbo, Japanese).
9. M.Kashiwara, Lectures at Nagoya Univ. 1974, Kokyuroku 226, RIMS. Kyoto
    Univ., 167(1975)(Notes by T.Kimura, Japanese); and to appear in
    Kokyuroku (1975)(Notes by T.Miwa, Japanese).
10. T.Suzuki, Master thesis, Nagoya Univ. 1975 (Applications of 9;Japanese).
11. F.Pham, Talk at Nice Conf. on Hyperfunctions and Physics 1973.
12. D.I.Olive, Talk at Nice Conf. on Hyperfunctions and Physics 1973.
13. R.J.Eden-P.V.Landshoff-D.I.Olive -J.C.Polkinghorne, The analytic
    S-matrix, Cambridge Univ. Press (1966).
14. D.Iagolnitzer, Introduction to S-matrix theory (1973).
15. N.Nakanishi, Graph theory and Feynman integrals, Gordon & Breach(1971).
16. M.Sato, Lectures on microanalyticity of the S-matrix, Kyoto Univ. 1973,
    Kokyuroku 209, RIMS.Kyoto Univ., 11(1974)(Notes by T.Kawai, Japanese).
17. M.Kashiwara, in this Proceedings.
18. T.Kawai-H.P.Stapp, in this Proceedings.
19. L.M.Brown, Nuovo Cimento 22, 178(1961).
20. H.P.Stapp, J.Math. Phys. 8, 1606(1967).
21. K.Nishijima, Phys. Rev. 119, 485(1960).
22. M.Sato-M.Kashiwara-T.Miwa, Microlocal study of infinite systems,
    to appear in Kokyuroku, RIMS. Kyoto Univ. (1975).
5*. A.Cerezo-J.Chazarain-A.Piriou, in Proc. Nice Conf. 1973.

M. Kashiwara
Department of Mathematics, Nagoya University
Nagoya, Japan

MICRO-LOCAL CALCULUS

Introduction

The methods of micro-local calculus are to characterize a function by the system of differential equations which it satisfies and to analyze the micro-local structure of the system. We are mostly concerned with holonomic systems, because they are the systems whose solutions form a vector space of finite dimension. Almost all functions which we encounter in Physics or Mathematics are solutions of holonomic systems.

In this paper an outline of the theory of holonomic systems is given and the micro-local properties of Feynman integrals are discussed as its application.

In section 1 we explain some basic notions concerning holonomic systems such as holonomy diagram, geometric and analytic interaction, order and principal symbol.[*] In section 2 we give explicit forms of functions determined by holonomic systems. Then in the last section we apply our theory to the Feynman integrals.

1. *Holonomic systems and holonomy diagram.*

First we review a micro-local calculus of holonomic systems.

Consider a system of differential (or micro-differential) equations on a manifold X with an unknown function $u(x)$;

$$\mathcal{M} : P_1(x, D)u(x) = \cdots = P_N(x, D)u(x) = 0$$

$P_j(x, D)$ $(j = 1, \cdots, N)$ are differential (resp. micro-differential) operators. Denoting by $\mathcal{D}$ (resp. $\mathcal{E}$) the sheaf of differential (resp. micro-differential) operators, we write by $\mathcal{J}$ the ideal of $\mathcal{D}$ (resp. $\mathcal{E}$) generated by $P_1(x, D), \cdots, P_N(x, D)$, so that $\mathcal{D}u = \mathcal{D}/\mathcal{J}$ (resp. $\mathcal{E}u = \mathcal{E}/\mathcal{J}$).

---

[*] As for details of holonomic systems, see S-K-K [1], [4], [5], where we have used "maximally overdetermined system" and "Lagrangean manifold" in place of "holonomic system" and "holonomic manifold", respectively.

We analyze its structure micro-locally, that is, locally on the cotangent bundle T*X of X.

The *symbol ideal* J is the ideal of the sheaf $\mathcal{O}_{T^*X}$ of holomorphic functions on the cotangent vector bundle T*X of X, which is generated by principal symbols of operators in $\mathcal{J}$. The common zeros Λ of all functions in the symbol ideal J is said to be the *character- istic variety* of the system $\mathcal{M}$. It is known that the codimension of the characteristic variety does not exceed the dimension of X, say n. A *holonomic system* is, by definition, a system whose characteristic variety has codimension n. In this case, its characteristic variety Λ is holonomic, that is, the fundamental 1-form $\omega = \Sigma\xi_j dx_j$ of T*X vanishes on it. Let $\Lambda = \cup \Lambda_j$ be a decomposition of Λ into irredu- cible components. Λ_j is called a holonomic component. Λ_j is the conormal bundle of its image Y_j by the projection to X.

If the symbol ideal J is reduced, in other words, if J coin- cides with the ideal of functions which vanish on Λ, then we say $\mathcal{M}$ is *very good*. If J is reduced at generic points of Λ_j's, then $\mathcal{M}$ is said to be *simple*.

When there exists an irreducible analytic set S of dimension (n-1) contained in Λ_j and Λ_k (j ≠ k), we say Λ_j and Λ_k have a *geometric interaction* at S. If, moreover, $\mathcal{M}$ is not a direct sum of two $\mathcal{E}$ -Modules whose characteristic varieties are Λ_j and Λ_k respec- tively in a neighborhood of a generic point of S, then we say that Λ_j and Λ_k have an *analytic interaction*.

A geometric interaction is important because no other interaction between irreducible components occurs by virtue of the following fact; if two holonomic systems $\mathcal{M}_1$ and $\mathcal{M}_2$ are isomorphic outside of (n+2) co-dimensional analytic set, then they are globally isomorphic. By these data, we write a *holonomy diagram*, which consists of dots and segments joining them. Dots represent the holonomic components Λ_j of characteristic variety and segments represent the analytic interactions between the holonomic components.

We explain the order of a generator of a simple holonomic system on each irreducible component Λ_j. There is a micro-differential operator $P(x, D) = P_m(x, D) + P_{m-1}(x, D) + \cdots$ of order m belonging to $\mathcal{J}$ such that $dP_m(x, \xi) \equiv \varphi\omega$ mod J for some function φ on Λ_j. Here

$\omega = \Sigma \xi_j dx_j$ is the fundamental 1-form. It is shown that

$$\varphi^{-1}[P_{m-1} - \frac{1}{2} \sum_{i=1}^{n} \frac{\partial^2 P_m}{\partial x_i \partial \xi_i}] - \frac{m-1}{2}$$ is a constant function on Λ_j indepen-

dent of the choice of P and called the *order* of u on Λ_j. The holonomy diagram with orders on each component gives us a fundamental information on the holonomic system.

A microfunction solution $u(x)$ of a given holonomic system on each Λ_j can be written $u(x) = P(x, D)\delta(x_1, \cdots, x_r)$ by a suitable elliptic micro-differential operator P, if we take a local coordinate system $(x_1, \cdots, x_n)$ so that Λ_j is a conormal bundle of a submanifold defined by $x_1 = \cdots = x_r = 0$. $(2\pi)^{-r/2}\sigma(P)|_{\Lambda_j} (d\xi_1 \cdots d\xi_r dx_{r+1} \cdots dx_n / dx_1 \cdots dx_n)^{1/2}$ is said to be the *principal symbol* of $u(x)$. Here, $dx_1 \cdots dx_n$ and $d\xi_1 \cdots d\xi_r dx_{r+1} \cdots dx_n$ are regarded as volume elements on X and Λ_j, respectively. It was shown that this is invariant by a coordinate transformation. The order of a generator u coincides with the homogeneous degree of the principal symbol with respect to the fiber coordinate ξ. The principal symbol $\sigma(u)$ satisfies the following system of differential equations of order 1 on Λ_j so that it can be characterized as its solution up to constant multiple: $(H_{P_m} +$

$$(P_{m-1} - \frac{1}{2} \Sigma \frac{\partial^2 P_m}{\partial x_i \partial \xi_i}))[\sigma(u)(dx_1 \cdots dx_n)^{1/2}] = 0 \quad \text{for any} \quad P = P_m(x, D)$$

$+ P_{m-1}(x, D) + \cdots$ in $\mathcal{J}$. Here, $H_{P_m} = \Sigma(\frac{\partial P_m}{\partial \xi_i} \frac{\partial}{\partial x_i} - \frac{\partial P_m}{\partial x_i} \frac{\partial}{\partial \xi_i})$.

Microfunction solutions of a simple holonomic system at a generic point of holonomic components form a vector space of one dimension, and the principal symbol characterizes them.

At a generic point of Λ_j, a simple holonomic system is determined by its order. That is, if two systems of equations $\mathcal{M}_\nu = \mathcal{E} u_\nu$ ($\nu=1,2$) with the same order are given, then there exists locally an elliptic operator $Q(x, D)$ of *order* 0 such that $u_1 = Qu_2$ gives an isomorphism between $\mathcal{M}_1$ and $\mathcal{M}_2$.

If Λ_j and Λ_k intersect regularly at an (n-1)-dimensional analytic set S (that is, in a neighborhood of generic point of S, Λ_j and Λ_k are non singular and the tangent space of S is the intersection of those of Λ_j and Λ_k) and $\mathcal{M}$ is very good, we call it the regular interaction. Then, $\mathcal{M}$ is determined by the orders e_j, e_k on Λ_j and Λ_k. If $e_j - e_k - \frac{1}{2} \neq 0, 1, 2, \cdots$, then the support of any non zero microfunction solution is not contained in Λ_j. That is, the solution propagates from Λ_k to Λ_j. On the contrary, if two systems $\mathcal{M}_\nu = \mathcal{E} / \mathcal{J}_\nu$ ($\nu = 1,2$) are given with the same orders e_j and e_k, $\mathcal{J}_1 = \mathcal{J}_2$ on Λ_j implies $\mathcal{J}_1 = \mathcal{J}_2$ on $\Lambda_j \cup \Lambda_k$, under the same condition $e_j - e_k - \frac{1}{2} \neq 0, 1, \cdots$. That is, the system itself propagates from Λ_j to Λ_k. Therefore, if the difference is not a half integer, they propagate in both directions. In the real domain, S divides Λ_j (resp. Λ_k) into two components. Microfunction solutions at a generic point of S form a vector space of two dimentions, and two relations among the four principal symbols of a microfunction solution can be written down.   This leads us to the global determination of principal symbols of microfunction solutions by means of analytic interactions. (See [5].)

Example

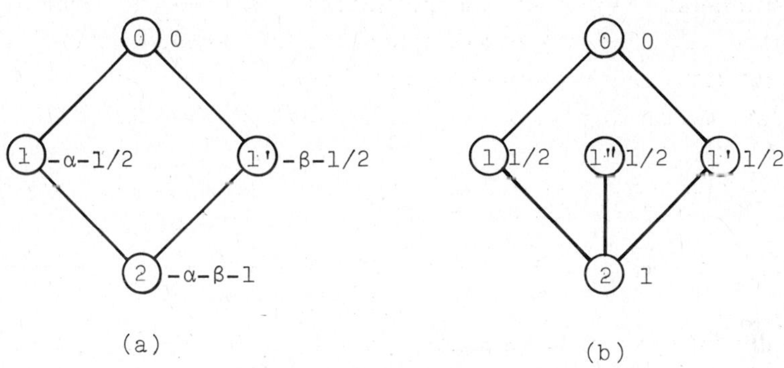

(a)   (b)

The number enclosed with a circle signifies the codimension of the image of holonomic components to the base manifold. The number beside a circle is the order on the corresponding holonomic manifold. Example (a) : $X = \mathbb{C}^2$, $(xD_x - \alpha)u = (yD_y - \beta)u = 0$, ⓪, ①①', ② are the conormal bundles of X, $x = 0$, $y = 0$, $x = y = 0$, resp. Example (b) : $X = \mathbb{C}^2$, $(D_x - D_y)xy(x + y)u = (D_x - D_y)D_x xu = (D_x - D_y)D_y yu = (D_x x + D_y y)u = 0$, ⓪, ①, ①', ①", ② are the conormal bundles of X, $x = 0$, $y = 0$, $x+y = 0$, $x = y = 0$, resp. ; In this case, ⓪ and ①" have a geometric interaction but no analytic one.

2. *Explicit forms of functions with simple singular spectrum.*

When a function $u(x)$ satisfies a simple holonomic system and the characteristic variety is not complicated, we can give an explicit form of functions. (In this section, the conormal of the total manifold is omitted.)

a) In the case the characteristic variety is the conormal bundle of a hypersurface $f(x) = 0$, $u(x)$ can be written as $u(x) = \varphi(x)f(x)^{\alpha} + \psi(x)$ where $\varphi(x)$ and $\psi(x)$ are functions analytic near the hypersurface. The order is $-\alpha-\frac{1}{2}$. When α is a non negative integer, $f(x)^{\alpha}$ must be replaced by $f(x)^{\alpha}\log f(x)$. If $u(x)$ is considered as hyperfunction, $f(x)^{\alpha}$ is interpreted as $(f(x)\pm i0)^{\alpha}$.

b) In the case the characteristic variety is the union of the conormal bundle of hypersurface $f(x) = 0$ and the conormal bundle of $f(x) = g(x) = 0$, $u(x)$ can be written as $\varphi(x)g(x)^{\alpha}\delta^{(m)}(f(x)) + \psi(x)$.

c) In the case the characteristic variety is the union of conormal bundles of two hypersurfaces $f(x) = 0$ and $g(x) = 0$ which have a contact of order 2, by modifying $f(x)$ and $g(x)$, we can set $g(x) = f(x) - h(x)^2$ with some function $h(x)$. Thus, we can take a local coordinate system $(x_1, \cdots, x_n)$ such that $f(x) = x_1$ and $h(x) = x_2$. Then, $u(x)$ satisfies micro-locally the following system of equations; $(x_1 D_1 + \frac{1}{2}x_2 D_2 - \alpha)v = 0$, $((D_2 + 2x_2 D_1)D_2 - 4\beta D_1)v = 0$, $D_3 v = \cdots = D_n v = 0$ by setting $u(x) = R(x, D)v(x)$ for a micro-differential operator R. If we denote by e_0, e_1, the orders of $u(x)$ at the conormal bundles of $f(x)=0$ and $g(x)=0$, we can say $u(x)$ is of the following form $u(x) = \varphi(x)v(x) + \psi(x)D_2 D_1^{-1}v(x) + \phi(x)$ for some analytic functions $\varphi(x)$, $\psi(x)$ and $\phi(x)$. Solving the above equation, $v(x)$ and $D_2 D_1^{-1}v(x)$ is given by

$$v(x) = f(x)^{-\frac{e_0+1}{2}} g(x)^{-\frac{e_1}{2}-\frac{1}{4}} Q_{e_0}^{e_1+1/2}(h(x)/\sqrt{f(x)})$$

and

$$D_2 D_1^{-1}v(x) = f(x)^{-\frac{e_0}{2}} g(x)^{-\frac{e_1}{2}-\frac{1}{4}} Q_{e_0-1}^{e_1+1/2}(h(x)/\sqrt{f(x)}).$$

Here, Q is the associated Legendre function. Especially, if $e_0 = 0$ and $e_1 = -\frac{1}{2}$,

$$u(x) = \varphi(x)f(x)^{-1/2}\log\frac{h(x) - f(x)^{1/2}}{h(x) + f(x)^{1/2}} + \psi(x)\log g(x) + \phi(x).$$

In the case $e_0 = -1$ and $e_1 = -\frac{1}{2}$,

$$u(x) = \varphi(x)f(x)^{1/2}\log\frac{h(x) - f(x)^{1/2}}{h(x) + f(x)^{1/2}} + \psi(x)f(x)^{1/2} + \phi(x).$$

The preceding discussion on holonomic system is applied to a microlocal study of Feynman integrals.

3. *Feynman Integrals.*

Consider a Feynman diagram D which consists of n external lines, n' internal lines joining n" vertices (parametrized by r, ℓ and j, resp.). Internal lines have directions. $j^{\pm}(\ell)$ signifies the end and the source of an internal line r. The incidence number $[j, \ell] = 1, -1, 0$ according to whether $j = j^{+}(\ell)$, $j^{-}(\ell)$ or else. The set of $(p_1, \cdots, p_n; u_1, \cdots, u_n)$ in the cotangent bundle $T^*(R^{\nu})^n$ of momentum space satisfying the following *Landau equation* is called *positive Landau holonomic manifold* and denoted by Λ_D^+ ; The relations $\sum\limits_{r\to j} p_r + \sum\limits_{\ell}[j : \ell]k_\ell = 0$ and $u_r = v_j$, where $j = 1, \cdots n"$ and r runs external lines ending at j, hold for some ν-vectors k_ℓ, v_j and scalars α_ℓ $(\ell = 1, \cdots, n'; \ j = 1, \cdots, n")$ which satisfy

$$k_\ell^2 = m_\ell^2, \quad v_{j^{+}(\ell)} - v_{j^{-}(\ell)} = \alpha_\ell k_\ell, \qquad \alpha_\ell > 0.$$

A set of $(p_1, \cdots, p_n ; u_1, \cdots, u_n)$ which satisfies the same equation replacing the last positivity condition $\alpha_\ell > 0$ with $\alpha_\ell \neq 0$ is denoted by Λ_D and called *Landau holonomic manifold* for D. The projection of Landau holonomic manifold on the momentum space $(R^{\nu})^n$ is called *Landau manifold* and denoted by L_D.

The Feynman integral $F_D(p_1, \cdots, p_n)$ is given by

$$\int_{} (\prod_{\ell}(m^2-k_\ell^2-i0))^{-1} \ \prod_{j}\delta^{\nu}(\sum_{r\to j} p_r + \sum_{\ell}[j : \ell]k_\ell)\prod_{\ell}d^{\nu}k_\ell.$$

This integral satisfies a holonomic system. It is shown that the singular spectrum of a Feynman integral $F_D(p_1, \cdots, p_n)$ is the union of the positive Landau holonomic manifolds of graphs obtained by contractions of internal lines of D.

For a diagram D, set cd(D) = # internal lines $+\nu$(# vertices - # internal lines).

Theorem. Let D' be a diagram obtained by contracting internal lines of D. If, for (p, u) in $\Lambda_{D'}^{+}$, we can solve uniquely (possibly finite) k_ℓ and α_ℓ from the Landau equation for D'. Then, the order of F_D on $\Lambda_{D'}$ is a half of $cd(D')$. Especially, if all vertices are external, the above condition holds and this formula is true.

This theorem is obtained by the formula concerning the order of an integral : the order of an integral is the order of the integrand minus a half of the number of the integration variables.

Theorem. If D' is a diagram obtained by contractions of internal lines of D and D'' is obtained by contraction of *one* internal line, say ℓ_0, from D', and if the Landau equation for D' in which $k_{\ell_0}^2 = m_{\ell_0}^2$ is replaced by $\alpha_{\ell_0}(k_{\ell_0}^2 - m_{\ell_0}^2) = 0$ gives α_ℓ and k_ℓ from $(p_1, \cdots, p_n, u_1, \cdots, u_n)$, then $\Lambda_{D'}$ and $\Lambda_{D''}$ have a regular interaction. The above condition is again satisfied when all vertices are external.

In this case, the difference of the orders on $\Lambda_{D'}$ and $\Lambda_{D''}$ is one half. Therefore, there exists a non trivial microfunction solution whose support is contained in $\overline{\Lambda}_{D'}$, but no solutions with support in $\overline{\Lambda}_{D''}$ in a neighborhood of $\overline{\Lambda}_{D'} \cap \overline{\Lambda}_{D''}$. That is, any solution on $\overline{\Lambda}_{D'}$ can be uniquely continued to some solution on $\overline{\Lambda}_{D''}$.

In the sequel, we assume all vertices are external. We say D is a *large diagram* if $cd(D') \geq (\nu+1)b_0$ for any subgraph D' of D. Here b_0 is the number of connected components of D'. The codimension (in $R^{\nu n}$) of the Landau manifold L_D of a large diagram D coincides with $cd(D)$. For example, the holonomy diagram for the Feynman integral of one loop with n-vertices is as follows.

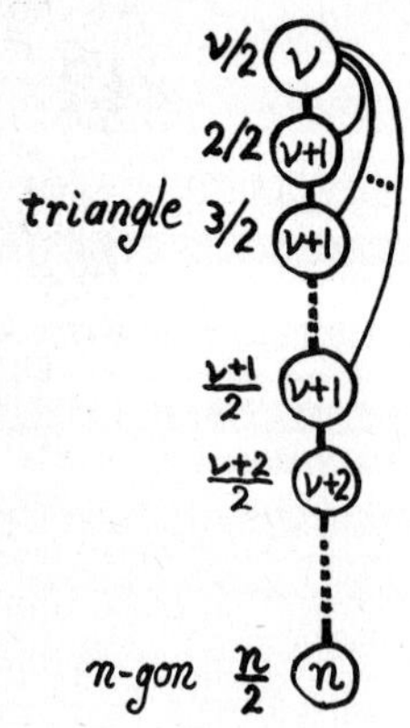

Each circle denotes the Landau holonomic manifold, and the number beside a circle indicates the order. Just one route from the full diagram is shown here. The full diagram contains many routes corresponding to the way of contracting internal lines successively. In the neighborhood of the intersections between Landau surfaces, the situation in 2 c) occurs.

On the Landau holonomic manifold of a large diagram D' obtained by contraction of D, the Feynman integral F_D is micro-locally equal to a δ-function on the Landau manifold $L_{D'}$ multiplied by an analytic function. This amplitude can be obtained by solving only an algebraic equation. In fact, the amplitude of the Feynman integral is determined by its principal symbol, and the principal symbol of F_D on $\Lambda_{D'}$ is given by :

$$\frac{\text{const.}}{\prod\limits_{\ell''} (m^2_{\ell''}-k^2_{\ell''})} \sqrt{\frac{\prod\limits_{\ell'} d\alpha_{\ell'}\, \Pi d^\nu p_j\, \prod\limits_{j'} d^\nu v_{j'}\, \prod\limits_{\ell'} d^\nu k_{\ell'}\, \prod\limits_{\ell''} d^\nu k_{\ell''}}{\prod\limits_{\ell'} d(m^2_{\ell'}-k^2_{\ell'})\, \prod\limits_j d^\nu(p_j+\Sigma[j:\ell]k_\ell)\, \prod\limits_{\ell'} d^\nu(\alpha_{\ell'} k_{\ell'}-\Sigma[j:\ell]v_j)}}$$

where ℓ', j' are internal lines and vertices on D' and ℓ'' are contracted internal lines.

References

[1] M. Sato, T. Kawai, M. Kashiwara : in Hyperfunctions and Pseudo-
       differential Equations, Lecture Notes in Mathematics 287,
       Springer Verlag.
[2] M. Sato : Contributions to this conference.
[3] T. Kawai and H. P. Stapp : Contribution to this conference.
[4] M. Kashiwara and T. Kimura : Surikaisekikenkyusho Kokyuroku 226,
       (1975) in Japanese.
[5] M. Kashiwara and T. Miwa : Micro-local Calculus and Fourier Trans-
       forms of Relative Invariants of Pre-homogeneous Vector Spaces,
       to appear in the Surikaisekikenkyusho Kokyuroku, in Japanese.

<u>MICROLOCAL STUDY OF S-MATRIX SINGULARITY STRUCTURE</u>

Takahiro Kawai[*]
Department of Mathematics
University of California
Berkeley, California 94720

and

Research Institute for Mathematical Sciences
Kyoto University
Kyoto, Japan

and

Henry P. Stapp[†]
Lawrence Berkeley Laboratory
University of California
Berkeley, California 94720

Abstract

Support is adduced for two related conjectures of simplicity of the analytic structure of the S-matrix and related function; namely, Sato's conjecture that the S-matrix is a solution of a maximally over-determined system of pseudo-differential equations, and our conjecture that the singularity spectrum of any bubble diagram function has the conormal structure with respect to a canonical decomposition of the solutions of the relevant Landau equations. This latter conjecture eliminates the open sets of allowed singularities that existing procedures permit.

1. <u>Introduction</u>

Two topics concerning the S-matrix singularity structure are discussed. The first is Sato's conjecture (1) that the S-matrix is a solution of a maximally over-determined system of pseudo-differential equations whose characteristic variety is given by the Landau equations. This property has already been established for a large class of Feynman integrals, and was used to investigate the character of the singularities associated with contracted diagrams (2). Sato's conjecture, if true, would provide a powerful tool for the determination of the analytic properties of scattering amplitudes.

Sato proposed, as a first test of his conjecture, a check of its consistency with the S-matrix discontinuity formulas. Some positive results along this line are reported here.

The second topic concerns the singularity structure of bubble diagram functions. These functions arise in the derivation of the S-matrix discontinuity formulas from unitarity and analyticity; both unitarity and the discontinuity formulas are expressed in terms of them. Bubble diagram functions are represented by bubble dia-

---

[*] Supported by Miller Institute for Basic Research in Science.
[†] Supported by ERDA.

grams, and are formed by integrating products of scattering amplitudes and complex conjugate amplitudes over the mass-shell variables corresponding to the internal lines of these diagrams.

The singularities of bubble diagram functions are controlled by the structure theorem (3). This theorem places conditions on the complete singularity spectrum of bubble diagram functions. That is, it limits the allowed real positions of singularities, and for each allowed real position it specifies an allowed set of imaginary directions from which the real point can be approached, staying in the domain of analyticity. This information combined with unitarity yields the S-matrix discontinuity formulas (4).

The structure theorem derived in (3) allows singularities corresponding to a certain degenerate case. This case is defined by an equation $u = 0$, and the corresponding allowed locations of singularities are called $u = 0$ points. These points cannot occur when the positive-α condition is imposed. But to study nonpositive-α singularities the $u = 0$ points must be considered.

We show here that in all cases known to us where $u = 0$ points cover open sets the actual singularities are confined to sets of lower dimension. This general result does not, however, specify the locations of these sets of lower dimension. This poses the problem of formulating and proving a generalized version of the structure theorem that restricts the allowed singularities to specified surfaces of codimension one or more.

A general theorem of this kind is not proved here. However, a number of special cases have been examined, and these conform to the following rule: A bubble diagram function $F^B(p)$ is singular at p only if p lies on a codimension-one or more <u>component</u> of the Landau surface associated with F^B. The components of this Landau surface are obtained by first considering for each Landau diagram D associated with F^B the solutions (p,k) of the corresponding Landau equations. Here k is the set of variables associated with the internal lines of D. These solutions define a variety in (p,k) space, which can be canonically decomposed into a set of analytic manifolds. The p-space images of these manifolds can then be canonically decomposed into analytic manifolds. These latter manifolds are the components of the Landau surface. In the cases we have examined the components of the Landau surface that cover open sets in p space are devoid of singularities, and the singularity spectrum has the conormal structure with respect to the remaining manifolds. That is, the cotangential component of the singularity spectrum at a point p lying on component N_i is confined to the conormal space to N_i at p. (The cotangential component of the singularity spectrum at p determines the allowed directions of approach to p.)

In the course of the analysis of $u = 0$ points another problem is encountered and resolved. This is the problem of $\mathcal{M}_o$ points. The set $\mathcal{M}_o$ is the part of the mass shell where two or more initial momentum-energy vectors are parallel or two or more final momentum-energy vectors are parallel. The information on the singularity structure of scattering functions near $\mathcal{M}_o$ points obtained by Iagolnitzer and Stapp (5) from macrocausality is insufficient for the present work. By introducing also the constraints of Lorentz invariance we obtain a more detailed characterization of the singularity structure at $\mathcal{M}_o$ points. The singularity spectrum has, in particular, the conormal structure demanded by Sato's conjecture and needed in our work on $u = 0$ points.

We shall now describe these results in more detail, together with some new technical results upon which they are based. The proofs will be given elsewhere.

The results are based on the general theory of microfunctions discussed by Sato, Kawai, and Kashiwara (6). Thus the singularity spectrum of a hyperfunction defined on a real analytic manifold M of dimension n is regarded as the support in $\sqrt{-1}\, S^*M$ of the hyperfunction regarded as a microfunction. The space $\sqrt{-1}\, S^*M$ consists of points $(p, \sqrt{-1}\, u)$ where p lies in M and u is a nonzero real n vector defined modulo real positive (multiplicative) factors. The vector u lies in the cotangent space and determines allowed directions in the dual tangent space.

2. Structure Theorem, $\mathcal{M}_o$ Points, and Generalized Landau Equations

An important concept is that of the space-time Landau equations. Consider a diagram D. This is a topological structure consisting of n external lines L_r and n' internal lines L_ℓ connected at n'' vertices V_j. Each line has a direction, and the incidence numbers $[j,\ell]$ (or $[j,r]$) are $+1, -1,$ or 0, according to whether V_j lies on the front-end, back-end, or no end of L_ℓ (or L_r). The indices $j^+(\ell)$ and $j^-(\ell)$ are uniquely defined: $[j^\pm(\ell), \ell] = \pm 1$. They label the vertices lying on the ends of internal line L_ℓ. Each L_ℓ (or L_r) is associated with a mass μ_ℓ (or m_r), and each line L_ℓ of some (possibly empty) subset of the internal lines carries a sign σ_ℓ.

$\underline{\text{Definition 1}}$. A set $(p_1, \cdots, p_n;\ u_1, \cdots, u_n) \equiv (p, u)$ consisting of n real four-vectors p_r and n real four vectors u_r is said to be a solution of the Landau equations corresponding to Landau diagram D if and only if there are a four vector a, sets of real four vectors $k_\ell\ (\ell = 1, \cdots, n')$ and $v_j\ (j = 1, \cdots, n'')$, and scalars $\alpha_\ell\ (\ell = 1, \cdots, n')$ and $\beta_r\ (r = 1, \cdots, n)$ such that the following equations hold:

$$
(1)\left\{
\begin{array}{lll}
\sum_r [j:r]p_r + \sum_\ell [j:\ell]k_\ell = 0 & \text{for } j = 1, \cdots, n'' & (1.a) \\[2ex]
p_r^{\,2} = m_r^{\,2}, \qquad p_r^{\,0} > 0 & \text{for } r = 1, \cdots, n & (1.b) \\[2ex]
k_\ell^{\,2} = \mu_\ell^{\,2} \qquad k_\ell^{\,0} > 0 & \text{for } \ell = 1, \cdots, n' & (1.c) \\[2ex]
v_{j+}(\ell) - v_{j-}(\ell) = \alpha_\ell k_\ell & \text{for } \ell = 1, \cdots, n' & (1.d) \\[2ex]
u_r + \beta_r p_j = -[j(r):r]\left(v_{j(r)} + a\right) & \text{for } r = 1, \cdots, n & (1.e)
\end{array}
\right.
$$

$$(1)\begin{cases} \sigma_\ell \alpha_\ell \geqq 0 & \text{for all signed lines } \ell \qquad (1.\text{f}) \\[2ex] \alpha_\ell \neq 0 & \text{for some } \ell \qquad (1.\text{g}) \end{cases}$$

Note that the vector $u \equiv (u_1, \cdots, u_n)$ is not uniquely determined by these equations: given one solution another is generated by a common positive scale change of the v_j's, α_ℓ's, β_r's, and u_r's. Also, if (p,u) is a solution then so is $\left(p,\, u+u_o(p)\right)$ where $u_o(p)$ is any $4n$ vector of the form $u_o = (\beta_1' p_1 + a, \beta_2' p_2 + a, \cdots, \beta_n' p_n + a)$. Thus u at p is defined modulo vectors of the form $u_o(p)$, and modulo positive scale changes.

The restricted mass shell $\mathcal{M}^r$ is defined by $\{p;\ p_r^2 = m_r^2$ for $r = 1, \cdots, n$, $\sum_r \big[j(r),r\big] p_r = 0$, at least two p_r are not parallel$\}$. The solutions $(p, \sqrt{-1}\, u)$ of (1) define a variety $\mathcal{L}(D)$ in $\sqrt{-1}\, S^* \mathcal{M}^r$. This variety is called the Landau variety and its projection to the base space $\mathcal{M}^r$ is the Landau surface $L(D)$.

The Landau equations associated with a bubble diagram function F_r^B (which is F^B with the conservation law δ-function factor removed) is the set of Landau equations (1) corresponding to all diagrams $D \subset \bar{B}$. The set $\bar{B}$ is the set of Landau diagrams that can be constructed by replacing each bubble b of B by some Landau diagram D_b, where the internal lines of D_b all carry the sign of b. {See (3), (4), or (5) for further details.}

The functions F^B are defined on the (unrestricted) mass shell $\{p;\ p_r^2 = m_r^2$, $r = 1, 2, \cdots, n\}$. The corresponding equations are (1) with $a = 0$.

<u>Definition 2</u>. A point p is said to be a $u = 0$ point if there is a solution $(p,u) = (p,0)$ of the Landau equations (1).

<u>Theorem 1</u>. If the singularity spectrums of the bubble diagram functions $F_r^{B_1}(p_1, \cdots, p_s, p_{s+1}, \cdots, p_{n_1})$ and $F_r^{B_2}(p_{s+1}, \cdots, p_{n_1}, p_{n_1+1}, \cdots, p_n)$ are confined to solutions of the associated space-time Landau equations except possibly at $u = 0$ points, then the bubble diagram function F_r^B corresponding to the bubble diagram B obtained by joining B_1 and B_2 with respect to $p_{s+1}, \cdots, p_{n_1}$ has the same property.

The usual structure theorem (3) follows from a repeated application of Theorem 1 starting from bubble diagrams B_1 and B_2 consisting of single bubbles. For these simplest bubble diagrams the only $u = 0$ points are the $\mathcal{M}_o$ points $\big(\,(7)\,\big)$.

To prove Theorem 1 we first prove a corresponding theorem for the (non-reduced) bubble diagram functions, i.e., bubble diagram functions which contain the overall δ-function. This is easily done by the successive application of Corollary 2.4.2 and Theorem 2.3.1 in Chapter I of (6). In fact, Corollary 2.4.2 guarantees that the integrand appearing in the definition of F^B is well defined under the $u \neq 0$ assumption and estimates its singularity spectrum. Therefore Theorem 2.3.1 immediately applies to estimate the singularity spectrum of F^B. Next we apply Theorem 2.1.8 in Chapter III of (6) to estimate the singularity spectrum of F_r^B, the function obtained by factorizing out the overall δ function from F^B.

That the theory of microfunctions would yield a simple proof of the structure theorem was noted by Professor Pham several years ago. (Private communication from Professor Iagolnitzer to HPS.)

Theorem 1 estimates the singularity spectrums of bubble diagram functions and is used in the derivation of the S-matrix discontinuity formula. ((4)) However, the limitation to $u \neq 0$ points is quite serious, since, unlike $u \neq 0$ points, the $u = 0$ points may cover open sets.

To overcome this difficulty we do two things. The first is to obtain more information on the singularity spectrum of the S matrix at $\mathcal{M}_o$ points, by making use of the Lorentz invariance property of the S matrix. The second is to take account of the specific form of the singularities. In fact, the troubles at $u = 0$ points come from the fact that the multiplication procedure needed to define the integrand at such points cannot be legitimate unless the singularities enjoy special properties. On the other hand, solutions of maximally overdetermined systems enjoy properties that allow their products to be defined even at $u = 0$ points.

In connection with the $\mathcal{M}_o$ problem we introduce a generalized version of the Landau equations.

Definition 3. The <u>generalized</u> space-time Landau equations corresponding to a scattering function $S_r(p)$ or its conjugate $S_r^\dagger(p)$ are the same as the original space-time Landau equations (1) for these functions except that for every set $\{L_r; r\epsilon\Gamma\}$ of external lines that all originate (or all terminate) on a single vertex, there is an alternative to the set of equations (1.d) and (1.e) associated with these lines. This alternative set consists of the equations

$$
(2)\begin{cases}
\dfrac{p_r}{p_r^{\ 0}} = \dfrac{p_{r'}}{p_{r'}^{\ 0}} & \text{if } r\epsilon\Gamma \text{ and } r'\epsilon\Gamma & (2.a)\\[2ex]
\text{and} & \\[1ex]
u_r + \beta_r p_r = -\big[j(r),r\big](v_r + \eta_r) & \text{for } r\epsilon\Gamma & (2.b)\\[2ex]
\text{where the } \eta_r \text{ satisfy} & \\[1ex]
\displaystyle\sum_{r\epsilon\Gamma}(p_r^{\ 0}\eta_r^{\ \nu} - p_r^{\ \nu}\eta_r^{\ 0}) = 0 \ . & & (2.c)
\end{cases}
$$

Also, in place of (1.g) one imposes on the complete solution the condition

$$(3) \qquad u \neq 0 \ (\text{mod } u_o) \ .$$

All solutions $(p; u)$ of the form $(p_1,p_2,\cdots,p_n; \beta_r'p_1+a, \beta_2'p_2+a, \cdots, \beta_n'p_n+a)$ are eliminated by (3).

Theorem 2. The macroscopic causality and Lorentz invariance properties of the S matrix entail that the singularity spectrums of $S_r(p)$ and $S_r^\dagger(p)$ be confined to solutions of the corresponding generalized Landau equations.

For a general bubble diagram function $F_r^{\ B}$ the generalized Landau equations

are obtained from the ordinary equations (1) by allowing for the variables associated
with each individual bubble b of B the options allowed by (2) and (3). That is,
if D_b is the part of some $D \subset \bar{B}$ that is associated with a bubble b of B, then
for each vertex v of D_b that is an external vertex of D_b considered alone, one
allows the option described by (2), but excludes, for the variables u_i associated
with all the external lines of D_b considered alone, all solutions excluded by (3).
{Geometrically, the option (2) allows certain parallel displacements of the space-
time trajectories corresponding to a set of explicit lines of B that all begin
or end on a common vertex v of D, provided these trajectories are all parallel.
Condition (3) excludes any solution in which the trajectories associated with the external lines of any single D_b considered alone, all pass through a common point.
The geometrical interpretation of the Landau equations as a classical space-time scattering diagram is discussed in (3), (4), and (5). The generalized Landau equations
allow the vertices to go, in effect, to infinity, subject to conservation-law constraints.}

Definition 4. A point p is said to be a generalized u = 0 point of F_r^B
if (p,u) = (p,0) satisfies the corresponding generalized Landau equations.

Theorem 3. The singularity spectrum of $F^B(p)$ is confined to solutions of
the generalized Landau equations except possibly at generalized u = 0 points.

A simple case covered by Theorem 3, but not by Theorem 1, is illustrated in
Fig. 1.

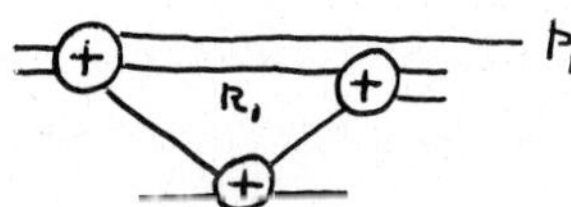

Fig. 1. Bubble diagram for a case covered by Theorem 3.

If for some p the integration region contains a point k where p_1 is parallel
to k_1 then the original Landau equations have a u = 0 solution. Such points p
cover an open set, and Theorem 1 gives no information there. However, the parallel-
ness of k_1 and p_1 does not lead to a u = 0 solution of the generalized Landau
equations. The generalized u = 0 points do not cover open sets, in this case, and
the singularities allowed by Theorem 3 are confined to sets of lower dimension.

3. Sato's Conjecture

Our further results depend on the form of the singularity itself, not merely
its location, and are restricted to simple cases where we have determined this form.
A principal limitation arises from the exclusion of three-particle thresholds; the
form of the singularity at such thresholds is very complicated and has not yet been
determined.

Two-particle threshold points are analyzable.

Definition 5. A two-particle threshold point is an argument $p = (p_1, \cdots, p_n)$
of $S_r(p)$ such that: (1) no three initial p_i are parallel, (2) no three final p_i

are parallel, (3) there is a (multisheeted) complex neighborhood ω of p such that $S_r(p)$ has no singularities in ω except those that coincide with the two-particle thresholds corresponding to the various pairs of parallel initial and final p_i, and (4) each of these two-particle thresholds lies below the lowest three-particle threshold in its channel.

From Zimmermann's result (8) on the square-root nature of singularities lying below the lowest three-particle threshold in elastic scattering amplitudes, combined with the S-matrix discontinuity formulas (4), one may obtain the following result:

<u>Theorem 4</u>. In a neighborhood of a two-particle threshold point p the S-matrix $S_r(p)$ is simply a product of normal-threshold factors

$$\left[\left((p_i + p_j)^2 - (m_i + m_j)^2 + i0 \right)^{\frac{1}{2}} \phi_{ij} + \psi_{ij} \right] \; , \tag{4}$$

where the ϕ_{ij} and ψ_{ij} are analytic.

This result validates Sato's conjecture in the neighborhood of any two-particle threshold point. It is used in conjunction with the discontinuity formulas to validate Sato's conjecture at points where the discontinuity functions involve scattering amplitudes evaluated at two-particle threshold points.

A positive-α diagram is a Landau diagram each line of which carries a positive sign σ_ℓ. A diagram subject to this condition is written D^+. The union of the $\mathcal{L}(D^+)$ is $\mathcal{L}^+$, the union of the $L(D^+)$ is L^+. Nonbasic D^+ are ignored (4).

<u>Definition 6</u>. A point $(p; \sqrt{-1}\, u)$ of $\mathcal{L}^+$ is <u>invertible</u> if and only if the following three conditions hold:

(5) There is a unique D^+ such that $(p; \sqrt{-1}\, u)$ lies in $\mathcal{L}(D^+)$.

(6) The complexification of $\mathcal{L}(D^+)$ is an analytic submanifold near $(p; \sqrt{-1}\, u)$.

(7) If x represents a set of local coordinates of $\mathcal{L}(D^+)$, then there is a unique set of analytic functions $k_\ell(x)$, $\alpha_\ell(x)$, and $v_j(x)$ such that for each point x in some complex neighborhood of the point $x_0 =$ image of $(p; \sqrt{-1}\, u)$ the unique solution of the Landau equations that define $\mathcal{L}(D^+)$ is $p = p(x)$, $u = u(x)$, $k = k(x)$, $\alpha = \alpha(x)$, and $v = v(x)$.

<u>Theorem 5</u>. Suppose $(p; \sqrt{-1}\, u)$ of $\mathcal{L}^+$ is invertible. Suppose the corresponding D^+ has the property that at most two lines connect any pair of vertices. Suppose each of the scattering functions that occurs in the discontinuity formula at p is evaluated at a two-particle threshold point. Then the S-matrix $S_r(p)$ satisfies near $(p; \sqrt{-1}\, u)$ a maximally over-determined system of pseudo-differential equations. This system is simple, in the sense of (6) Chapter II §4, except for some analytic varieties. Its order is $\alpha = 2n'' - \frac{3}{2} n'$, where n' is the number of internal lines and n'' is the number of vertices of D^+.

<u>Remark 1</u>. The excluded subvarieties in Theorem 6 correspond to the zeros of the scattering amplitude. In other words, the S matrix is a solution of a simple maximally overdetermined system except for a multiplicative factor, possibly vanishing.

Remark 2. To obtain these results we have used a micro-local version of the discontinuity formula (4). This version makes sense even when L^+ is not a hypersurface. The arguments in (4) establish, in effect, also this generalization.

Remark 3. The result stated in Theorem 5 arises from a very special property of the S-matrix discontinuity formula. It would not hold if that function were replaced by the similar bubble diagram function F^B corresponding to the B obtained by simply replacing the vertices of D^+ by plus bubbles. The S-matrix is thus particularly simple, from the point of view of maximally over-determined systems.

Definition 7. A point $p\epsilon L^+$ is _simple_ if and only if

(8) L^+ near p is a codimension-one analytic submanifold $\{p\epsilon\mathcal{M}^r;\ \phi(p) = 0\}$, and

(9) The point $(p;\ \sqrt{-1}\ u) = \left(p;\ \sqrt{-1}\ \mathrm{grad}_p\ \phi(p)\right)$ in $\mathcal{L}^+$ is invertible.

Theorem 6. If p is a simple point of L^+ then the scattering amplitude near p has the following form (15):

$$(10) \begin{cases} h_1(p)\left(\phi(p) + i0\right)^{-\alpha+3/2} + h_2(p) & \text{if } -\alpha + \tfrac{3}{2} \text{ is neither a positive integer} \\ & \qquad\qquad \text{nor zero.} \\[2mm] \text{or} \\[2mm] h_1(p)\ \phi(p)^{-\alpha+3/2}\ \log\left(\phi(p) + i0\right) + h_2(p) & \text{if } -\alpha + \tfrac{3}{2} \text{ is a positive integer} \\ & \qquad\qquad \text{or zero.} \end{cases}$$

Here $h_1(p)$ and $h_2(p)$ are analytic.

To fully analyze the S-matrix singularity structure we must study it also at singular points of the Landau variety. In view of Theorem 1 of (9), and the remark following it, the most important singular points are the points where two irreducible components of the Landau variety cross normally along some subvariety of codim 1 in the Landau variety. This is the situation which appears if some single internal line is contracted. To delineate this case we introduce the following definition.

Definition 8. A point $(p;\ \sqrt{-1}\ u)$ of $\mathcal{L}^+$ is a _simple contraction point_ if and only if the following conditions are satisfied:

(11) There are exactly two basic diagrams D_1^+ and D_2^+ that satisfy $p\epsilon\mathcal{L}(D_i^+)$.

(12) D_2^+ is obtained by contracting exactly one line of D_1^+.

(13) The complexifications $\mathcal{L}_1$ and $\mathcal{L}_2$ of $\mathcal{L}(D_1^+)$ and $\mathcal{L}(D_2^+)$ are both analytic submanifolds near $(p;\ \sqrt{-1}\ u)$, and each satisfies the invertibility condition (7).

(14) $\mathcal{L}_1 \cap \mathcal{L}_2$ is an analytic submanifold near $(p;\ \sqrt{-1}\ u)$, and $\mathcal{L}_1$ and $\mathcal{L}_2$ intersect transversally at $(p;\ \sqrt{-1}\ u)$.

Theorem 7. Suppose $(p;\ \sqrt{-1}\ u)$ in $\mathcal{L}^+$ is a simple contraction point. Suppose both D_1^+ and D_2^+ have the properties required in Theorem 5 of D^+, except for the pole singularity associated with the contracted line. Then the conclusions of Theorem 5 still hold, except for the simplicity of the system.

The explicit form of the scattering amplitude can also be given in this case under some moderate conditions on D_1, by making use of a canonical form of the system of equations it must satisfy.

The results stated in Theorems 5 and 7 verify Sato's conjecture in nontrivial cases.

4. <u>Analysis of Generalized $u = 0$ Points</u>

All open sets of generalized $u = 0$ points known to us arise from the occurrence within a diagram D of a symmetric hammock part. A <u>hammock part</u> is a part that is connected to the rest of the diagram at precisely two vertices, has other vertices, and has no external lines. Two examples are shown in Fig. 2.

Fig. 2. Two hammock parts.

The dark vertices at the ends are the only vertices connected to the rest of the diagram D, which is not shown. If a hammock diagram is symmetric with respect to a vertical center-line, apart from a reversal of the signs σ_ℓ, then for every solution of the Landau equations for the half diagram there is a corresponding solution for the whole diagram. This solution is symmetric, in the sense that corresponding vertices from the two half diagrams coincide. In particular the two end points coincide. Hence a solution of the Landau equations for the entire diagram D can be obtained by setting to zero all α_ℓ's corresponding to lines outside the hammock part. This solution is a $u = 0$ solution, and it usually generates open sets of generalized $u = 0$ points. Any bubble diagram with some pair of bubbles connected by more than two lines has open sets of generalized $u = 0$ points of this kind.

Consider, for example, the bubble diagram B of Fig. 3.

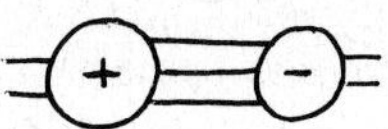

Fig. 3. The bubble diagram B.

Both diagrams of Fig. 2 are contained in $\bar{B}$, and each gives an open set of generalized $u = 0$ points that covers the entire region where F^B is nonzero. Thus the (generalized) structure theorem gives no information about the singularity spectrum of F^B. Special assumptions were needed in (4) to exclude effects of $u = 0$ contributions.

Points $k = (k_1, k_2, k_3)$ in the integration domain where the only singularities of the integrand are those associated with Fig. 2a lead to no singularities of F^B. The integrand contains a factor $(\phi + i0)^{\frac{1}{2}}(\phi - i0)^{\frac{1}{2}}$ which is not defined by the rules for products of microfunctions. (The product is too similar to $\left(\delta(\phi)\right)^2$.) However, for products of representatives of Hilbert space kernels one has, for $\lambda \geq 0$,

$$(15) \quad (\phi + i0)^{\lambda}(\phi - i0)^{\lambda} = \phi_+^{2\lambda} + \phi_-^{2\lambda}$$

where ϕ is the distance below threshold, and $\phi_+^{2\lambda}$ and $\phi_-^{2\lambda}$ are the two Heaviside resolutions. Using (15) one can effectively reduce the diagram of Fig. 2a to a simple three-particle normal threshold diagram, which yields no singularities above the three-particle threshold.

The same argument eliminates most of the singularities of F^B coming from Fig. 2b. However, there are two values of $(p_1 + p_2)^2 = (p_3 + p_4)^2$ where the singularities do not have the required form $(\phi \pm i0)^\lambda$, and the argument fails. These two values of $(p_1 + p_2)^2$ are those such that the triangle diagram singularity of the half diagram, considered as a surface in k space that changes as p is changed, either (1) touches the two-particle normal threshold surface corresponding to Fig. 2a, or (2) touches the point where k_3 is parallel to $p_1 + p_2$. (Here k_3 is the variable associated with the lowest line in Fig. 2b.) When either of these cases is reached the singularity surface in k-space suffers an abrupt change of topological structure, and a singularity of $F^B(p)$ is expected, barring cancellations.

The above argument rests on (15), and works at points p such that all the points (p,k) on the singularity surface of the half diagram are simple points, since the forms of the singularities are then $(\phi \pm i0)^\lambda$. The result, however, continues to hold when the singularity surface in (p,k) space includes also simple contraction points:

<u>Theorem 8</u>. Consider any bubble diagram function $F^B(p)$ and an argument $\bar{p}$ that satisfies the following conditions:

(16) There is a unique D such that $\bar{p}$ lies on $L(D')$ for $D' \subset \bar{B}$ only if D' is D, or a diagram obtained from D by contracting some signed lines.

(17) This unique diagram D is a symmetric hammock diagram D_h with half diagrams D_r and D_ℓ.

(18) Every solution (p,k) of the Landau equations for any $D \subset \bar{B}$ is obtained from coincident points of D_r and D_ℓ, or from coincident simple points of contractions of D_r and D_ℓ, or from coincident simple contraction points of D_r and D_ℓ.

(19) p lies on a codimension zero component of $L^B \equiv \bigcup_{D \subset \bar{B}} L(D)$.

Then F^B is not singular at $\bar{p}$.

The components of L^B are constructed by decomposing the union of the solutions of the corresponding Landau equations into analytic manifolds in the canonical fashion, then further decomposing these manifolds into manifolds that are analytic over p space, in the sense that the local coordinates of the p space manifold can be used as a subset of the local coordinates of the (p,k) space manifold lying over it. This construction gives a well-defined decomposition of L into analytic manifolds, which are called its components. If the analytic properties are controlled by the geometry of the Landau surfaces, as they are in the simple examples studied above, then any single codimension zero component would be expected to be completely free of singularities or completely filled with singularities, since the geometry is analytically uniform over each component.

That no codimension zero component is filled with singularities is suggested by the following result.

<u>Theorem 9</u>. Let the phase space functions F^D be defined by the same rules as the bubble diagram functions F^B, except that whereas the bubbles of B are associated with scattering amplitudes, the vertices of D are associated with constants, which are all taken to be unity. Let D_h be a hammock diagram. Then F^{D_h} is analytic except on a locally finite set of analytic manifolds of codimension one or more. Moreover, its singularity spectrum enjoys the conormal structure.

This result is proved by first reducing out the delta functions in the phase space functions, and then applying the result of (10).

The singularities of the bubble diagram functions are, in a certain sense, generated by the singularities of phase space functions. Hence it seems unlikely that the former could give open sets of singularities if the latter do not.

<u>References</u>

1. M. Sato, Contribution to this conference.
2. M. Sato, T. Kawai, and M. Kashiwara, Pseudo-differential equation theoretic approach to the Feynman integral, in preparation.
3. H. P. Stapp, J. Math. Phys. <u>9</u>, 1548 (1968) and in Ref. 4; D. Iagolnitzer, The Structure Theorem in S-matrix Theory, Saclay DPh-T/74-91, to be published in Mathematical Physics.
4. J. Coster and H. P. Stapp, J. Math. Phys. <u>11</u>, 2743 (1970) and LBL-2427 to be published in J. Math. Phys.
5. D. Iagolnitzer and H. P. Stapp, Comm. Math. Phys. <u>14</u>, 15 (1969).
6. M. Sato, T. Kawai, and M. Kashiwara, in Hyperfunctions and Pseudo-Differential Equations, Lecture Notes in Mathematics 287, Springer-Verlag.
7. See References 3, 4, and 5.
8. W. Zimmermann, Nuovo Cimento <u>21</u>, 249 (1961).
9. M. Kashiwara, T. Kawai, and T. Oshima, Proc. Japan Acad. <u>50</u>, 420 (1974).
10. M. Kashiwara and T. Kawai, Micro-local properties of $\prod_{j=1}^{n} f_{j+}^{s_j}$, to be published in Proc. Japan Acad.

CONVOLUTORS FOR ULTRAHYPERFUNCTIONS

Mitsuo MORIMOTO
Department of Mathematics
Sophia University

Tokyo, 102 Japan

In 1958, Sebastião e Silva [7] introduced the tempered ultrahyperfunction and the Fourier ultrahyperfunction in the name of tempered ultradistribution and of ultradistribution of exponential type. He studied
mainly the 1-dimensional case and Hasumi [2] generalized the notion of
tempered ultrahyperfunction in the case of n-dimension. Park-Morimoto
[5] then studied the Fourier ultrahyperfunction in the n dimensional
case. The author considered the (cohomological) ultrahyperfunction in
[3].

The ultrahyperfunction can be thought of an analytic functional with
not necessarily compact porter. It generalizes the notion of hyperfunction on R^n but can not be localized as hyperfunctions. In this paper
we will recall very briefly the definitions of the various types of
ultrahyperfunctions and announce some new results concerning the convolutors and the multiplicators for ultrahyperfunctions.

§1. <u>Definitions.</u> For a subset A of R^n, we will denote by $T(A) =
R^n \times iA \subset C^n$ the tubular set with base A. For a convex compact set K of
R^n, $\mathcal{O}_b(T(K))$ is, by definition, the space of all continuous functions
φ on $T(K)$ such that φ is holomorphic in the interior of $T(K)$ and that
$\zeta^p \varphi(\zeta)$ is bounded in $T(K)$ for any multiindex p, where $\zeta^p = \zeta_1^{p_1} \zeta_2^{p_2} \cdots \zeta_n^{p_n}$
for $\zeta = (\zeta_1, \zeta_2, \ldots, \zeta_n)$ and $p = (p_1, p_2, \ldots, p_n)$. With the seminorms

$$\|\varphi\|^{K,k} = \sup\left\{|\zeta^p \varphi(\zeta)| \; ; \; \zeta \in T(K), \; |p| \leq k\right\}, \quad k = o, 1, 2, \ldots,$$

$\mathcal{O}_b(T(K))$ is a Fréchet space. If $K_1 \subset K_2$ are two convex compact sets, we
have the canonical injection:

(1) $\qquad \mathcal{O}_b(T(K_2)) \hookrightarrow \mathcal{O}_b(T(K_1))$.

Let O be a convex open set in R^n. We define

(2) $\qquad \mathcal{O}(T(O)) = \lim \text{proj} \; \mathcal{O}_b(T(K_1))$,

where K_1 runs through the convex compact sets contained in O and the
projective limit is taken following the restriction mappings (1). Let
K be a convex compact set in R^n. We define

(3) $\qquad \mathcal{O}(T(K)) = \lim \text{ind} \; \mathcal{O}_b(T(K_1))$,

where K_1 runs through the convex compact sets such that K is contained in
the interior of K_1 and the inductive limit is taken following the restriction mappings (1).

Definition 1.   A continuous linear functional on $\mathcal{H}(T(R^n))$ is, by definition, a tempered ultrahyperfunction(or tempered ultradistribution of Sebastião e Silva).

We will put $\mathcal{H} = \mathcal{H}(C^n) = \mathcal{H}(T(R^n))$. The dual space of $\mathcal{H}$ is denoted by $\mathcal{H}'$. As we know, $\mathcal{H}$ is dense in $\mathcal{H}(T(0))$ and in $\mathcal{H}(T(K))$, we can consider $\mathcal{H}'(T(0)) \subset \mathcal{H}'$ and $\mathcal{H}'(T(K)) \subset \mathcal{H}'$. Hasumi proved that $\mathcal{H}$ is the Fourier image of the space $H(R^n)$ of exponentially decreasing C^∞ functions and that $\mathcal{H}'$ is the Fourier image of the space $H'(R^n)$ of distributions of exponential growth.

Let K' be a convex compact set of R^n. We put

(4) $h_{K'}(x) = \sup\{\langle x, \eta\rangle ; \eta \in K'\}$.

Let $K \subset R^n$ be a convex compact set. We denote by $Q_b(T(K); K')$ the space of all continuous functions f on T(K) which are holomorphic in the interior of T(K) and satisfy the estimate:

(5) $\sup\{\exp(h_{K'}(x))|f(z)| ; z \in T(K)\} < \infty$,   where x = Re z.

Endowed with the norm(5), the space $Q_b(T(K); K')$ is a Banach space. If $L \supset K$ and $L' \supset K'$, then the canonical mapping

(6) $Q_b(T(L); L') \longrightarrow Q_b(T(K); K')$

is continuous.

Let 0 and 0' be convex open sets of R^n. We define

(7) $Q(T(0); 0') = \lim \text{proj} \, Q_b(T(K); K')$,

where K and K' run through the convex compact sets contained in 0 and 0' respectively and the projective limit is taken following the canonical mappings (6). Let K and K' be convex compact sets in R^n. We define

(8) $Q(T(K); K') = \lim \text{ind} \, Q_b(T(K_1); K_1')$,

where K_1 (resp. K_1') runs through the convex compact sets such that K (resp. K') is contained in the interior of K_1(resp. K_1') and the inductive limit is taken following the canonical mappings (6).

Definition 2.   A continuous linear functional on $Q(T(R^n); R^n)$ is, by definition, a Fourier ultrahyperfunction (or ultradistribution of exponential type of Sebastião e Silva).

It is known that the space $Q(T(R^n); R^n)$ is invariant under the Fourier transformation.

The relation of various spaces are as follows:

$$
\begin{array}{cccc}
Q(T(R^n); R^n) & \subset \mathcal{H}(T(R^n)) & Q'(T(R^n); R^n) & \supset \mathcal{H}'(T(R^n)) \\
\cap & \cap & \cup & \cup \\
H(R^n) & \subset \mathcal{S}(R^n), & H'(R^n) & \supset \mathcal{S}'(R^n),
\end{array}
$$

(9)

where $\mathcal{S}(R^n)$ denotes the space of rapidly decreasing C^∞ functions on R^n and $\mathcal{S}'(R^n)$ is the space of tempered distributions. Two diagrams in (9) are in duality.

§2. Cohomological representation of ultrahyperfunctions.

For an open set W of C^n, $Q_0(W)$ is the space of all holomorphic functions ψ such that for any positive number ε, there exist a multiindex p and a constant $C \geqslant 0$ such that

(10) $\quad |\psi(\zeta)| \leqslant C (1 + |\zeta^p|) \quad$ for $\zeta \in C^n \setminus (C^n \setminus W)_\varepsilon$,

where $(C^n \setminus W)_\varepsilon$ denotes the ε-neighbourhood of $(C^n \setminus W)$. $W \mapsto Q_0(W)$ forms a presheaf of vector spaces over C^n. Let K be a convex compact set of R^n. Choose open half spaces E_j of R^n such that E_j, $j = 1, 2, \ldots$ cover $R^n \setminus K$. Then $\mathcal{U} = \{T(E_j); j = 1, 2, \ldots\}$ is an open covering of $C^n \setminus T(K)$.

Theorem 1. (Hasumi) We have the following linear isomorphism:

(11) $\quad \mathcal{y}'(T(K)) \cong H^{n-1}(\mathcal{U}; Q_0) \quad$ for $n > 1$.

If $n = 1$, then we have

(11') $\quad \mathcal{y}'(T(K)) \cong Q_0(C \setminus T(K)) / Q_0(C)$.

The inner product of $\mathcal{y}(T(K)) \times H^{n-1}(\mathcal{U}; Q_0)$ giving the above isomorphisms is defined by the Cauchy-Weil formula for the tubular domain.

As an example we consider the simplest case where $K = \prod_{j=1}^{n} [-a_j, a_j]$. In this case we may take $\mathcal{U} = \{T(E_j); j = 1, 2, \ldots, 2n\}$, $E_j = \{y \in R^n; y_j > a_j\}$ and $E_{n+j} = \{y \in R^n; y_j < -a_j\}$ for $j = 1, 2, \ldots, n$. An element of $H^{n-1}(\mathcal{U}; Q_0)$ is represented by 2^n functions $\psi_\sigma(z) \in Q_0(\{z \in C^n; z=x+iy, \sigma_j y_j > a_j$ for $j = 1, 2, \ldots, n\})$, $\sigma = (\sigma_1, \sigma_2, \ldots, \sigma_n)$, $\sigma_j = \pm 1$. The inner product is given by

(12) $\langle [\psi_\sigma], \varphi \rangle = \sum_\sigma \sigma_1 \cdots \sigma_n \int \cdots \int_{R^n} \psi_\sigma(x_1+i\sigma_1(a_1+\varepsilon), \ldots, x_n+i\sigma_n(a_n+\varepsilon)) \cdot$
$\qquad \cdot \varphi(x_1+i\sigma_1(a_1+\varepsilon), \ldots, x_n+i\sigma_n(a_n+\varepsilon)) \, dx_1 dx_2 \ldots dx_n$.

See for the details Hasumi [2]. (12) can be generalized for any convex compact set K.

We can formulate the edge of the wedge theorem for the tempered ultrahyperfunctions. Among the various versions of the theorem, the following is one of the simplest versions:

Theorem 2. Let $a_j \geqslant 0$, $j = 1, 2, \ldots, n$. We put $\Gamma = \{y \in R^n; y_j > a_j$ for $j = 1, 2, \ldots, n\}$ and $K = \{y \in R^n; |y_j| \leqslant a_j$ for $j = 1, 2, \ldots, n\}$. Suppose that $\psi_1 \in Q_0(T(\Gamma))$ and $\psi_2 \in Q_0(T(-\Gamma))$ are given and they satisfy the following equality:

(13) $\quad \int_{R^n+i(a+\varepsilon I)} \psi_1(z) \varphi(z) \, dz = \int_{R^n-i(a+\varepsilon I)} \psi_2(z) \varphi(z) \, dz$

for any $\varphi \in \mathcal{y}(T(K))$, where $a = (a_1, a_2, \ldots, a_n)$, $I = (1, 1, \ldots, 1)$ and $\varepsilon > 0$ is sufficiently small. Then ψ_1 and ψ_2 can be continued analytically each other and define a polynomial.

The similar results hold for the Fourier ultrahyperfunctions.

More generally we define the (cohomological) ultrahyperfunction as follows:

Definition 3. (Morimoto [3]) Let $\mathcal{O}$ be the sheaf of germs of holomorphic functions over C^n. For a convex compact set K of R^n, we define the n-th group of cohomology of C^n with support in $T(K)$ and with coefficients in the sheaf $\mathcal{O}$, $H^n_{T(K)}(C^n; \mathcal{O})$, to be the space of (cohomological) ultrahyperfunctions with porter in $T(K)$.

As we have by the Leray theorem,

$$(14) \quad H^n_{T(K)}(C^n; \mathcal{O}) = H^{n-1}(C^n \setminus T(K); \mathcal{O}) = H^{n-1}(\mathcal{U}; \mathcal{O}) \quad \text{for } n > 1$$

and

$$(14') \quad H^1_{T(K)}(C; \mathcal{O}) = \mathcal{O}(C \setminus T(K))/\mathcal{O}(C),$$

a tempered ultrahyperfunction can be considered as a (cohomological) ultrahyperfunction. So is a Fourier ultrahyperfunction.

The space of (cohomological) ultrahyperfunctions over C^n is defined to be the space

$$(15) \quad \lim \text{ind } H^n_{T(K)}(C^n; \mathcal{O}),$$

where K runs through the convex compact sets of R^n and the inductive limit is taken following the canonical mappings

$$(16) \quad H^n_{T(K)}(C^n; \mathcal{O}) \longrightarrow H^n_{T(L)}(C^n; \mathcal{O}), \quad K \subset L.$$

It is known that (16) is injective.

Put for $k > 0$

$$(17) \quad T_k = T([-k, k]^n) = \left\{ z = x+iy; \; |y_j| \leqslant k \; \text{ for } j = 1, 2, \ldots, n \right\}$$

and

$$(18) \quad C^n \# T_k = \bigcup_\sigma \left\{ z = x+iy; \; \sigma_j y_j > k \; \text{ for } j = 1, 2, \ldots, n \right\},$$

where $\sigma = (\sigma_1, \sigma_2, \ldots, \sigma_n)$, $\sigma_j = \pm 1$. The ultrahyperfunction T is represented by a holomorphic function $\psi \in \mathcal{O}(C^n \# T_k)$, k depending on T. We will note it by $T = [\psi]$.

§ 3.   <u>Multiplicators and Convolutors.</u>   L. Schwartz [6] determined the multiplicators and the convolutors for the tempered distributions.

$\mathcal{O}_M(\mathcal{S}, \mathcal{S})$ denotes the space of multiplicators of $\mathcal{S}(R^n)$:

$$\alpha \in \mathcal{O}_M(\mathcal{S}, \mathcal{S}) \Leftrightarrow \left\{ \begin{array}{l} \alpha \in C^\infty(R^n); \text{ for any multiindex } p, \text{ there exists a} \\ \text{polynomial } P \text{ such that } |D^p \alpha| \leqslant |P| \text{ on } R^n. \end{array} \right.$$

One refers the elements of $\mathcal{O}_M(\mathcal{S}, \mathcal{S})$ as C^∞ functions slowly increasing at infinity.

$\mathcal{O}'_C(\mathcal{S}', \mathcal{S}')$ denotes the space of convolutors for $\mathcal{S}'(R^n)$:

$$T \in \mathcal{O}'_C(\mathcal{S}', \mathcal{S}') \Leftrightarrow \left\{ \begin{array}{l} T \in \mathcal{S}'(R^n); \text{ for any integer } k, \; (1+\|x\|^2)^{k/2} T(x) \\ \text{is a bounded distribution.} \end{array} \right.$$

One refers the elements of $\mathcal{O}'_C(\mathcal{S}', \mathcal{S}')$ as the distributions rapidly decreasing at infinity.

<u>Theorem 3.</u> [6]   The Fourier transformation $\mathcal{F}$ gives a topological linear isomorphism:

$$(19) \quad \mathcal{O}_M(\mathcal{S}, \mathcal{S}) \xrightarrow{\mathcal{F}} \mathcal{O}'_C(\mathcal{S}', \mathcal{S}').$$

If $\alpha \in \mathcal{O}_M(\mathcal{S}, \mathcal{S})$ and $T \in \mathcal{S}'(R^n)$, then $\mathcal{F}\alpha \in \mathcal{O}'_C(\mathcal{S}', \mathcal{S}')$ and $\mathcal{F}T \in \mathcal{S}'(R^n)$ and we have

(20) $\qquad \mathcal{F}(\alpha T) = (\mathcal{F}\alpha) * (\mathcal{F}T).$

Hasumi [2] determined the multiplicator for $\mathcal{G}$ and the convolutor for H'. (See also Zieleźny[9].)

$O_M(\mathcal{G}, \mathcal{G})$ denotes the space of multiplication operators for the space $\mathcal{G}$:

$$\alpha \in O_M(\mathcal{G}, \mathcal{G}) \Leftrightarrow \begin{cases} \alpha \in O(C^n), \text{ for any } h > 0 \text{ there exists a multiindex } p \\ \text{such that } |\alpha(z)|(1 + |z^p|)^{-1} \text{ is bounded on } T_h. \end{cases}$$

One refers the elements of $O_M(\mathcal{G}, \mathcal{G})$ as the entire functions slowly increasing in any horizontal bands.

$O_C'(H', H')$ denotes the space of convolution operators for the space H' of distributions of exponential growth:

$$T \in O_C'(H', H') \Leftrightarrow \begin{cases} T \in \mathcal{D}'(R^n), \text{ for any } a \in R^n, \exp\langle a, x\rangle T(x) \text{ is} \\ \text{a bounded distribution on } R^n. \end{cases}$$

One refers the elements of $O_C'(H', H')$ as the distributions very rapidly decreasing at infinity.

<u>Theorem 4.</u> The Fourier transformation $\mathcal{F}$ gives an isomorphism:

(21) $\qquad O_M(\mathcal{G}, \mathcal{G}) \xrightarrow{\mathcal{F}} O_C'(H', H').$

If $\alpha \in O_M(\mathcal{G}, \mathcal{G})$ and $T \in \mathcal{G}'$, then $\mathcal{F}\alpha \in O_C'(H', H')$ and $\mathcal{F}T \in H'$ and we have (20).

The space $O_M(H, H)$ of multiplicators for H is given as follows:

$$\alpha \in O_M(H, H) \Leftrightarrow \begin{cases} \alpha \in C^\infty(R^n); \text{ for any multiindex } p, \text{ there exist an integer} \\ k \text{ and } C \geq 0 \text{ for which } |D^p \alpha(x)| \leq C\, e^{k|x|}, \ x \in R^n. \end{cases}$$

We refer the elements of $O_M(H, H)$ as the C^∞ functions exponentially increasing at infinity.

The space $O_C'(\mathcal{G}', \mathcal{G}')$ of convolutors for $\mathcal{G}'$ is given as follows:

$$T \in O_C'(\mathcal{G}', \mathcal{G}') \Leftrightarrow \begin{cases} T \text{ is an ultrahyperfunction defined by } \varphi \in O(C^n \# T_{k_0}) \\ \text{such that for any multiindex } p \text{ there exists } k > k_0 \text{ such} \\ \text{that } |\zeta^p \varphi(\zeta)| \text{ is bounded in } C^n \# T_k. \end{cases}$$

We refer the elements of $O_C'(\mathcal{G}', \mathcal{G}')$ as the ultrahyperfunctions rapidly decreasing at real infinity.

<u>Theorem 5.</u> The Fourier transformation $\mathcal{F}$ defines an isomorphism:

(22) $\qquad O_M(H, H) \xrightarrow{\mathcal{F}} O_C'(\mathcal{G}', \mathcal{G}').$

If $\alpha \in O_M(H, H)$ and $T \in H$, then $\mathcal{F}\alpha \in O_C'(\mathcal{G}', \mathcal{G}')$ and $\mathcal{F}T \in \mathcal{G}'$ and we have (20).

We will abbreviate as follows: $Q = Q(T(R^n); R^n)$ and $Q' = Q'(T(R^n); R^n)$. Q' is the space of the Fourier ultrahyperfunctions. The space $O_M(Q, Q)$ of multiplicators for the space Q is given as follows:

$$\alpha \in O_M(Q, Q) \Leftrightarrow \begin{cases} \alpha \in O(C^n); \text{ for any } h > 0, \text{ there exist } K_0 > 0 \text{ and } C \geq 0 \text{ such} \\ \text{that } |\alpha(z)| \leq C \exp(k_0|z|) \text{ for } z \in T_h. \end{cases}$$

We refer the elements of $O_M(Q, Q)$ as the entire functions exponentially increasing in any horizontal bands.

The space $O'_C(Q', Q')$ of convolutors for Q' is given as follows:

$$T \in O'_C(Q', Q') \Leftrightarrow \begin{cases} T \text{ is an ultrahyperfunction defined by } \varphi \in O(C^n \# T_{k_0}); \\ \text{for any } h > 0 \text{ there exists } k > k_0 \text{ such that } |e^{h|z|} \varphi(z)| \\ \text{is bounded in } C^n \# T_k. \end{cases}$$

We refer the elements of $O'_C(Q', Q')$ as the ultrahyperfunctions very rapidly decreasing at real infinity.

<u>Theorem 6.</u> The Fourier transformation $\mathcal{F}$ establishes an isomorphism:

$$(23) \qquad O_M(Q, Q) \xrightarrow{\mathcal{F}} O'_C(Q', Q').$$

If $\alpha \in O_M(Q, Q)$ and $T \in Q$, then $\mathcal{F}\alpha \in O'_C(Q', Q')$ and $\mathcal{F}T \in Q'$ and we have (20).

Summing up the results, we write down the tables of various spaces of multiplicatos and convolutors:

$$(24) \qquad \begin{array}{ccc} O_M(Q, Q) & \subset & O_M(H, H) \\ \cup & & \cup \\ O_M(\mathcal{G}, \mathcal{G}) & \subset & O_M(\mathcal{S}, \mathcal{S}), \end{array} \qquad \begin{array}{ccc} O'_C(Q', Q') & \subset & O'_C(\mathcal{G}', \mathcal{G}') \\ \cup & & \cup \\ O'_C(H', H') & \subset & O'_C(\mathcal{S}', \mathcal{S}'). \end{array}$$

The Fourier transformation $\mathcal{F}$ maps the first diagram onto the second one.

<u>References</u>

[1] Carmichael, R.D. : Distributions of exponential growth and their Fourier transforms, Duke Math. J., <u>40</u>(1973), 765-783.

[2] Hasumi, M. : Note on the n-dimensional tempered ultradistributions, Tôhoku Math. J., <u>13</u>(1961), 94-104.

[3] Morimoto, M. : Sur les ultradistributions cohomologiques, Ann. Inst. Fourier, <u>19</u>(1970), 129-153.

[4] Morimoto, M. : Theory of tempered ultrahyperfunctions, I. to appear in Proc. Japan Acad.

[5] Park, Y.S. and M. Morimoto : Fourier ultrahyperfunctions in the Euclidean n-space, J. Fac. Sci. Univ. Tokyo Sec. IA, <u>20</u>(1973), 121-127.

[6] Schwartz, L. : Théorie des Distributions, Hermann, Paris, 1967.

[7] Sebastião e Silva, J. : Les fonctions analytiques commes ultra-distributions dans le calcul opérationnel, Math. Ann., <u>136</u>(1958), 58-96.

[8] Sebastião e Silva, J. : Les séries de multipôles des physicians et la théorie des ultradistributions, Math. Ann., <u>174</u>(1967), 109-142.

[9] Zieleźny, Z. : On the space of convolution operators in K'_1, Studia Math., <u>31</u>(1968), 111-124.

MASS SINGULARITIES OF FEYNMAN AMPLITUDES

Toichiro Kinoshita*
Laboratory of Nuclear Studies
Cornell University, Ithaca, New York 14853
and
Akira Ukawa
Department of Physics
University of Tokyo, Tokyo 113, Japan

Singularities of Feynman amplitudes associated specifically with vanishing internal masses appear frequently in various applications of field theory. The most well-known example is the infrared divergences in QED and non-Abelian gauge theories. In examining the large momentum behaviour of Feynman amplitudes, we also encounter with such singularities since large momentum limit with fixed masses may be translated to certain zero mass limit with fixed momenta. It is our ultimate purpose to construct a general method for treating these so-called mass singularities[1]. Here we give a brief account of the results obtained thus far. The details will be published elsewhere[2].

We illustrate the salient features of mass singularities choosing the box diagram G (fig.1) as an example. If we ignore spin, the corresponding Feynman amplitude F_G, which is regarded as a function of internal masses $\underline{m} = \{m_1,\ldots,m_4\}$ as well as external momenta $\underline{P} = \{P_1,\ldots,P_4\}$, is written using Feynman parameters $\underline{z} = \{z_1,\ldots,z_4\}$ as[3]

$$F_G(\underline{m},\underline{P}) = \int \frac{\delta(1-\sum_{i=1}^{4} z_i) \prod_{j=1}^{4} dz_j}{[V(\underline{z},\underline{m},\underline{P})-i\varepsilon]^2} , \quad (1)$$

where

$$V(\underline{z},\underline{m},\underline{P}) = \frac{1}{2} \sum_{i,j=1}^{4} v_{ij} z_i z_j , \quad (2)$$

$v_{ij} (= v_{ji})$ being given by

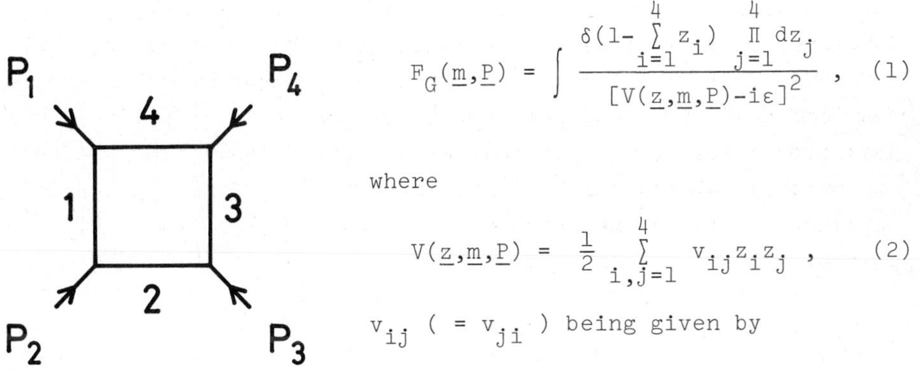

Fig.1. The box diagram G

$$((v_{ij})) = \begin{bmatrix} 2m_1^2 & , & m_1^2+m_2^2-P_2^2 & , & m_1^2+m_3^2-(P_2+P_3)^2 & , & m_1^2+m_4^2-P_1^2 \\ & & 2m_2^2 & , & m_2^2+m_3^2-P_3^2 & , & m_2^2+m_4^2-(P_3+P_4)^2 \\ & & & & 2m_3^2 & , & m_3^2+m_4^2-P_4^2 \\ & & & & & & 2m_4^2 \end{bmatrix} \qquad (3)$$

The Feynman amplitude (1) develops a singularity when the singularity of the $\underline{\text{integrand}}$ given by $V(\underline{z},\underline{m},\underline{P})=0$ becomes unavoidable by any distortion of the hypersurface of integration $\sum_{i=1}^{4} z_i=1$. The corresponding singularity of F_G is classified into two types according to whether the subhypersurface of $\sum_{i=1}^{4} z_i=1$ along which V vanishes depends on (i) $\underline{\text{all}}$, or (ii) $\underline{\text{only some}}$ of the Feynman parameters $z_1,\ldots,z_4$. We find that the former gives rise to ordinary threshold singularities. They are of no interest to us since it is not necessary for their appearance that any of the internal masses vanish. On the other hand, some of the internal masses inevitably vanish in the latter case (see below). In this sense, we call singularities of the latter type as $\underline{\text{mass singularities}}$.

To analyze their structure, suppose, for instance, that $V(\underline{z},\underline{m},\underline{P})$ vanishes identically on $\underline{\text{the domain boundary D}}$ defined by $z_1=z_3=0,z_2+z_4=1$. We easily find from (2) and (3) that

$$V(\underline{z},\underline{m},\underline{P})\Big|_{z_1=z_3=0} = m_2^2 z_2^2 + \frac{1}{2}[m_2^2+m_4^2-(P_3+P_4)^2]z_2 z_4 + m_4^2 z_4^2 , \qquad (4)$$

which agrees with the V function for the reduced diagram M shown in fig.2. We thus find two necessary conditions

$$(\alpha)\quad m_2 = m_4 = 0 , \qquad (\beta)\quad (P_3+P_4)^2 = (P_1+P_2)^2 = 0$$

for a mass singularity to occur. The first condition (α) justifies our definition of mass singularity. The second condition (β) says that the external momenta must be exceptional [4].

The behaviour of $F_G(\underline{m},\underline{P})$ near the singularity depends strongly on how fast V vanishes at the boundary D. To examine this problem, consider

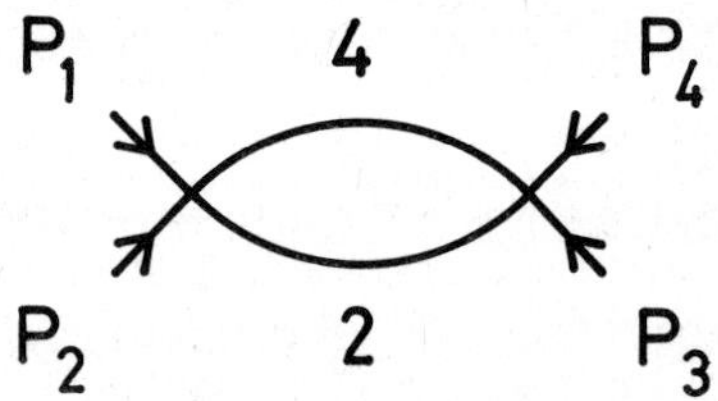

Fig.2. The reduced diagram M

the limit $\delta \to 0$ where z_1, $z_3 = O(\delta)$ and z_2, $z_4 = O(1)$, $z_2 + z_4 = 1 - O(\delta)$. The leading term of V in this limit is given by (4), which of course vanishes at the singularity. After substituting (α) and (β) into v_{ij}, we find that the $O(\delta)$ term is given by

$$(m_1^2 - P_2^2)z_1 z_2 + (m_1^2 - P_1^2)z_1 z_4 + (m_3^2 - P_3^2)z_3 z_2 + (m_3^2 - P_4^2)z_3 z_4 \quad . \qquad (5)$$

Since (5) does not necessarily vanish, V behaves as $O(\delta)$ in the general case. If $\underline{m}$ and $\underline{P}$ satisfy the further condition

$$(\gamma) \quad m_1^2 = P_1^2 = P_2^2 \qquad \text{and} \qquad m_3^2 = P_3^2 = P_4^2$$

in addition to (α) and (β), however, (5) vanishes identically and V behaves as $O(\delta^2)$, leading to an <u>enhancement</u> of mass singularity. Note that the three conditions (α), (β) and (γ) agree with those for <u>the infra-red singularity</u> which appears in the forward Coulomb scattering amplitude in QED.

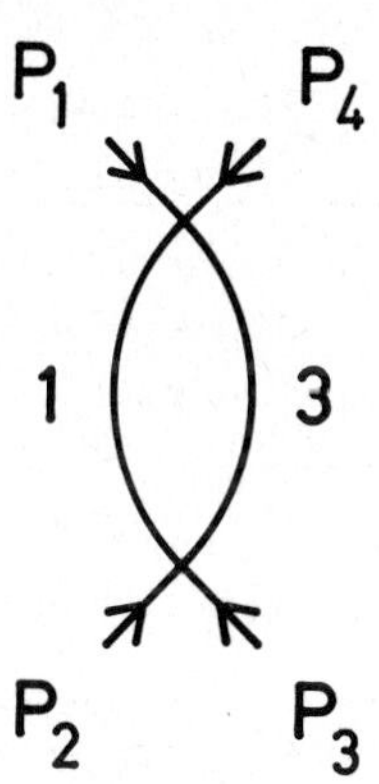

Fig.3. The reduced diagram $\tilde{M}$

The singularities we have examined above do not represent the most general ones associated with vanishing masses. In fact, z_1 and z_3 may make <u>pinches</u> instead of sticking to the boundary. If this happens, z_1 and z_3 have to satisfy

$$\partial V / \partial z_i = \sum_{j=1}^{4} v_{ij} z_j = 0 \quad (i=1,3) \ , \quad (6)$$

<u>for any values of z_2 and z_4.</u> It follows that $v_{ij} = 0$ ($i=1,3$, $j=2,4$) and $v_{11}v_{33} - v_{13}^2 = 0$. The latter gives

$$(\delta) \quad (P_2 + P_3)^2 = (m_1 + m_3)^2 \quad \text{and} \quad m_1 z_1 - m_3 z_3 = 0$$

which are the threshold conditions for the reduced diagram $\tilde{M}$ shown in fig.3. Substituting these back into V and demanding it to vanish for arbitrary z_2 and z_4, we find that the four conditions (α), (β), (γ) and (δ) are all necessary. In a rough sense, this singularity may be interpreted as <u>a product</u> of the enhanced mass singularity at M (fig.2) and the threshold singularity of $\tilde{M}$ (fig.3).

All these considerations can be generalized in a straightforward way to arbitrary diagrams. In parallel with the simple example treated above, we find that mass singularities in general can be classified schematically as shown in fig.4 (in this figure, $\overline{M}'$ ($\overline{M}$) is a subdiagram (proper subdiagram) of $\overline{M}$ (G) and $G/\overline{M}$ etc. denotes the reduced diagram obtained from G by shrinking the lines of $\overline{M}$ to points, etc.).

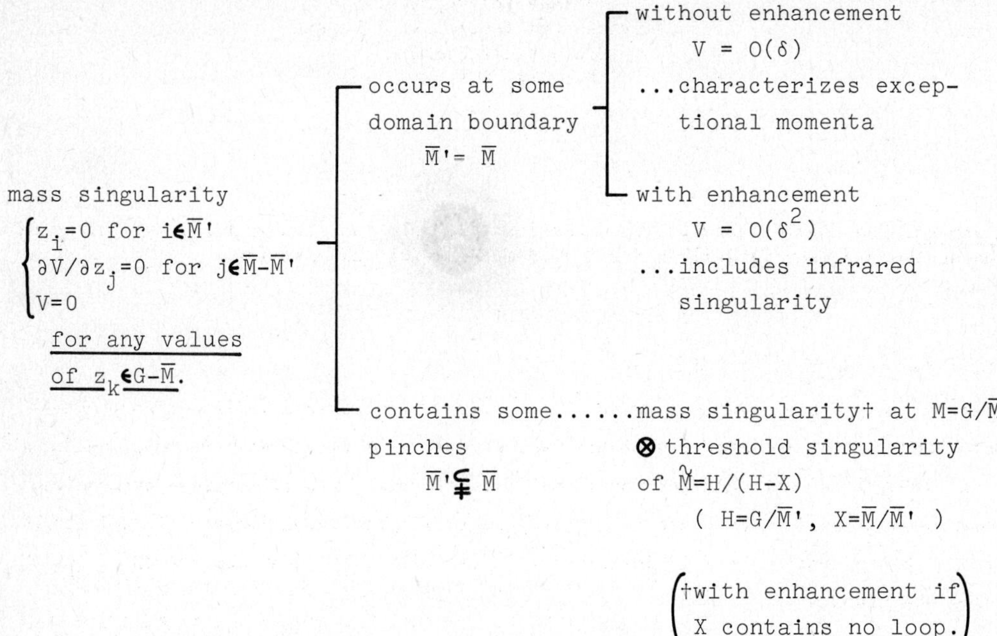

Fig.4.   Classification of mass singularities.

Our future program is to establish general power counting rules for various types of mass singularities. Complications arising from ultraviolet divergences and their renormalization will be analyzed choosing $\lambda\phi^4$ theory as an example. Non-Abelian gauge theories will also be examined in connection with their infrared properties.

References

* Work supported in part by the U. S. National Science Foundation.

[1] T. Kinoshita, J. Math. Phys. 3 (1962) 650.
[2] T. Kinoshita and A. Ukawa, in preparation.
[3] P. Cvitanovic and T. Kinoshita, Phys. Rev. D10 (1974) 3978.
[4] K. Symanzik, Commun. Math. Phys. 23 (1971) 49.

LARGE MOMENTUM BEHAVIOUR OF FEYNMAN AMPLITUDES

Klaus Pohlmeyer
II. Institut für Theoretische Physik
Universität Hamburg
Hamburg, Germany

Abstract: The complete asymptotic expansion of Feynman amplitudes for large values of
the scale parameter is derived for Euclidean and Minkowski metrics.

I. Formulation of the Problem

Consider an arbitrary Feynman graph G^o occurring in a Lagrangian field theory which
describes the self-interaction of one sort of particles with mass $m \neq 0$. G^o consists
of a collection $\mathcal{V}$ of vertices and a collection $\mathcal{L}$ of L internal lines (no tadpoles,
no external lines!). Let N stand for the number of independent loops of G^o and let $\mathcal{U}$
denote the collection of all U external vertices of G^o. The four-momenta $(p_u)_{u \in \mathcal{U}} = \underline{p}$
enter the graph G^o.

Consider the renormalized Feynman amplitude associated with G^o: $\widetilde{\mathcal{C}}(\underline{p};m)$. Replace $\underline{p}$ by
$\Lambda\underline{p}$, $\Lambda \in \mathbb{R}_+$. Show that the following asymptotic expansion for large values of the scale
parameter Λ holds in the sense of distributions

$$\widetilde{\mathcal{C}}(\Lambda\underline{p};m) = \sum_{\mu=n}^{M} \sum_{\varkappa=0}^{m_\mu - 1} [\Lambda^2]^{-\mu-2} [\ln\Lambda^2]^{\varkappa} \, \widetilde{\mathcal{C}}_{\mu,\varkappa}(\underline{p};m) + \widetilde{\mathcal{R}}_M$$

where M is an arbitrary integer, larger than or equal to n, and set up rules how to
calculate the numbers n, m_μ , the coefficient distributions $\widetilde{\mathcal{C}}_{\mu,\varkappa}$ and the remainder distribution $\widetilde{\mathcal{R}}_M$.

Without loss of generality, G^o may be assumed to be irreducible (I) i.e. one-particle-
irreducible and one-vertex-irreducible. Otherwise, $\widetilde{\mathcal{C}}(\underline{p};m)$ would factorize into a product of renormalized Feynman amplitudes corresponding to the maximal irreducible com-
ponents of G^o and propagators.

The numbers n and m_n were determined by S. Weinberg [1] and J. P. Fink [2] respectively
for the case of convergent graphs and Euclidean metrics. Under the same restrictions
D. A. Slavnov [3] proved the validity of the general form of the asymptotic expansion as
written down by us. Finally, M. C. Bergère and Y-M. P. Lam [4] determined the coefficients of the polynomial in $(\ln \Lambda^2)$ accompanying the leading inverse power of Λ^2, again

assuming Euclidean metrics. We have formulated the problem in the distribution theoretical frame as opposed to the pointwise discussion of the above mentioned authors. For Minkowski metrics this turns out to be both adequate and helpful (ensuring uniform convergence at a later stage). The difference between distribution theoretical and pointwise discussion may be illustrated by the simple mathematical example: $\frac{\varepsilon}{\varepsilon^2 + p^2}$, $\varepsilon > 0$, p a real scalar variable

	distribution theoretical	pointwise
$\dfrac{\varepsilon}{\varepsilon^2 + (\Lambda p)^2} =$	$\Lambda^{-1}\pi\,\delta(p) + O(\Lambda^{-2})$	$\begin{cases} \Lambda^{-2}\cdot\dfrac{\varepsilon}{p^2} + O(\Lambda^{-4}), & p\neq 0 \\ \varepsilon^{-1} = \text{const}, & p=0 \end{cases}$

II. Resolution of the Ultraviolet Singularities and Renormalization

As to the renormalization scheme, we decide in favour for Speer's analytical renormalization [5] since the theory of analytical functions will be employed anyway: to every line $1 \in \mathcal{L}$ we associate a complex variable λ_1: $\underline{\lambda} = (\lambda_1)_{1\in\mathcal{L}}$ (and a Feynman parameter α_1: $\underline{\alpha} = (\alpha_1)_{1\in\mathcal{L}}$) such that the renormalized Feynman amplitude $\mathcal{C}(\underline{p};m)$ is obtained by evaluating the analytical continuation of the analytically regularized Feynman amplitude $\widetilde{\mathcal{C}}_{\underline{\lambda}}(\underline{p};m)$, $\underline{\lambda} \in \Omega_2$, $\Omega_2 = \{ \underline{\lambda} \, / \, \mathcal{Re}\,\lambda_1 > 2$ for every $1\in\mathcal{L} \}$ at the point $\underline{\lambda} = \underline{1}$:

$$\widetilde{\mathcal{C}}(\underline{p};m) = \mathcal{W}_L\,\widetilde{\mathcal{C}}_{\underline{\lambda}}(\underline{p};m)$$

with the help of Speer's generalized evaluator $\mathcal{W}) = \{\mathcal{W}_L \, / \, L = 1, 2, \ldots\}$. In Ω_2 $\widetilde{\mathcal{C}}_{\underline{\lambda}}$ is holomorphic. To achieve the analytical continuation from Ω_2 to $\underline{1}$, $\widetilde{\mathcal{C}}_{\underline{\lambda}}$ is represented as a Feynman parameter integral which for the case of scalar particles in the absence of derivative coupling – we shall restrict our discussion to this case for the sake of simplicity – reads as follows

$$\widetilde{\mathcal{C}}_{\underline{\lambda}}(\underline{p};m) = \lim_{\varepsilon\downarrow 0}\,\delta\Big(\sum_{u\in\mathcal{U}}p_u\Big)\Gamma(\nu)\Big[\prod_{\ell\in\mathcal{L}}\int_0^\infty d\alpha_\ell\,\alpha_\ell^{\lambda_\ell - 1}\Big]\,d(\underline{\alpha})^{-\lambda}\cdot$$

$$\cdot \exp\Big\{-i\Big[A_{\underline{\alpha}}(\underline{p},\underline{p}) + m^2\sum_{\ell\in\mathcal{L}}\alpha_\ell\Big]\Big\}\,;\quad \lambda\in\Omega_2,\,\nu = \sum_{\ell\in\mathcal{L}}(\lambda_\ell - 1) + L - 2N,$$

$A_{\underline{\alpha}}(\underline{p},\underline{p})$ being a homogeneous quadratic form in $\underline{p}$, $d(\underline{\alpha})$ denoting the Feynman determinant. Both $A_{\underline{\alpha}}(\underline{p},\underline{p})$ and $d(\underline{\alpha})$ depend polynomially on $\underline{\alpha}$.

The vanishing of $d(\underline{\alpha})$ when some or all of the α-parameters are zero gives rise to a loss of integrability in the course of the analytical continuation in $\underline{\lambda}$, the ultraviolet divergences. In order to get the complete information about the analytical structure of $\widetilde{\mathcal{C}}_{\underline{\lambda}}$ outside of Ω_2, the intersection of the zero-surfaces of $d(\underline{\alpha})$ at $\underline{\alpha} = 0$ has to be resolved. In differential topology this problem is known under the name "resolution of a singularity". The outcome of the analyzis there – suitably adapted to the present situation – is the existence of a covering of the α-space $\mathbb{R}_+^L$ by a minimal number of sectors $\mathcal{D}$ such that the union of all sectors makes up the entire α-space, the intersections of any two different sectors have Lebesgue measure zero and such that for every sector $\mathcal{D}$ there exists a parametrization $\underline{\alpha} : \mathcal{D} = \underline{\alpha}(\Delta)$

$$\underline{\alpha} = \underline{\alpha}((t),(\beta)) \quad \text{real analytical for} \quad ((t),(\beta)) \in \Delta$$

with

$$d(\underline{\alpha}) = \prod_1^N t_i^{N_i} \cdot \hat{d}((t),(\beta))$$

where

$$\hat{d}(t),(\beta)) \geqslant 1 \quad \text{for} \quad ((t),(\beta)) \in \Delta$$

and where the powers N_i are non-negative integers.

This resolution of the ultraviolet singularities has been accomplished by Speer with the help of the concept of a labeled singularity family. To explain this concept, along with the graph G^o we consider its subgraphs G consisting of a collection $\mathcal{V}(G)$ of vertices and a collection $\mathcal{L}(G)$ of $L(G)$ internal lines. Let $N(G)$ stand for the number of independent loops of G.

A singularity family $\mathcal{E}$ is a maximal collection of non-trivial irreducible non-overlapping subgraphs G of G^o such that if $G',G \in \mathcal{E}$, $G' \subsetneqq G$ there exists at least one line $1 \in \mathcal{L}(G), \notin \mathcal{L}(G')$.

A labeled singularity family $(\mathcal{E},\sigma)$ arises from $\mathcal{E}$ by distinguishing for every $G \in \mathcal{E}$ one line $1 = \sigma(G) \in \mathcal{L}(G), \notin \mathcal{L}(G')$ for any $G' \in \mathcal{E}$, $G' \subsetneqq G$, the line whose α-parameter is largest. Thus σ is a map from $\mathcal{E}$ into $\mathcal{L}$.

A labeled singularity family leads to a sector $\mathcal{D} = \mathcal{D}(\mathcal{E},\sigma)$ in α-space

$$\mathcal{D} = \{ \alpha \ / \ \alpha_1 \geqslant 0 \ \text{ for every } 1 \in \mathcal{L}, \ \alpha_1 \leqslant \alpha_{\sigma(G)} \ \text{ for every } 1 \in \mathcal{L}(G), G \in \mathcal{E} \ \}$$

characterized by a partial ordering of the α-parameters. This partial ordering is made explicit by the parametrization

$$\alpha_1 = \begin{cases} \displaystyle\prod_{G \subset G' \in \mathcal{E}} t_{G'} & \text{if} \quad 1 = \sigma(G) \quad \text{for some} \quad G \in \mathcal{E} \\[2em] \beta_1 \cdot \displaystyle\prod_{G(1) \subset G' \in \mathcal{E}} t_{G'} & \text{if} \quad 1 \in \mathcal{L} \smallsetminus \sigma(\mathcal{E}) \ , \end{cases}$$

$$0 \leqslant t_{G^o} < \infty, \quad ((t_G)_{G \in \mathcal{E} \smallsetminus \{G^o\}}, (\beta_1)_{1 \in \mathcal{L} \smallsetminus \sigma(\mathcal{E})}) = (\underline{t},\underline{\beta}) \in I^{L-1} \quad \text{where} \quad I = [0 , 1]$$

and

$$G(1) = \text{minimal element of the set} \ \{G \ / \ G \in \mathcal{E} \ , \ 1 \in \mathcal{L}(G)\} \ .$$

On the sector $\mathcal{D} = \mathcal{D}(\mathcal{E},\sigma)$

$$d(\underline{\alpha}) = \prod_{G\in\mathcal{E}} t_G^{N(G)} \cdot \hat{d}(\underline{t},\underline{\beta}) \qquad ; \qquad \hat{d}(\underline{t},\underline{\beta}) \geq 1 \quad \text{for} \quad (\underline{t},\underline{\beta}) \in I^{L-1}.$$

The analytically renormalized Feynman amplitude then takes the following form

$$\tilde{\mathcal{C}}_\lambda(\underline{p};m) = \sum_{\mathcal{D}} \tilde{\mathcal{C}}_\lambda^{\mathcal{D}}(\underline{p};m)$$

with

$$\tilde{\mathcal{C}}_\lambda^{\mathcal{D}}(\underline{p};m) = \delta\left(\sum_{u\in\mathcal{U}} p_u\right)\left[\prod_{G\in\mathcal{E}} \Gamma(\nu(G))\right]\left[\prod_{\ell\in\mathcal{L}\cdot\sigma(\mathcal{E})} \int_0^1 d\beta_\ell\, \beta_\ell^{\lambda_\ell-1}\right]\cdot$$

$$\cdot\left[\prod_{G\in\mathcal{E}-\{G^0\}} \Gamma'(\nu(G))^{-1}\int_0^1 dt_G\, t_G^{\nu(G)-1}\right]\left[D_{\underline{t},\underline{\beta}}(\underline{p},\underline{p}) + m^2\sum_{\ell\in\mathcal{L}}\frac{\alpha_\ell}{\alpha_{\sigma(G^0)}} - i0\right]^{-\nu}\hat{d}(\underline{t},\underline{\beta})^{-2}$$

and

$$\nu(G) = \sum_{1\in\mathcal{L}(G)}(\lambda_1 - 1) + L(G) - 2N(G), \qquad \nu = \nu(G^0).$$

The Γ-functions have been distributed in such a way that the λ-singularities are
entirely contained in the first factor. This is so because the massive Feynman denominator $[\ldots\ldots]^{-\nu}$ is an entire function of its negative power ν and an infinite
differentiable function in the t- and β-variables.

III. Simultaneous Resolution of the Ultraviolet and Infrared Singularities

If we replace $\underline{p}$ by $\Lambda\underline{p}$ and let Λ tend to plus infinity, we are dealing essentially with
an inhomogeneous quadratic form with an ever decreasing inhomogeneous term raised to
the power $-\nu$:

$$\left[D_{\underline{t},\underline{\beta}}(\Lambda\underline{p},\Lambda\underline{p}) + m^2\sum_{\ell\in\mathcal{L}}\frac{\alpha_\ell}{\alpha_{\sigma(G^0)}} - i0\right]^{-\nu} = \left[\Lambda^2\right]^{-\nu}\cdot\left[D_{\underline{t},\underline{\beta}}(\underline{p},\underline{p}) + \frac{m^2}{\Lambda^2}\sum_{\ell\in\mathcal{L}}\frac{\alpha_\ell}{\alpha_{\sigma(G^0)}} - i0\right]^{-\nu}$$

In the limit of an infinitely large Λ the inhomogeneous term disappears altogether
and the situation changes drastically: the massless Feynman denominator is no longer
an entire function of ν nor is it an infinite differentiable function of the t-variables. Instead, it has simple poles as a function of ν and branching singularities in
the t-variables arising whenever and wherever the quadratic form $D_{\underline{t},\underline{\beta}}(\underline{p},\underline{p}) / \sum_{u\in\mathcal{U}} p_u = 0$
in the 4(U-1) p-variables degenerates.

To illustrate the latter point we consider the following parameter dependent distri-
bution of the real four-vector variables p_i i = 1,2,3 : $\left[- p_1^2 - \Theta\cdot p_2^2 - \Theta\cdot\theta\cdot p_3^2 - i0\right]^{-\nu}$,
$\Theta, \theta \in I$, $\nu \in \mathbb{C}$, $\mathcal{R}e\,\nu < 6$, p^2 either the Euclidean or the Minkowski length of p,
as a mathematical example for a massless Feynman denominator. Its singularities as a
distribution valued function of Θ and θ are exhibited by the following representation

$$\left[- p_1^2 - \Theta\cdot p_2^2 - \Theta\cdot\theta\cdot p_3^2 - i0\right]^{-\nu} = F_\nu(\underline{p};\Theta,\theta) + \Theta^{2-\nu}\cdot G_\nu(\underline{p};\Theta,\theta) + \Theta^{2-\nu}\cdot\theta^{4-\nu}\cdot H_\nu(\underline{p},\Theta,\theta)$$

where F_ν, G_ν and H_ν as distribution valued functions of Θ and θ are infinite differentiable in the unit square I^2 and where the sum on the right hand side depends holomorphically on ν for $\mathcal{R}e\,\nu < 6$.

Thus, in order to keep track of the formation of these singularities we have

  i) to resolve the intersection of the zero-surfaces of the Feynman determinant $d(\underline{\alpha})$ and the Jacobian of the transformation $\underline{p} \to (D_{\underline{t},\underline{\beta}}(\underline{p},\underline{p})$, coordinates on the surfaces $D_{\underline{t},\underline{\beta}}(\underline{p},\underline{p}) = $ constant)

ii) extract explicitly the branching singularities in the t-variables from the Feynman denominator.

The first task is accomplished by the introduction of a coarser concept of irreducibility: the so-called irreducibility in view of infinity (I_∞). In order to define the property (I_∞) we embed the graph G^o into a bigger graph G^o_∞ by adding to $\mathcal{V}$ one more vertex v_∞, the infinite vertex, and joining every external vertex $u \in \mathcal{U}$ to v_∞ by one separate line 1_u. Similarly, we embed the subgraphs G of G^o into subgraphs G_∞ of G^o_∞.

Definition: A subgraph G of G^o possesses the property (I_∞) if G is irreducible (I) or if G_∞ is one-particle irreducible and one-vertex-irreducible with the possible exception of the infinite vertex v_∞.

Examples:

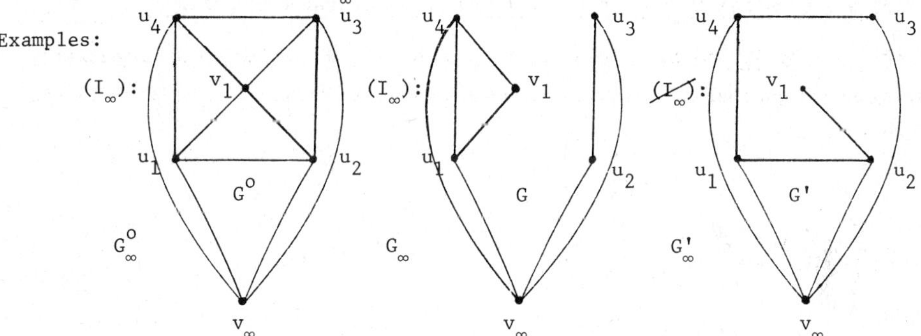

With the help of the new irreducibility concept (I_∞) in complete analogy to Speer, we define new singularity families $\mathcal{E}_\infty$, labeled singularity families $(\mathcal{E}_\infty,\sigma_\infty)$, sectors $\mathcal{D}_\infty = \mathcal{D}_\infty(\mathcal{E}_\infty,\sigma_\infty)$ and appropriate parametrizations. However, now we have to distinguish two subfamilies $\mathcal{F}_\infty$ and $\mathcal{H}_\infty$ which are disjoint and which make up $\mathcal{E}_\infty$:

$G \in \mathcal{F}_\infty$ iff the removal of the line $\sigma_\infty(G)$ from $\mathcal{L}(G)$ opens a loop, $G \in \mathcal{H}_\infty$ iff the removal of the line $\sigma_\infty(G)$ from $\mathcal{L}(G)$ dissolves a connection between external vertices.

The number of elements in $\mathcal{F}_\infty$ is equal to N, the number of elements in $\mathcal{H}_\infty$ equal to $U - 1$.

The second task is accomplished in two steps:

First, the quadratic form $D_{\underline{t},\underline{\beta}}(\underline{p},\underline{p})$ is diagonalized

$$D_{\underline{t},\underline{\beta}}(\underline{p},\underline{p}) = E_{\underline{t},\underline{\beta}}\left(\frac{q(\underline{t},\underline{\beta})}{}, \frac{q(\underline{t},\underline{\beta})}{}\right)$$

$$E_{\underline{t},\underline{\beta}}(q,q) = -t_{G^0}^{-1}\cdot \sum_{H\in\mathscr{H}_\infty}\left(\prod_{H\subset G\in\mathscr{E}_\infty}t_G\right)\cdot\hat{e}_H(\underline{t},\underline{\beta})\, q_H^2$$

$\hat{e}_H(\underline{t},\underline{\beta}) \geqslant 1$ for all $H\in\mathscr{H}_\infty$ and $((\underline{t}),(\underline{\beta}))\in I^{L-1}$

with the help of an invertible linear transformation $T_{\underline{t},\underline{\beta}}$ which as a function of $\underline{t}$

and $\underline{\beta}$ is infinite differentiable [6]: $\left(\underline{q(t,\beta)}, \sum_{u\in\mathscr{U}}p_u\right) = \left((q_H(\underline{t},\underline{\beta}))_{H\in\mathscr{H}_\infty}, \sum_{u\in\mathscr{U}}p_u\right) = T_{\underline{t},\underline{\beta}}\,\underline{p}$.

Second, the additive occurrence of $\dfrac{m^2}{\Lambda^2}\cdot\sum_{\ell\in\mathscr{Z}}\dfrac{\alpha_1}{\alpha_{\sigma(G^0)}}$ and the $(U-1)$ terms of $E_{\underline{t},\underline{\beta}}$ in the

Feynman denominator is converted into a multiplicative appearance in the Feynman para-

meter integral by means of an $(U-1)$-fold Mellin transformation.

The points i) and ii) being settled in this way, we arrive at the following represen-

tation for $\tilde{\mathcal{C}}_{\underline{\lambda}}(\Lambda\underline{p};m)$:

$$\tilde{\mathcal{C}}_{\underline{\lambda}}(\Lambda\underline{p};m) = \sum_{\mathscr{D}_\infty}{}' \tilde{\mathcal{C}}_{\underline{\lambda}}^{\mathscr{D}_\infty}(\Lambda\underline{p};m)$$

$$\tilde{\mathcal{C}}_{\underline{\lambda}}^{\mathscr{D}_\infty}(\Lambda\underline{p};m) = \delta\left(\sum_{u\in\mathscr{U}}p_u\right)\left[\prod_{H\in\mathscr{H}_\infty}\frac{1}{2\pi i}\int_{-\gamma_H-i\infty}^{-\gamma_H+i\infty}ds_H\,\Gamma(-s_H)\,\Gamma(2+s_H)\right]\cdot$$

$$\left[\Lambda^2\right]^{\sum\limits_{H\in\mathscr{H}_\infty}s_H}\left[\prod_{G\in\mathscr{E}_\infty}\Gamma\!\left(\nu(G)+\sum_{G\subset H\in\mathscr{H}_\infty}s_H\right)\right]\cdot h_{\underline{\lambda},\underline{s}}(\underline{p};m)$$

where $h_{\underline{\lambda},\underline{s}}(\underline{p};m)$ is an entire function of $\underline{\lambda}$ and $\underline{s} = (s_H)_{H\in\mathscr{H}_\infty}$, $0 < \gamma_H < 1$, $H\in\mathscr{H}_\infty$.

Note the presence of simple poles contained in the factors $\Gamma(2+s_H)$. The residues of

these infrared poles bring about δ-functions: $\delta(\sum'p_u)$, where $\sum'$ stands for a partial

sum of external momenta, or derivatives of them.

The integrations converge uniformly in $\underline{\lambda}$ in the topology of $\mathscr{S}'$ for Euclidean as well

as for Minkowski metrics. It is at this point that the distribution theoretical dis-

cussion pays off.

As $\underline{\lambda}$ tends to $\underline{1}$, poles in s_H move in from the left of the contours. Together with the

fixed poles on the right they may or may not pinch the contours. In the latter case

we pick up extra contributions. In view of the uniform convergence, the generalized

evaluator may be applied under the integration signs. Thus we obtain the following

form for the renormalized amplitude

$$\tilde{\mathcal{C}}(\Lambda\underline{p};m) = \frac{1}{2\pi i}\int_{g-i\infty}^{g+i\infty}dz\,\left[\Lambda^2\right]^{z-2}\tilde{\mathcal{C}}_{\underline{w}}^z(\underline{p};m),\quad L-2N<g<L-2N+1$$

where $\tilde{\mathcal{C}}_{\underline{w}}^z(\underline{p};m)$ is a meromorphic function of z. The localization and order of its poles

can be read off directly from the graph G^{o} [7]. Its asymptotic behaviour in z is entirely known.

The complete asymptotic expansion is now obtained by shifting the z-contour to the left
and picking up contributions from the poles according to their order.

References

1) Weinberg S.: Phys. Rev. $\underline{118}$, 838 (1960)
2) Fink J.P.: Journ. Math. Phys. $\underline{9}$, 1389 (1968)
3) Slavnov D.A.: Teor. Mat. Fiz. $\underline{17}$, 169 (1973)
4) Bergère M.C. and Lam Y-M.P.: Commun. Math. Phys. $\underline{39}$, 1 (1974) and
                                Freie Universität Berlin preprint 74/9 (1974)
5) Speer E.R.: Generalized Feynman Amplitudes, chapter III,
                Princeton University Press, Princeton 1969
6) Trute H.: Desy report 74/44 (1974)
7) Pohlmeyer K.: Desy report 74/36 (1974)
8) Rübsamen R.: forthcoming Desy report

Discussion

A. Ukawa (question): Can you treat exceptional momenta cases by your method?

K. Pohlmeyer (answer): Yes, you can. If you introduce the restriction $\sum' p_u = 0$,
$\sum'$ denoting a partial sum of external momenta, right from the beginning you have to
resolve the intersection of the zero-surfaces of the Feynman determinant $d(\underline{\alpha})$ and the
Jacobian of the transformation $\underline{p}/\sum'_{p_u=0} \rightarrow (D_{\underline{t},\underline{\beta}}(\underline{p},\underline{p})/\sum'_{p_u=0}$, coordinates on the
surfaces: $D_{\underline{t},\underline{\beta}}(\underline{p},\underline{p})/\sum'_{p_u=0} =$ constant). This can be done by introducing one more infinite vertex with the help of which an appropriate irreducibility concept - finer than
(I_∞) but still coarser than (I) - is defined [8].

SPECTRAL DEPENDENCE OF THE ANALYTICITY DOMAIN OF
LOCAL VERTEX FUNCTIONS

G. Sommer

Department of Theoretical Physics

University of Bielefeld

Germany

ABSTRACT

A method is outlined which allows to construct the envelope of holomorphy for general local vertex functions with arbitrary mass spectrum.

0.

Analyticity properties of the momentum space three point function V are considered in the frame-work of axiomatic field theory. More precisely, we are studying the Fourier transform of the vacuum expectation value of the retarded product of three in general different Wightman fields or currents.

The problem is to find the envelope of holomorphy $\mathcal{H} = \mathcal{H}(\mathcal{B}_\nu)$ of the primitive analyticity domain $\mathcal{B}_\nu$ of V , in its dependence on the spectral assumptions of the theory. We disregard non-linear properties of field theory, so the results are independent of an appearance of Martin catastrophies.

Since we can disregard one-particle poles in V , the boundary $\partial\mathcal{H}$ of $\mathcal{H}$ depends only on the lower limits of the continuous spectra of the mass operator in the different channels (i.e. the multi-particle thresholds M_k , k = 1,2,3): $\partial\mathcal{H}(\mathcal{B}_\nu(M_k))$ $\equiv \partial\mathcal{H}$. The primitive domain in the variables $z_k = p_k^2$, k = 1,2,3 with $\sum p_k = 0$ is

$$\mathcal{B}_\nu(M_k) = \bigcup_{k=1}^{3} \left\{ (z_k, z_l, z_m) \; ; \; 0 \leqslant z_k < M_k^2 \, , \, (z_l, z_m) \in \mathbb{C}^2, \, (z_k \pm i\varepsilon, z_l, z_m) \in \mathcal{B}_\nu(0) \; \forall \varepsilon > 0 \right\} \cup \mathcal{B}_\nu(0)$$

where (k,l,m) denotes a cyclic permutation of (1,2,3) and $\mathcal{B}_\nu(0)$ is the Källen - Wightman domain [1].

The procedure of constructing $\partial\mathcal{H}$ runs through following steps [2]:

1) choice of suitable variables,
2) determination of edges of $\partial\mathcal{H}(\mathcal{B}_\nu(M_k))$,
3) choice of a suitable parametrization of a family of smooth hypersurfaces $\mathcal{S}_\lambda$ which interpolate between these edges,
4) geometrical properties of $\mathcal{H}$,

5) characterization of the requested pseudoconvexity properties of $\partial \mathcal{H}$,

6) introduction of a constraint on $\mathcal{S}_\lambda$ through a differential equation (PSC-equation) on the defining functions of $\mathcal{S}_\lambda$ which is sufficient for $\mathcal{S}_\lambda$ to have the desired pseudoconvexity,

7) proof of the existence and uniqueness of a solution $\overset{\circ}{\lambda}$ of the resulting system of boundary value problems,

8) proof that the resulting unique hypersurface $\mathcal{S}_{\overset{\circ}{\lambda}}$ is the boundary of a (natural) domain of holomorphy,

9) construction of disks to establish $\partial \mathcal{H}(\mathcal{B}_v(M_k)) = \mathcal{S}_{\overset{\circ}{\lambda}}$.

1.

For the choice of variables we first remark that in the special cases where $M_k \to 0$ $\forall\, k$ or $M_k \to \infty\, \forall k$ the holomorphy envelope of V is invariant under the two transformation groups

$$v \to R_\varphi\, v \qquad \text{(simultaneous rotations: SR)}$$
$$v \to D_x\, v \qquad \text{(simultaneous dilatations: SD)}$$

where

$$v := \begin{pmatrix} v_1 \\ v_2 \\ v_3 \\ v_4 \end{pmatrix} \equiv \begin{pmatrix} j_1\,(p_{10} - p_{11}) - j_2\,(p_{20} - p_{21}) \\ j_1\,(p_{20} - p_{21}) - j_2\,(p_{10} - p_{11}) \\ j_1\,(p_{10} + p_{11}) - j_2\,(p_{20} + p_{21}) \\ j_1\,(p_{20} + p_{21}) - j_2\,(p_{10} + p_{11}) \end{pmatrix}$$

$$R_\varphi := \begin{pmatrix} e^{i\frac{2\pi}{3}\varphi} & & & \bigcirc \\ & e^{-i\frac{2\pi}{3}\varphi} & & \\ & & e^{i\frac{2\pi}{3}\varphi} & \\ \bigcirc & & & e^{-\frac{2\pi}{3}\varphi} \end{pmatrix}$$

$$D_x := x \cdot \mathbb{1}$$

with momenta (p_{k0}, p_{k1}), $k = 1,2,3$, $\sum_k p_{k0} = \sum_k p_{k1} = 0$ in a two-dimensional Minkowski space, with $j_k = e^{i\frac{2\pi}{3}k}$, $k = 1,2,3$ and with $\varphi \in \mathbb{Z}$, $x \in \mathbb{R}_+$. For $M_k \to \infty$ the holomorphy envelope is even invariant under a larger group, namely $\varphi \in \mathbb{R}$. In general both symmetries are broken if $0 < M_k < \infty$ for at least one k, i.e. if at least one of the multiparticle thresholds is finite. However, if the multiparticle thresholds in the three channels are equal, one has still invariance under SR, but not under SD.

Because we are interested in the dependence of the analyticity domain on M_k it is desirable to choose instead of z_k three other Lorentz invariants x, y, α in such a way that only one of them is not SD-invariant (X) respectively not SR-invariant (α). Such a choice is possible :

$$x := (v_1 v_2 v_3 v_4)^{-1/2}, \quad y := v_1 v_4 x, \quad \alpha := v_1 v_3 x.$$

Then the Källén-Wightman manifolds C_k and $F'_{k\ell}$ have similar representations: With

$$\Psi_k(\tau, t) := \frac{\tau \alpha}{j_k} + \frac{j_k}{\tau \alpha} - \frac{y}{\tau} - \frac{\tau}{y} + t x$$

both C_k and $F'_{k\ell}$ can be represented by $\Psi_k(\tau, t) = 0$ with $t \in \mathbb{R}$ and different choices of τ, $\tau \in S_2$.

Furthermore for fixed $y \in \mathbb{C}$ the singularity region of V is restricted to the union of the 3 cut manifolds C_k and a region $\triangle(y) \subset \mathbb{C}^2$, $\triangle(y) := \{ (x, \alpha) ; x \in \Omega(y), \alpha \in d(y,x) \}$, which is finite of $M_k \neq 0 \ \forall \ k$. Here $\Omega(y)$ depends analytically on y in the sense that it can be represented by

$$f(x,y; \sigma, \sigma') = 0 \qquad (\sigma, \sigma') \in \Omega_1$$

where f is holomorphic in x and y and $C^{(\infty)}$ in σ, σ' in Ω_1 with

$$\Omega_\varrho := \left\{ (\sigma, \sigma') \in \mathbb{R}^2; \ (\sigma \cdot \min_k M_k^2 \varrho) \in [0,1], \ \sigma'^2 \in [0,1], \ \varrho \text{ fixed} \in \mathbb{R} \right\}.$$

2.

By construction of perturbation theoretical examples it can be shown that certain portions of C_k, $F'_{k\ell}$ are edges E_ϱ of $\partial \mathcal{H}$, that for arbitrary $y_0 \in \mathbb{C}$ and arbitrary $x_0 \in \partial \Omega(y_0)$ the union of these edges has a non-vanishing intersection with $\{x, y, \alpha ; x = x_0, y = y_0, \alpha \in d(y_0, x_0)\}$ and that the edges are analytic in y.

3.

One can provide that a family $\{ \mathcal{J}_\lambda \}$ of hypersurfaces exists which for $(\sigma, \sigma') \in \partial \Omega_\varrho$ $\varrho \in \mathbb{R}$ pass through the edges E_ϱ of $\partial \mathcal{H}$, are quasi-analytic of rank 1 and can be represented by

$$\mathcal{J}_\lambda : \begin{cases} f(x,y; \sigma, \sigma') = 0 \\ \lambda(x,y, \alpha ; \varrho, \sigma, \sigma') + \wedge(x,y, \alpha ; \varrho, \sigma, \sigma') = 0 \end{cases}$$

where f is the special function characterizing the domain $\Omega(y)$ and λ, $\wedge$ are holomorphic in (x,y,α) for $y \in \mathbb{C}$, $(x, \alpha) \in \triangle(y)$ and twice continuously differentiable in $(\varrho, \sigma, \sigma')$ for $\varrho \in \mathbb{R}$, $(\sigma, \sigma') \in \Omega_\varrho$. The function $\wedge$ can be constructed in such a way that the hypersurface $f = \wedge = 0$ interpolates between the edges E_ς of $\partial\mathcal{H}$ and that E_ς is approached on $f = \wedge = 0$ in the limit dist $\left[(\sigma, \sigma'), \partial\Omega_\varsigma\right] \to 0$.

For this aim the similarity of the representations of C_k and $F'_{k\ell}$ in (x,y,α)-space can be exploited: The desired interpolation between the edges E_ς is established by an ansatz

$$\wedge : = s^{-1} \prod_{k=1}^{3} \psi_k (\tau, t_k)$$

where $S = S(x,y,\alpha; \varsigma, \sigma, \sigma')$ never vanishes in $\triangle(y)$ and where $\tau = \tau(\varsigma,\sigma,\sigma')$, with $|\tau| = 1$, and $t_k = t_k(\varsigma, \sigma, \sigma')$, $t_k \geqslant 0$ are suitably chosen, twice continuously differentiable ($\varsigma \in \mathbb{R}$, $(\sigma, \sigma') \in \overline{\Omega}_\varsigma$) and depend on the lower limits M_k, $k = 1,2,3$, of the mass spectrum in the three channels. (For details see ref. 2). In the case when we still have invariance under SR (i.e. in the equal mass case),

$$t_k \equiv t \quad \forall k .$$

The hypersurfaces $\mathcal{S}_\lambda$ yield interpolations of the desired type for arbitrary λ with $\lambda = 0$ on $\partial\Omega_\varsigma$.

4.

Using the fact that a common scaling of the three mass spectra results in a scaling of x with y, α unchanged (more precisely: of σ with y, α, σ' unchanged) the boundary $\partial\mathcal{H}$ of $\mathcal{H}$ must be such that on $\partial\mathcal{H}$ the coordinate σ is uniquely determined, $\sigma = \sigma_0(y, \alpha, \sigma')$, by y, α, σ'. Thus we may use the notion: below (above) $\partial\mathcal{H}$, for characterizing points with $\sigma < \sigma_0$ ($\sigma > \sigma_0$).

5.

From the geometrical property of $\mathcal{H}$ just stated a necessary condition for a hypersurface to be in $\partial\mathcal{H}$ is that it be Cartan pseudoconvex from below.

6.

If a hypersurface $\phi = 0$, $\phi : \mathbb{C}^3 \to \mathbb{R}$, $\phi \in C^{(2)}$ possesses in a neighbourhood of some point P supporting analytic manifolds of lower dimension (and if the rank of the Hesse form restricted to the analytic tangent space fulfils a further condition, both conditions being satisfied for quasi-analytic hypersurfaces of rank 1) then pseudoconvexity in the sense of Cartan and pseudoconvexity in the sense of Hesse forms are equivalent. Thus for pseudoconvexity from below of the hypersurface $\mathcal{S}_\lambda$ at some point P semi-definiteness of the corresponding Hesse form at P (instead of definiteness in the general case) and $\lambda + \Lambda \in C^{(2)}$ with respect to ς , σ , σ ' (instead of $C^{(4)}$ in the general case) are already sufficient. In particular a system of second order elliptic differential equations (pseudoconvexity equation) for λ in the variables σ , σ ' can be stated:

$$(PSC) \qquad \frac{\partial^2}{\partial\sigma^2}\lambda + \frac{\partial^2}{\partial\sigma'^2}\lambda + a\,(y,\alpha,\varsigma\,;\,\sigma,\sigma'\,;\,\frac{\partial}{\partial\sigma}\lambda,\frac{\partial}{\partial\sigma'}\lambda) = 0$$

which is sufficient for $\mathcal{S}_\lambda$ to be pseudoconvex from below (for any smooth solution λ with $\frac{\partial}{\partial\sigma}\lambda + \frac{\partial}{\partial\sigma}\Lambda \neq 0$ on $\mathcal{S}_\lambda$).

7.

The system of boundary value problems consisting of (PSC) and the boundary condition $\lambda \equiv 0$ on $\partial\Omega_\varsigma$ can be shown by use of the Leray-Schauder theorem (and of Lady-zhenskaya-Ural'tseva estimates) to have a solution $\overset{\circ}{\lambda}$ which is unique and which, moreover, is analytic in y, α , ς with $y \in \mathbb{C}$, $\alpha \in d\,(y,x(y,\sigma,\sigma'\,))$, $\varsigma \in \mathcal{N}(\mathcal{R})$.

8.

The uniquely determined hypersurface $\mathcal{S}_{\overset{\circ}{\lambda}}$ is pseudoconvex from below at all points where $\frac{\partial}{\partial\sigma}\overset{\circ}{\lambda} + \frac{\partial}{\partial\sigma}\Lambda \neq 0$. From the boundary condition and the properties of the solution $\overset{\circ}{\lambda}$ the assumption that $\frac{\partial}{\partial\sigma}\overset{\circ}{\lambda} + \frac{\partial}{\partial\sigma}\Lambda = 0$ somewhere on $\mathcal{S}_{\overset{\circ}{\lambda}}$ implies the existence of points where $\mathcal{S}_{\overset{\circ}{\lambda}}$ is convex from above and $\frac{\partial}{\partial\sigma}\overset{\circ}{\lambda} + \frac{\partial}{\partial\sigma}\Lambda \neq 0$ what leads to a contradiction by the fact that (PSC) is a sufficient condition for pseudoconvexity from below at this point. It follows that $\mathcal{S}_{\overset{\circ}{\lambda}}$ is the boundary of <u>some</u> domain of holomorphy.

9.

The last step is to show that $\mathcal{S}_{\overset{\circ}{\lambda}}$ coincides with part of $\partial\mathcal{H}$ by use of the continuity theorem. In the proof the construction of disks is based on the fact that the rank 1 quasi-analytic hypersurface $\mathcal{S}_{\overset{\circ}{\lambda}}$ has some analyticity properties: through any of its points P passes a 1-dimensional analytic manifold completely belonging to $\mathcal{S}_{\overset{\circ}{\lambda}}$ in a full neighbourhood of P . The boundary condition imposed on $\overset{\circ}{\lambda}$ on $\partial\Omega_\varsigma$ allows to construct a one-parameter sequence of disks $\Delta(\kappa)$, $\kappa \in [0,1]$ in such a way that for arbitrary $\kappa \in [0,1]$ the large α region on $\Delta(\kappa)$ and for $\kappa = 1$ the whole of Δ is in the analyticity domain.

lo.

To conclude, for $(x,\alpha) \in \triangle(y)$, $y \in \mathbb{C}$ the envelope of holomorphy of the general local vertex function for arbitrary mass spectrum is bounded by a quasi-analytic hypersurface $\mathcal{S}_{\mathring{\lambda}}$ of rank 1. Here $\mathring{\lambda}$ is the unique solution of the boundary value problem consisting of (PSC) and the boundary condition $\mathring{\lambda} = 0$ on $\partial\Omega_\rho$ and can be determined numerically. The dependence of $\mathcal{S}_{\mathring{\lambda}}$ on the mass spectrum is displayed by the functions $\mathring{\lambda}$ and $\wedge$.

REFERENCES

1) G. Källén , A. Wightman, Math. Fys. Skv. Dan. Vid. Selsk. <u>1</u>, no.6 (1958).
   For further references see 2).

2) G. Sommer, On the envelope of holomorphy of the general local vertex function in
            momentum space (with arbitrary mass spectrum),
            University of Bielefeld preprint, Bi-73/16.

DISCUSSION

V.S. Vladimirov (question):

Is it possible to calculate an explicit form of the surfaces at the boundary of $\mathcal{H}$ in p-space, as in x-space (Källén-Wightman), at least for equal masses?

G. Sommer (answer):

Unfortunately not. The dependence of the pseudoconvexity equations on the three massparameters as well as on the variables σ , σ' and the other parameters is still rather involved. This is,because the hypersurface has to interpolate between the sets given by the boundary conditions, and these are complicated. All this gets only slightly simplified in the equal mass case. The structure of the pseudoconvexity equations remains unchanged, because essentially the only simplification in this case is that all three t_k in the parametrization of $\wedge$ are equal.

ASYMPTOTICS AND LIGHT-CONE SINGULARITIES
IN QUANTUM FIELD THEORY

by

E. Brüning and P. Stichel

Department of Theoretical Physics
University of Bielefeld
Germany

Abstract:

For a local amplitude we prove a one-to-one correspondence between
properly defined scaling, the leading light-cone singularity and the
asymptotic behaviour of the corresponding Jost-Lehmann spectral function
in the sense of distribution theory.

1. Introduction

It has been claimed very often that Bjorken scaling of the imaginary part of the forward scattering amplitude for the virtual Compton
process is explained in terms of the light-cone (LC-) singularities of
the Fourier transform of that amplitude [1]. This claim has led to the
development of the light-cone physics [1].

It is the aim of the present paper to discuss the connection
between LC-singularities, Bjorken scaling and related asymptotics [+)].
In particular we will show that there is a one-to-one correspondence
between scaling, light-cone singularities and asymptotic behaviour of
the Jost-Lehmann (JL) spectral function, if all these limits are understood as limits of sequences of distributions [2,3].

In order to avoid complications due to the photon spin we consider
a model-amplitude

+) For comments on earlier attempts we refer to [2].

$$\widetilde{T}(x,p) := \langle p | j(\tfrac{x}{2}) j(-\tfrac{x}{2}) | p \rangle$$

with a real scalar current $j(x)$ and a state $|p\rangle$ of one scalar particle
of momentum p ($p^2 = 1$). Our assumptions are those of general quantum
field theory, e.g. [4]

(A) $T(q,p)$ satisfies

    A1) Lorentz-invariance:

$$T(\Lambda q, \Lambda p) = T(q,p) \qquad \forall \Lambda \in L_+^\uparrow$$

    A2) Spectrum:

$$\operatorname{supp} T(q,p) \subset \{(q,p) \,|\, q \cdot p \geq 0,\ q^2 + 2q \cdot p \geq 0\}$$

    A3) Locality:

$$\widetilde{T}(x,p) - \widetilde{T}(-x,p) = 0 \quad \text{for} \quad x^2 < 0 \ \forall p$$

    A4)[†] Temperedness:

$$T(q,p) \in \mathscr{S}'(\mathbb{R}^4_q) \quad \text{for fixed } p,\ p^2 = 1.$$

(B) Positivity:

$\widetilde{T}(x,p)$ is of positive type, e.g.

$$\langle \widetilde{T}(x-y,\ p), \varphi(x)\varphi^*(y) \rangle \geq 0 \quad \forall \varphi \in \mathscr{S}(\mathbb{R}^4) \forall p \text{ fixed.}$$

Let us comment on condition A4). We think A4) to be natural in
our context: On one hand, it is well known from the perturbative
treatment of renormalizable field theories that the n-point-functions
are tempered distributions. On the other hand, we don't know of any
example for a nontrivial ordinary function $\widetilde{T}(x,p)$ which satisfies A2)
A3) and (B)[††]. Thus, in the general case, we cannot expect both $T(q,p)$
and $\widetilde{T}(x,p)$ to be functions, and are thus forced to introduce a notion

---

†) In the usual axiomatic framework temperedness of matrix elements of
    field operators holds if these are taken with respect to proper
    (wave packet) states. For reasons of simplicity we restrict ourselves
    to improper (plane wave) states.

††) In the special case supp $T(q,p) \subset \overline{V^+}$ it can be shown (Borchers
    [5]) by A3) and (B) that $\widetilde{T}(x,p)$ cannot be an ordinary function
    different from a constant.

of asymptotic behaviour of (tempered) distributions for a characterization of Bjorken scaling.

At first it seems to be most natural to define the asymptotic behaviour of a distribution $F \in \mathscr{S}'(\mathbb{R}^1)$ via the asymptotic behaviour of its regularization [6], e.g.

Definition 1:

We say, $F \in \mathscr{S}'(\mathbb{R}^1)$ behaves asymptotically like $x^{-\gamma}$ (is of asymptotic degree $\gamma \in \mathbb{R}^1$) if

(i) $\lim\limits_{x \to +\infty} x^\gamma \ (F * \varphi)(x) \ \exists = c_\gamma(\varphi) \quad \forall \varphi \in \mathscr{D}(\mathbb{R}^1)$

(ii) $c_\gamma(\varphi) \neq 0$ for at least one φ, if $F \neq 0$.

But we prefer another definition [7]:

Definition 2:

We say, $F \in \mathscr{S}'(\mathbb{R}^1)$ behaves asymptotically like $x^{-\gamma}$ (is of asymptotic degree $\gamma \in \mathbb{R}^1$) if

(i) $\lim\limits_{\lambda \to +\infty} \lambda^\gamma F(\lambda x) \ \exists = F_\gamma(x)$ in $\mathscr{S}'(\mathbb{R}^1)$

(ii) $F_\gamma \neq 0$ if $F \neq 0$

There are some essential differences between both definitions:

(a)   The asymptotic degree of $F \in \mathscr{S}'(\mathbb{R}^1)$ may be well-defined in the sense of both definitions but may differ: take F of compact support, then the asymptotic degree γ of F in the sense of definition 1 is $\gamma_1 = +\infty$, but $\gamma = \gamma_2$ is finite in the sense of definition 2.

(b)   The asymptotic degree may be defined in the sense of one definition but not in the other:

$F(x) = e^{ix}$ has the asymptotic degree $\gamma_2 = +\infty$ in the sense of definition 2 but is not defined in the sense of definition 1.

(c)   There are classes of tempered distributions on which both definitions agree, for instance on those homogeneous distributions which don't have compact support.

(d)   Both definitions agree especially for

$$F \in \mathcal{C}(\mathbb{R}^1), \quad \lim_{x \to \pm\infty} x^\gamma F(x) \, \exists = c_\gamma \neq 0, \quad \gamma < 1. \tag{1}$$

Such a situation we want to cover: If we identify q and p
with the photon and nucleon momentum in the virtual Compton process,
we expect $T(q,p)$ to be a continuous function at least for $q^2 \leq 0$,
because it is measurable there. The experimentally observed
behaviour of $T(q,p)$, $q^2 \leq 0$, suggests a structure for $T(q,p)$
which can be expressed via a condition like (1).

We prefer definition 2, because

(α)   we think it to be sufficiently good adapted to the experimental
      situation (see above).
(β)   it is not far away from the most general situation we have in
      quantum field theory, e.g. each tempered distribution is almost
      power-behaved in the sense of definition 2:

<u>Lemma 1</u>   (Steinmann [7]):

There exists a function $\gamma : \mathscr{S}'(\mathbb{R}^1) \to (-\infty, +\infty]$ such that for
all $F \in \mathscr{S}'(\mathbb{R}^1)$:

(i) $\lim\limits_{\lambda \to +\infty} \lambda^\delta F(\lambda x) \, \exists = 0 \qquad \forall \, \delta < \gamma(F)$

(ii) $\lim\limits_{\lambda \to +\infty} \lambda^\delta F(\lambda x) \, \nexists \qquad \forall \, \delta > \gamma(F)$

$\left. \right\}$ in $\mathscr{S}'(\mathbb{R}^1)$

(γ)   Definition 2 is rather symmetric with respect to Fourier trans-
      formation:

$$\lim_{\lambda \to +\infty} \lambda^\gamma F(\lambda p) \, \exists \quad \text{in} \quad \mathscr{S}'(\mathbb{R}^1_p) \quad \Leftrightarrow$$

$$\lim_{\lambda \to +\infty} \lambda^{\gamma-1} \tilde{F}(x/\lambda) \, \exists \quad \text{in} \quad \mathscr{S}'(\mathbb{R}^1_x)$$

and thus seems to be an appropriate notion for characterizing
Bjorken scaling of

$$V(q,p) := T(q,p) - T(-q,p) \tag{2}$$

<u>and</u> the LC-singularity structure of $\tilde{V}(x,p)$.
We say that V satisfies condition A (B) if T does.

2. Asymptotics in Minkowski Space

In the following we want to define for V the properties of strong scaling[†] and weak scaling respectively in accordance with definition 2 and lemma 1.

Due to condition A1), V is a $0_+(3)$-invariant tempered distribution in the rest system $p = (1,\underline{0})$. We define

$$V(q): = V(q, (1,\underline{0})) \qquad\qquad (3)$$

Hence $V(q) \in \mathscr{S}'(\mathbb{R}^4, 0_+(3))$.

Due to the topological isomorphism between $\mathscr{S}'(\mathbb{R}^3, 0_+(3))$ and $\mathscr{S}'(\mathbb{R}^1_+)$ we may define uniquely a tempered distribution V_1 of the $0_+(3)$-invariants only:

$$\langle V_1(q_0,\rho), \phi f(q_0,\rho)\rangle = \langle V(q), f(q)\rangle \;\; \forall f \in \mathscr{S}(\mathbb{R}^4) \qquad (4)$$

with $\quad \phi f(q_0,\rho): = (4\pi)^{-1} \int\limits_{|\underline{\omega}|=1} d\underline{\omega}\, f(q_0,\sqrt{\rho}\underline{\omega}).$

We have [8] $\phi f \in \mathscr{S}(\mathbb{R}^1 \times \mathbb{R}^1_+)$ and therefore $V_1(q_0,\rho) \in \mathscr{S}'(\mathbb{R}^1 \times \mathbb{R}^1_+)$.

There exists a unique extension of V_1 to a distribution $V_2 \in \mathscr{S}'(\mathbb{R}^2)_-$[††]:

$$\langle V_2(q_0,w), \varphi(q_0,w)\rangle = \langle V_1(q_0,\rho), \frac{1}{\sqrt{\rho}}\, \varphi(q_0,\sqrt{\rho})\rangle \qquad (5)$$

$$\forall \varphi \in \gamma(\mathbb{R}^2)_-$$

and

$$V_2(q_0,w) = 0 \quad \text{on} \quad \mathscr{S}(\mathbb{R}^2) \setminus \mathscr{S}(\mathbb{R}^2)_-.$$

For the following it is advantageous to define a tempered distribution $F(u,v)$ by means of the relation[†††]

---

†) Here and in the following 'scaling' always means 'Bjorken scaling'.

††) By $\mathscr{S}(\mathbb{R}^n)$ we denote the space consisting of testfunctions from $\mathscr{S}(\mathbb{R}^n)$ which are antisymmetric in all variables.

†††) There is a slight difference in the definition of (u,v) compared to [2].

$$F(u,v) := V_2(v+u, v-u). \tag{6}$$

From the antisymmetry of V_2 we obtain the symmetry relations

$$F(u,v) = -F(v,u), \tag{7a}$$

$$F(u,v) = F(-u, -v). \tag{7b}$$

We are now prepared to state the scaling condition in two alternative forms.

<u>Definition 3</u>: (Strong scaling)

$V(q)$ shows strong scaling of degree β, if there exists a real constant β, such that

$$\lim_{\lambda \to +\infty} \lambda^{\beta-1} F(\lambda u, v) = F_\beta(u,v) \qquad \text{in } \mathscr{S}'(\mathbb{R}^2),$$

$$F_\beta \neq 0 \quad \text{if} \quad F \neq 0.$$

We conclude that $F_\beta(u,v)$ is a homogeneous distribution of degree $1-\beta$ on $\mathbb{R}^1$ with respect to u. With that, the symmetry relation (7b) and the support of F we obtain [8]

$$F_\beta(u,v) = u_+^{1-\beta} F_\beta(v) + u_-^{1-\beta} F_\beta(-v) \tag{8}$$

$$\text{for} \quad \beta \neq 2,3,4,\ldots \quad \text{with} \quad \text{supp } F_\beta(v) \subset [-1/2, +\infty).$$

Later on we will show that (8) may be continued into the points $\beta = 2,3,4,\ldots$ by means of locality leading to

$$F_n(u,v) = u^{1-n} F_n(v) \tag{9}$$

We note that the form of the scaling limit (8), (9) is just the naively expected result.

It is well known that strong scaling is violated for renormalizable interactions by logarithmic terms in finite order of perturbation theory. Recent investigations even suggest that this is true for the exact amplitude in asymptotic free theories.

In this situation we consider in accordance with lemma 1 a weak form of the scaling limit which always exists. Weak scaling of degree

β is a consequence of strong scaling of degree β, but not vice versa.

Our formulation of strong scaling agrees with the point limit of F for $F \in \mathcal{C}(\mathbb{R}^2)$ if $\beta < 2$ as shown in the introduction. With the usual variables (ν,ω) we have the asymptotic correspondence $u \to \nu$ $\omega: = -q^2/2\nu \to -2\nu$.

Similarly we treat the leading LC-behaviour in the following. We define a distribution $\widetilde{V}^1$ of the $0_+(3)$-invariants x_0 and $\underline{x}^2$ by means of the chain of relations $(\forall f \in \mathscr{G}(\mathbb{R}^4))$

$$\langle \widetilde{V}^1(x_0,\sigma), \phi f(x_0,\sigma)\rangle = \langle \widetilde{V}(x_0,\underline{x}), f(x_0,\underline{x})$$
$$= (2\pi)^4 \langle V(q_0,\underline{q}), f(q_0,\underline{q})\rangle = (2\pi)^4 \langle V_1(q_0,\rho), \phi f(q_0,\rho)\rangle. \tag{10}$$

In exactly the same way as in case of $V_1(q_0,\rho)$ we extend $\widetilde{V}^1(x_0,\sigma)$ to a distribution $\widetilde{V}^2 \in \mathscr{G}'(\mathbb{R}^2)_-$. By means of eqs. (5) and (10) we obtain

$$\widetilde{V}^2(x_0,\boldsymbol{\varkappa}) = -2\pi i \int dq_0 \int dw \, \sin q_0 x_0 \cdot \sin w \, \boldsymbol{\varkappa} \, V_2(q_0,w). \tag{11}$$

Due to crossing (2) and locality A3) we may define a distribution $\overline{V} \in (\mathscr{G}(\mathbb{R}^1_+) \otimes \mathscr{G}(\mathbb{R}^1)_-)'$ by

$$\widetilde{V}^2(x_0,\boldsymbol{\varkappa}) = \mathcal{E}(x_0)\overline{V}(x_0^2 - \boldsymbol{\varkappa}^2,\boldsymbol{\varkappa}). \tag{12}$$

Two alternative forms of leading LC-behaviour are defined in terms of $\overline{V}$ now.

<u>Definition 4</u>:   (Strong LC-behaviour)[+)]

$\overline{V}(\eta,\boldsymbol{\varkappa})$ shows strong LC-behaviour of degree γ, if there exists a real constant γ, such that

$$\lim_{\lambda \to +\infty} \lambda^{\gamma-2}\overline{V}(\eta/\lambda,\boldsymbol{\varkappa})\exists = \overline{V}_\gamma(\eta,\boldsymbol{\varkappa}) \text{ in } (\mathscr{G}(\mathbb{R}^1_+) \otimes \mathscr{G}(\mathbb{R}^1)_-)',$$

$$\overline{V}_\gamma \neq 0 \quad \text{if} \quad \overline{V} \neq 0.$$

Definition 4 implies that $\overline{V}_\gamma(\eta,\boldsymbol{\varkappa})$ is a homogeneous distribution of degree $\gamma - 2$ on $[0,\infty)$ with respect to η , i.e. [8]

+) Compare ref. [3].

$$\bar{V}_\gamma(\eta,\varkappa) = \frac{(\eta)_+^{\gamma-2}}{\Gamma(\gamma-1)} \, g_\gamma(\varkappa). \tag{13}$$

This result is equal to the usually assumed form of the leading LC-singularity.

With the same arguments as given above and in accordance with lemma 1 we introduce the notion of weak LC-behaviour of degree γ.

Now we give a general representation formula for a distribution $\bar{V}$ satisfying strong LC-behaviour.

<u>Lemma 2</u>:[†]

A distribution $\bar{V}(\eta,\varkappa)$ which satisfies condition A shows strong LC-behaviour of degree γ if and only if there exists a natural number n with $n + \gamma - 2 \geq 0$, such that

$$\bar{V}(\eta,\varkappa) = D_\eta^n \, \eta^{n+\gamma-2} \, \bar{V}^0(\eta,\varkappa), \tag{14}$$

where $\bar{V}^0(\eta,\varkappa)$ exhibits the following properties:

(i) it is continuous and polynomially bounded in
$$(\eta,\varkappa) \in \mathbb{R}_+^1 \times \mathbb{R}^1,$$

(ii) it is an odd entire function of exponential type 1 in
$$\varkappa \in \mathbb{C}, \quad \forall \eta \in \mathbb{R}_+^1, \quad \eta \ \text{fixed}.$$

<u>Corollary</u>

Sufficient for the validity of (14) is strong LC-behaviour of degree γ of $\bar{V}(\eta,\varkappa)$ on $\mathbb{R}_\varepsilon^1$, $\forall \varepsilon > 0$,

$$\mathbb{R}_\varepsilon^1 := \{\varkappa \mid \, |\varkappa| > \varepsilon > 0\}$$

3. Equivalence of Scaling and Leading LC-behaviour

The equivalence of scaling and leading LC-behaviour is stated in our following main theorem 1 by means of definitions 3 and 4 given in section 2.

[†] For all proofs we refer to ref. [2].

Theorem 1

For a distribution $V(q)$ which satisfies condition A, we have:

(a) Strong scaling of degree β implies strong LC-behaviour of degree $\beta \quad \forall \quad \beta \in \mathbb{R}^1$.

(b) Strong LC-behaviour of degree $\beta > 0$ implies strong scaling of degree β.

In addition we may derive some interesting properties of the distribution $F_\beta(v)$ and $g_\beta(\varkappa)$ which occur in the asymptotic forms of F and $\bar{V}$ (eqs. (8), (13)) respectively.

Lemma 3:

The distribution $g_\beta(\varkappa)$ defined by means of eq. (13) is an odd entire function of exponential type 1 which is polynomially bounded for $\text{Im}\varkappa = 0$.

Lemma 3 is an immediate consequence of the representation eq. (14) for $\bar{V}$.

Lemma 4:[†]

The distribution $F_\beta(v)$ (eq. (8)) and $g_\beta(\varkappa)$ are related to each other according to ($\beta > 0$)

$$F_\beta(v) = -i\pi^{-1}\int_0^\infty d\varkappa\, g_\beta(\varkappa)\varkappa^{\beta-2}2^{\beta-1}\cos\left[2v\varkappa + \frac{\pi}{2}(\beta-1)\right] \qquad (15)$$

$$\text{for} \quad v \geq -\tfrac{1}{2}.$$

From the spectrum condition we know supp $F_\beta \subset [-1/2,\infty)$. On the other hand the r.h.s. of eq. (15) is symmetric (antisymmetric) under the substitution $v \to -v$, if β is equal to an odd (even) integer. Therefore, we obtain from eq. (15) the well-known fact, that the scaling function F_β has bounded support for integer $\beta = n$

$$\text{supp} \quad F_n \subset [-\tfrac{1}{2}, \tfrac{1}{2}].$$

This result proves eq. (9) as the continuation of eq. (8) to integer β.

---

†) Compare Gatto, Menotti - ref. [1] and ref. [9].

Our theorem 1 may be extended immediately to the cases of weak scaling and weak LC-behaviour respectively.

Corollary 1a

For a distribution $V(q)$ which satisfies condition A, we have:
(a) Weak scaling of degree $\beta \geq 0$ implies weak LC-behaviour of degree β.
(b) Weak LC-behaviour of degree $\beta > 0$ implies weak scaling of degree β.

Corollary 1b

Singularities in the interior of the light cone do not contribute to the scaling limit.

4. Equivalence of Leading LC-behaviour and JL-asymptotic [3]

As a consequence of condition A the $0_+(3)$-invariant tempered distribution $V(q)$ satisfies a JL-representation.

Lemma 5 (Jost-Lehmann [10]):

A distribution $V(q)$ satisfies conditions A2) – A4), if and only if there exists an $0_+(3)$-invariant tempered distribution $\psi(s,u)$ with

$$\text{supp }\underline{\psi} \subset \{(s,\underline{u}) \mid |\underline{u}| \leq 1, s \geq s_0(u) = (1 - \sqrt{1-\underline{u}^2})^2\}$$

such that

$$\langle V(q), f(q) \rangle = \langle \underline{\psi}(s,\underline{u}), Tf(s,\underline{u}) \rangle \; \forall \, f \in \mathscr{S}(\mathbb{R}^4) \tag{16}$$

with

$$Tf(s,\underline{u}) := \int d^4q \, f(q) \, \varepsilon(q_0) \, \delta(q_0^2 - (\underline{q}-\underline{u})^2 - s)$$

By means of the Fourier transform of ψ with respect to $\underline{u}$ we introduce a distribution $\psi_2 \in (\mathscr{S}(\mathbb{R}^1_+) \otimes \mathscr{S}(\mathbb{R}^1)_-)'$ of the $0_+(3)$-invariants only:

$$\langle \tilde{\underline{\psi}}_2(s,\varkappa), \varkappa_\phi \tilde{f}(s,\varkappa^2)\rangle = (2\pi)^3 \langle \underline{\psi}(s,\underline{u}), f(s,\underline{u})\rangle. \tag{17}$$

With that and the definition of $\bar{V}$ we obtain as an alternative form of the JL-representation

$$\langle \bar{V}(\eta,\varkappa), g(\eta,\varkappa)\rangle = \langle \tilde{\underline{\psi}}_2(s,\varkappa), \mathcal{P}_g(s,\varkappa)\rangle, \tag{18}$$

where the mapping

$$g(\eta,\varkappa) \mapsto \mathcal{P}_g(s,\varkappa) := i(2\pi)^2 \int_0^\infty d\eta J_0(\sqrt{\eta s}) \frac{\partial}{\partial \eta} g(\eta,\varkappa) \tag{19}$$

is a topological isomorphism of $\mathcal{G}^\nu(\mathbb{R}_1^+) \otimes \mathcal{G}^\nu(\mathbb{R}^1)_-$.

In terms of $\tilde{\underline{\psi}}_2$ we may formulate the JL-asymptotic now.

Definition 5: (Strong JL-asymptotic)

$\tilde{\underline{\psi}}_2(s,\varkappa)$ shows strong JL-asymptotic of degree δ, if there exists a real constant δ, such that

$$\lim_{\lambda\to+\infty} \lambda^\delta \tilde{\underline{\psi}}_2(\lambda s,\varkappa) \exists = \tilde{\underline{\psi}}_\delta(s,\varkappa) \quad \text{in } (\mathcal{G}^\nu(\mathbb{R}_+^1) \otimes \mathcal{G}^\nu(\mathbb{R}^1)_-)',$$

$$\tilde{\underline{\psi}}_\delta \neq 0 \quad \text{if} \quad \tilde{\underline{\psi}}_2 \neq 0.$$

Definition 5 implies that $\tilde{\underline{\psi}}_\delta(s,\varkappa)$ is a homogeneous distribution of degree $-\delta$ on $[0,\infty)$ with respect to s, i.e. [8]:

$$\tilde{\underline{\psi}}_\delta(s,\varkappa) = \frac{(s)_+^{-\delta}}{\Gamma(1-\delta)} \mathcal{P}_\delta(\varkappa), \tag{20}$$

Again we may introduce a weak form of JL-asymptotic in accordance with lemma 1.

From the properties of the mapping (19) we obtain the following theorem:

Theorem 2: (Zavialov [3])

Suppose $V(q)$ satisfies condition A. Then $\bar{V}(x^2,\varkappa)$ shows strong LC-behaviour of degree β if and only if $\tilde{\underline{\psi}}_2(s,\varkappa)$ shows strong JL-asymptotic of degree β, $\forall \beta \in \mathbb{R}^1$.

Corollary 2

Theorem 2 may be extended to the cases of weak LC-behaviour and weak JL-asymptotic respectively.

5. Equal-time Limits and LC-singularities

The connection between leading LC-behaviour and equal-time commutation, supposed by many authors[+], can be put on a rigorous mathematical basis. Let us consider the most important example, the so-called "Schwinger term sum rule". By means of the representation eq. (14) and Lebesque's bounded convergence criteria one easily proves the following lemma:

Lemma 6:

Suppose $\bar{V}(x^2,\varkappa)$ shows strong LC-behaviour of degree 1, then

$$\lim_{\lambda\to+\infty} \lambda\frac{\partial}{\partial x_0}\ \widetilde{V}(\frac{x_0}{\lambda},\ \underline{x})\,] = \frac{g_1(\varkappa)}{\varkappa}\Bigg|_{\varkappa=0}\ \delta(\underline{x}) \qquad \text{in } \mathscr{P}'(\mathbb{R}^4). \tag{21}$$

Inverting eq. (15) we obtain the desired sum rule

$$\frac{g_1(\varkappa)}{\varkappa}\Bigg|_{\varkappa=0} = -2i \int_{-1/2}^{+1/2} dv\ F_1(v),$$

where the integral on the r.h.s. has to be understood as a regularized one if necessary.

By the same method other equal-time sum rules including their generalization for $\beta \neq 1$ may be derived.

But the argumentation leading to lemma 6 cannot be reversed. The equal-time limit might exist in the sense of (21) even for an arbitrary singular behaviour of $\bar{V}$ on the light-cone [11].

6. Conclusions

The combination of theorems 1 and 2 leads to the supposed equivalence between scaling, leading LC-behaviour and JL-asymptotic in a region $\beta > 0$ which contains the physical relevant interval $1 \leq \beta \leq 2$. We suppose that zero is not a natural lower bound for β in this context, because quite recently Zavialov proved the same equivalence for a some-

---

[+] Compare the first two papers of ref. [1].

what different definition of scaling without any restriction on β [3].

REFERENCES

1. R. Jackiw, R. Van Royen, G.B. West: Phys. Rev. D2, 2473 (1970)
   H. Leutwyler, J. Stern: Nucl. Phys. B20, 77 (1970)
   R. Brandt, G. Preparata: Nucl. Phys. B27, 541 (1971)
   R. Gatto, P. Menotti: Phys. Rev. D5, 1493 (1972)
   Y. Frishman: Schladming Lectures, 1971, Acta Phys. Austr. 34, 351
   (1971)
   P. Stichel: Lectures given at the "Ecole Internationale de la Physi-
   que des Particules Elementaires", Basko Polje-Makarska (Yougoslavie),
   September 1971

2. E. Brüning, P. Stichel: Comm. Math. Phys. 36, 137 (1974)

3. B.I. Zavialov: TM 17, 178 (1973); TM 19, 163 (1974)

4. R.F. Streater, A.S. Wightman: "Die Prinzipien der Quantenfeldtheorie"
   , BI-Hochschultaschenbücher, Bd. 435/435a

5. H. Borchers: Lectures on quantum field theory, held at the University
   of Göttingen, winter 1966/67

6. H. Leutwyler, P. Otterson in "Scale and Conformal Symmetry in Hadron
   Physics", ed. by R. Gatto, John Wiley & Sons, New York, 1973
   M. Magg: Comm. Math. Phys. 38, 225 (1974)
   J. Kühn: Nuovo Cimento 23A, 420 (1974)

7. O. Steinmann: Lecture Notes in Physics, Vol. 11, Springer Verlag,
   Berlin-Heidelberg-New York, 1971

8. I.M. Gelfand, G.E. Schilov: Verallgemeinerte Funktionen (Distri-
   butionen), VEB Deutscher Verlag der Wissenschaften, Berlin 1967,
   Vol. I,II

9. C.A. Orzalesi: Phys. Rev. D7, 488 (1973)

10. F.J. Jost, H. Lehmann: Nuovo Cimento 5, 1598 (1957)
    F.J. Dyson: Phys. Rev. 110, 1460 (1958)
    A.S. Wightman in "Relations de dispersions et particules elementaires
    ", ed. by C. de Witt, R. Omnes, Hermann, Paris, 1960

11. B. Schroer, P. Stichel: Comm. Math. Phys. 3, 258 (1966)

STRUCTURE OF SINGULARITIES OF WIGHTMAN-DISTRIBUTIONS AND THE
WILSON-ZIMMERMANN-EXPANSION RESPECTIVELY LIGHTCONE-EXPANSION

Siegfried Schlieder

Max-Planck-Institut für Physik und Astrophysik

München, Germany

Abstract: For a relativistic quantum field theory of WIGHTMAN type necessary and sufficient conditions are formulated for the existence of a WILSON-ZIMMERMANN-expansion (short distance expansion). One can deduce for those fields, which fulfil our sufficient conditions from the short-distance expansion a lightcone expansion.

1. Introduction

A field theory in which the principle of contact interaction shall be valid needs for the definition of the interaction terms products of field quantities taken at the same position.

If one intends to apply this principle also in a relativistic quantum field theory one has to define products of field operators at the same position. The difficulties which arise in this procedure are very well known. They have their origin in the distributive character of the field operators.

In the case of the free field $A_o(x)$ with $x = (x_o, x_1, x_2, x_3)$ however it is well known how one can get $A_o^n(x)$, n = 2, 3, ... in the form of WICK products of the field. In the last years a heuristic ansatz for the products of field-operators at the same position in the case for interacting fields has been proposed especially by WILSON and ZIMMERMANN [1, 2, 3, 4] and by BRANDT [5]. For instance for a real scalar field WILSON and ZIMMERMANN assume the expansion

$$A(x+\chi_1) A(x+\chi_2) \cdots A(x+\chi_n) = \sum_{j=1}^{m} s_j(\chi_1, \chi_2, \cdots \chi_n) B_j(x) + R_m(x; \chi_1, \chi_2, \cdots, \chi_n)$$

The $B_j(x)$ are field operators relatively local to $A(x)$, the $s_j(\chi_1, \cdots, \chi_n)$ functions which become in general singular for $\chi_j \to 0$, j = 1, 2, ..., n, so that

$$\lim_{x_1, \cdots x_n \to 0} \frac{S_{j+1}(x_1, \cdots, x_n)}{S_j(x_1, \cdots, x_n)} = 0 .$$

With respect to the operator $R_m(x; x_1, \cdots, x_n)$ it is assumed that

$$\lim_{x_1, \cdots x_n \to 0} \frac{R_m(x; x_1, \cdots, x_n)}{S_m(x_1, \cdots, x_n)} = 0 .$$

Thereby the operators $B_j(x)$ are candidates for $A^n(x)$. ZIMMERMANN [2] and BRANDT [5] have proved the validity of such an expansion for certain examples in perturbation theory, WILSON and ZIMMERMANN [3] gave conditions under which the expansion is valid and they discussed several consequences of it; furthermore LOWENSTEIN [6] has proved the expansion rigorously for the THIRRING model.

The aim of the main part of this talk is to study the question of the existence of a short distance expansion (synonym for WILSON-ZIMMERMANN expansion) for the class of relativistic quantum field theories fulfilling WIGHTMAN's conditions. To keep the formalism as simple as possible we restrict our attention to the cases of a real scalar field A(x) and the products of only two operators.

We intend to use the characterisation of the field theory by WIGHTMAN distributions

$$\left(\Omega, A(x_1) \cdots A(x_j) A(x_{j+1}) \cdots A(x_n) \Omega \right)$$
$$= \mathcal{W}_n(x_1, \cdots, x_j, x_{j+1}, \cdots, x_n) = W_n(\xi_1, \cdots, \xi_j, \cdots, \xi_{n-1})$$

with $\xi_j = x_{j+1} - x_j$ $j = 1, 2, \ldots n - 1$.

If a short distance expansion exists, then the structures of the singularities for the n-point functions should show a certain uniform behaviour, independent of n and independent of j for $\xi_j \to 0$. This is the content of section 2. In section 3 sufficient conditions are given for an uniform structure of the singularities. Section 4 connects a set of WIGHTMAN distributions showing an uniform singular behaviour with the WILSON-ZIMMERMANN expansion. Relations between a short distance expansion and a lightcone expansion are the content of section 5.

2. Necessary Conditions for the Existence of the Short-Distance Expansion

According to the remarks in the introduction one can show that the assumption of the existence of the WILSON-ZIMMERMANN expansion implies that singularities appearing for $\xi_j \to 0$ in

$$W_n(\xi_1, \cdots, \xi_j, \cdots, \xi_{n-1})$$

have their counterparts in singularities appearing in $W_4(\xi_1, \xi_2, \xi_3)$ for $\xi_3 \to 0$

For the proof of this statement one has to assume that the expansion is an expansion for operators. If it is an expansion which yields for $\xi_j \to 0$ only a sum of bilinear forms, then the singularities appearing for $\xi_j \to 0$ in W_n must be found again for $\xi_1, \xi_3 \to 0$ simultaneously in $W_4(\xi_1, \xi_2, \xi_3)$. The proofs can be found in [7] .

3. Conditions for an uniform structure of the singularities

The problem is to show, that the singularities which arise if in

$$W_n\left(\xi_1, \cdots, \xi_j, \cdots \xi_{n-1}\right)$$

$\xi_j \to 0$ are essentially not dependent on n and on j. More precisely: It is intended in the following to give such conditions, that in the limit above no other singularities can appear, than those which arise in

$$W_4\left(\xi_1, \xi_2, \xi_3\right)$$

for $\xi_3 \to 0$. Regarding the cluster properties of the n-point-functions, one recognises easily that for n even, n > 4, at least the same singluarities as in the case W_4 and $\xi_3 \to 0$ must appear. What sufficient conditions, however, can be given so that no other and probably more serious singularities appear if n is growing?

Before formulating them it may be of advantage to give at first an idea of the procedure:

Clearly the singularity structure of WIGHTMAN distributions is characterized by the behaviour of the analytic WIGHTMAN functions near the boundary. In the forward tubes $\mathcal{T}_n^+$ these holomorphic functions can be interpreted as scalar-products of holomorphic states: For instance [see 8]

$$W_4\left(\xi_1, \xi_2, \xi_3\right) = \left(\phi_2(\bar{z}_1, \bar{z}_0), \phi_2(z_2, z_3)\right)$$

with
$$\xi_i = z_i - z_{i-1} \, , \quad \xi_i \in \mathcal{T}^+$$

and
$$\phi_2(z, z') = A(z)\, A(z')\, \Omega = \int e^{i(pz + p'z')}\, \tilde{A}(p)\, \tilde{A}(p')\, dp\, dp' \cdot \Omega$$

with
$$z' - z \in \mathcal{T}^+ \, , \quad z \in \mathcal{T}^+ .$$

Now one makes the assumption, that W_4 can be written as

$$W_4\left(\xi_1, \xi_2, \xi_3\right) = \sum_{k=1}^{\ell} G_{4,k}\left(\xi_1, \xi_2, \xi_3\right) u_k\left(\xi_3\right) \quad \text{with} \quad \xi_j \in \mathbb{C}^4$$

where the u_k carry the singularities at $\xi_3 = 0$, and the $G_{4,k}$ are not only holomorphic for $\{\xi_1, \xi_2, \xi_3\} \in \mathcal{T}_3^+$ but also in $\{\xi_1, \xi_2\} \in \mathcal{T}_2^+, \xi_3 \in \mathcal{T}^+ \cup \mathcal{U}_r(0)$ with $\mathcal{U}_r(0)$ as a real neighbourhood of 0.

Under certain conditions it is possible to regard also the $G_{4,k}$ $k = 1, 2, \ldots, l$ as scalar-products of states which are holomorphic even at $\zeta_3 = 0$.

$$W_n\left(\zeta_1, \zeta_2, \cdots, \zeta_{n-1}\right)$$ then can be regarded as a sum of l terms, where each again can be interpreted as a scalar-product of states. The states on the right in these scalar-products can be identified with those, which appear in $G_{4,k}$, $k = 1, 2, \ldots, l$, to the right. So a corresponding expansion of W_n into l terms

$$W_n\left(\zeta_1, \cdots, \zeta_{n-1}\right) = \sum_{k=1}^{l} G_{n,k}\left(\zeta_1, \cdots, \zeta_{n-1}\right) u_k\left(\zeta_{n-1}\right)$$

with $\zeta_j \in \mathbb{C}^4$ and u_k as before is possible.

$G_{n,k}$ are holomorphic for $\{\zeta_1, \cdots, \zeta_{n-1}\} \in \tau_{n-1}^{+}$ and also holomorphic for $\{\zeta_1, \cdots, \zeta_{n-2}\} \in \tau_{n-2}^{+}$ and $\zeta_{n-1} \in \tau^{+} \cup U_r(0)$. One can generalise this procedure by allowing that the $G_{4,k}$ respective the $G_{n,k}$ are not holomorphic, however, meromorphic with the condition that a pole at $\zeta_3 = 0$ respectively $\zeta_{n-1} = 0$ is excluded. This procedure shows that the structure of singularities appearing in W_n for $\zeta_{n-1} \to 0$ must appear also in W_4 for $\zeta_3 \to 0$.

After that one can try to use the invariance of the WIGHTMAN function $W_n\left(z_1, \cdots, z_n\right)$ under permutation of its arguments to get a similar structure of the singularities for the case that not the last variable $\zeta_{n-1} \to 0$ however an arbitrary variable $\zeta_j \to 0$; $j = 1, 2, \ldots n-1$.

Now I wish to give conditions under which the described procedure works. At first the most simple case is treated:

(A) Let us assume that $W_4\left(\zeta_1, \zeta_2, \zeta_3\right)$ is such, that a function $r(\zeta) \neq 0$ exists with the following properties:

a) r is holomorphic in τ^{+}

b) $r(\zeta) = \overline{r(-\bar{\zeta})}$

c) r is invariant under the (homogeneous) Lorentz group

d) $W_4\left(\zeta_1, \zeta_2, \zeta_3\right) r(\zeta_3)$ has a (locally unique) analytic continuation to the points $\{\zeta_1, \zeta_2\} \in \tau_2^{+}$, $\zeta_3 \in U_r(0)$ where $U_r(0)$ is a real neighbourhood of 0 independent of $\{\zeta_1, \zeta_2\}$ and $W_4\left(\zeta_1, \zeta_2, 0\right) r(0) \neq 0$

e) $r(\zeta) \in D'\left(U_r(0)\right)$. ($D(U_r(0))$ test-functions with support in $U_r(0)$).

<u>Theorem 1.</u>: Assuming the conditions (A) $W_n(\mathfrak{z}) \uparrow (\mathfrak{z}_j)$ has an (locally unique) analytic continuation to the points $\mathfrak{z}_\flat \in \tau^+_{n-2}$, $\mathfrak{z}_j = 0$.

For the proof of the theorem see 7 .

Examples for this simple case characterized by the conditions (A) are:

The free scalar field with mass 0: $\Box A_0(x) = 0$

Wickproducts of this field

Exponentials of the free field with arbitrary mass in 2-dimensions.

For the more general case, where expansion of the 4-point-functions contains a finite number of terms it is suitable to split the assumptions into three groups. (A') replaces (A):

(A') It is assumed that an expansion of W_4

$$W_4(\mathfrak{z}_1,\mathfrak{z}_2,\mathfrak{z}_3) = \sum_{k=1}^{l} G_{4,k}(\mathfrak{z}_1,\mathfrak{z}_2,\mathfrak{z}_3)\, u_k(\mathfrak{z}_3)$$

exists with the following properties:

a_1.) $u_k(\mathfrak{z})$, $k = 1,2,\cdots,l$ are holomorphic in τ^+, they are Lorentzinvariant and

$$u_k(\mathfrak{z}) = \overline{u_k(-\bar{\mathfrak{z}})}$$

a_2.) $G_{4,k}(\mathfrak{z}_1,\mathfrak{z}_2,\mathfrak{z}_3)$ are holomorphic in τ^+_3 and can be continued as holomorphic functions to the points $\{\mathfrak{z}_1,\mathfrak{z}_2\} \in \tau^+_2$, $\mathfrak{z}_3 = 0$ (in a locally unique manner).

a_3.) $\{u_1, u_2, \cdots, u_l\}$ are linearily independent over the field of the meromorphic functions at $\mathfrak{z} = 0$.

The second group of conditions (B') shall guarantee, that an expansion

$$\Phi_2(z, z+\mathfrak{z}) = \sum_{k=1}^{l} \Phi_{2,k}(z, z+\mathfrak{z})\, u_k(\mathfrak{z})$$

exists where the $\Phi_{2,k}$ are vectorstates holomorphic for $\{z,\mathfrak{z}\} \in \tau^+ \times [\tau^+ \cup U_r(0)]$. Since (B') is more or less of technical nature, it will not be given here explicitly (see [9]).

From (A') and (B') one derives

<u>Theorem 2.</u>: There exists an expansion $W_n(\mathfrak{z}_1,\cdots,\mathfrak{z}_{n-1}) = \sum_1^{l} G_{n,k}^{(n-1)}(\mathfrak{z}_1,\cdots,\mathfrak{z}_{n-1}) u_k(\mathfrak{z}_{n-1})$

with $G_{n,k}^{(n-1)}(\mathfrak{z}_1,\cdots,\mathfrak{z}_{n-1})$ holomorphic in $\{\mathfrak{z}_1,\cdots,\mathfrak{z}_{n-2}\} \in \tau^+_{n-2}$, $\mathfrak{z}_{n-1} \in [\tau^+ \cup U_r(0)]$

To get the corresponding expansion for an arbitrary argument $\mathfrak{z}_j$

$$W_n(\mathfrak{z}_1,\cdots,\mathfrak{z}_j,\cdots,\mathfrak{z}_{n-1}) = \sum_{k=1}^{l} G_{n,k}^{(j)}(\mathfrak{z}_1,\cdots,\mathfrak{z}_j,\cdots,\mathfrak{z}_{n-1})\, u_k(\mathfrak{z}_j)$$

with $G_{n,k}$ holomorphic in $\mathfrak{z}_j \in [\tau^+ \cup \mathcal{U}_r(0)]$ one can formulate the following sufficient conditions:

(C') It exists an expansion

$$W_n(\mathfrak{z}_1,\cdots,\mathfrak{z}_j,\cdots,\mathfrak{z}_{n-1}) = \sum_{k=1}^{\infty} H_{n,k}^{(j)}(\mathfrak{z}_1,\cdots,\mathfrak{z}_j,\cdots,\mathfrak{z}_{n-1})\, w_k(\mathfrak{z}_j)$$

with

c_1.) $\quad w_k(\mathfrak{z}_j) = \overline{w_k(-\bar{\mathfrak{z}}_j)}\quad$, $k = 1,2,3,\cdots$

c_2.) $\quad H_{n,k}^{(j)}\quad$ holomorphic for $\mathfrak{z}_j \in [\tau^+ \cup \mathcal{U}_r(0)]$, $\vec{\mathfrak{z}} \in \tau_{n-2}^+$.

Remark: All known examples of WIGHTMAN distributions fulfill the conditions (A') (B') and (C'). For the details see $[9]$.

4. Short Distance Expansion

Given a theory as described before one is able to use the holomorphy of $\{G_{n,k}^{(j)}\}$ at $\mathfrak{z}_j = 0$ to formulate a short distance expansion for $A(x_j)A(x_{j+1})$ with $x_{j+1}-x_j - 0$ and express the singularities by $u_k(\mathfrak{z}_j)$.

Technically there are different versions of the short distance expansion possible ($[7]$, $[9]$). I will give here the simplest one:

The WILSON-ZIMMERMANN expansion

$$A(x+\chi)\, A(x-\chi) = \sum_{j=1}^{m} s_j(\chi)\, B_j(x) + R_m(x,\chi)$$

exists as an expansion in bilinear forms over $\bar{J} \times J$, where J is the linear hull of the vectors

$$\{\Omega,\ \{\Phi_1(z_1')\},\ \{\Phi_2(z_1'',z_2'')\},\cdots,\{\Phi_n(z_1^{(h)},\cdots,z_n^{(n)})\},\cdots\}\ ;\ J\ \text{is}$$

dense in H.

5. Lightcone Expansion

From the Lorentzinvariance of the $\{u_k(\mathfrak{z}_j)\}$ and their independence it follows the Lorentzinvariance of the $\{G_{n,k}^{(j)}(\mathfrak{z}_1,\cdots,\mathfrak{z}_j,\cdots\mathfrak{z}_{n-1})\}$. Using this property one is able to enlarge the assumed holomorphy domain of the $\{G_{n,k}^{(j)}\}$. The interesting part of this enlargement is, that the real neighbourhood $\mathcal{U}_r(0)$ of the origin goes over into a real neighbourhood of the lightcone. Using the holomorphy of

$G^{(j)}_{n,k}(\zeta_1, \cdots, \zeta_j, \cdots, \zeta_{n-1})$ for ζ_j in this neighbourhood one is able to formulate an expansion of $A(x+\chi)A(x-\chi)$ with χ on the lightcone (the $\{u_k(\zeta_j)\}$ are again responsible for the singularities) (see [9], [10], [11]).

<u>References</u>:

1. Wilson, K.: Cornell Report (1964); Phys. Rev. <u>179</u>, 1499 (1969).

2. Zimmermann, W.: Commun. math. Phys. <u>6</u>, 161 (1967); <u>10</u>, 325 (1968).

3. Wilson, K., Zimmermann, W.: Commun. math. Phys. <u>24</u>, 87 (1972).

4. Otterson, P., Zimmermann, W.: Commun. math. Phys. <u>24</u>,107 (1972).

5. Brandt, R.A.: Ann. Phys. (N.Y.) <u>44</u>, 221 (1967), <u>52</u>, 122 (1969).

6. Lowenstein, J.H.: Commun. math. Phys. <u>16</u>, 265 (1970).

7. Schlieder, S., Seiler, E.: Commun. math. Phys. <u>31</u>, p 137-159 (1973).

8. Jost, R.: The General Theory of Quantized Fields. Providence (AMS) 1965.

9. Baumann, K.: Thesis 1974 and Prepr.: Januar 1975.

10. Kühn, J., Seiler, E.: Commun. math. Phys. <u>33</u>, 253-257 (1973) and

11. Kühn, J.: Thesis 1974.

$$\underline{\text{CONFORMAL INVARIANCE IN MINKOWSKIAN QUANTUM FIELD THEORY}}$$
$$\underline{\text{AND GLOBAL OPERATOR EXPANSIONS}}$$

Bert Schroer

Institut für Theoretische Physik
Freie Universität Berlin

The two main problems in conformal-invariant QFT on which there has been some recent progress[1-3] are

1. The construction of the correct global conformal transformation of fields consistent with the Einstein causality structure.
2. The validity of global conformal operator expansions on and away from the vacuum.

It is evident that then two problems are intimately related, since the precise formulation of global transformation is a necessary prerequisite for the criterion of global conformal invariance of global operator expansions. So the first question is in what field theoretical frame-work should one analyze the problem. Traditionally physicists have studied this problem in the euclidean formulation[4]. Indeed by using methods of euclidean field theory[5] one can show that the conformal euclidean invariance will lead to a representation of the (∞ sheeted) universal covering group on physical states[6]. Due to the distribution theoretical pathologies in the connection of euclidean and minkowskian theories, this method is not so good if one asks question concerning fields (i.e. basic and composite fields), for example the problems 1) and 2).

Another popular method, namely the Green's function technique for minkowskian field theories[7], is not very suitable either, because the very concept of time ordering is somewhat alien to conformal invariance.

Consider for example the two dimensional conformal group which is known to factorize. The variables appropriate to this factorization are the light cone variables:

$$u = t + x, \quad v = t - x \tag{1a}$$

$$u = \frac{u}{1-\beta u}, \quad v = \frac{v}{1-\gamma v} \tag{1b}$$

A time ordered function, for example:

$$\frac{1}{x^2+i\epsilon} = \frac{1}{u \cdot v + i\epsilon} \tag{2}$$

is <u>not</u> adapted to this factorization, whereas the unordered function:

$$\frac{1}{x^2-i\epsilon x} = \frac{1}{(u-i\epsilon)(v-i\epsilon)} \tag{3}$$

is adapted.

So the most suitable framework is the Wightman framework[8].

In order to get the flavour of the problem consider first the free field case in D-dimensions

$$A(x) = A^{(+)}(x) + A^{(-)}(x) \tag{4}$$

It can be easily demonstrated[9] that there is a unitary operator $U(b)$ which implements the conformal transformation:

$$x' = \frac{x-bx^2}{\sigma(b,x)}, \quad \sigma(b,x) = 1 - 2bx + b^2x^2$$

$$U(b)A^{(\pm)}(x)U^+(b) = \left(\frac{1}{\sigma_\pm(b,x)}\right)^{d_A} : A^{(\pm)}(x') \tag{5}$$

where: $\quad d_A = \frac{D-2}{2}$

is the canonical dimension of the field A and

$$\sigma_\pm(b,x) = 1-2(b_0 \pm i\epsilon)(x_0 \pm i\epsilon) + 2\vec{b}\vec{x} + ((b_0 \pm i\epsilon)^2 - \vec{b}^2)((x_0 \pm i\epsilon)^2 - \vec{x}^2)$$

are appropriate boundary values of σ.

Expressed in terms of the local field we have

$$U(b)A(x)U^+(b) = \frac{1}{|\sigma|^{d_A}} e^{-id_A N \arg \sigma} A(x') e^{+id_A N \arg \sigma} \tag{6}$$

$\arg \sigma = 0$ or $\pm\pi$. N = number operator.

i.e. a transformation law, which through the appearance of N (a conserved quantity which is not the space integral of a local conserved

Noether current) is non-local.

Consider for example the case $D = 3$ with $d_A = \frac{1}{2}$. In this case the two point function has a square root cut leading to the violation of Huygens principle[10]:

$$[A(x), A(y)] \neq 0 \quad \text{for} \quad \xi_0^2 - \vec{\xi}^2 > 0 \tag{7}$$

Therefore a unitary conformal transformation which would just transform the local fields at x,y spacelike into the local fields at x' and y' time-like would violate causality. The non-local transformation law (6) saves the consistency: the transformed field is <u>not</u> the local field at the transformed point.

In order to maintain the formal language of causality it is convenient to say that the transformed non-local field is not the local field at Minkowski-point x' but rather at a point p' which is above of below the point x' in another sheet of the universal conformal covering of Minkowski-space. One can show that the concept of Einstein causality can be formally generalized to the covering of Minkowski-space in such a way the notion of "spacelike" within one sheet agrees with the usual concept[11].

Of course in special cases, as for D-dimensional free fields, one does not need the full universal covering, but by studying for example generalized free fields one can come to situation which necessitates the universal covering.

Introducing central elements:

$$Z(C_2, C_1) = U^{-1}(C_2 C_1) U(C_2) U(C_1) \tag{8}$$

one can easily see that[1]

$$Z(C_2, C_1) = e^{-i d_A N \varphi} \tag{9}$$

with $\varphi = \arg \sigma_+ (b_2, L_2 e^{-\delta_2} a_1)$ (10)

where $C_i = (D_i, e^{-\delta_i}, L_i, a_i) = $ general conformal element.

The central elements depend on the group only via the phase φ.

We therefore introduce the special operator

$$Z \equiv Z(\, \varphi = \pi) \tag{11}$$

Now we try to understand the <u>general interacting</u> situation. For proper conformal transformations we would expect:

$$U(b)A(x)U^+(b) = \frac{1}{|\sigma|^{d_A}} Z(b,x)A(x')Z^+(b,x) \tag{12}$$

where $Z(b,x)$ is a phase operator which only depends via $\varphi = \arg \sigma_+$ on the conformal transformation. This Z brings in the non-local aspect necessary to get reconciliation with Einstein causality.

Reducing out the phase operator by defining[1] (mod 1)

$$A(x) = \sum_{n=-\infty}^{\infty} Z^n A(x) Z^{+n} e^{in\pi(d_A - 2\xi)} \tag{13}$$

which formally is connected with the local field by

$$A(x) = \oint A^{\xi}(x) d\xi \tag{14}$$

we obtain for the "phase fields" $A^{\xi}(x)$ the simple transformation law:

$$U(b)A(x)U^+(b) = \frac{1}{\sigma_+^{d-\xi}\sigma_-^{\xi}} A(x') \tag{15}$$

Because of the spectrum condition the vacuum only sees the $\xi = 0$ component:

$$A(x)\,|\,0> \;=\; A^{\xi=0}(x)\,|\,0> \tag{16a}$$

and
$$<0\,|\,A(x) \;=\; <0\,|\,A^{\xi=d_A}(x) \tag{16b}$$

The possible eigenvalues ξ of the phase operators are obtained from the 3-point function:

$$<C(x)A(y)B(z)> = \text{const}\left(\frac{1}{-(x-y)_-^2}\right)^{\delta_3}\left(\frac{1}{-(x-z)_-^2}\right)^{\delta_2}\left(\frac{1}{-(y-z)_-^2}\right)^{\delta_1}$$

$$= <C^{d_c}(x)A^{\xi}(y)B^0(y)>, \quad \delta_1 = \frac{d_a + d_b - d_c}{2}$$

$$\delta_2, \; \delta_3: \text{cycl.}$$

with $\xi = \frac{1}{2}(d_a + d_b - d_c)$.

The statement that the ξ-spectrum obtained from all 3-point functions of the theory exhausts all possible values of ξ is related to the statement that the totality of all composite local fields applied to the vacuum generate a dense set of states i.e.

$$\overline{\{ \mathcal{O}(x) \mid 0 > \}} = \mathcal{H} \tag{17}$$

if $\mathcal{O}$ runs through all local (composite) fields.
This requirement is in turn related to the validity of operator product expansion. It has been checked in the Thirring model and Thirring-like models. The completeness of local composite states (17) is a property which transcends conformal invariance and holds presumably in renormalized perturbation theory.

We will come back to this problem at the end of this talk.

In all solvable two dimensional models the question whether the 3-point functions give all ξ spectrum can be answered positively[1].

We asked a group theoretical question and ended up with a dynamical problem: in order to know the conformal transformation property of a minkowskian field A, we have to find the total dimensional spectrum of all composites which communicate with A. Let us give an example of some dimensional "trajectories" which are different from the free field linear trajectories (no binding dimension for free fields).

A. Thirring model[12]

Denote with

$$\mathcal{O}^{(n_1,n_2,m_1,m_2)} = P^{(n_1,n_2,m_1,m_2)}(\psi_1,\psi_2,\psi_1^+,\psi_2^+) \tag{18}$$

the composite field with the smallest dimension occurring in the operator short-distance expansion of a product involving n_i-factors ψ_i i=1, 2 and m_i factors ψ_i^+. The dimensional trajectory is:

$$d_{\mathcal{O}} = \lambda(Q_1^2 + Q_2^2) + \mu Q_1 \cdot Q_2 \tag{19}$$

where:

$$Q_i = m_i - n_i$$

and λ and μ are parameters related to the dimension and the spin (which may have also an anomalous contribution) of ψ. The higher dimensional fields in the operator expansion have dimensions which are bigger from (19) by integers.

B. A^4 coupling in $D = 2$ dimensions[13]

The following results rely on the assumption that the $D = 2$ A^4 coupling and the Lenz-Ising model belong to the same "long distance universality class" (i.e. have the same scale-invariant fixed points).

The underlying field (auxiliary field) is a free two component <u>Majorana</u> (selfconjugate) spinor field ψ.

Whereas the even powers of A are rather simple expressions in $D=2$ and lead to [13]

$$\dim A^n = \frac{n^2}{4} , \quad n \text{ even} \tag{20a}$$

the odd powers have a very complicated connection with ψ and the formula

$$\dim A^n = -\frac{1}{8} + \frac{n^2}{4} , \quad n \text{ odd} \tag{20b}$$

has up to now not been obtained by direct computation but rather by consisting arguments. We think that we will be able to construct the Wightman functions explicitly or to give an operator solution in terms of the Majorana fields.

C. Baxter-Model, $A_1^2 \cdot A_2^2$-coupling in $D = 2$[14]

Again the underlying scale invariant field theory is best described in terms of Majorana fields. This time we have two Majorana fields coupled by a 4 fermion-coupling. Again the even polynomials have a rather simple relation to the auxiliary Majorana fields and their dimensional trajectory turns out to be of the Thirring type. We have not investigated the problem of odd polynomials.

Let us now come to our second point, the validity of <u>global operator</u>

expansions.

If the local composites in a QFT are complete, one should be able to write:

$$A(x)B(y) = \int K(x,y,z)C(z)d^D z + \cdots \tag{21}$$

where the sum extends over all (composite) local fields and as a typical contribution we have exhibited a scalar (composite) field C. In a conformally invariant minkowskian QFT we would expect a simple kernel formula in terms of the nonlocal "phase"-fields $A(x)$:

$$A^{\xi_a}(x_1)B^{\xi_b}(x_2) = \int K^{(\xi)}(x_1,x_2,x_3)C^{\xi_c}(x_3)d^D x_3 + \cdots \tag{22}$$

$$\xi_c = \xi_a + \xi_b + \frac{d_a+d_b-d_c}{2} \quad \mathrm{mod}\,(1) \tag{23}$$

$$K^{(\xi)}(x_1,x_2,x_3) = h_{ABC}(\xi_a,\ \xi_b)\cdot \prod_i \prod_j \left[\frac{1}{-(x_i-x_j)^2}\right]^{(\lambda_{ij},\,\xi_{ij})} \tag{24}$$

with $\quad \left[\dfrac{1}{-\xi^2}\right]^{(\lambda,\xi)} = \left[\dfrac{1}{-\xi_+^2}\right]^{\lambda-\xi}\left[\dfrac{1}{-\xi_-^2}\right]^{\xi}$

and $^{3)}\quad \lambda_{ij},\ \xi_{ij} = \mathrm{fct.}(d_a,\ d_b,\ d_c;\ \xi_a,\ \xi_b) \tag{25}$

In particular on the vacuum we have:

$$A^{\xi_a}(x_1)B^0(x_2)|0\rangle = \int K(x_1 x_2,x_3)C^0(x_3)d^D x |0\rangle \tag{26}$$

with $\quad \xi_a = 0 = \dfrac{d_a+d_b-d_c}{2} \quad \mathrm{mod}\,(1) \tag{27}$

and $\quad \xi_{23} = 0,\ \xi_{12} = \lambda_{12},\ \xi_{13} = \xi_a - \lambda_{12} \quad \mathrm{mod}\,(1) \tag{28}$

$$\lambda_{12} = \frac{d_a+d_b-d_c^*}{2}\ ,\ \mathrm{cycl.} \tag{29}$$

and $\quad d^* = D - d \tag{30}$

The following statements concerning global expansions have been established.

1. Validity of global operator expansions for free fields on and off the vacuum. On the vacuum they have the form (26). Off the vacuum the expansion does not have the conformal invariant form (22) as can best be seen by studying a product of two different free fields$^{3)}$:
$A^{(+)}(x)B^{(-)}(y)$:

The principles which determine the coefficients of these expansions
have not been completely understood.

2.   The vacuum-expansion (26) reproduces the 3-point function i.e.

$$<C(x)A(x_1)B(x_2)> = \int K^{(\xi)}(x_1,x_2,x_3)<C(x)C(x_3)>d^Dx_3 \qquad (31)$$

For the general expansion such consistency relations are under investigation.

Finally some general (more speculative) remarks are in order.

a)   Global conformal operator expansions (GCOE) clearly prefer the
     unordered expectation values and therefore are at the heart of the
     Wightman-formulation which is very close to operators. Any time-
     ordering would mass up the mathematical discussion.

b)   GCOE express a kind of field theoretical duality[16], since in the
     4-point function:

$$<A(x_1)B(x_2)C(x_3)D(x_4)> \qquad (32)$$

the vacuum expansion has to match e.i. the expansion in the middle,
at least if one sums over all phase indices. (This is presumably
in part a determining principle for the off vacuum expansions.)
Instead of particles (which do not exist in a nontrivial conformally
invariant theory) one sums over all so called type I[15] conformal
tensor fields. In contrast to dual S-matrix models where the higher
order destroys the duality of the Born approximation, we expect to
get duality for the full field theory.

c)   Global operator expansions (GOPE) without conformal invariance are
     most probably true in renormalized perturbation theory. They could
     be a valuable tool (together with C.S.-type partial differential
     equations) to control the problem of highly exceptional momentum
     configurations. In case the high energy on shell configurations
     belong to this set, GOPE's could provide a new basis for a field
     theoretical derivation of Regge-behaviour.
     Making a Taylor expansion in momentum space at $p = 0$ for $C(p)$,
     each term in the GOPE leads to infinitely many terms in the short-
     distance (Wilson) expansion. This short distance expansion is in

general asymptotically convergent, only for particular matrix-
elements such expansions become convergent. The global expansions
represent the correctly sum and up short distance expansions.

References

1)   B. Schroer and J.A. Swieca, Phys.Rev. D10, 480(1974)

2)   M. Lüscher and G. Mack, "Global Conformal Invariance in Quantum
     Field Theory", University of Bern preprint 1974

     G. Mack, Lectures presented at the 1974 Bonn summer school

3)   B. Schroer, J.A. Swieca and A.H. Völkel, "Global Operator Expansions
     in Conformally Invariant Relativistic Quantum Field Theory", FU
     Berlin preprint Oct.74/13 to be published in Phys.Rev. D May 15th
     1975

4)   A.A. Migdal, Phys.Lett. 37B, 98(1971); 37B, 386(1971)

     A.M. Poliakov, Zh. Eksp. Teor. Fiz. Pis'ma Red. 12, 538(1970) [Sov.
     Phys. JETP Lett. 12, 381(1970)]

     G. Mack, Proceedings of the Aix en Provence Conference 1973

5)   K. Osterwalder and R. Schrader, Commun.Math.Phys. 31, 83(1973) and
     to be published in Commun.Math.Phys.

6)   See last reference in 4) and reference 2)

7)   S. Ferrara, R. Gatto, A. Grillo and G. Parisi, Springer Tracts in
     Modern Physics, vol. 67, Heidelberg, Springer Verlag 1973, and
     literature quoted therein

8)   R.F. Streater and A.S. Wightman, PCT, Spin and Statistics and All
     That (W.A. Benjamin, New York, 1964)

9)   J.A. Swieca and A.H. Völkel, Commun.Math.Phys. 29, 319(1973)

10) M. Hortaçsu, R. Seiler and B. Schroer, Phys.Rev. D5, 2519 (1972)

11) For the geometrical concepts see reference 2) and literature quoted
     therein

12) Conformal invariance in the Thirring model and Thirring-like models
     have been carried out in J. Kupsch, W. Rühl and B.C. Yunn,
     Kaiserslautern preprint 1974 and reference 1)

13) These results have been obtained in collaboration with B. Berg,
     as yet unpublished

14) In field theoretical language the Baxter model consists in two
     superimposed A^4 couplings (i.e. Lenz-Ising models) with an energy-
     energy interaction $A_1^2 \cdot A_2^2$. See remark in reference 13)

15) L. Bonora, Nuovo Cimento Letters 3, 548 (1973)

16) The duality idea has been first discussed in euclidean field theory
     in G. Mack, Proceedings of the Aix en Provence Conference 1973.
     In this discussion one finds a reference to A.M. Poliakov who dis-
     cussed it first in the Green function language.

Discussion

Di Castro (question): Could you further comment on the fact that $\beta \geq 0$ in the Baxter model?

Schroer (answer): The scale invariant limit of the D=2 Baxter-model is described by a Lagrangian of two (either commuting or anticommuting) Majorana fields coupled by a marginal 4-fermion coupling. As in the Thirring model the marginality of the interaction is preserved to every order due to the symmetry $SO_2 \times SO_2$ ($\beta=0$). The interaction just leads to a shift of dimensions depending on the coupling strength. The difficult aspect (compared with the Thirring model) is the complicated relation between the spinor-fields and odd powers of the original scalar fields. For a detailed discussion, see a forthcoming publication of B. Berg and myself.

Suzuki (question): Would you give us more explanation on some applications of the conformal invariance to statistical mechanics of critical phenomena?

Schroer (answer): According to ideas of Kadanoff, Poliakov, Migdal, Wilson et al., critical phenomena are described by scale-invariant field theories. There are rather general arguments saying that scale invariant field theories, in particular those of the A^4 type, are also conformal invariant. This can be partially demonstrated also for the Onsager solution of the D=2 Lenz Ising model. Here the correlation functions on one line are asymptotically ($T=T_c$) conformal invariant. Conformal invariance (for example the bootstrap equations of Migdal and Poliakov) can be used as the starting point for computing critical in indices. This method has the same shortcoming as many other methods: one has to resort to ε-expansions ($D=4-\varepsilon$).

ON SOME MASSLESS SUPERRENORMALIZABLE AND NONRENORMALIZABLE THEORIES

K. Symanzik

Deutsches Elektronen-Synchrotron DESY

Hamburg, Federal Republic of Germany

Massless superrenormalizable and nonrenormalizable theories have in common that they cannot be constructed in perturbation theory. Examples of the first theories are massless $(\phi^4)_2$ and $(\phi^4)_3$, while (massive or massless) $(\phi^4)_5$ and $(\phi^4)_6$ are nonrenormalizable. It is here advantageous to consider massless $(\phi^4)_{4-\varepsilon}$, $\varepsilon>0$ ("IR case"), and (massive or massless) $(\phi^4)_{4+\varepsilon}$, $\varepsilon>0$ ("UV case"), with ε also noninteger /1/, and even complex, since thereby not only several cases are treated simultaneously, but for generic ε degeneracies are lifted that otherwise would require the consideration of logarithms, which are less easily handled than powers, in the formulae below.

We first discuss the IR case. The Lagrangean is

$$(1) \qquad L = \frac{1}{2}\partial_\mu\phi\partial^\mu\phi \;-\; \frac{1}{2}m_B^2\,\phi^2 \;-\; \frac{1}{24}g\phi^4 \;.$$

It is well known how to adjust, in perturbation theory, m_B^2 such that e.g. the physical mass is m, and we consider this mass renormalization carried out with $m>0$, such that the IR divergences are regularized. The connected amputated one-particle-irreducible parts of the Green's functions

$$2^{-\ell}<T\phi(x_1) \; \ldots \; \phi(x_{2n}) \; \phi^2(y_1) \; \ldots \; \phi^2(y_\ell)>$$

are called vertex functions, and their Fourier transforms, with a factor $(2\pi)^{4-\varepsilon}\,\delta(\Sigma p + \Sigma q)$ taken out, we denote as $\Gamma(p_1\ldots p_{2n},q_1\ldots q\;;\;m^2,g,\varepsilon)\equiv$

$$\equiv \Gamma((2n),(\ell);\;m^2,g,\varepsilon).$$

These functions have, as functions of m, in perturbation theory an expansion in powers of m, which, for nonexceptional momenta (e.g., for Euclidean momenta, no partial sum of an even number of p and any number of q must vanish), is

$$(2) \quad \Gamma((2n),(\ell);\;m^2,g,\varepsilon) = f_{oo}((2n),(\ell);\;g,\varepsilon) + \sum_{j=1}^{\infty}\;\sum_{k=o}^{\infty}\;m^{2j-\varepsilon k}f_{jk}((2n),(\ell);\;g,\varepsilon)$$

where for diagrams with L loops only terms with $k \leq L$ appear. f_{oo} is the function obtained in perturbation theory by using bare propagators $i(p^2 + iO)^{-1}$ and bare vertices $-ig$, computing /1/ for $0 < \varepsilon < \frac{2}{L}$ and continuing termwise analytically in ε from there. f_{oo} has (in sufficiently high order in g) singularities at all positive rational ε. The singularities at $\varepsilon = R$, say, on the R.H.S. of (2) must, because the L.H.S. has no such singularities, be cancelled identically in g and the momenta by singularities in all the f_{jk} for which $2j-Rk=0$. Similarly, the singularities, if any, in all other f_{jk}, at $\varepsilon = R$, must cancel among those f_{jk} with equal $2j-Rk$.

Since the intended limit $m \rightarrow 0$ in meaningless for the R.H.S. of (2), partial summations at least are necessary. It turns out that it suffices to sum for constant j. We describe $j = 1$ only. One finds /2/

$$(3) \quad \sum_{k=o}^{\infty} m^{2-\varepsilon k} f_{1k}((2n),(\ell); g,\varepsilon) = -iA(g,m,\varepsilon) \, \underline{\Gamma}((2n),(\ell)0; g,\varepsilon) +$$

$$+ B(g,m,\varepsilon) \, \underline{\Gamma}'((2n)00,(\ell); g,\varepsilon)$$

where $\underline{\Gamma}$ and $\underline{\Gamma}'$ are functions with similar properties as $f_{oo}((2n),(\ell); g,\varepsilon)$ and constructible in perturbation theory. Furthermore,

$$(4a) \quad A(g,m,\varepsilon) = \begin{array}{c} \text{anal.contin.} \\ \text{from small } \varepsilon \end{array} \int_{o}^{m^2} dx \, \rho(gx^{-\frac{1}{2}\varepsilon},\varepsilon)$$

and

$$(4b) \quad B(g,m,\varepsilon) = \begin{array}{c} \text{anal. contin.} \\ \text{from small } \varepsilon \end{array} i \int_{o}^{m^2} dx \, \rho(gx^{-\frac{1}{2}\varepsilon},\varepsilon) \, \Gamma(,00;x,g,\varepsilon)$$

where

$$(5) \quad \rho(gm^{-\varepsilon},\varepsilon) = \frac{\partial m_B^2}{\partial m^2} = 1 + \sum_{k=1}^{\infty} (gm^{-\varepsilon})^k \rho_k(\varepsilon)$$

and

$$(6) \quad \Gamma(,00; m^2,g,\varepsilon) = m^{-\varepsilon} \sum_{k=o}^{\infty} (gm^{-\varepsilon})^k \chi_k(\varepsilon),$$

with computable $\rho_k(\varepsilon)$ and $\chi_k(\varepsilon)$ holomorphic in ε outside $(-\infty,0]$. (4a) yields

$$\text{(7a)} \quad A(g,m,\varepsilon) = m^2 \left[1 + \sum_{k=1}^{\infty} (gm^{-\varepsilon})^k \, \rho_k(\varepsilon)(1-k\tfrac{\varepsilon}{2})^{-1} \right] =$$

$$= \operatorname*{anal.contin.}_{\text{from } \varepsilon>2} \left\{ m^2 + \int_{\infty}^{m^2} dx \left[\rho(gx^{-\frac{1}{2}\varepsilon},\varepsilon) - 1 \right] \right\} =$$

$$= \operatorname*{anal.contin.}_{\text{from } \varepsilon>2} \; m_B^{\,2} \equiv m_B^{\,2} \, ,$$

and (4b) yields

$$\text{(7b)} \quad B(g,m,\varepsilon) = \operatorname*{anal.contin.}_{\text{from } \varepsilon>2} i \int_{\infty}^{m^2} dx \, \rho(gx^{-\frac{1}{2}\varepsilon},\varepsilon) \, \Gamma(,00; x,g,\varepsilon) =$$

$$= \operatorname*{anal.contin.}_{\text{from } \varepsilon>2} \left\{ -\tfrac{1}{2} <\phi^2> \right\} \equiv -\tfrac{1}{2} <\phi^2> \, .$$

Now (7a) yields

$$\text{(8a)} \quad A(g,0,\varepsilon) = \operatorname*{anal.contin.}_{\text{from } \varepsilon>2} g^{\frac{2}{\varepsilon}} \int_{\infty}^{0} dx \left[\rho(x^{-\frac{1}{2}\varepsilon},\varepsilon) - 1 \right] =$$

$$= g^{\frac{2}{\varepsilon}} \left\{ \sum_{k=1}^{\infty} \rho_k(\varepsilon)(1-k\tfrac{\varepsilon}{2})^{-1} + \int_{1}^{0} dx \left[\rho(x^{-\frac{1}{2}\varepsilon},\varepsilon) - 1 \right] \right\} \equiv g^{\frac{2}{\varepsilon}} \mu(\varepsilon)$$

and similarly (7b)

$$\text{(8b)} \quad B(g,0,\varepsilon) = \operatorname*{anal.contin.}_{\text{from } \varepsilon>2} i \, g^{\frac{2}{\varepsilon}-1} \int_{\infty}^{0} dx \, \rho(x^{-\frac{1}{2}\varepsilon},\varepsilon) g_o \Gamma(,00; xg_o^{\frac{2}{\varepsilon}},g_o) \equiv g^{\frac{2}{\varepsilon}-1} \nu(\varepsilon)$$

where $\nu(\varepsilon)$ has an expansion similar to $\mu(\varepsilon)$ and $g_o > 0$ is arbitrary.

Using (8a) and (8b) in (3) shows how from the $j=1$ series in (2), for
which the termwise $m \to 0$ limit is meaningless, a sensible limit can be extracted by
virtue of factorizations. The explicit singularities in $\mu(\varepsilon)$ and $\nu(\varepsilon)$ are precisely
those that cancel via (3) the ones in $f_{oo}((2n),(\ell); g,\varepsilon)$ that arise due to
coincidences in (2) with the $j=1$ terms, i.e. at $2-\varepsilon k = 0$, as discussed earlier.
At such ε, terms with factors $g^N \ln g$ will arise, with N positive integer, however,
even to a low order in g coefficients will arise, such as $a_k \equiv \frac{\partial}{\partial \varepsilon} (2-\varepsilon k)\mu(\varepsilon)\big|_{\varepsilon = \frac{2}{k}}$,

that can in principle be calculated in perturbation theory only by an infinite
summation. That these coefficients are indeed finite is strongly suggested by the
success of renormalization group methods in the theory of critical phenomena /3/.

Details of the method described here, which can be extended to higher j in (2), can be found in ref. /2/. It leads to a pseudo-perturbation expansion that proceeds not only in integer powers of g but also in powers of logarithms. The "physical" interpretation of the occurence of $g^{\frac{2}{\epsilon}}, g^{\frac{4}{\epsilon}}$ etc. for generic ϵ is in terms of a perturbation theoretical construction using bare propagators involving the true bare mass squared $g^{\frac{2}{\epsilon}} \mu(\epsilon)$ of the zero-mass theory.

We turn to the UV case, i.e. $(\phi^4)_{4+\epsilon}$, $\epsilon > 0$. Here an UV-regularization of the Lagrangean is called for, and we choose in place of (1)

$$(9) \qquad L = -\frac{1}{2}\phi \Box (1 + \Lambda^{-2} \Box)\phi - \frac{1}{2} m_B^2 \phi^2 - \frac{1}{24} g\phi^4$$

where

$$(10a) \qquad m_B^2 = m_{Bo}^2 + \Delta m_B^2$$

with

$$(10b) \qquad m_{Bo}^2 = \sum_{k=1}^{\infty} g^k \Lambda^{\epsilon k} \sigma_k(\epsilon)$$

the bare mass squared of the zero-mass theory. (There are no IR problems now.) The vertex functions have, as functions of Λ , in perturbation theory an expansion in powers of Λ , which is

$$(11) \qquad \Gamma_\Lambda((2n), (\ell); \Delta m_B^2, g, \epsilon) = \sum_{j=o}^{\infty} \sum_{k=o}^{\infty} \Lambda^{-2j+\epsilon k} h_{jk}((2n),(\ell); \Delta m_B^2, g, \epsilon)$$

whereby, as in (2), diagrams with L loops only contribute to $k \leq L$. $h_{oo}(\ldots;0,g,\epsilon)$ is the analytic continuation of $f_{oo}(\ldots;g,-\epsilon)$ of (2) from small negative to small positive ϵ, and $h_{oo}(\ldots; \Delta m_B^2, g, \epsilon)$ is similarly related to $\Gamma((2n),(\ell);m^2,g,-\epsilon)$ in (2). h_{oo}, for $\Delta m_B^2 = 0$ or $\neq 0$, has singularities in ϵ at positive rational ϵ which, for $\epsilon < 4$, are cancelled on the R.H.S. of (11) in complete analogy as in (2), since the L.H.S. in (11) has no such singularities.

The terms in (11) with $j = 0$, $k > 0$ can be removed by "renormalization" i.e. appropriate substitutions for g and Δm_B^2, and multiplication by a suitable factor. Hereby one may use e.g. the usual mass-shell renormalization conventions, or the more convenient intermediate ones /4/, carrying Λ along as a parameter, in order not to make substitutions that introduce singularities for $0 < \epsilon < 4$ previously absent in Γ_Λ. Having achieved a form analogous to (2), one analyses the $j = 1$ terms with the help of Schwinger's action principle and Zimmermann's normal product formalism /5/, and again finds that they can be written as a sum of products of two factors, with the Λ dependence in one and the momenta dependence in the other factor. The calculations are rather more involved than in the IR case, however, and no firm conclusions on the plausability of a $\Lambda \to \infty$ limit have been obtained yet. In particular, for the $\Lambda \to \infty$ limit there is so far no independent evidence analogous to the one that exists for the $m \to 0$ limit /3/ in (2). However, it will be possible to determine what characteristic form the pseudo-perturbation expansion will take if the limit $\Lambda \to \infty$ does exist, and to which extent a dependence on the manner of introducing the cutoff may remain.

References

/1/ K.G. Wilson, Phys. Rev. D7, 2911 (1973), App.;
    G. 't Hooft, M. Veltman, CERN 73-9, Lab. I, Theor. Study Div.
/2/ K. Symanzik, Lett. Nuovo Cimento 8, 771 (1973);
    DESY 73/58 (1973).
/3/ K.G. Wilson, Report at the Topical Meeting on Gauge Theories,
    C.N.R.S. Marseille, June 1974;
    E. Brézin, J.C. le Guillou, J. Zinn-Justin, DPhy-T/74/100, CEN Saclay.
/4/ J.D. Bjorken, S.D. Drell, Relativistic Quantum Fields,
    MacGraw-Hill: New York 1965.
/5/ W. Zimmermann, Ann. Phys. (N.Y.) 77, 536, 570 (1973).

Discussion

B. Schroer: Is there any restriction on the internal symmetry structure of the D=2 4-fermion coupling in order that your method works?
K. Symanzik: The β-function of the strictly renormalizable theory, i.e. the one for $\varepsilon = 0$, must not vanish identically, as it does not e.g. for the Gross-Neveu model.

THE LOCALITY CONDITION IN PARAFERMI FIELD THEORY

Y. Ohnuki and S. Kamefuchi*

Department of Physics, Nagoya University, Nagoya
*Department of Physics, Tokyo University of Education, Tokyo

We consider a system consisting of a single parafermi field $\psi(x)$ of order p, which satisfies the usual, trilinear commutation relations (C.R.).[1] The field operator $\psi(x)$ is known to be expressible in terms of the Green-component fields $\psi^{(\alpha)}(x)$ ($\alpha = 1,2,\cdots,p$) satisfying the anomalous case of bilinear C.R.,i.e., $\hat{\psi}(x) = \sum_{\alpha=1}^{p}\hat{\psi}^{(\alpha)}(x)$, where $\hat{\psi}$ stands for either ψ or $\psi^{\dagger}$. The basic C.R. of $\hat{\psi}$ then lead us to the so-called decomposition theorem[2]: Any monomial consisting of equal-time opera-tors $\hat{\psi}(1)\hat{\psi}(2) \cdots \hat{\psi}(n)$ can be written as a finite sum of terms such as

$$\{\hat{\psi}(i_1),\hat{\psi}(i_2),\cdots,\hat{\psi}(i_a)\}' [\hat{\psi}(j_1),\hat{\psi}(k_1)] \cdots [\hat{\psi}(j_b),\hat{\psi}(k_b)] \qquad (1)$$
$$(n \geqslant a + 2b,\ a \leqslant p),$$

with
$$\{\hat{\psi}(1),\hat{\psi}(2),\cdots,\hat{\psi}(a)\}' \equiv a! \sum \hat{\psi}^{(\alpha_1)}(1)\hat{\psi}^{(\alpha_2)}(2) \cdots \hat{\psi}^{(\alpha_a)}(a), \qquad (2)$$

where the summation $\sum$ is taken over all different values of $\alpha_1,\alpha_2,\cdots,\alpha_a$. Conversely any expression of the form (2) can be expressed as a polynomial of degree $m(m \leqslant a)$ in $\hat{\psi}$.

For bracketed quantities there exist the following relations:

$$[[\hat{\psi}(x_1),\hat{\psi}(x_2)],\hat{\psi}(y)] = 0,\quad [\{\hat{\psi}(x_1),\hat{\psi}(x_2),\cdots,\hat{\psi}(x_p)\}',\ \hat{\psi}(y)]_+ = 0,$$
$$[\{\hat{\psi}(x_1),\hat{\psi}(x_2),\cdots,\hat{\psi}(x_a)\}',\ \{\hat{\psi}(y_1),\hat{\psi}(y_2),\cdots,\hat{\psi}(y_b)\}']_\pm \neq 0 \qquad (3)$$
$$(1 \leqslant a,b < p),$$

where the two spatial domains V and V' ($x_i \in V$, $y_i \in V'$) are spacelike to each other (the relationship being denoted hereafter by $V \sim V'$).

Let F(V) be an observable defined for a spatial domain V. The locality condition for the present theory can then be given in either of the following forms:

$$[F(V),\ \hat{\psi}(y)] = 0 \qquad \text{for}\ \ y \in V' \sim V, \qquad (4.1)$$
$$[F(V),\ F'(V')] = 0 \qquad \text{for}\ \ V \sim V', \qquad (4.2)$$

which we shall refer to, in what follows, as the strong and weak forms, respectively. Obviously, the former implies the latter, but the converse is not necessarily true. As an immediate consequence of the above theorem and (3) we can make the following statement as to the structure of observables $F(V)$ in general:[3] i) Any $F(V)$ that satisfies (4.1) should be a functional of $[\hat{\psi}, \hat{\psi}]$'s; ii) Any $F(V)$ that satisfies (4.2) should be a functional of $[\hat{\psi}, \hat{\psi}]$'s ($[\hat{\psi}, \hat{\psi}]$'s and $\{\hat{\psi}(1), \hat{\psi}(2), \cdots, \hat{\psi}(p)\}'$) for the case $p = $ odd (even). As to the locality condition we therefore find that for the case $p = $ odd (even) (4.1) implies, and is (not necessarily) implied by, (4.2).

As an example of the field theory for which only the locality condition of the weak form holds true, we consider a parafermi field $\psi(x)$ of $p = 2$ with the Hamiltonian:[2,4,5]

$$H_\kappa = \frac{1}{2}\int d^3x([\psi^\dagger(x), D_m\psi(x)] + \kappa\{\psi^\dagger(x), \beta\psi(x)\}'), \tag{5}$$

where $D_m \equiv -i\vec{\alpha}\cdot\vec{\nabla} + m\beta$. Owing to the second term the locality condition here holds true only in the weak form. It is then an easy matter to see that the Heisenberg equation of motion for $\hat{\psi}$ takes a nonlocal form. This undesirable feature of the theory is in fact related to the cluster property of the system. To see this let us denote by $H_\kappa(V)$ the Hamiltonian (5) whose integration is taken over a sufficiently large, but finite, spatial domain V and require that $H_\kappa(V)|0\rangle = 0$ for the vacuum state $|0\rangle$. Next we consider, for example, a one-particle state $\hat{\psi}(\vec{x},0)|0\rangle$, where $\vec{x}$ is assumed to be sufficiently far away from V. As a consequence of (3) the following relation is obtained: $H_\kappa(V)\hat{\psi}(\vec{x},0)|0\rangle = \hat{\psi}(\vec{x},0)H_{-\kappa}(V)|0\rangle$ for $\vec{x} \sim V$. Naively speaking, the above expression should be expected to vanish in general, but $H_{-\kappa}(V)|0\rangle$ cannot offhand be put equal to zero since $H_{-\kappa}$ is no longer the Hamiltonian of the system. The situation is essentially the same for similar state vectors $|-\rangle$ in the odd sector, whereas no such difficulty arises for state vectors $|+\rangle$ in the even sector.

Our basic assumption to overcome this difficulty is that any observable F takes a sector-dependent form $F|\pm\rangle = F_\pm|\pm\rangle$ such that[4]

$$F_\pm = F([\hat{\psi},\hat{\psi}], \pm\{\hat{\psi},\hat{\psi}\}'). \tag{6}$$

Thus for the Hamiltonian H in particular we have $H_\pm = H_{\pm\kappa}$. This assumption about H necessitates a corresponding modification of the Heisenberg equation for $\hat{\psi}$:

$$i\frac{\partial\hat{\psi}(x)}{\partial t}\Big|\pm\rangle = [\hat{\psi}(x), H]\,|\pm\rangle$$
$$= (\hat{\psi}(x)H_\pm - H_\mp\hat{\psi}(x))|\pm\rangle, \tag{7}$$

which, when use is made of C.R., results in

$$(\gamma \cdot \partial + m \pm \kappa)\psi(x)|\pm> = 0, \quad (\gamma \cdot \partial + m \mp \kappa)\psi^{\dagger}(x)|\pm> = 0. \tag{7'}$$

This implies that a single field ψ describes two species of particles with masses $m \pm \kappa$.

We now introduce three kinds of number operators N, N_1 and N_2. While the operator N for the total number of particles is defined as usual, for both sectors, by

$$N = \frac{1}{2}\int d^3x[\psi^{\dagger}(x), \psi(x)] = N_1 + N_2, \tag{8}$$

the operators $N_j (j = 1,2)$ are defined according to (6) by

$$N_{j\pm} = \frac{1}{4}\int d^3x([\psi^{\dagger}(x), \psi(x)] \mp (-)^j\{\psi^{\dagger}(x), \psi(x)\}'). \tag{9}$$

When use is made of C.R., the following relations are obtained for equal-time quantities:

$$[N_1, \psi]|+> = -\psi|+>, \quad [N_1, \psi]|-> = 0,$$
$$[N_2, \psi]|+> = 0, \quad [N_2, \psi]|-> = -\psi|->, \tag{10}$$
$$[N, \psi] = -\psi, \quad [N_1 - N_2, \hat{\psi}\hat{\psi}] = 0 .$$

Combining (10) with (7') we find that N_1 (N_2) is the number operator for particles with mass $m + \kappa$ ($m - \kappa$). Furthermore the last relation in (10) implies that ($N_1 - N_2$) commutes with any observables and therefore obeys a superselection rule. This leads us immediately to a theorem such that if any one of the operators N, N_1 and N_2 is conserved, so are the other two as well.

The present theory is of further specific features: i) The charge conjugation cannot be defined as a unitary transformation for the operators $\hat{\psi}$, and hence the C- and CPT- symmetries do not hold good in the conventional sense; ii) all state vectors are subject to constraints of a nonlocal character such that $N_1|+> = N_2|+>$, $(N_1 + 1)|->$ $= N_2|->$. It may be of interest to identify the two species of particles with the muon and electron (To incorporate the neutrinos is straightforward).

The theory takes a more familiar form when the $\psi^{(\alpha)}$'s are Klein-transformed as $\psi^{(\alpha)}(\vec{x},0) = (-i)^{\alpha-1}K\phi^{(\alpha)}(\vec{x},0)$, where the Klein operator K is defined, at $t = 0$, in such a way that the field operators $\phi^{(\alpha)}$ ($\alpha = 1,2$) satisfy the normal case of bilinear C.R. for ordinary fermi fields. In contrast with the case of a non-relativistic field or of a relativistic field satisfying the locality condition of the strong form[3] a number of complications arise here owing to $[K, H] \neq 0$ or $KH_{+\kappa}K = H_{-\kappa}$. It is nevertheless possible to establish a one-to-one correspondence between the parafermi and fermi theories that are described, respectively, by the ψ and $\phi^{(\alpha)}$'s.[4] The latter theory is then found to be invariant under those transformations that correspond to the gauge group SO(2):

$$\phi^{(1)} \to \cos\theta \cdot \phi^{(1)} + \sin\theta \cdot \phi^{(2)}, \quad \phi^{(2)} \to -\sin\theta \cdot \phi^{(1)} + \cos\theta \cdot \phi^{(2)}, \quad (11)$$

the Casimir operator C being given by $C = N_1 - N_2$. This explains the reason why the operator $(N_1 - N_2)$ obeys a superselection rule.

We conclude this paper by making the following remarks. i) The redefinition of F, (6), leads us to the relation $[F(V), \psi(y)] | \pm >$ $= (F_{\mp}(V)\psi(y) - \psi(y)F_{\pm}(V)) | \pm > = 0$ for $y \sim V$ since $\psi F_{\pm} = F_{\mp}\psi$. Thus, by introducing the redefinition we are in fact changing the locality condition of the weak form into that of the strong form. This makes us suspect that any theory which somehow premises the existence of field operators or such operators as to connect the two sectors might not be consistent unless the locality condition of the strong form is adopted. ii) Our method developed above for the case $p = 2$ can be so generalized as to apply to a parafermi field of $p = 2k$ $(k > 1)$,[6] in which case, however, there arises no mass splitting. iii) Without introducing the redefinition of F, (6), we can alternatively formulate the theory by restricting ourselves only to the even sector and by exclusively using 'current operators' or bilinear expressions in $\hat{\psi}$. In the case $p = 2$,[4] possible representations of the current operators are shown to be consistent with those of the $\hat{\psi}$'s obtained above, except for certain trivial cases.

References

1) H. S. Green, Phys. Rev. **90**(1953), 270.

2) Y. Ohnuki and S. Kamefuchi, Phys. Rev. **170**(1968), 1279.

3) K. Druhl, R. Haag and J. E. Roberts, Comm. Math. Phys. **18**(1970), 204;
   Y. Ohnuki and S. Kamefuchi, Ann. Phys. (N.Y.) **78**(1973), 64;
   Prog. Theor. Phys. **50**(1973), 258.

4) Y. Ohnuki and S. Kamefuchi, Prog. Theor. Phys. **50**(1973), 1696.

5) H. S. Green, Prog. Theor. Phys. **47**(1972), 1400.

6) Y. Ohnuki and S. Kamefuchi, Prog. Theor. Phys. **51**(1974), 1206.

BOUND STATE NATURE
AND DEEP INELASTIC STRUCTURE FUNCTIONS

Masao Ninomiya and Keiji Watanabe
Physics Department
University of Nagoya

Nagoya, Japan

It is one of the recent interests to study the short distance behavior of the current product via applied field theoretic approach. The purpose of this note is to discuss the effect of bound state on the deep inelastic structure functions by taking composite model for hadrons in a _bona fide_ relativistic field theory.

To investigate the high energy limit of the structure functions we apply the Wilson expansion for the product of currents.[1] One may get the structure function $F_2(\omega,q^2)$ by means of the inverse Mellin transformation,[1]

$$F_2(\omega,q^2) = \frac{1}{2\pi i} \int_{c-i\infty}^{c+i\infty} dn\,\omega^{n-1} \sum_\alpha f_\alpha^n(q^2) A_n^\alpha \tag{1}$$

with $c>0$, where q is the virtual photon momentum, ω is the scaling variable given by $\omega=2Pq/-q^2$, α denotes superscript for the possible operators and f_α^n are the Fourier transform of the coefficient functions in the Wilson expansion. A_n^α are given by a proton expectation value of the twist two operators: $\langle P|O_{\mu_1\ldots\mu_n}(0)|P\rangle = P_{\mu_1}\cdots P_{\mu_n} A_n^\alpha + \text{terms containing } g_{\mu_i\mu_j}$. According to this treatment the dynamical structure of the elementary constituents is reflected in A_n^α which we calculate by using bound state model. As an illustration we take on a charged spin 0 boson with self interaction. Here we calculate a diagram for the matrix element by extending Nishijima and Mandelstam's method,[2] which is expressed in terms of the Bethe-Salpeter (B-S) amplitude $\Phi(P,q)$:

$$\langle P|O_{\mu_1\cdots\mu_n}(0)|P\rangle = -2i\int dq\,\Phi(P,q-\tfrac{P}{2})\,\Phi(P,q-\tfrac{P}{2})\,[(P-q)^2-m^2]\times q_{\mu_1}\cdots q_{\mu_n}. \qquad (2)$$

with the boson mass m. The matrix element is calculated by assuming that the operators are defined according to Zimmermann's normal ordering.[1] For the B-S amplitude we apply the Deser-Gilbert-Sudarshan-Ida representation[3] which is given by $\Phi(P,q)=\mathcal{Y}_\ell^m(q)\int_0^\infty dt\int_{-1}^1 dz\,g(z,t)(-q^2+Pqz+m^2-M^2/4+t)^{-\ell-2}$. Here ℓ and M are the angular momentum and mass of the bound state and $\mathcal{Y}_\ell^m(q)$ is the associated Legendre function. The spectral function $g(z,t)$ is a function of z and t, and its boundary values which are assumed to be $g(z,t)\sim(1-z^2)^{\delta_\ell}\,(z\to\pm1)$ and $t^{-\alpha_\ell}\,(t\to\infty)$ have been discussed by Menotti[4] by applying the Wilson expansion to the product of two boson field in B-S amplitude. It has been shown that the boundary values are connected to the high energy limit of the B-S amplitude[2],[4] and the relations, $\delta_\ell=\alpha_\ell+\ell-1$ and $\alpha_\ell=1+\min_{n\geq\ell}(-\tfrac{1}{2}\gamma_n)=1-\tfrac{1}{2}\gamma_\ell$, are obtained,[4] because $-\gamma_n$ is non-decreasing function of n as was proven by Nachtman.[5] Here γ_ϕ and $2\gamma_\phi-\gamma_n$ are the anomalous dimensions of the boson field ϕ and the local operator $O_{\mu_1\cdots\mu_n}$.

Now let us discuss the threshold behavior of $F_2(\omega,q^2)$ for ω near 1. As $F_2(\omega,q^2)$ is expressed by Eq.(1), we may evaluate large n limit of the matrix element denoted by Eq.(2). As to the large n limit of A_n, the boundary value of $g(z,t)$ near $z=\pm1$ dominate and we get $A_n\to\text{const.}n^{-2\delta_\ell-2}+O(n^{-2\delta_\ell-3})$. Applying the technique of the renormalization group[1] to $f_n(q^2)$, we get $f_n(q^2)=(\mu^2/-q^2)^{\gamma_\phi-1/2\gamma_n}$ with $\gamma_n=\text{const.}g^2[n(n+1)]^{-1}$ and $\gamma_\phi=\text{const.}g^2$ up to the second order in the coupling constant. Here μ^2 is a scale parameter. Thus $F_2(\omega,q^2)$ is obtained to be near the threshold $\omega=1$,

$$F_2(\omega,q^2)\sim\text{const.}\,(\mu^2/-q^2)^{\gamma_\phi}\,(\omega-1)^{2\delta_\ell+1}. \qquad (3)$$

To make the model somewhat more realistic we consider a bound state model for the nucleon so called Drell-Lee model[6] -the bound state of spin 0 and spin 1/2 constituents. In the ladder approximation we obtain $A_n\sim\text{const.}n^{-4}+O(n^{-5})$ for large n. Approximating the function $f_n(q^2)$ in the lowest order in the coupling constant, we obtain $F_2(\omega,q^2)\to\text{const.}(\omega-1)^3$. The result agrees with the present experiment.[7] Let us now refer to the non-Abelian gauge theory for the strong interaction.[8] Taking $A_n^\alpha=\text{const.}n^{-p}+O(n^{-p-1})$ we get $F_2(\omega,q^2)$ near $\omega=1$ as $\{\Gamma(\text{const.}\ln(1+t)+p)\}^{-1}(1+t)^{-\text{const.}}(\omega-1)^{\text{const.}\ln(1+t)+p-1}$. Here g is the gluon quark coupling constant and $t=\text{const.}g^2\ln(-q^2/\mu^2)$. In the non-Abelian gauge theory the threshold behavior may be broken up into two aspects: Firstly, due to the effect of the bound state, the structure function goes to $(\omega-1)^{p-1}$ and secondary $F_2(\omega,q^2)$ shows a shrinkage. Considering that the quark gluon coupling constant has been estimated to be fairly small in an asymptotically free theory,[8] we may take the present experimental result for the threshold behavior as a manifestation of the bound state nature of the nucleon.

Next we shall discuss the connection between $F_2(\omega, q^2)$ and the elastic electromagnetic form factor $G(q^2)$. We remark that the so called Drell-Yan (D-Y) relation[9] does not generally hold, because the high energy limit of $G(q^2)$ depends upon α_ℓ;[4] $G(q^2)$ becomes $(-q^2)^{-\ell-2\alpha_\ell+1}$ for large $-q^2$. Comparing $G(q^2)$ with Eq.(3), we see that the D-Y relation is broken down by γ_ℓ which is supposed to be fairly large from the present data.[10] However in some models the D-Y relation holds, for example the ϕ^4 theory without anomalous dimensions and the Drell-Lee model in the ladder approximation.

Finally let us go into large ω limit. Recently De Rujula et al.[11] have shown that the rightmost singularity of $f_\alpha^n A_n^\alpha$ as to n determines the large ω behavior of $F_2(\omega, q^2)$. Actually in the ϕ^4 theory γ_n has an essential singularity at $n=0$ if it is calculated with recourse to the perturbation theory. In the bound state model A_n^α are regular at $n=0$. Accordingly the effect of the bound state is shielded in the large ω limit, and we have

$$F_2(\omega, q^2) \sim \mathrm{const.}\,(-q^2/\mu^2)^{-b}\, z^{\frac{1}{2}(\beta+2)}\, (\ln \omega)^{-\frac{1}{4}(\beta+3)} \times \exp[2(z \ln \omega)^{\frac{1}{2}}]$$

where $z = a \ln(-q^2/\mu^2)$ and we have set $\gamma_\phi - \frac{1}{2}\gamma_n \approx -\frac{a}{n} + b$, $A_n \approx n^\beta$ $(\beta > 0)$ for small n.

The authors wish to express their gratitude to Professor T. Takabayasi and Professor H. Ezawa.

References

1) N. Christ et al., Phys. Rev. D6 (1972), 3543.

2) M. Ciafaloni, Phys. Rev. 176 (1968), 1898.

3) S. Deser et al., Phys. Rev. 115 (1959), 731.
    M. Ida, Prog. Theor. Phys. 23 (1960), 1151.

4) P. Menotti, Phys. Rev. D9 (1974), 2767.

5) O. Nachtman, Nucl. Phys. B63 (1973), 237.

6) T.D. Lee and S.D. Drell, Phys. Rev. D5 (1972), 1738.

7) G. Miller et al., Phys. Rev. D5 (1972), 528.

8) H.D. Politzer, Physics Report 14C (1974), 129, for a review.

9) S.D. Drell and T.M. Yan, Phys. Rev. Letters 24 (1970), 181.

10) O. Nachtman, Nucl. Phys. B78 (1974), 455.

11) A.De Rujula et al., Phys. Rev. D10 (1974), 1649.

THE PHYSICAL CONTENT OF THE STATISTICAL
BOOTSTRAP AND HIGH ENERGY HADRON INTERACTION

V.P. Shelest

Steclov Mathematical Institute, USSR Ac. of Sci,
Moscow, USSR

The essential results of the statistical bootstrap model (SBM) are:
a) a level density linearly exponential in total energy and b) an
asymptotically bounded average energy per secondary. It could be shown
however that these results follow from much more general considerations
than the usual assumptions of the statistical bootstrap model. Namely,
once the concept of coordinate space volume is modified such as to
justify a free gas picture, the linearly exponential level degeneracy
appears. This leads to the generalized statistical bootstrap model [1],
the basic postulates of which are the following:

I) Any hadronic system at rest decaying into N constituents can
be represented as a free gas contained in a volume NV_0, with $V_0 \sim \mu^{-3}$
determined by the range of strong interactions.

II) The level density $\bar{\tau}(M^2)$ for any hadronic system at rest and
of mass M is for $M > \mu$ equal to the mass spectrum $\bar{\rho}(M^2)$ of its
constituents

$$\bar{\tau}_1(M^2) = \bar{\rho}(M^2) - \delta(M^2 - \mu^2) \ .$$

It is also possible to use an alternative approach to obtain the
fundamental results a) and b): the composition picture. A hadron here
is not a system of free constituents as in the decay picture of SBM;
nevertheless, introducing the Regge pattern of resonance distribution,
we get again the a)-result. From this point of view, the SBM is a
Fermi model in Regge (M^2) space. [2]

The problem now arises, in the framework of the dynamical description provided by the dual resonance model (DRM), to study under what
conditions a statistical approach can be valid. It turns out to be that
a purely statistical description and in particular the implications of
an exponential mass spectrum can be completely altered through the
introduction of interaction dynamical features. Namely, the formation
dynamics decouples the initial state from a wide class of in principle
possible final states ("dynamical selection rules" [3]).

The first generation system produced in the multiparticle production process in hadron-hadron collisions has then a memory of its formation, a memory leading eventually to an anisotropic jet structure decay. The corresponding calculations show that this first generation system is highly non-equilibrium and anisotropic. [2]

We are led thus to ask if fireballs are at all produced in the DRM or in any similar dynamical model of hadron-hadron collisions? This is possible in the proposed two-stage picture of multihadron production: a non-equilibrium stage of fireball formation is followed by an equilibrium stage of statistical fireball decay.

The interesting problem of the ultimate temperature T_0 existence [4] in this picture is then solved as follows: for the equilibrium (fireball decay) stage of the process where the SBM approach is justified such ultimate temperature indeed exists. But it must be considered not as a general upper bound to temperature in all hadronic interactions but as a specific ultimate temperature for statistical decay of clusters formed, and is therefore reached from above as a result of cooling a certain dynamical system. [2]

It is possible also to get an explanation for the experimental evidence of anomalously high creation probabilities of particles with large transverse momenta. To this end in our two-stage picture we propose some generalization of Landau hydrodynamical model [5], making an assumption that the secondary particles can appear not only on the decay stage $(T=T_0)$ but also due to the evaporation of some amount of particles on the hydrodynamical expansion stage, $T_0 \ll T \leq T_{in}$ ("leakage process"). The details of these calculations will be published elsewhere.

References

[1] M. I. Gorenstein, V.A. Miransky, V.P. Shelest, G.M. Zinoviev,
     Phys. Lett $\underline{45B}$ (1973), 475.

[2] M.I. Gorenstein, V.A. Miransky, V.P. Shelest, G.M. Zinoviev, H.Satz,
     Nucl. Phys. $\underline{B76}$ (1974), 453.

[3] M.I. Gorenstein, V.A. Miransky, V.P. Shelest, B.V. Struminsky,
     G.M. Zinoviev, Nuovo Cim. Lett $\underline{6}$ (1973), 325.

[4] R. Hagedorn, Nuovo Cim. Suppl. $\underline{3}$ (1965), 147.

[5] L.D. Landau, Izvest. Akad. Nauk Ser. Fiz. $\underline{17}$ (1953), 51.

FIELD THEORETICAL APPROACH TO COMPOSITE PARTICLE REACTIONS

Kisei Kinoshita
Department of Physics
Kyushu University

Fukuoka, 812 Japan

The purpose of our research[1~3] is to obtain unified understanding of high energy hadron phenomena from the composite model of hadrons, on the basis of urbaryon field theory.   A brief report of recent progress and conjectures are given here.   More details are described in references 2) and 3).

1. <u>Whole region formula</u>.   A formula is obtained which unifiedly describes inclusive reactions $ab \to cX$ in the entire kinematical region, including two-body reactions $ab \to cd$ at all angles as the exclusive limit of inclusive ones.   The formula has the following properties:

	small p_T	large p_T
$ab \to cX$:	interpolation of Mueller-Regge forms,	$p_T^{-2N} \cdot f(x_+, x_-)$
$ab \to cd$:	Regge-like behaviour,	$s^{2-n_a-n_b-n_c-n_d} f(\cos\theta)$,

where $x_\pm = (E_c \pm P_{\|c})/\sqrt{s}$ in the C.M.S., and n_i is the valence number of the hadron i.   The formula consists of coupling constant g, dimensional factor F and scaling factor f as

$$E_c d\sigma/d^3 p_c = \sum_\nu g_\nu F_\nu (m_T^2) \, f_\nu (x_+, x_-), \qquad m_T^2 = p_T^2 + m_c^2,$$

where ν denotes a set $\{n_{ij}\}$ of the number of valence urbaryon lines n_{ij} connecting i and j.   At large p_T, we have $F_\nu (m_T^2) \to p_T^{-2N_\nu}$ and

$$f_\nu \to \frac{1}{1-\bar{x}} \left[\frac{x_+^{n_{ab}+n_{ac}} x_-^{n_{ab}+n_{bc}} (1-\bar{x})^{n_{aX}+n_{bX}+n_{cX}-1}}{(1-x_+)^{n_{bX}-1} (1-x_-)^{n_{aX}-1}} \right]^2, \qquad \bar{x} = x_+ + x_- .$$

At small p_T, the above form persists with the reduction to "effective number" $n_{ij} \to n_{ij} (p_T^2)$ in the formula.

In the degenerate region $m_T^2/s \sim 1$ with large p_T, the scaling factor can be approximated by $f_\nu \sim (m_T^2/s)^{2n_{ab}} F_{c/a}(x_+) F_{c/b}(x_-)$, where $F_{c/a}(x_+) = x_+^{2n_{ac}} (1-x_+)^{2(n_{aX}+n_{cX})-1}$ representing the effect of a on c.

2. <u>Model of urbaryon-hadron amplitude</u>.   As a tentative basis
of the formula, we consider manifestly covariant, non-perturbative
model of urbaryon-hadron amplitude $k+p \to k'+p'$, where k and k' are
(anti-)urbaryons off-shell, and p and p' denote hadrons on-shell.
This amplitude determines the structure function of the hadron, and
appropriate combinations of this give inclusive cross sections for
$ab \to cX$.   We assume that ordinary urbaryon mass is around 1 GeV, and
explicitly specify the amplitude, distinguishing whether k annihilates
with a valence in a hadron or not.   For example, the urbaryon-meson
amplitude with $n_{pk}=n_{p'k'}=n_{pp'}=1$ is written as

$$T = \frac{g^2}{M^2-s} + \int_{s_L}^{\infty} \frac{\beta(t,k^2,k'^2)s'^{\alpha(t)}}{s'-s}\, ds'$$

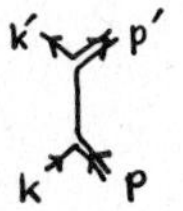

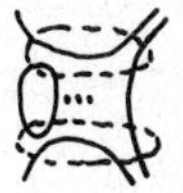

The off-shell dependence of the residue function is related to the
threshold behaviour of the structure function, and also to $x_+ \sim 1$ beha-
viour of $F_{c/a}(x_+)$.   We assume $\beta(t=0,\ k=k') \sim (k^2)^{-2n_\beta}$, where $n_\beta+1$ equals
to the minimum number of urbaryons in the intermediate state.   This
assumption matches with the dimensional counting of the form factor
and also with the "Born" argument of hard interaction in renormalizable
model of urbaryon interaction.   Simpler way of justifying our assumption may be super-renormalizable theory of urbaryon-hadron interaction
with $Z_{hadron}=0$.   Perturbative results of such a theory gives desired
behaviour of urbaryon-hadron amplitude.

3. <u>Three-step structure of strong interaction</u>.   On the basis
of the results in 2.   We may speculate a profile of strong interaction
field theory to be the following three-step structure:

A. The basic interaction of urbaryons (and gluons) is renormalizable and asymptotic free.

B. The urbaryon-hadron interaction is super-renormalizable with
$Z_{hadron}=0$.

C. Effective interactions among hadrons are "ultra-renormalizable",
in the sense that they are very convergent, allowing tree approximations in terms of hadrons and Reggeons in most situations.

It is to be stressed that the step B theory is very important
for full understang of strong interaction.

1) K.Kinoshita and H.Noda, Proceedings of International Symposium on
   High Energy Physics, Tokyo, July 23-27(1973), p.403.
2) K.Kinoshita and H.Noda, Prog. Theor. Phys. <u>52</u>(1974), 1622.
3) K.Kinoshita, Y.Kinoshita and Y.Myozyo, Prog. Theor. Phys. <u>54</u>(1975),
   to appear (preprint KYUSHU-74-HE-16).

PARTICLES AND BOUND STATES AND PROGRESS

TOWARD UNITARITY AND SCALING

James Glimm[1]
Rockefeller University
New York, N.Y.   10021

Arthur Jaffe[2]
Harvard University
Cambridge, MA   02138

Abstract

We present a survey of recent developments in constructive quantum field theory.

Introduction. The program of constructive quantum field theory starts with an approximate field theory whose existence is known and constructs a Lorentz covariant limit as the approximations are removed [50]. Frequently it has been convenient to work in the path space world of imaginary time [14,15,42]. The Osterwalder-Schrader axioms [34,35] give sufficient conditions on the Euclidean Green's functions (i.e. the path space theory in the case of bosons) to allow analytic continuation back to real (Minkowski) time and a verification of all Wightman axioms.

This program has been carried out in a number of models in space-time dimension $d < 4$. Once a model has been constructed, the interesting questions involve its detailed properties and how these properties depend on the parameters. For space-time dimension $d = 2$, considerable insight has been obtained into several models including the Sine-Gordon equation [11]. For $d = 3$, recent work yields the first nontrivial model [9]. We describe some of these recent results below, but first we propose a program for $d = 4$.

Consider a lattice ϕ_4^4 model, with lattice spacing ε . Using correlation inequalities of Lebowitz [18,30,47] or the Lee Yang theorem [32,33], we can bound the n-point Schwinger functions

---

[1] Supported in part by the National Science Foundation under Grant MPS 74-13252.

[2] Supported in part by the National Science Foundation under Grant MPS 73-05037.

$$S^{(n)}(x_1,\ldots,x_n) = \int \phi(x_1)\ldots\phi(x_n)dq\,,$$

defined as moments of the measure

$$dq = \frac{e^{-V(\phi)}\prod\limits_{x} d\phi(x)}{\int e^{-V(\phi)}\prod\limits_{x} d\phi(x)}\,,$$

$$V(\phi) = \sum_{x}[\frac{1}{2} : (\nabla\phi)^2 + m_0^2\phi^2 : + \lambda : \phi^4(x) :]\,.$$

In fact $S^{(n)}$ is bounded by a sum of products of two point functions. Thus a sufficient condition for the existence of a ϕ^4 field theory (by the method of compactness and subsequences) is a bound, uniform in the lattice spacing ε, on the lattice approximation two point function. In order to prevent the identical vanishing of all n-point functions in the limit $\varepsilon \to 0$, we hold the physical mass fixed, and since $d = 4$, we perform a field strength renormalization.

Field strength renormalization assumes the existence of a one particle pole in the Fourier transform of the two point function, of strength $Z = Z(\varepsilon) \neq 0$. We define

$$S_{ren}^{(n)} = Z^{-n/2}S^{(n)}\,,$$

and as above $S_{ren}^{(n)}$ is bounded in terms of $S_{ren}^{(2)} = Z^{-1}S^{(2)}$. Thus the essential missing steps are (i) existence of the one particle pole, so that Z is defined and not zero, for $\varepsilon > 0$, and (ii) control over $Z(\varepsilon)$ as $\varepsilon \to 0$. Since one expects $Z(\varepsilon) \to 0$ (infinite field strength renormalization), it is necessary to show that the rest of the mass spectrum in $S^{(2)}$ has a spectral weight converging to zero (in each bounded mass interval) with ε, so that $S_{ren}^{(2)}$ remains bounded.

The coupling constant renormalization, $\lambda = \lambda(\varepsilon) \to \infty$ as $\varepsilon \to 0$, should not be required for existence of the $\varepsilon \to 0$ limit, but if it is not performed, the limit should be a free field, for $d = 4$. See [24] for further discussion.

In the case $d = 2,3$, we can apply these same ideas to construct the "scaling limit" of (superrenormalizable) ϕ^4 models. In this case we do not require a cutoff for the approximate theories, but vary the bare parameters so $Z \to 0$ and $\lambda \to \infty$, while the physical mass remains fixed and the dimensionless charge approaches its critical value (characterized by the onset of symmetry breaking). We now consider the case $d = 2$ in more detail.

<u>The scaling limit in</u> ϕ_2^4 [24]. Consider an interaction Lagrangian

$$L^{Int} = -\frac{1}{2}\,\sigma\int :\phi^2: dx - \lambda\int :\phi^4: dx\,.$$

Here σ is related to the bare mass m_0, but does not equal m_0^2, since the free Lagrangian will contain a mass parameter also; see the appendix of [20] for a discussion. It is expected that a critical value $\sigma_c = \sigma_c(\lambda)$ of σ exists so that as $\sigma \downarrow \sigma_c$, the physical mass m goes to zero. The scaling limit combines the limit $\sigma \downarrow \sigma_c$ with an infinite scale transformation,

$$\lambda \to s\lambda, \qquad \sigma \to s\sigma, \qquad m^2 \to sm^2,$$

so that m is held fixed. The correct choice of s is given by the conditions

$$\lambda/\sigma \to \lambda/\sigma_c(\lambda) = (\lambda/\sigma(\lambda))_c\,,$$

and in this limit $\lambda \to \infty$, $\sigma_c(\lambda) \to \infty$. Note that $\lambda/\sigma_c(\lambda)$ is dimensionless, hence a pure number independent of λ. In this limit all unrenormalized correlation functions converge (after passage to a subsequence). In taking the scaling limit, we also perform a field strength renormalization, so the one particle pole has residue 1. We expect $Z(\sigma) \to 0$ in the scaling limit.

As before, control over the renormalized two point function is sufficient for the existence of the scaling limit, and again the missing steps are (i), (ii) above. Other sufficient conditions are (α) negativity of the six point vertex function $\Gamma^{(6)} \leq 0$, or (β) better than one particle decay of the inverse propagator $-\Gamma^{(2)}(x)$ (which for $x \neq 0$ equals the proper self energy part $\Pi(x)$) or (γ) absence of level crossings and control over CDD zeros of the momentum space two point function [24], and also [23]. Thus (α)-(γ) are central open issues.

A numerical analysis of the ϕ_1^4 model (anharmonic oscillator) is consistent with the validity of parts of (γ) [28]. An explicit calculation shows that $\Gamma^{(6)} < 0$ in the one-dimensional Ising model [39]. A related correlation inequality $S_T^{(6)} \geq 0$ (positivity of the connected part of $S^{(6)}$) has been announced by Cartier; recently Percus [38] and Sylvester [47] have given proofs. Computer studies of the Ising model by Sylvester indicate furthermore that $(-1)^{n+1} S_T^{(2n)} \geq 0$, [48]. The question of correlation inequalities will be discussed at greater length in the lecture of Simon.

In addition to its relation to the problem of constructing ϕ_4^4, the scaling limit is important in the renormalization group approach to the study of critical exponents; see for example [37]. In this connection we note that a number of critical exponents have been bounded from below by their canonical (mean field) values [17]. Furthermore we prove a priori bounds on renormalized coupling constants [21].

The problems (i) and (ii) above arise also in the scaling limit of the d-dimensional Ising model, $2 \leq d \leq 4$. For I_2, asymptotic calculations using Toeplitz determinants indicate that the required bounds are in fact satisfied [31, 49]. The model I_1 is already scale invariant. A formal interchange of limits identifies the scaling limit of I_d with the infinitely scaled ϕ_d^4 model. This statement provides a basis for the idea that spin $1/2$ Ising model critical exponents are independent of the details of the lattice structure and are equal to those defined by a ϕ^4 field theory. It also suggests that $\phi^6, \phi^8, \ldots$ tri and multi-critical points are associated with higher spin Ising model tri and multi-critical points.

<u>Particles and Unitarity</u> [25,26]. Let $M = (H^2 - \vec{P}^2)^{\frac{1}{2}}$ be the mass operator. For weakly coupled $P(\phi)_2$ models, the spectrum of M is completely known in the interval $[0, 2m-\varepsilon]$. At zero, M has a simple eigenvalue with eigenvector Ω, the vacuum state. At m, M has an isolated eigenvalue, and the corresponding eigenspace, the space of one particle states, carries an irreducible representation of the Poincaré group. M has no other spectrum below $2m-\varepsilon$; here $\varepsilon \to 0$ and $m \to m_0$ (the bare mass) as the coupling tends to zero. The existence of an isometric S matrix and n-particle in and out states then follows from the Haag-Ruelle scattering theorem.

These results about particles (spectrum of M) are proved using a "cluster expansion," similar to the high temperature expansions in statistical mechanics [25,26]. These expansions, furthermore, give spectral information about the mass (generalized) eigenvectors on any bounded spectral interval, for λ sufficiently small. The main consequence of these expansions is the fact that the Euclidean correlation functions decay (become uncorrelated) as the points separate into clusters, and rate of decay is exponential in the separation distance. Let

$$X = x_1, \ldots, x_n, \qquad Y = y_1, \ldots, y_m$$

$$\Phi(X) = \prod_{i=1}^{n} \phi(x_i), \qquad \Phi(Y) = \prod_{j=1}^{m} \phi(y_j).$$

A typical estimate is

$$(1) \qquad \int \Phi(X)\Phi(Y)d\phi \; - \int \Phi(X)d\phi \int \Phi(Y)d\phi \; = \; O(e^{-md})$$

where $m > 0$ is independent of X, Y and

$$d = \text{dist}(X, Y)$$

is the separation distance. The subtraction in (1) can be recognized as a Euclidean vacuum subtraction. Let $\Omega_E \equiv 1$ be the Euclidean vacuum state and let $< , >$ be the Euclidean inner product defined by the measure $d\phi$. Let P_0 be the orthogonal projection onto Ω_E. Then (1) can be written

$$(2) \qquad <\Phi(X), (I-P_0)\Phi(Y)> \; = \; O(e^{-md}) \; .$$

In fact, the exponential decay rate in (1),(2) is related to the spectrum of M. Define m_1 as the infimum over the exponential decay rates for different choices of X, Y. Then m_1 is the one particle mass (mass gap) in the spectrum of M.

We generalize (2) by replacing the projection $I-P_0$ with the projection onto the orthogonal complement of the subspace spanned by polynomials of degree $< n$. We then expect decay rates $m_n > m_1$. Let P_n denote the orthogonal projection onto the subspace spanned by Euclidean vectors

$$|X_n> \; \equiv \; (I - \sum_{i=0}^{n-1} P_i)\phi(x_1)\ldots\phi(x_n)\Omega \; .$$

The projection P_n can be written in terms of a kernel $P_n(X,Y)$, and in Dirac notation

$$(3) \qquad P_n \; = \; \int dX dY \, |X_n> P_n(X,Y) <Y_n| \; ,$$

where the integration extends over $R^{nd} \times R^{nd}$ and where the kernel $P_n(X,Y)$ is the inverse (in the sense of integral operators) to the kernel $<X_n|Y_n>$.

The kernel

$$P_1(X,Y) \; = \; -K_1(X,Y) \; = \; -\Gamma^{(2)}(x-y)$$

is the familiar inverse propagator, while the kernel $P_2(X,Y)$ is related to the Bethe-Salpeter kernel $K_2(X,Y)$, namely

$$P_2(X,Y) = -K_2(X,Y) + \frac{1}{2} K_1 \otimes K_1 .$$

In terms of Feynman graphs, K_2 is the connected part of $-P_2$. More generally, we can define an n-body Bethe-Salpeter kernel $K_n(X,Y)$ as the connected part of $-P_n$. In fact, P_n can be written as a sum of tensor products of Bethe-Salpeter kernels, generalizing the expressions for P_1, P_2 above:

$$(4) \qquad P_n(X,Y) = \sum_{r=1}^{n} \sum_{|\alpha_1|+\ldots+|\alpha_r|=n} (-1)^r \binom{n}{\alpha}^{-2} K_{|\alpha_1|} \otimes \ldots \otimes K_{|\alpha_r|} ,$$

where $\binom{n}{\alpha} \equiv n! \left(\prod_{i=1}^{r} |\alpha_i|! \right)^{-1}$ and the sum extends over all cluster decompositions into elastic channels. For instance, the term with $r=1$ is just $-K_n$, while the term with $r=n$ equals $(-1)^n (n!)^{-1} \Gamma \otimes \ldots \otimes \Gamma$.

The existence of the Bethe-Salpeter kernel K_n, for $n=1,2,3$ is proved in [19,45,23]. We will publish an analysis of (4). A much deeper question than existence of K_n is the magnitude of the exponential decay rates associated with these kernels. Let

$$|X_{n,r}\rangle \equiv (I - \sum_{i=0}^{n-1} P_i)\phi(x_1)\ldots\phi(x_r)\Omega .$$

We seek estimates generalizing (2) of the form

$$(5) \qquad \langle X_{n,r} | Y_{n,r} \rangle = O(e^{-m_n d}),$$

i.e. Euclidean cluster properties, or decay estimates

$$(6) \qquad K_n(X,Y) = O(e^{-\overline{m}_n d}) .$$

In order to analyze the decay of K_n, Spencer has derived a new cluster expansion [45] which generalizes [26] by giving higher particle subtractions. Explicitly in the case $n=2$, for weak coupling even $P(\phi)_2$ models, Spencer proves that $\overline{m}_2 \geq 4m(1-\varepsilon)$, where $\varepsilon \to 0$ as the coupling tends to zero. He and Zirilli expect that the decay rate for the two body Bethe-Salpeter kernel will provide information related to asymptotic completeness, up to the threshold $\overline{m}_2$, i.e. for $M \leq 4m(1-\varepsilon)$ [46]. Similar methods give the decay rate for the part of $S_T^{(4)}$ that is two particle irreducible in each channel [2]; this amplitude is obtained by a second Legendre transformation.

If one imagines an extension of this structure analysis to arbitrary n, it appears that the allowed size of the weak coupling region would be n-dependent, and tend to zero as n tends to infinity. The question of dealing with the particle structure away from weak coupling is also of great importance. In this case we only have results for the ϕ^4 interaction, which is repulsive, as described in the following section.

Bound States. For $d = 2,3$, bound states should occur for weak as well as strong attractive forces; they should be missing for repulsive forces. For single phase ϕ^4 models, it is known that no even bound states can occur [44,7,26]. We expect that odd bound states are also missing, cf [24]. For weak $(\phi^6-\phi^4)_2$ models, it is known that mass spectrum occurs in the bound state interval $(m,2m)$ [26]. The Bethe-Salpeter equation combined with improved decay estimates above should allow a complete analysis of the bound state problem for weak coupling $P(\phi)_2$ models; partial results have been obtained [46]. An interesting question is whether bound states occur in ϕ^4 models with symmetry.

Fermions and Many Body Systems. The original construction of the infinite volume limit for the d=2 Yukawa model was given by Schrader [41]; see also [14]. Some portions of this construction have been derived in a Euclidean formalism [43,1]. Aside from the increased simplicity which may accompany a covariant Euclidean construction, the Euclidean formalism is important as a natural framework for a cluster expansion and a study of particles.

Federbush [5,6] has simplified the Dyson-Lenard proof of the stability of matter. The methods were suggested in part by constructive field theory techniques, including techniques he previously employed for the Yukawa_2 model. [6] contains a cluster expansion, which should have a number of applications.

Phase Transitions. Fröhlich [10] has shown that the Euclidean decomposition of the measure $d\phi$ into time translation invariant components coincides with the direct integral decomposition of the associated quantum fields into pure phases.

Three Dimensions. The original semiboundedness proof for the ϕ^4_3 Hamiltonian [16] and related Schwinger function bounds [8,36] were given in a finite volume. A cluster expansion for ϕ^4_3 has been established by Feldman and Osterwalder [9], yielding the first nontrivial d=3 model of the Osterwalder-Schrader axioms. We can hope that further progress will soon bring ϕ^4_3 to the level of understanding we have for d=2 .

REFERENCES

1. D. Brydges, Boundedness below for fermion model theories. Preprint.

2. C. Burnap, private communication.

3. J. Dimock, The $P(\phi)_2$ Green's functions: smoothness in the coupling constant.

4. J.-P. Eckmann, J. Magnon and R. Seneor, Decay properties and Borel summability
   for Schwinger functions in $P(\phi)_2$ theories. Commun. Math. Phys. To appear.

5. P. Federbush, A new approach to the stability of matter I,II. J. Math Phys.
   to appear.

6. ___________, The semi-Euclidean approach in statistical mechanics I. Basic
   expansion steps and estimates II. The custer expansion, a special example.
   Preprint.

7. J. Feldman, On the absence of bound states in the $\lambda\phi_2^4$ quantum field model
   without symmetry breaking. Canadian J. Phys. $\underline{52}$, 1583-1587 (1974).

8. _________, The $\lambda\phi_3^4$ field theory in a finite volume, Commun. Math. Phys. $\underline{37}$,
   93-120 (1974).

9. J. Feldman and K. Osterwalder, The Wightman axioms and mass gap for ϕ_3^4, these
   proceedings.

10. J. Fröhlich, Schwinger functions and their generating functionals, II. Adv. Math,
    to appear.

11. __________, The quantized "Sine-Gordon" equation with a nonvanishing mass term in
    two space-time dimensions. Preprint.

12. J. Glimm, The mathematics of quantum field theory. Adv. Math. To appear.

13. ________, Analysis over infinite dimensional spaces and applications to quantum
    field theory. Proceedings Int. Congress Math., 1974.

14. J. Glimm and A. Jaffe, Quantum field models, in: Statistical mechanics and quantum
    field theory, ed. by C. de Witt and R. Stora, Gordon and Breach, New York,
    1971.

15. ________________, Boson quantum field models, in: Mathematics of contempo-
    rary physics, ed. by R. Streater, Academic Press, New York, 1972.

16. ________________, Positivity of the ϕ_3^4 Hamiltonian, Fort. d. Physik, $\underline{21}$,
    327-376 (1973).

17. ___________________, ϕ^4 quantum field model in the single phase region:
    Differentiability of the mass and bounds on critical exponents, Phys. Rev. D$\underline{10}$,
    536-539 (1974).

18. ___________________, A remark on the existence of ϕ_4^4 . Phys. Rev. Lett. $\underline{33}$,
    440-442 (1974).

19. ___________________, The entropy principle for vertex functions in quantum
    field models, Ann. l'Inst. H. Poincaré, $\underline{21}$, 1-26 (1974).

20. ___________________, Critical point dominance in quantum field models, Ann.
    l'Inst. H. Poincaré, $\underline{21}$, 27-41 (1974).

21. J. Glimm and A. Jaffe, Absolute bounds on vertices and couplings, Ann. l'Inst. H. Poincaré, $\underline{22}$, to appear.

22. _______________________, On the approach to the critical point, Ann. l'Inst. H. Poincaré, $\underline{22}$, to appear.

23. _______________________, Two and three body equations in quantum field models, Preprint.

24. _______________________, On three-particle structure of ϕ^4 and the infinite scaling limit. Preprint.

25. J. Glimm, A. Jaffe and T. Spencer, The Wightman axioms and particle structure in the $P(\phi)_2$ quantum field model. Ann. Math. $\underline{100}$, p. 585–632 (1974).

26. _______________________________, The particle structure of the weakly coupled $P(\phi)_2$ model and other applications of high temperature expansions, in: Constructive quantum field theory, Ed. by G. Velo and A. Wightman, Springer-Verlag, Berlin, 1973.

27. F. Guerra, L. Rosen and B. Simon, Correlation inequalities and the mass gap in $P(\phi)_2$ III. Mass gap for a class of strongly coupled theories with nonzero external field. Preprint

28. D. Isaacson, Private communication.

29. A. Jaffe, States of constructive field theory. Proceedings of 17th International Conference on high energy physics, London, 1974. J.R. Smith, editor, pp. I-243 to I-250.

30. J. Lebowitz, GHS and other inequalities. Commun. Math. Phys. $\underline{35}$, 87–92 (1974).

31. B. McCoy and T. Wu, The two dimensional Ising model. Harvard University Press, Cambridge, 1973.

32. C. Newman, Inequalities for Ising models and field theories which obey the Lee-Yang theorem. Commun. Math. Phys. To appear.

33. _________, Moment inequalities for ferromagnetic Gibbs distributions. Preprint.

34. K. Osterwalder and R. Schrader, Axioms for Euclidean Green's functions, I. Commun. Math. Phys. $\underline{31}$, 83–112 (1973).

35. _______________________________, Axioms for Euclidean Green's functions, II. Preprint.

36. Y. Park, Lattice approximation of the $(\lambda\phi^4 - \mu\phi)_3$ field theory in a finite volume. Preprint.

37. G. Parisi, Field theory approach to second order phase transitions in three and two dimensional systems. Cargèse Summer School, 1973.

38. J. Percus, Correlation inequalities for Ising spin lattices. Preprint.

39. J. Rosen, Private communication.

40. J. Rosen and B. Simon, Fluctuations in $P(\phi)_1$ processes. Preprint.

41. R. Schrader, Yukawa quantum field theory in two space time dimensions without cutoff. Ann. Phys. $\underline{70}$, 412–457 (1972).

42. B. Simon, The $P(\phi)_2$ Euclidean quantum field theory. Princeton University Press, Princeton, 1974.

43. E. Seiler, Schwinger functions for the Yukawa model in two dimensions with space-time cutoff.

44. T. Spencer, The absence of even bound states in ϕ_2^4. Commun. Math. Phys., 39, 77-79 (1974).

45. __________, The decay of the Bethe Salpeter kernel in $P(\phi)_2$ quantum field models. Preprint.

46. T. Spencer and F. Zirilli, private communication.

47. G. Sylvester, Representations and inequalities for Ising model Ursell functions, Commun. Math. Phys., to appear.

48. __________, private communication.

49. C. Tracey and B. McCoy, Neutron scattering and the correlation functions of the Ising model near T_c. Phys. Rev. Lett. 31, 1500-1504 (1973).

50. A. Wightman, Introduction to some aspects of the relativistic dynamics of quantized fields, in: 1964 Cargèse Summer School Lectures, Ed. by M. Lévy, Gordon and Breach, New York (1967), p. 171-291.

DISCUSSION

Masuo Suzuki (comment): I hope that by using your inequalities you can obtain such qualitative results as the dependence of critical exponents upon the dimensionality and potential-range parameter in your ϕ^4 model. In the ferromagnetic Ising model, I proved inequalities such as $\gamma(d) \geq \gamma(d+1) \geq \ldots$ and $\nu(d) \geq \nu(d+1) \geq \ldots$, using Griffiths' inequalities. (Physics Letters 38A (1972) 23.)

FIELD THEORY OF RELATIVISTIC STRINGS

K. Kikkawa
Department of Physics
College of General Education
Osaka University

Toyonaka 560, JAPAN

§1. Introduction

Recent development of the dual resonance model is reviewed[1].

First, the quantum mechanics of a relativistic free string is presented by putting the emphasis on the gauge invariance and the Poincaré invariance. The invariances are guaranteed only when the space-time dimension D is 26 (D=10 in another model) and the Regge intercept α_0 equals 1.

Second, the second quantized formalism of the string model is demonstrated. By this we mean the theory of many pieces of strings interacting each other, which is analogous to the Dyson's formalism in the local field theory. The basic interactions between (closed and/or open) strings are determined from the continuity condition of the world sheets swept out by the string motion. The five basic interactions are shown to be necessary.

So far, the second quantized formalism is possible only in a special gauge, i.e., in the light cone gauge.

Third, we show the equivalence of the string theory to the dual resonance model. The scattering amplitude corresponding to each Feynman diagram can be mapped onto the amplitude in the dual resonance model. The five basic interactions introduced from the geometrical reason are shown to be necessary and sufficient to reproduce the covariant dual resonance model.

§2. Quantum Mechanics of a String

<u>Classical Theory</u>: The motion of a classical string is governed by the action

$$I = \int_\Delta \mathcal{L}\, d\sigma\, d\tau = (1/\alpha')\int_\Delta \sqrt{(\dot{X}X')^2 - \dot{X}^2 X'^2}\; d\sigma\, d\tau \tag{2.1}$$

where $X^\mu(\tau, \sigma)$ represents a point on a world sheet swept out by a string motion in a D-dimensional space-time ($\mu = 0, \cdots, D-1$). The set of parameters (σ, τ) is an arbitrary coordinate on the world sheet Δ, and $\dot{X}^\mu = \partial X^\mu/\partial\tau$, $X^{\mu\prime} = \partial X^\mu/\partial\sigma$ and $X^2 = \sum_{\mu=0}^{D-1} X^\mu X_\mu$ [Fig.1]. The integrand $\mathcal{L}\, d\sigma\, d\tau$ is the infinitesimal surface element of Δ. The action principle implies that the classical motion of string is determined in such a way that the area of world sheet is made minimum, a generalized Fermat's principle.

The action (2.1) has an invariance under the reparametrization

$$\{\sigma,\tau\} \longrightarrow \{\sigma'(\sigma,\tau), \tau'(\sigma,\tau)\} \tag{2.2}$$

where σ' and τ' are arbitrary continuous functions. In such a case, the velocities $\dot{X}^\mu(\sigma,\tau)$ cannot be solved with respect to the canonical momenta

$$P^\mu(\sigma,\tau) \equiv \frac{\partial \mathcal{L}}{\partial \dot{X}_\mu(\sigma,\tau)} \ . \tag{2.3}$$

The momenta, however, satisfy a set of "gauge conditions",

$$P^2(\sigma,\tau) + X'^2(\sigma,\tau) = 0 \ , \quad P(\sigma,\tau) X'(\sigma,\tau) = 0 \ . \tag{2.4}$$

Owing to (2.4), the mechanical system cannot be solved unless we specify a special gauge. The gauge we choose is the light cone gauge:

$$P^+ \equiv (P^0 + P^{D-1})/\sqrt{2} = 1 \ , \quad X^+ \equiv (X^0 + X^{D-1})/\sqrt{2} = \tau \tag{2.5}$$

which, when combined with (2.4), determine other components;

$$P^- \equiv (P^0 - P^{D-1})/\sqrt{2} = (\vec{P}^2 + \vec{X}'^2)/2 \ , \quad X^{-\prime} \equiv (X^{0\prime} - X^{D-1\prime})/\sqrt{2} = \vec{P}\cdot\vec{X}' \ . \tag{2.6}$$

Consequently, independent variables for string are the transverse components of coodinates $\vec{X}(\sigma,\tau) = (X^i(\sigma,\tau)\,|\,i=1,2,\cdots D-2)$, the zero frequency modes of light cone components x^-, and their canonical conjugates. The equation of motions, therefore, should be solved for $\vec{X}(\sigma,\tau)$ under the boundary condition

$$\vec{X}'(\sigma,\tau)\big|_{\sigma_\Delta} = 0 \tag{2.7}$$

for an open string, or the periodicity condition

$$\vec{X}(\sigma,\tau) = \vec{X}(\sigma+\sigma_0,\tau) \tag{2.8}$$

for a closed string.

<u>Quantum Theory</u>: In going to the quantum mechanics, we first assume that X^- except for the zero-modes, are functions of other independent quantities due to (2.5) and (2.6). The state vector which describes the string motion must be a functional of $\vec{X}$, and x^-, the zero-modes of $X^-(\sigma,\tau)$: $\phi[\vec{X}, x^+, x^-]$. For later convenience we will introduce the following Fourier components of ϕ with respect to x^-:

$$\Phi[\vec{X}, x^+, x^-] = \int_0^\infty dp^+ [\,\Phi_{p^+}[\vec{X}, x^+] e^{ip^+x^-} + c.c.\,] \ . \tag{2.9}$$

We have already imposed all the conditions (2.4)–(2.6) but the zero-mode condition of (2.6), which should be imposed as an equation of motion on ϕ. In the Schödinger representation, the condition turns out to be

$$[\, i\,\partial/\partial x^+ - (1/2)\int_0^{\pi d} d\sigma : \{-\delta^2/\delta \vec{X}(\sigma)^2 + \vec{X}'^2(\sigma)\} : -\frac{\alpha_0}{\alpha}\,]\,\Phi_{p^+} = 0 \ , \tag{2.10}$$

where we have used

$$P^- = i\,\partial/\partial x^+ \ , \quad \vec{P}(\sigma) = -i\,\delta/\delta\vec{X}(\sigma) \ , \text{ and } \alpha \equiv 2p^+ \ .$$

Fig.1.   The world sheet.

A great advantage of the approach presented above consists in the ghost elimination. The time component coordinate $X^0(\sigma, \tau)$ has a negative norm if it is quantized covariantly. The trouble has been overcome by taking advantage of gauge freedom, because X^0 has been eliminated by (2.5), and (2.6). This method is analogous to the Coulomb gauge quantization in the usual quantum electrodynamics.

Contrary to the ghost elimination, the Lorentz covariance, or the Poincaré covariance is not manifest in the above treatment. GGRT[2] was the first who confirmed the covariance by the explicit construction of the Poincaré generators. From the geometrical meaning of $X^\mu(\sigma, \tau)$ and $P^\mu(\sigma, \tau)$ one can guess that

$$P^\mu = \sqrt{1/2\pi} \int_0^{\pi\alpha} P^\mu(\sigma)\, d\sigma$$
$$M^{\mu\nu} = 1/4 \int_0^{\pi\alpha} [\, \{X^\mu(\sigma), P^\nu(\sigma)\} - \{X^\nu(\sigma), P^\mu(\sigma)\} \,] \ . \tag{2.11}$$

In constructing P^μ, and $M^{\mu\nu}$, the explicit solutions (2.5) and (2.6) has to be used with careful attention to the normal ordering for operators. Against expectation, GGRT found that the Poincaré invariance is possible only when $D=26$, and $\alpha_0=1$, where α_0 is the intercept of the Regge trajectory. The mass spectrum is given by

$$M^2 \equiv P_\mu^2 = \sum_{\ell=1}^{\infty} \ell\, \vec{a}_\ell^+ \vec{a}_\ell - 1 \ . \tag{2.12}$$

The reason why $D=26$ can be trace back to the normal ordering of operators in (2.10) and (2.11) when the system is quantized. The operators $\vec{a}_\ell^+$ and $\vec{a}_\ell$ are quantized coefficients in normal mode expansions of $\vec{X}(\sigma, \tau)$ and $\vec{P}(\sigma, \tau)$.

Finally, we note that a similar discussion is possible for a closed string with the cyclic boundary condition (2.8).

§3. The Second Quantization

In the previous section we discussed the quantum mechanics of a single string, and showed that the motion can be described in the Hilbert space spanned by x^μ, $\vec{X}(\sigma, \tau)$ and $\vec{P}(\sigma, \tau)$.

In order to construct a theory of interacting strings, one has to consider an infinite direct products of these Hilbert spaces:

$$\mathcal{H} = \prod_{i=1}^{\infty} \mathcal{H}_i \tag{3.1}$$

to each of which a string belongs. The purpose in this section is to formulate the perturbation theory for the many strings interacting each other.

Free Strings: Let us begin with the second quantization of interaction free strings. The wave function Φ_{p+} introduced in (2.9) now should be quantized as a field operator. The Lagrangian which provides (2.16) is given by

$$L_0 = \int_0^\infty dP_+ \int_0^{\pi\alpha} d\sigma \int \mathcal{D}\vec{X}\ \Phi_{p+}[X] \left\{ i\frac{\partial}{\partial x_+} - : \frac{1}{2}\left(-\delta^2/\delta\vec{X}^2(\sigma) + \vec{X}'^2(\sigma)\right) : - \frac{\alpha_0}{\alpha} \right\} \Phi_{p+}[X] \ . \tag{3.2}$$

The field Φ_{p+}, then, has the following expansion:

$$\Phi_{p+}[X] = (2\pi)^{-(D-2/2)} \int d\vec{p} \sum_{\{n_\ell\}} A_{\vec{p},\,p+,\,\{n_\ell\}} \, \xi_{\vec{p},\,p+,\,\{n_\ell\}}[X] \qquad (3.3)$$

where the annihilation operator $A_{\vec{p},\,p+,\,\{n_\ell\}}$ (creation operator $A^+_{\vec{p},\,p+\,\{n_\ell\}}$) destructs (creates) a whole string with excitation $\{n_\ell\}$.

The propagator for the string is, then, defined as

$$G_{p+}[X,Y] = \langle\langle 0| \xi_{p+}[X] \, \xi^+_{q+}[Y] |0\rangle\rangle. \qquad (3.4)$$

The Fourier transformation of (3.4) turns out to be

$$G_{p+}[P,Q]$$
$$= \int_{\sigma,\tau} \pi \, \mathcal{D}\vec{X}(\sigma,\tau) \exp\left\{ - \int_{\Delta_p} \mathcal{L}\, d\sigma d\tau + i\int_{\tau=\tau_1} \vec{P}(\sigma)\vec{X}(\sigma,\tau_1)d\sigma + i\int_{\tau=\tau_2} \vec{Q}(\sigma)\vec{X}(\sigma,\tau_2)d\sigma \right\} \qquad (3.5)$$

where

$$\mathcal{L} = (1/2)\left[\, \dot{\vec{X}}^2(\sigma,\tau) - \vec{X}'^2(\sigma,\tau)\,\right] \qquad \text{and} \qquad x^+ = -i\tau, \qquad (3.6)$$

and the integration region Δ_p is given in Fig.2. Note that the propagator (3.4) is now expressed as a path integral on the two dimensional domain Δ_p, which can be interpreted as an image of the world sheet on a complex plane.

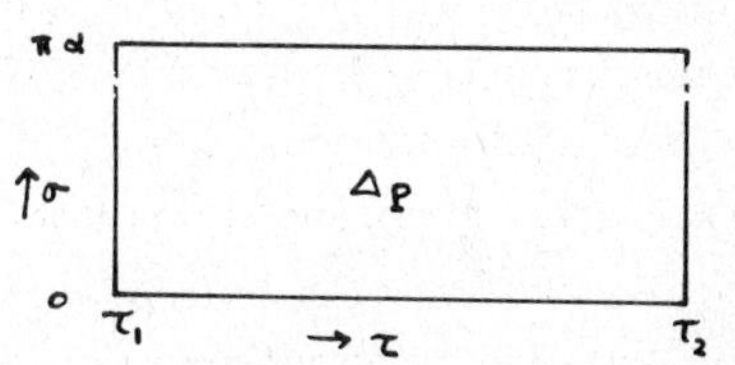

Fig.2. The propagator.

Interactions

The string picture has another advantage in determining the interactions between strings. Following geometrical observations provide us with an intuitive method to obtain the five basic interactions, which turn out to be sufficient to reproduce DRM.

(i) Suppose a piece of string $\vec{X}_1(\sigma_1, \tau)$ flies in the space time, and splits into two pieces $\vec{X}_2(\sigma_2, \tau)$ and $\vec{X}_3(\sigma_3, \tau)$ at $ix^+ = \tau$. The world sheet swept out by this process will be the one shown in Fig.3. In the classical picture, the continuity of the world sheet requires that, at $ix_+ = \tau$,

$$\vec{X}_1(\sigma_1) - \vec{X}_2(\sigma_2) = 0 \qquad , \quad \text{if} \ \ 0 \le \sigma_1 \le \pi\alpha_2$$
$$\vec{X}_1(\sigma_1) - \vec{X}_3(\sigma_3) = 0 \qquad , \quad \text{if} \ \ \pi\alpha_2 \le \sigma_1 \le \pi(\alpha_2+\alpha_3) \qquad (3.7)$$

Similary, the momentum density distributed on the string must flow smoothly along the world sheet;

$$\vec{P}_1(\sigma_1) + \vec{P}_2(\sigma_2) = 0 \qquad , \quad \text{if} \ \ 0 \le \sigma_1 \le \pi\alpha_2$$
$$\vec{P}_1(\sigma_1) + \vec{P}_2(\sigma_2) = 0 \qquad , \quad \text{if} \ \ \pi\alpha_2 \le \sigma_1 \le \pi(\alpha_2+\alpha_3), \qquad (3.8)$$

where the direction of momentum is taken to be positive for inward.

In quantum mechanics, the above condition should be satisfied when the left hand sides of (3.7) and (3.8) are operated on the vertex function, provided $\vec{P}_i = -i\delta/\delta\vec{x}_i$. The relations are then considered to be a set of homogeneous differential equations to the vertex. The

solution is unique except for a normalization constant. The formal
solution turns out to be

$$L_1 = g \int \prod_{i=1}^{3} \delta \vec{X}_i \; \Phi^+_{P_1}[X_1] \Phi_{P_2}[X_2] \Phi_{P_3}[X_3] \; \prod_i \frac{d P_{+i}}{\sqrt{2 P_{+i}}} \; \delta(P_{+1} - P_{+2} - P_{+3})$$

$$\times \prod_\sigma \delta[\vec{X}_1(\sigma_1) - \Theta(\sigma_1 - \pi d_2)\vec{X}_3(\sigma_3) - \Theta(\pi d_2 - \sigma_1)\vec{X}_2(\sigma_2)] + h.c. \,, \qquad (3.9)$$

$$\prod_\sigma \delta[\vec{X}_1 - \Theta\vec{X}_3 - \Theta\vec{X}_2] = g \lim_{\tau_i \to \tau} \int \delta\vec{X}_1 \cdots \delta\vec{X}_3 \, exp\{-\int_{\Delta_V} \mathcal{L} \, d\sigma d\tau\} \prod_{i=1}^{3} \prod_\sigma \delta[\vec{X}(\sigma, \tau_i) - \vec{X}_i(\sigma)]$$

$$(3.10)$$

The last expression is the path integral formula and $\mathcal{L}$ in the exponent
is given by (3.6). The domain Δ_V is shown in Fig.3.

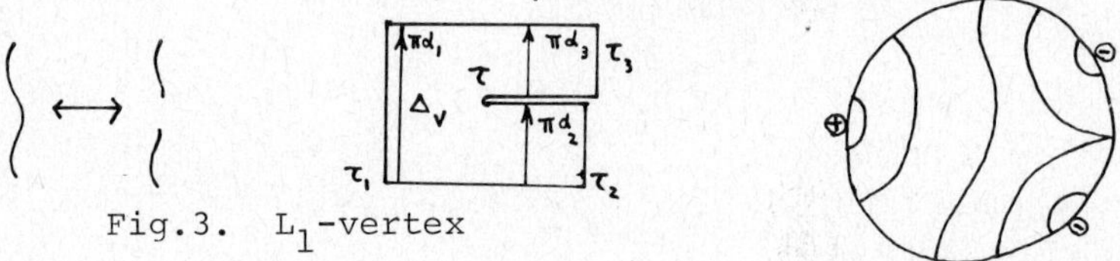

Fig.3. L_1-vertex

(ii) Another possible interaction will be the case when two pieces
of strings collide at their intermediate points and undergo a rearrangement [Fig.4]. The continuity condition of the world sheet again
determines the interaction Lagrangian, whose explicit form will be
found in Ref.3.

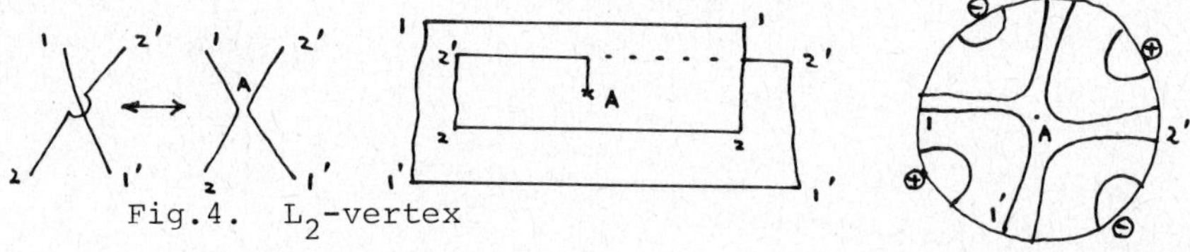

Fig.4. L_2-vertex

Assuming that the local structure of the string interactions is
either the type of L_1 or L_2, we look for all other possibilities.
Then, we have following three interactions.

(iii) L_3: The open-to-closed transition. Fig.5.

(iv) L_4: The open-to-an-open-and-a-closed transition. Fig.6.

(v) L_5: The closed-to-two-closed transition. Fig.7.

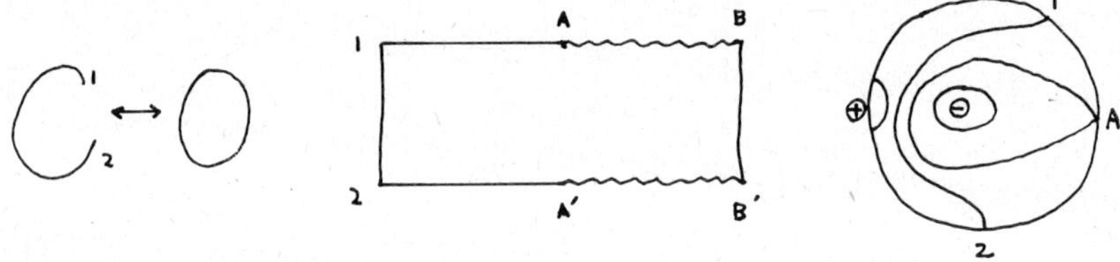

Fig.5. L_3-vertex. On the light cone sheet, AB is identified
with A'B'. The equi-potential line 1-2 in the right fig. is
mapped onto the vertical line 1-2 in the middle.

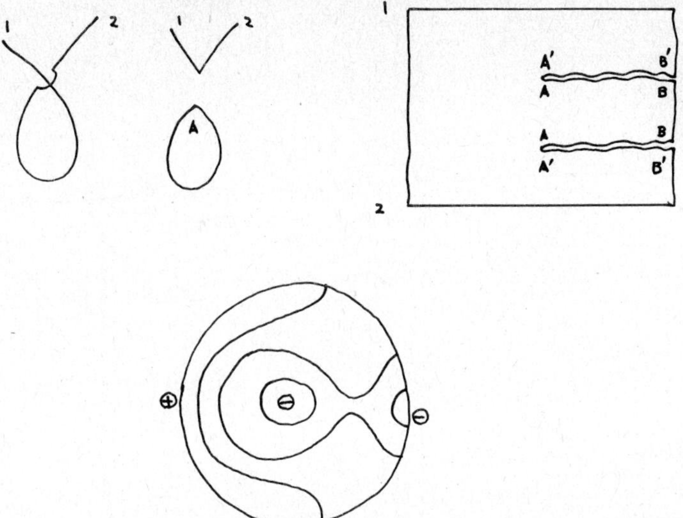

Fig.6. L_4-vertex. The two lines AB(A'B') are identified each other.

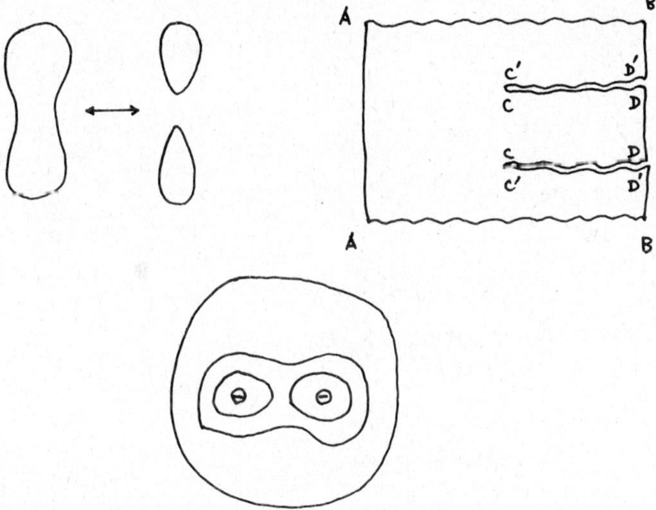

Fig.7. L_5-vertex. The two lines AB(CD and C'D') are identified so that any vertical line forms a loop in the right top fig.

Corresponding interaction Lagrangians can be written down from the continuity condition. The explicit forms are found in Ref.3.

Scattering matrix

The Lagrangian for interacting strings is now given by

$$L = L_o + L_c + \sum_{i=1}^{5} L_i \quad ,$$

(3.11)

where L_o and L_c are the interaction free parts of the open and the closed strings, respectively.

We introduce the interaction representation operators as

$$V_i(\tau) = - e^{-H_o \tau} L_i e^{H_o \tau}$$

(3.12)

where H_o is the unperturbed Hamiltonian determined from $L_o + L_c$ in (3.11). It will be worthwhile to point out that the interaction Lagrangians L_i in (3.11) are all local with respect to the time x^+, and with no derivative. This is the great advantage of the light cone gauge. In covariant gauges the interaction term becomes generally non-local, which destroys the possibility of using the canonical quantization method.

The transition matrix is now given by

$$U(\tau_f, \tau_i) = \sum_{n=0}^{\infty} \int_{\tau_i}^{\tau_f} d\tau_1 \int^{\tau_1} d\tau_2 \cdots \int^{\tau_{m-1}} d\tau_m \prod_{i=1}^{n} V(\tau_i)$$

(3.13)

where

$$V(\tau) = \sum_{i=1}^{5} V_i(\tau) \quad .$$

As is well known in the usual local field theory the Dyson-Wick contraction automatically provides us with any matrix element. The results will be summerlized as follows.

1) Each way of contractions of operators gives a light cone Feynman diagram.

2) Associated with each internal line propagator, we obtain

$$G_{p+}[P_1, P_2] = \boxed{\quad \Delta_P \quad}$$

(3.14)

where, by the rectangle, we mean the path integral over the domain Δ_p defined in (3.5).

3) Associated with each 3-vertex L_1, we obtain

$$\Gamma[P_1, P_2, P_3] = \lim_{\tau_i \to \tau} {}^{\tau_1}\boxed{\begin{matrix} \Delta_V \\ \tau \end{matrix}}{}^{\tau_3}_{\tau_2}$$

(3.15)

where the polygon again means the path integral over the domain Δ_V. Similarly, associated with other interactions L_2-L_5, we obtain corresponding path integrals.

4) Associated with each incoming external line, we obtain (for a lowest excited state)

$$G_{p+}[P_1, P_2] e^{\tau_i E\{P_1, P_1, \ell_0\}} = \boxed{\quad \Delta_P \quad} e^{\tau_i E\{P_1, P_1, \ell_0\}}$$

(3.16)

As for the outgoing external line, the Hermitian conjugate of (3.16) appears.

5) When each factor in 1)-4) are combined according to the Feynman
diagram, we obtain a path integral over the region, say,

$$e^{\tau_i E_1} \boxed{} e^{-\tau_f E_3}$$
$$e^{\tau_i E_2} \qquad\qquad\Delta\qquad\qquad e^{-\tau_f E_4}$$

$$= e^{\sum_{j=1}^{4} \tau_j P_j^-} \int \exp\left\{ -\int_\Delta \mathcal{L}\, d\sigma d\tau - i \int \sum_j \vec{X}_j(\sigma_j, \tau_j) \vec{P}_j(\sigma_j)\, d\sigma_j \right\} \mathcal{D}\vec{X}(\sigma, \tau) \qquad (3.17)$$

where we used a new notation; $P_j^- = E_j$ for incomming strings and $P_j^- = -E_j$
for outgoing strings. The second factor in (3.17) can be explicitly
performed if the Newmann function for the given domain Δ is known. As
for the first factor we use the following formula:

$$\tau_j = 2 \sum_i \int N_\Delta(\sigma_j, \tau_j\,;\, \sigma_i, \tau_i)\, P_i^+\, d\sigma_i$$
$$= 2 \oint_{\partial\Delta} N_\Delta(\rho, \rho')\, P^+(\rho')\, d\rho' \qquad (3.18)$$

where $N_\Delta(\rho, \rho')$ with $\rho = \tau + i\sigma$ is the Neumann function on the domain Δ.

6) Substituting these Neumann function expressions into (3.17), we
will finally obtain a <u>formally</u> covariant expression for the scattering
matrix element

$$T = \sum_{\substack{over\\ contractions}} \int_{-\infty}^{\infty} d\tau_1 \int_{-\infty}^{\tau_1} d\tau_2 \cdots \int_{-\infty}^{\tau_{m-1}} d\tau_m\; J_\Delta(\tau_1 \cdots \tau_m) \cdot$$
$$\cdot \exp\left\{ \oint_{\partial\Delta} \oint P_r(\rho) N_\Delta(\rho, \rho') P^\mu(\rho')\, d\rho d\rho' \right\} \qquad (3.19)$$

where N_Δ, of course, depends on τ_i. Although we have not discussed
the path integration measure factor, from which $J(\tau_1 \cdots \tau_n)$ is determined, the explicit forms are known for simple scattering diagrams.
$J_\Delta(\tau_1 \cdots \tau_n)$ does depend on the space-time dimension D and the intercept α_0 also, and the theory is consistent only when D=26, and $\alpha_0 - 1$.

§4. General Mapping

One will notice that the formula (3.19) is quite similar to the
analogue model representation of DRM. In fact, a DRM amplitude is
given by (3.19), provided, however, that the integration region Δ is
not the light cone world sheet but that Δ is a certain simply or multiply connected compact domain in a complex z-plane. All the conformally equivalent domains are supposed to be equivalent.

What we should prove are, first, the existence of mapping of DRM
amplitudes to (3.19), and second, that all the DRM amplitudes are
reproduced by the combination of the five basic interactions L_1, L_2,
$\cdots$, and L_5 introduced in §3.

We shall outline the proof below.

The mapping of the Neumann function defined above to the one in
(3.19) is rather well known in fluid dynamics. Let $N_R(z, z')$ be a
Neumann function for a given (simply or multiply) connected domain R.
For a given source of the light cone momentum $P^+ = \frac{1}{2}(P^0 + P^{D-1})$:

$$P^+(z) = \sum_i p_{+i}\, \delta(z - z_i) \qquad (4.3)$$

we can make an harmonic function

$$U_R(z) = (1/2\pi) \oint_{\partial R} N_R(z.z') P^+(z') dz' \tag{4.4}$$

Associated with $U_R(z)$, let us define a complex harmonic function on R by

$$\rho_R = Z_R(z) = U_R(z) + i V_R(z) \tag{4.5}$$

where V_R is a function determined by the Canchy-Riemann relations

$$\partial U_R(z)/\partial x = \partial V_R(z)/\partial y$$
$$\partial U_R(z)/\partial y = -\partial V_R(z)/\partial x \ . \tag{4.6}$$

In the fuild dynamics U_R and V_R are known as the potential and the velocity potential, respectively.

The mapping we are looking for is given by (4.5). In fact an equi-potential line in R is mapped by (4.5) onto a vertical line in the light cone diagram. A stream line in R is mapped onto a horizontal line in the light cone diagram.

Finally, we show that DRM can be reproduced with the five basic interactions L_1, $\cdots L_5$. The necessity can be easily understood by looking the diagrams Fig.3-Fig.7. Since the mapping (4.5) transforms an equipotential line to a vertical string line on the light cone Feynman diagram, one will see the necessity of the four string interaction in Fig.5.

Other interactions are explained in similar ways. The sufficiency is given as follows. Suppose, for example, a six string interaction existed. Owing to the conformal invariance, three out of six z_i's (source position in (4.3)) can be fixed at any point. Therefore, the other threes z_4, z_5, and z_6 should be integration parameters in DRM amplitude. By the mapping the contact point P is mapped on the interaction point τ in the light cone diagram, which now is the new integration parameter. Since this point τ has only two degrees of freedom, the Jacobian for $(z_4, z_5, z_6) \rightarrow \tau$, must be zero on a non-zero measure. Since the mapping is conformal, (4.5) must be a constant. This is a contradiction.

For other string interactions, a similar argument works. The five interactions $L_1 \cdots L_5$, therefore, are sufficient to reproduce DRM.

References

1) Recent reviews for dual resonance models are, for example, J. Scherk, NYU preprint (to be published in Rev. Mod. Phys.) and S. Mandelstam "Dual-Resonance Models" Physics Report, <u>13C</u> no.6 (1974).

2) J. Goldstone, P. Goddard, C. Rebbi and C. Thorn, Nucl. Phys. <u>B56</u>, 109 (1973).

3) M. Kaku and K. Kikkawa, Phys. Rev. <u>10D</u>, 1110 (1974) and ibid.
    <u>10D</u>, 1823 (1974).

Discussions

Matsuda: You have discussed here the mesonic strings with your five
    fundamental string interactions. What about the baryonic strings?
    In particular, how many topological interactions do you have for
    the latter? And also, can you bring the classical topology of the
    baryonic strings directory into the quantized version consistent-
    ly?

Kikkawa: For the first two questions, see X. Artru, Nucl. Phys. <u>B85</u>,
    442 (1975). As for the last, the answer is that the field theo-
    retical formulation is not simple, because we have an infinite
    number of strings which are topologically different.

Takabayashi: Is your interaction theory Lorentz invariant?

Kikkawa: Yes. See S. Mandelstam, Nucl. Phys. <u>B83</u>, 413 (1974).

RESULTS AND PROBLEMS IN IRREVERSIBLE STATISTICAL MECHANICS OF OPEN SYSTEMS

Klaus Hepp
Physics Department E.T.H.
CH-8049 Zürich, Schweiz

§ 1. Introduction

Rigorous results about the foundations of equilibrium statistical mechanics are extremely difficult to obtain. The beautiful proof of the Bernoulli property of billiards with smooth convex scatterers [S1] , the geometry of the energy surface for large closed systems [R3] , and the justification of the Boltzmann equation for hard spheres in the Grad limit [L1] are illustrative examples, which will caution us from expecting rapid progress in this difficult field.

The study of open systems does not really help to overcome these technical problems. It does, however, open new possibilities which are in a sense complementary to collision mixing in closed systems. One can rather easily prove that small quantum systems coupled to suitable reservoirs of a diffusive type (e.g. an ideal gas in thermal equilibrium) via suitable interactions strongly approach equilibrium. Conceptually, however, open systems are important for describing stationary nonequilibrium situations. Here the proof of the Onsager relations [O1] is more natural than in linear response theory, and the main input is the KMS property [H1] of the reservoir rather than time reversal invariance. Most interesting are open systems far from thermal equilibrium with the possibility of a wide range of non-equilibrium phase transitions (see [G2]) .

In my review I shall not treat the classical case and deal only with a very special class of models, which have come up frequently in my joint work with E.H. Lieb, but which belong to the folklore of the yet unwritten general theory of irreversible statistical mechanics.

§ 2. Open Systems of Finitely Many Degrees of Freedom

Small quantum systems of finitely many degrees of freedom (for instance any number of particles in a box) coupled to suitable (but not very physical) infinite quantum systems, called reservoirs, can show a very strong approach to equilibrium.

The small _system_ S has a Hilbert space $\mathcal{H}^S$ with the C^*-algebra $\mathcal{A}^S$ of all bounded operators on $\mathcal{H}^S$ as observables. The states of S are all positive normalized trace class operators ω^S on $\mathcal{H}^S$, and the time evolution α_t^S is generated by a self-adjoint Hamiltonian $H^S = H^{S*}$ with $\mathrm{Tr}\, \exp{-\beta H^S} < \infty$ for all inverse temperatures $\beta > 0$. For simplicity of exposition we shall make the inessential assumption that $\dim \mathcal{H}^S < \infty$. The _reservoir_ R is composed of r parts $R^1, .. R^r$, each with a C^*-algebra of observables $\mathcal{A}^\varrho$, and a continuous 1-parameter group of time evolutions $\alpha_t^\varrho \in \mathrm{Aut}\, \mathcal{A}^\varrho$. We set $\mathcal{A}^R = \mathcal{A}^1 \otimes \cdots \otimes \mathcal{A}^r$ and $\alpha_t^R = \alpha_t^1 \otimes \cdots \otimes \alpha_t^r$.

The combined system has then the algebra $\mathcal{A} = \mathcal{A}^R \otimes \mathcal{A}^S$. An _interaction_ is any $V = V^* \in \mathcal{A}$, and this gives rise to a time evolution α_t of $\mathcal{A}$ [A1], [R2], which is the perturbation of $\alpha_t^R \otimes \alpha_t^S$. In an equilibrium state ω^R for α_t^R, where one has a GNS Hamiltonian H^R, the generator of α_t is $H^S + H^R + V$.

The reservoir (R, α^R) _drives_ the system (S, α^S) _strongly into equilibrium_ through the interaction V, if for a dense set of $A \in \mathcal{A}$ in the strong topology of $\mathcal{A}$

$$\mathrm{n\text{-}lim}_{t \to \pm\infty} \alpha_{-t}^R \circ \alpha_t(A) = \alpha_\pm(A) \qquad (2.1)$$

exists and $\alpha_\pm(A) \in \mathcal{A}^R$.

Wave epimorphisms $\alpha_\pm$ have been studied by Robinson in the return to equilibrium of closed systems [R2]. It is not necessary that $\alpha_+ = \alpha_-$, as we shall see in theorem 2.4. The $\alpha_\pm$ intertwine between α_t and α_t^R: $\alpha_\pm \circ \alpha_t = \alpha_t^R \circ \alpha_\pm$. For many purposes, the following weaker property is sufficient: The reservoir (R, α^R, ω^R) in the α^R equilibrium state ω^R drives (S, α^S) _weakly_ into equilibrium via V, if for a dense set of $A \in \mathcal{A}$

$$\mathrm{s\text{-}lim}_{t \to \pm\infty} \pi \circ \alpha_{-t}^R \circ \alpha_t(A) = A_\pm \qquad (2.2)$$

exists in the strong topology in the GNS representation space $\mathcal{H}^R \otimes \mathcal{H}^S$ for ω^R, and if $A_{\pm} \in \pi(\mathcal{O}^R)''$. The following obvious result motivates our definition:

<u>Theorem 2.1</u>: If (R, α^R, ω^R) drives (S, α^S) weakly via V , then for every initial state ω^S of S

$$\lim_{t \to \pm\infty} \omega^R \otimes \omega^S(\alpha_t(A)) = \omega_{\pm}(A) \qquad\qquad (2.3)$$

exists on $\mathcal{O}$ and is an equilibrium state with respect to α_t . If ω^R is a thermal equilibrium (KMS) state w.r.t. α^R with inverse temperature β , then $\omega_{\pm}$ are KMS states w.r.t. α for the same β .

We are interested in reservoirs $R^1, \ldots R^r$ which are in KMS states $\omega^{\beta_1}, \ldots \omega^{\beta_r}$ at different temperatures $\beta_1, \ldots \beta_r$ and which are coupled to S by $V = \sum_{\mathcal{g}=1}^{r} V^{\mathcal{g}}$, where $V^{\mathcal{g}} = V^{\mathcal{g}*} \in \mathcal{O}^{\mathcal{g}} \otimes \mathcal{O}^S$. For such a coupling, there is still some ambiguity in defining energy flux observables $J^{\mathcal{g}}$ from $R^{\mathcal{g}}$ to S . We require $J^{\mathcal{g}} = J^{\mathcal{g}*} \in \mathcal{O}^{\mathcal{g}} \otimes \mathcal{O}^S$ and

$$\frac{d}{dt} \alpha_t(H^S) \Big|_{t=0} = \dot{H}^S = \sum_{\mathcal{g}=1}^{r} J^{\mathcal{g}} . \qquad\qquad (2.4)$$

Obviously, $J^{\mathcal{g}} = i[V^{\mathcal{g}}, H^S] + J_o^{\mathcal{g}}$, where $J_o^{\mathcal{g}} = J_o^{\mathcal{g}*} \in \mathcal{O}^S$ and $\sum J_o^{\mathcal{g}} = o$.

<u>Problem</u>: Suppose that $(R, \alpha^R, \omega^{\beta_1} \ldots \omega^{\beta_r})$ drives (S, α^S) weakly via $\sum V^{\mathcal{g}}$ for an open neighborhood of $\beta_1 = \ldots \beta_r = \beta$. When do there exist energy fluxes $J^1, \ldots J^r$ (independent of $\beta_1 \ldots \beta_r$) such that the <u>Onsager relations</u> hold:

$$\omega_{\pm}(J^{\mathcal{g}}) \Big|_{\beta} = 0 \qquad\qquad (2.5)$$

$$\frac{\partial}{\partial \beta_{\tau}} \omega_{\pm}(J^{\mathcal{g}}) \Big|_{\beta} = \frac{\partial}{\partial \beta_{\mathcal{g}}} \omega_{\pm}(J^{\sigma}) \Big|_{\beta}$$

for $\beta_1 = \ldots \beta_r = \beta$ and all $1 \leq \mathcal{g}, \sigma \leq r$?

<u>Problem</u>: Suppose that (R, α^R, ω^R) drives (S, α_λ^S) weakly via V for all $-\lambda_o \leq \lambda \leq \lambda_o > 0$, where α_λ^S is generated by $H^S + \lambda A$ with $A = A^* \in \mathcal{O}^S$. Under what conditions is the equilibrium stable under the "mechanical perturbation" A , and when does it define a <u>linear response</u> to A by

$$\omega_{\pm}^{\lambda}(B) = \omega_{\pm}(B) + i\lambda \int_{\mp\infty}^{o} ds \, \omega_{\pm}([\alpha_s(A), B]) + o(\lambda)$$

$$\text{for } \lambda \to 0 \text{ and all } B \in \mathcal{O} \text{ ?} \qquad\qquad (2.6)$$

Up to now the most satisfactory results on the drive to equilibrium are known in the weak coupling (van Hove [H11]) limit. Let $\{\phi_1,\ldots\phi_n\}$ be an orthonormal basis of $\mathcal{H}^S$ with $H^S\phi_i = \varepsilon_i\phi_i$ and $H^S \neq \varepsilon\mathbb{1}$. The reservoirs $R^1,\ldots R^r$ are taken in KMS states $\omega^{\beta_1},\ldots\omega^{\beta_r}$, defining a representation π of $\mathcal{O}$ and a Hamiltonian H^R. We take $V = \sum V^s$, $V^s = V^{s*} \in \mathcal{O}^s \otimes \mathcal{O}^S$ and set $H_o = H^R + H^S$ and (suppressing π)

$$H_\lambda = H_o + \lambda V \tag{2.7}$$

In this framework, the following result by Davies [D2] and Pulè [P1] is relevant:

<u>Theorem 2.2:</u> Assume that for all $1 \leq s \leq r$

$$\omega^R(V^s) = o \qquad \text{and} \tag{2.8}$$

$$\sup_t \| \omega^R(V^s e^{iH_o t} V^s e^{-iH_o t}) \| (1 + |t|)^{1+\varepsilon} < \infty \tag{2.9}$$

for some $\varepsilon > o$, and let either ω^R be quasifree, or assume bounds of type (2.9) for all truncated multi-time correlation functions of k V-operators, with rapid decrease of the suprema in k. Then for every $B \in \mathcal{O}^S$ and every ω^S

$$\lim_{\substack{\lambda \to o \\ \lambda^2 t = \tau}} \omega^R \otimes \omega^S(e^{-iH_o t} e^{iH_\lambda t} B e^{-iH_\lambda t} e^{iH_o t}) = \omega^S(\tau)(B) =$$

$$= \lim_{\substack{\lambda \to o \\ \lambda^2 t = \tau}} \omega^R \otimes \omega^S(e^{iH_\lambda t} e^{-iH_o t} B e^{iH_o t} e^{-iH_\lambda t}) \tag{2.10}$$

exists and defines a density matrix in $\mathcal{H}^S$. For many interactions the non-diagonal elements $\omega^S_{ij}(\tau)$ of $\omega^S(\tau)$ in the energy basis decrease purely exponentially for $\tau \geq o$, $\omega^S_{ij}(\tau) = \exp\lambda_{ij}\,\omega^S_{ij}$, $\mathrm{Re}\lambda_{ij} < o$, while the weak coupling time evolution defines on the diagonal elements a strongly continuous Markov semigroup with the <u>master equation</u>

$$\frac{d}{d\tau}\omega^S_{ii}(\tau) = \sum_{j=1}^n K_{ij}(\beta_1,\ldots\beta_r)\,\omega^S_{jj}(\tau) \tag{2.11}$$

$$K_{ij}(\beta_1,\ldots\beta_r) = \sum_{s=1}^r \left\{ W^s_{ij} - \delta_{ij}\sum_{k=1}^n W^s_{ki} \right\},$$

where the W^s_{ij} satisfy <u>detailed balance</u>

$$W^s_{ij}\,\exp-\beta_s\varepsilon_j = W^s_{ji}\,\exp-\beta_s\varepsilon_i . \tag{2.12}$$

Generically for H^S, this Markov process is ergodic. Then the unique equilibrium state $\bar{\omega}^S(\beta_1,\ldots\beta_r)$ is for $\beta_1 = \ldots = \beta_r = \beta$ the Gibbs state

$$\bar{\omega}_{ij}^S(\beta,\ldots\beta) = Z^{-1}\exp(-\beta\varepsilon_i)\,\delta_{ij}\,, \quad Z = \sum \exp{-\beta\varepsilon_i}\,. \tag{2.13}$$

In thermal equilibrium the only manifestation of the reservoir is its temperature. There exists a large class of interactions, where $\bar{\omega}^S(\beta_1,\ldots\beta_r)$ for small deviations $|\beta_\delta - \beta|$ from thermal equilibrium can be computed by a convergent perturbation expansion. This gives a microscopic model for the stochastic approach to open systems by Bergmann, Lebowitz et al. [B2]. Obviously, the energy H^S of S satisfies

$$\lim_{\substack{\lambda\to 0 \\ \lambda^2 t = \tau}} \omega^R \otimes \omega^S\,(e^{iH_\lambda t}\,H^S\,e^{-iH_\lambda t}) = \omega^S(\tau)\,(H^S) \tag{2.14}$$

and the time evolution of $\omega^S(\tau)(H^S)$ is completely determined by (2.9). We take the simplest choice for the energy flux in the weak coupling limit:

$$\frac{d}{d\tau}\,\omega^S(\tau)(H^S) = \sum_{\delta=1}^{r}\omega^S(\tau)(J^\delta)\,, \tag{2.15}$$

$$J^\delta = \sum_{i,j=1}^{n} W_{ji}^\delta\,(\beta_\delta)(\varepsilon_j - \varepsilon_i)\,|\phi_i\rangle\langle\phi_i|\,.$$

Then the entropy production in $\omega^S(\tau)$

$$\sigma(\omega^S(\tau)) = -\frac{d}{d\tau}\,k\sum_{i=1}^{n}\omega_{ii}^S(\tau)\,\ell n\,\omega_{ii}^S(\tau)$$

$$- k\sum_{\delta=1}^{r}\beta_\delta\,\omega^S(\tau)\,(J^\delta) \tag{2.16}$$

is ≥ 0 with equality only for $\beta_1 = \ldots = \beta_r = \beta$ and $\omega_{ii}^S(\tau) = \bar{\omega}_{ii}^S(\beta,\ldots\beta)$ [B2]. An easy calculation [H8] leads to

<u>Theorem 2.3</u>: For the energy flows (2.15) through S weakly coupled to $R^1, \ldots R^r$ in KMS states $\omega^{\beta_1}, \ldots \omega^{\beta_r}$, the Onsager relations hold in the stationary state $\bar{\omega}^S(\beta_1,\ldots\beta_r)$ of (2.11):

$$\bar{\omega}^S(\beta_1,\ldots\beta_r)\,(J^\delta)\big|_\beta = 0 \tag{2.17}$$

$$L_{\delta\sigma} = \frac{\partial}{\partial\beta_\sigma}\,\bar{\omega}^S(\beta_1,\ldots\beta_r)\,(J^\delta)\big|_\beta = \frac{\partial}{\partial\beta_\delta}\,\bar{\omega}^S(\beta_1,\ldots\beta_r)\,(J^\sigma)\big|_\beta = L_{\sigma\delta}$$

for $\beta_1 = \ldots \beta_r = \beta$ and all $1 \leq \varrho, \sigma \leq r$. The assumptions about dim $\mathcal{H}^S$ can be relaxed for suitable V, retaining only $\operatorname{Tr} \exp{-\beta H^S} < \infty$ for $\beta > 0$. For interacting particles in an external magnetic field $\underline{B}$, our model leads to the symmetry

$$L_{\varrho\sigma}(\underline{B}) = L_{\sigma\varrho}(\underline{B}) \tag{2.18}$$

<u>without assuming time reversal invariance</u>. If in addition the latter holds for H^S and H_λ, then one obtains the standard form

$$L_{\varrho\sigma}(\underline{B}) = L_{\sigma\varrho}(-\underline{B}). \tag{2.19}$$

Not very much is known about the drive to equilibrium without taking the weak coupling limit. It is <u>not</u> true that closed infinite systems with a strong return to equilibrium can always serve as reservoirs for driving even very simple open systems to equilibrium. Let e.g. S be one fermion with creation and annihilation operators $a^{\#}$ ($\{a, a^*\} = 1$, $\{a, a\} = 0$) and

$$\mathcal{H}^S = \mathbb{C}^2 \quad , \quad H^S = \varepsilon a^* a \quad , \quad \varepsilon \geq 0. \tag{2.20}$$

Let $\mathcal{A}$ be the C^*-algebra generated by $a^{\#}$ and a Fermi field $A^{\#}(\underline{k})$, $\underline{k} \in \mathbb{R}^3$, with $\{A(\underline{k}), A^*(\underline{k}')\} = \delta(\underline{k} - \underline{k}')$ and $\alpha_t^R(A(\underline{k})) = \exp(-i|\underline{k}|^2 t) A(\underline{k})$. Take the interaction $V = a^* A(v) + A(v)^* a$, where $A(v) = \int d\underline{k}\, v(\underline{k})\, A(\underline{k})$ and $v = v^* \in \mathcal{S}(\mathbb{R}^3)$ and $\int d\underline{k}\, v(\underline{k})^2 / |\underline{k}|^2 > \varepsilon$. The full time evolution α_t acts in the test function space $\mathcal{L} = \mathbb{C} \oplus L^2(\mathbb{R}^3)$: For $\mathcal{A}(F) = \varphi a + A(f)$, $F = (\varphi, f) \in \mathcal{L}$, $\alpha_t(\mathcal{A}(F)) = \mathcal{A}(e^{-iHt} F)$, where $HF = (\varepsilon\varphi + \int d\underline{k}\, v(\underline{k})\, f(\underline{k})$, $v(\underline{k})\varphi + |\underline{k}|^2 f(\underline{k}))$. The existence of an eigenstate $F_d = (\varphi_d, f_d)$ of H with $\varphi_d \neq 0$ is responsible for the memory on the initial state ω^S in the approach to equilibrium: For $A \in \mathcal{A}$, $\omega^R \otimes \omega^S(\alpha_t(A))$ converges for $t \to \pm\infty$, but the limit depends on ω^S.

A good case for drive to equilibrium is found in a class of small Fermi systems coupled to Fermi field reservoirs, which we shall call <u>elementary open systems</u> (EOS) and which one can couple to <u>interacting open systems</u> (IOS).

An <u>EOS</u> is one fermion $a^{\#}$ with $\mathcal{H}^S = \mathbb{C}^2$, $H^S = \varepsilon a^* a$, coupled to r Fermi fields $A_\varrho^{\#}(\omega)$, $\omega \in \mathbb{R}^1$, $1 \leq \varrho \leq r$, with $\alpha_t^R(A_\varrho(\omega)) = e^{-i\omega t} A_\varrho(\omega)$ and $\mathcal{A}$ the CAR algebra generated by $\{a^{\#}, A_\varrho^{\#}(\omega)\}$. A family of <u>regular</u> interactions is given by $V^\tau = \sum_{\varrho=1}^{r} V_\varrho^\tau$, where $g_\varrho = g_\varrho^* \neq 0$, $1 \leq \tau < \infty$ and

$$V_\varrho^\tau = g_\varrho \int (1 + \omega^2/\tau^2)^{-1/2} (A_\varrho^*(\omega)\, a + a^*\, A_\varrho(\omega))\, d\omega \tag{2.21}$$

The interacting time evolution α_t^{τ} leads to a stochastic differential equation for $a^{\tau}(t) = \alpha_t^{\tau}(a)$:

$$\dot{a}^{\tau}(t) = -i\varepsilon a^{\tau}(t) - \sum_{\varsigma=1}^{r} \gamma_{\varsigma}\tau \int_0^t ds\; e^{-\tau(t-s)}\; a^{\tau}(s)$$

$$- i\sum_{\varsigma=1}^{r} g_{\varsigma} A_{\varsigma}^{\tau}(t) \quad , \quad a^{\tau}(o) = a \quad , \tag{2.22}$$

where $\pi g_{\varsigma}^2 = \gamma_{\varsigma}$ and $A_{\varsigma}^{\tau}(t) = \int d\omega\, A_{\varsigma}(\omega)\, e^{-i\omega t}\,(1+\omega^2/\tau^2)^{-1/2} \in \mathcal{OL}^R$.

From $a^{\tau}(t)$, $A_{\varsigma}^{\tau}(\omega,t) = \alpha_t^{\tau}(A_{\varsigma}(\omega))$ can be computed: $A_{\varsigma}^{\tau}(\omega,t) =$

$e^{-i\omega t} A_{\varsigma}(\omega) - ig_{\varsigma} \int_o^t ds\,(1+\omega^2/\tau^2)^{-1/2}\, e^{-i\omega(t-s)}\, a^{\tau}(s)$. One differentiation removes the retardation in (2.22), and the elementary solution shows exponential approach to equilibrium uniformly in $1 \leq \tau \leq \infty$. This property is not restricted to the Lorentzian kernel in (2.21). The <u>singular interaction</u> limit $\tau \to \infty$ is particularly simple: α_t^{∞} is still an automorphism of $\mathcal{OL}$, derived from the 1st order stochastic differential equation

$$\dot{a}^{\infty}(t) = -(i\varepsilon + \sum \gamma_{\varsigma})\, a^{\infty}(t) - i\sum g_{\varsigma} A_{\varsigma}^{\infty}(t) \quad , \tag{2.23}$$

with fermion "white noise" $A_{\varsigma}^{\infty}(t) = \int d\omega e^{-i\omega t} A_{\varsigma}(\omega)$ as input. For $t < o$, the γ_{ς} change to $-\gamma_{\varsigma}$.

An <u>IOS</u> consists of N EOS with uncoupled identical reservoirs, $a_n^{\#}$, $A_{\varsigma n}^{\#}(\omega)$ $(1 \leq n \leq N$, $1 \leq \varsigma \leq r)$, and any nonlinear interaction $H_I^S = H_I^{S*} \in \mathcal{OL}^S$:

$$H^S = \sum_{n=1}^{N} \varepsilon_n\, a_n^* a_n + H_I^S$$

$$V^{\tau} = \sum_{n=1}^{N} \sum_{\varsigma=1}^{r} V_{\varsigma n}^{\tau}$$

where $V_{\varsigma n}^{\tau}$ has the form (2.21) with $g_{\varsigma n} = g_{\varsigma}$ n-independent. In the singular limit, this system is a Hamiltonian model for the following non-linear dissipative stochastic differential equation $(t \geq o)$

$$\dot{a}_n^{\infty}(t) = -(i\varepsilon + \sum \gamma_{\varsigma})\, a_n^{\infty}(t) - i\sum_{\varsigma} g_{\varsigma} A_{\varsigma n}^{\infty}(t) \tag{2.24}$$

$$+ i\left[H_I^S, a_n\right]^{\infty}(t) \quad , \quad a_n^{\infty}(o) = a_n \quad .$$

As energy flow we take

$$J_{\mathcal{S}}^{\tau}(t) = i\left[V_{\mathcal{S}}^{\tau}, H^S\right]^{\tau}(t) . \tag{2.25}$$

For $\tau < \infty, J_{\mathcal{S}}^{\tau}(t) \in \mathcal{O}L$, while $J_{\tau}^{\infty}(t)$ is an operator-valued distribution which in Theorem 2.4 has continuous expectation values in $\omega^R \otimes \omega^S$:

<u>Theorem 2.4</u>: In every EOS and in every IOS with small non-linearity, $\| H_i^S \| < C_N \Sigma \gamma_{\mathcal{S}}$, the reservoirs (R, α^R) and interaction V^{τ} drive (S, α^S) strongly to equilibrium and uniformly for $1 \leq \tau \leq \infty$. In general $\alpha_+ \neq \alpha_-$. If the reservoirs $R^{\mathcal{S}}$ are in KMS states with respect to $\alpha^{\mathcal{S}}$ with temperatures $\beta_{\mathcal{S}}$, then in the equilibrium state $\omega_{\pm}^{\tau}$ of the open driven system the Onsager relations (2.5) hold and $\omega_{\pm}^{\tau}$ is stable in the sense of (2.6).

One proves this proposition for an IOS by an analysis of the time dependent perturbation expansion, which converges uniformly for $-\infty \leq t \leq +\infty$, if $\| H_i^S \| < C_N \Sigma \gamma_{\mathcal{S}}$ holds [H9]. In the weak coupling limit one recovers Theorem 2.3. Again, instead of assuming time reversal invariance, one uses the KMS property of the reservoir states $\omega^{\beta_{\mathcal{S}}}$.

§ 3. <u>The Limit of ∞-Many Degrees of Freedom</u>

Nothing dramatic happens in the weak coupling limit, if the temperatures of the reservoirs are far apart. If the generator $P(\tau)$ of the positive transition semigroup (2.11) on ℓ is (quasi)compact for some $\tau > 0$, then there exist a projection Π with finite dimensional range and strictly positive constants α and γ , such that $\| P(\tau) - \Pi \| < \alpha \exp -\gamma\tau$. Then zero is an isolated point of $\sigma(K)$, and the remainder of $\sigma(K)$ lies in the half-plane $\{ \mathrm{Re}\, \lambda \leq -\gamma \}$ [W1] . In particular there exist <u>no oscillatory stationary states</u> for these continuous time Markov chains even far from thermal equilibrium.

Similarly, the non-linear Langevin equations of the type (2.24) for finitely many degrees of freedom should not lead to sharp "dissipative structures" [G1] even for strong nonlinearities. A general theory is lacking, but the example of the classical equation

$$\dot{z}(t) = (i\omega + \mu)\, z(t) - |z(t)|^2\, z(t) + f(t) \tag{3.1}$$

with $z \in \mathbb{C}$, $\omega, \mu \in \mathbb{R}$ and $f(t)$ Gaussian white noise, should be typical.

Without f one has a Hopf bifurcation [H10] at $\mu = o$ from a stationary to a periodic attractor, while the stochastic equation has a unique equilibrium state which is analytic in μ across $\mu = o$ [R1].

Hence, as in the case of closed systems, one has to pass to the thermodynamic limit in order to obtain phase transitions. In this section we study the limit $N \to \infty$ for IOS of the type (2.24). With the present technology, only mean field non-linearities can be analysed in a conclusive way, and this has been done by Lieb and myself in a series of papers [H4-6], [L2] , guided from the findings in quantum optics [H3].

Mean field interactions between EOS are introduced by grouping the local fermion degrees of freedom into a large number N of groups each having M elements. The fermions with a_n^m , $a_{n'}^{m'}$ ($| \le m,m' \le M$, $| \le n,n' \le N$) interact for fixed m,m' with the same strength, irrespective of whether n and n' are close together or far apart. Such non-linearities occur in qualitatively successful models for lasers [H3] and superconductors [B1] and lead to H^S of the type

$$H^S_{(N)} = N \, P(\underline{\sigma}_{(N)}) \, , \qquad (3.2)$$

where P is a hermitean polynomial in the <u>intensive observables</u>

$$\underline{\sigma}_{(N)} = (\sigma^{ij}_{(N)} = N^{-1} \sum_{n=1}^{N} a_n^{i*} a_n^{j}) \, . \qquad (3.3)$$

It is helpful, but not necessary, to restrict oneself to Hamiltonians of the type (3.2). The advantage is algebraic: the $\sigma^{ij}_{(N)}$ are closed under commutation. Hence under the combined action of $H^S_{(N)}$ and $H^R_{(N)}$ (see (3.5)) one obtains the Heisenberg equations of motion for $t \ge o$:

$$\dot{\sigma}^{ij}_{(N)} (t) = - \gamma^{ij}(\sigma^{ij}_{(N)} (t) - \eta^{ij}) + i[H^S_{(N)}, \sigma^{ij}_{(N)}](t) + \psi^{ij}_{(N)}(t) \, . \qquad (3.4)$$

Hence $i[H^S_{(N)}, \sigma^{ij}_{(N)}] = Q^{ij}(\underline{\sigma}_{(N)})$ is again a polynomial in the components of $\underline{\sigma}_N = \underline{\sigma}_N(0)$. We have taken singular reservoirs and have suppressed $\tau = \infty$. For finite τ one needs a larger set of intensive observables for closure [H6]. We further choose ω^R to be a Fock state and double the number of reservoirs in order to obtain sufficiently many parameters:

$$H^R_{(N)} = \sum_{m,n} \int_{-\infty}^{+\infty} d\omega \, \omega (A_n^m(\omega)^* A_n^m(\omega) + B_n^m(\omega)^* B_n^m(\omega))$$

$$+ \sum_{m,n} \int_{-\infty}^{+\infty} d\omega \{ f^m A_n^m(\omega)^* a_n^m + g^m B_n^m(\omega)^* a_n^m + h.c. \} \qquad (3.5)$$

147

$$A_n^m(\omega)\,\Omega^R = B_n^m(\omega)^*\,\Omega^R = o \quad . \tag{3.6}$$

Then the damping and pumping constants in (3.4) become

$$\gamma^{ij} = \gamma^i + \gamma^j, \ \pi(|f^i|^2 + |g^i|^2) = \gamma^i \ , \tag{3.7}$$

$$\eta^{ij} = \delta^{ij}\,|g^i|^2\,(|f^i|^2 + |g^i|^2)^{-1}$$

The remarkable feature of (3.4) is that the influence of the reservoir in the Heisenberg equation of motion for the intensive observables splits into a friction and a pumping term and a <u>fluctuation force</u>

$$\varphi^{ij}_{(N)} = \varphi^{ij}_{+(N)} + \varphi^{ij}_{-(N)} \ , \ \varphi^{ij\,*}_{-(N)} = \varphi^{ji}_{+(N)} \tag{3.8}$$

$$\varphi^{ij}_{+(N)}(t) = \frac{i}{N} \sum_{n=1}^{N} \{ f^i A_n^i(t)^* a_n^i(t) + g^{j*} B_n^j(t)\, a_n^i(t)^* \} \ . \tag{3.9}$$

Due to (3.9) one has, quite independent of the solutions $a_n^m(t)$ of the microscopic system observables, for all $t \geq o$

$$\varphi^{ij}_{-(N)}(t)\,\Omega_R = \varphi^{ji}_{+(N)}(t)^*\,\Omega_R = o \ , \tag{3.10}$$

Furthermore, the $\varphi^{ij}_{+(N)}(t)$ are averages over the contributions of the individual degrees of freedom. At $t = o$ they are all independent, and the $A_n^m(t)^{\#}$, $B_n^m(t)^{\#}$ anticommute for all times. If the $a_n^m(t)^{\#}$ are only weakly dependent due to their mean field interaction, then one expects some kind of "law of large number" to make $\varphi^{ij}_{(N)}(t)$ vanish for $N \to \infty$. The $\sigma^{ij}_{(N)}(t)$ should become c-numbers, because of $[\sigma^{ij}_{(N)}(t) , \sigma^{kl}_{(N)}(t)] = N^{-1}\{ \delta^{jk} \sigma^{il}_{(N)}(t) - \delta^{il} \sigma^{kj}_{(N)}(t) \}$, and this is in fact true in sufficiently homogeneous states $\omega^S_{(N)}$, where the $\sigma^{ij}_{(N)}$ converge weakly for $N \to \infty$. $\omega^S_{(N)}$ is called <u>classical</u> w.r.t. $\sigma_{(N)}$, if there exists a probability measure μ on phase space

$$P = \{ \sigma \in \mathbb{C}^{M^2} \mid \sigma^{ij} = \sigma^{ji*} , \ \sum |\sigma^{ij}|^2 \leq M \} \tag{3.11}$$

such that for all monomials

$$\lim_{N \to \infty} \omega^S_{(N)}(\sigma^{ij}_{(N)} \cdots \sigma^{pq}_{(N)}) = \int \mu(d\sigma)\, \sigma^{ij} \cdots \sigma^{pq} \ . \tag{3.12}$$

<u>Theorem 3.1:</u> If $\omega^S_{(N)}$ is classical and $\omega^R_{(N)}$ satisfies (3.6), then for all monomials and positive times

$$\lim_{N \to \infty} \omega^R_{(N)} \otimes \omega^S_{(N)} \; (\sigma^{ij}_{(N)}(r) \ldots \sigma^{pq}_{(N)}(t))$$

$$= \int \mu(d\underline{\alpha}) \sigma^{ij}(r,\underline{\alpha}) \ldots \sigma^{pq}(t,\underline{\alpha}) \; , \tag{3.13}$$

where $\underline{\sigma}(t,\underline{\alpha})$ is the unique solution of

$$\dot{\sigma}^{ij}(t,\underline{\alpha}) = -\gamma^{ij} (\sigma^{ij}(t,\underline{\alpha}) - \eta^{ij}) + Q^{ij}(\underline{\sigma}(t,\underline{\alpha})), \tag{3.14}$$

$$\sigma^{ij}(0,\underline{\alpha}) = \alpha^{ij} \; .$$

The proof of this theorem as well as finer results on the fluctuations of the quantum observables $\sigma^{ij}_{(N)}(t)$ around the classical orbit $\sigma^{ij}(t)$ can be found in [H4]. In [H5] a different topology is used to deal with the unbounded boson observables, which occur in the laser. KMS reservoirs are studied in [D1].

One sees that in the limit $N \to \infty$ open systems built up from EOS with a strong mean field non-linearity can show very dramatic non-equilibrium phase transitions, which manifest themselves in the non-analytic change of the attractors of the corresponding classical "mean motion" equations when varying the parameters η^{ij} and γ^{ij} of the reservoirs. In the 1-mode laser the threshold can be explained by a Hopf bifurcation, which is a second-order phase transition from a time-independent to a periodic attractor [H4]. There exists a model for a Josephson oscillator with a first order non-equilibrium phase transition [H7]. In this framework we expect most of the "catastrophies" of classical dynamical systems as non-equilibrium phase transitions.

The next difficult and unsolved problem is the investigation of the thermodynamic limit of EOS with a local nonlinear interaction. The physically most interesting example is macroscopic quantum electrodynamics, e.g. the interaction of a continuum of modes of the radiation field with localized "pumped" M-level atoms. The semiclassical theory, where the electromagnetic field is treated classically, shows interesting nonlinear wave propagation phenomena [C2]. W. Braun and myself [B3] have studied the model of a nerve membrane built from EOS ("pores" for Na- and K-conductivity) and interacting through a classical potential. The resulting non-linear diffusion equation has a propagating pulse solution [C1] under conditions which are typical in real life (necessity of two ion species with different time constants and opposite concentration gradients).

§ 4.) <u>Conclusions</u>

The beauty of our approach to irreversible statistical mechanics
lies in the possibility of building microscopic models for dynamical
systems with considerable structure. One of the many unsatisfactory
aspects of the present state of the art is the very special class of
quasifree reservoirs and interactions (weak coupling limit or IOS),
for which a drive to equilibrium can be established. The validity of
many conclusions about the nature of non-equilibrium phase transitions
will remain doubtful, as long as non-markovian reservoirs and local
interactions cannot be treated rigorously. Open systems close to thermal equilibrium, however, might soon be understood from first principles, and the Onsager relations for heat flows be related to the
intrinsic stability properties of the KMS state [H2].

<u>References</u>

[A1] H. Araki, Ann. Sci.École Norm.Sup. <u>6</u>, 67(1973)

[B1] J. Bardeen, L.N. Cooper, J.R. Schrieffer, Phys.Rev. <u>108</u>, 1175(1957)
W. Thirring, A. Wehrl, Comm.Math.Phys. <u>4</u>, 303 (1967)

[B2] P.G. Bergmann, J.L. Lebowitz, Phys.Rev. <u>99</u>, 578 (1955)
J.L. Lebowitz, A. Shimony, Phys.Rev. <u>128</u>, 1945 (1962)
C.R. Willis, P.G. Bergmann, Phys.Rev. <u>128</u>, 391 (1962)

[B3] W. Braun, K. Hepp (unpublished)

[C1] G. Carpenter, Thesis, U. of Wisconsin, Madison (1974)

[C2] E. Courtens, in "Laser Handbook, Vol. II" (F.T. Arecchi,
E.O. Schulz-Dubois eds.), North Holland (1972);
R.E. Slusher in "Prog. Optics XII" (E. Wolf ed.) North Holland
(1974)

[D1] W. Daeppen, Diplomarbeit, ETH, Zürich (1975)

[D2] E.B. Davies, Comm.Math.Phys. <u>39</u>, 91 (1975)

[F1] K.O. Friedrichs, Comm.Pure Appl.Math. <u>1</u>, 361 (1948)

[G1] P. Glansdorff, I. Prigogine,"Thermodynamic Theory of Structure,
Stability and Fluctuations", Wiley, N.Y. (1971)

[G2] R. Graham, in "Prog.Optics XII" (E. Wolf ed.), North Holland (1974)

[H1] R. Haag, N.M. Hugenholtz, M. Winnink, Comm.Math.Phys. <u>5</u>, 215(1967)

[H2] R. Haag, D. Kastler, E.B.Trych-Pohlmeyer, Comm.Math.Phys. <u>38</u>, 173
(1974)

[H3] H. Haken, "Handbuch der Physik XXV 2c", Springer, Berlin (1957)

[H4] K. Hepp, E.H. Lieb, Helv.Phys. Acta <u>46</u>, 573 (1973)

[H5] K. Hepp, E.H. Lieb, in "Constructive Quantum Field Theory"
     (G. Velo, A.S. Wigthman eds.), Springer, Berlin (1973)

[H6] K. Hepp, E.H. Lieb, in "Battelle Rencontres 1974" (J.L. Lebowitz,
     J.K. Moser eds.), Springer, Berlin (1975)

[H7] K. Hepp, Ann.Phys. <u>90</u> (1975)

[H8] K. Hepp, Z. Phys. B (to appear)

[H9] K. Hepp, unpublished

[H10] E. Hopf, Ber.math.phys. Kl. Sächs.Akad.Wiss.Leipzig <u>94</u>, 3 (1942)

[H11] L. van Hove, Physica <u>21</u>, 517 (1955); <u>23</u>, 441 (1957)

[L1] O.E. Lanford, in "Battelle Rencontres 1974" (J.L. Lebowitz,
     J.K. Moser eds.), Springer, Berlin (1975)

[L2] E.H. Lieb, Physica <u>73</u>, 226 (1974)

[O1] L. Onsager, Phys.Rev. <u>37</u>, 405; <u>38</u>, 2265 (1931)

[P1] J. Pulè, Comm.Math.Phys. <u>38</u>, 241 (1974)

[R1] H. Risken in "Prog.Optics IIX" (E. Wolf ed.), North Holland (1970)

[R2] D.W. Robinson, Comm.Math.Phys. <u>31</u>, 171 (1973); in "C^*-Algebras and
     Quantum Statistical Mechanics", Varenna Lectures (1973)

[R3] D. Ruelle, J. Math. Phys. <u>6</u>, 201 (1965)
     O.E. Lanford, in "Battelle Rencontres 1971" (A. Lenard ed.),
     Springer, Berlin (1973)

[S1] Y. Sinai, Usp. Math. Nauk. <u>27</u>, 137 (1972)
     G. Gallavotti, D. Ornstein, Comm.Math.Phys. <u>38</u>, 83 (1974)

[W1] D. Williams, "Proc. 5th Berkeley Symp. on Math. Statistics and
     Probability" II, 187 (1966)

THE WIGHTMAN AXIOMS AND THE MASS GAP FOR WEAKLY COUPLED $(\Phi^4)_3$ QUANTUM FIELD THEORIES

Joel S. Feldman[+]
Konrad Osterwalder[+][*]

Departments of Mathematics and Physics
Harvard University
Cambridge, Massachusetts 02138, U.S.A.

Abstract

The weakly coupled $\lambda \Phi^4$ quantum field theory models in three space-time dimensions exist and satisfy all the Wightman axioms. The mass operator $M = (H^2 - \vec{P}^2)^{1/2}$ has an isolated simple eigenvalue at zero and no spectrum in the interval $(0, m)$ for some $m > 0$.

I. Introduction

The topic of this talk is the construction of quantum field theory models with a quartic boson selfinteraction in three dimensional space-time. These models are superrenormalizable, but they require a mass renormalization whose infinite part is logarithmically divergent in the ultraviolet cutoff and of second order in the coupling constant λ. The $(\lambda \Phi^4)_3$ models are much more singular than the $P(\Phi)_2$ models. Therefore the construction requires more work.

These models were first studied by Glimm [1] who constructed the ultraviolet limit

$$(1) \qquad H(\Lambda) = \lim_{\kappa \to \infty} H(\Lambda, \kappa)$$

[+] Supported in part by the National Science Foundation under Grant MPS 73-0537.

[*] A. Sloan Foundation Fellow.

of a sequence $\{H(\Lambda, \kappa)\}$ of Hamiltonians with fixed space cutoff Λ and momentum cutoff κ. The limit (1) has to be defined in a weak sense; while all the $H(\Lambda, \kappa)$ act on the Fock space $\mathfrak{F}_0$ of the free field, $H(\Lambda)$ is defined on a new space $\mathfrak{F}_\Lambda$ which is disjoint from $\mathfrak{F}_0$. In a natural way $\mathfrak{F}_\Lambda$ becomes the carrier space of a <u>non Fock representation</u> of the CCR or of the Weyl relations, which was studied first by Hepp [2] and Fabrey [3] and subsequently also by Eckmann, Osterwalder [4], Eckmann [5] and again by Fabrey [6].

The next step towards the construction of a $(\Phi^4)_3$ model without cutoffs was due to Glimm, Jaffe [7], who showed that the Hamiltonian $H(\Lambda)$ is bounded from below by a constant proportional to the volume Λ.

The success of Euclidean methods in the construction of $P(\Phi)_2$ models (and in the semiboundedness proof of Glimm, Jaffe [7]) suggested that for (Φ^4_3) too, it might be easier to construct the model through its Euclidean Green's functions.

A first step in that direction was taken by Feldman [8]. He used the methods and results of Glimm, Jaffe [7] to show that the ultraviolet limit $\kappa \to \infty$ of the space and momentum cutoff unnormalized Euclidean Green's functions $Z_{\Lambda, \kappa} S^{(n)}_{\Lambda, \kappa}$ exists, is C^∞ in the coupling constant λ and satisfies

$$\left| Z_\Lambda S^{(n)}_\Lambda (f_1 x \ldots x f_n) \right| \equiv \left| \int Z_\Lambda S^{(n)}_\Lambda (x_1 \ldots x_n) \prod f(x_i) dx_i \right|$$

$$\leq n! \prod |f_i|_s \cdot e^{O(1)|\Lambda|}$$

for some Schwartz norm $|\cdot|_s$. He also exhibited $Z_\Lambda S^{(n)}_\Lambda$ as the n'th moment of a unique measure on $\mathscr{S}'_R(R^3)$.

Park [9] showed that the lattice approximation of the Euclidean Green's functions converges as the lattice spacing tends to 0.

The main purpose of this talk is to report on the completion of the construction of the $(\lambda \Phi^4)_3$ model, for λ/m_0 sufficiently small (m_0 is the mass of the <u>free</u> Euclidean field in terms of which we define the cutoff Euclidean Green's functions). See Feldman, Osterwalder [10].

We summarize our principal results in four theorems:

<u>Theorem I</u>

For λ/m_0 sufficiently small, the normalized finite volume Euclidean Green's functions $S^{(n)}_\Lambda$ have a limit as $\Lambda \to R^3$. They are the Euclidean Green's functions of a uniquely determined Wightman quantum field theory.

Theorem II

This Wightman theory has a nonzero mass gap m. There is no mass spectrum in $(0, 2m)$ on the subspace of vectors that are even under the transformation $\varphi \rightarrow -\varphi$.

Theorem III

The Euclidean Green's functions are C^∞ in λ.

Perturbation theory provides an asymptotic expansion for the exact Green's functions.

Theorem IV

The Euclidean Green's functions are the moments of a (unique) probability measure on $\mathscr{S}'_R(R^3)$.

Remark: The corresponding results for $P(\Phi)_2$ models were first proven by Glimm, Jaffe, Spencer [11], (Theorem I, II), Spencer [12], (Theorem II), Dimock [13], (Theorem III), Fröhlich [14], (Theorem IV).

II.   The Euclidean approach

In this chapter we briefly describe the idea of constructing a relativistic quantum field model through its Euclidean Green's functions.

The Euclidean Green's (or Schwinger) functions are formally given by

$$S^{(n)}(x_1 \ldots x_n) = (\Omega, T\left[\prod_{i=1}^{n} e^{-x_i^0 H} \varphi(0, \vec{x}_i) e^{x_i^0 H} \right] \Omega),$$

(where T means time ordering) or, in terms of a functional integral, by

$$(2) \qquad S^{(n)}(x_1 \ldots x_n) = \frac{\int \Phi(x_1) \cdot \ldots \cdot \Phi(x_n) e^{-V(\Phi)} d\mu_0}{\int e^{-V(\Phi)} d\mu_0} .$$

Here $V(\Phi) = \lambda \int :\Phi^4(x): d^3x + \frac{1}{2} \delta m^2 \int :\Phi^2(x): dx$ is the renormalized __Euclidean action__ and $d\mu_0 \ "=" \ e^{-(\Phi, (-\Delta+m_0^2)\Phi)}[d\Phi]$ is the free measure, i.e. the Gaussian measure on $\mathscr{S}'_R(R^3)$ with mean zero and covariance $(-\Delta+m_0^2)^{-1}$.

More precisely we want to find Euclidean Green's functions $S^{(n)}$ which have the following properties.

(E0) $S^{(n)} \in \mathscr{S}'(R^{3 \cdot n})$ and for $f_1, \ldots, f_n \in \mathscr{S}(R^3)$

$$|S^{(n)}(f_1 \times f_2 \ \ldots \ f_n)| \leq a(n!)^\beta \prod_{i=1}^{n} |f_i|_s$$

for some a, β and some Schwartz norm $|\cdot|_s$.

(E1)   Euclidean invariance

(E2)   Positivity

(E3)   Symmetry

(E4)   (Exponential) clustering

For precise formulations of E1-E4 see [15].   Then the <u>reconstruction</u> <u>theorem</u> of Osterwalder, Schrader [15, 16] gives, as a consequence of E0-E4 the existence of a unique Wightman quantum field theory having the $S^{(n)}$ as its Euclidean Green's functions.   If the clustering of $S^{(n)}$ is exponential, then the Wightman quantum field theory has a nonzero mass gap.

<u>Remark</u>:   Postulate E0 could be relaxed considerably, (see Osterwalder, Schrader [16]) but that is not necessary for our present purposes.

It is clear how we will proceed to obtain the Euclidean Green's functions for the $(\lambda \Phi^4)_3$ models.   We first introduce a volume cutoff Λ and a momentum cutoff κ into the formal expression (2) for $S^{(n)}$.

$$S^{(n)}_{\Lambda, \kappa} = \frac{\int \Phi(x_1) \ldots \Phi(x_n) \, e^{-V_{\Lambda, \kappa}(\Phi)} \, d\mu_0}{\int e^{-V_{\Lambda, \kappa}(\Phi)} \, d\mu_0}$$

are well defined distributions.

Secondly we prove that the limits $\kappa \to \infty$, $\Lambda \to R^3$ exist and have properties E0-E4.   Euclidean invariance (E1) follows if the limit is independent of how the volume Λ tends to R^3.   All the other requirements should be established for the cutoff $S^{(n)}_{\Lambda, \kappa}$ uniformly in Λ and κ.

The $S^{(n)}_{\Lambda, \kappa}$ are obviously symmetric (E3).   They satisfy positivity (E2) for suitable forms of the cutoffs.   This follows from the Feynman-Kac formula.

This concludes the easy part of the argument.   In order to establish the distribution and cluster properties (E0) and (E4) and the convergence of the $S^{(n)}_{\Lambda, \kappa}$ we need estimates.   In the next section we outline how one gets them.

III.   Estimates: the main tools

We combine two types of expansion - the cluster expansion, (see Glimm, Jaffe, Spencer [11]), to prove exponential clustering and convergence of the infinite volume limit, and - the phase space cell expansion, (see Glimm, Jaffe [7]), to control the ultraviolet limit.

The cluster expansion was first used to establish exponential clustering for weakly coupled $P(\Phi)_2$ models, [11].   The idea is to introduce a lattice structure in R^3, e.g. Z^3, and to replace the covariance $(-\Delta + m_0^2)^{-1}$ in the Gaussian measure $d\mu_0$ by $(-\Delta_\Gamma + m_0^2)^{-1}$, where Γ is a set of unit squares whose corners are at lattice points in Z^3 and Δ_Γ is the Laplace operator with zero Dirichlet data on Γ.   We thus obtain a new Gaussian measure $d\mu_\Gamma$ which decouples across closed surfaces Γ.   If Γ contained all possible surfaces, then $d\mu_\Gamma$ would factor into a product measure over unit cubes.   Clustering of the Euclidean Green's functions would be complete and all estimates could be performed in a finite volume.   The art of the game is of course to introduce the boundary conditions on one unit square at a time and to control the error terms we obtain from changing the covariance.   The crucial formula is

$$\int F(\Phi, C_1) d\mu_{C_1} - \int F(\Phi, C_0) d\mu_{C_0}$$

$$= \int_0^1 ds \frac{d}{ds} \int F(\Phi, C_s) d\mu_{C_s}$$

$$= \int_0^1 ds \left\{ \frac{1}{2} (C_1(x, y) - C_0(x, y)) \frac{\delta^2}{\delta \Phi(x) \, \delta \Phi(y)} F(\Phi, C_s) \right.$$

$$\left. + \frac{d}{ds} F(\Phi, C_s) \right\} d\mu_{C_s} .$$

Here $d\mu_{C_s}$ is the Gaussian measure with covariance $C_s = sC_1 + (1 - s)C_0$ and F is a function of Φ and C_s.   Typically F is a polynomial in Φ, Wick ordered with respect to the covariance C_s.

For this method to work it is crucial that

a) the covariance $(-\Delta + m_0^2)^{-1}$ decouples exponentially (i.e.
$(-\Delta + m_0^2)^{-1}(x, y) \sim e^{-m_0|x-y|}$ for $|x-y|$ large) and

b) the Euclidean action $V(\Phi)$ is local, in the sense that $e^{-V(\Phi)} d\mu_\Gamma$ factors across the same surfaces as $d\mu_\Gamma$.

This is where the problems start in $(\Phi^4)_3$. We would like to do all our manipulations with the doubly cutoff Euclidean Green's functions $S_{\Lambda, \kappa}^{(n)}$. However the ordinary methods of cutting off the high momenta destroys both the local character of the action and the exponential decoupling of the covariance. Also, the kernel of the propagator $(-\Delta_\Gamma + m^2)^{-1}$ does not have a simple Fourier transform, so that the momentum space methods that have till now dominated the phase space cell expansion become unmanageable.

Our method is to use a cutoff which simultaneously suppresses high momenta, preserves locality of the interaction and exponential decay of the propagators and allows for the introduction of boundary conditions.

To this end we define a free field $\psi(t, x)$, $t \in (0, \infty)$, $x \in R^3$, whose two point function is given by

$$\langle \psi(t, x) \, \psi(s, y) \rangle_\Gamma = \delta(t - s) \, G_\Gamma(t, x, y)$$

where

$$G_\Gamma(t, x, y) = e^{-m_0^2 t} \int P_{xy}^t(d\omega) \, \chi_\Gamma(\omega).$$

Here $P_{xy}^t(d\omega)$ is the conditional Wiener measure on the set of all paths starting at x at time 0 and arriving at y at time t. The function χ_Γ is used to introduce boundary conditions in the cluster expansion and is always between 0 and 1. Typically for $\Gamma \subset R^3$,

$$\chi_\Gamma(\omega) = \begin{cases} 0 & \text{if } \omega(s) \in \Gamma \text{ for some } 0 \le s \le t \\ 1 & \text{otherwise,} \end{cases}$$

so that $G_\Gamma(x, y, t) = 0$ whenever x and y are on different sides of a closed surface Γ. This says that the Gaussian measure belonging to the field $\psi(t, x)$ factors across closed surfaces with zero Dirichlet boundary conditions.

We now define the cutoff fields by

$$\Phi(r, x) = \int_r^\infty \psi(t, x) dt$$

and the cutoff Euclidean action by

$$V_{\Lambda, t} = \int_{\Lambda} : \Phi(t, x)^4 : d^3x + \text{counter terms}.$$

To demonstrate that the t-cutoff does indeed suppress high momenta we evaluate the two point function (assume $r \le s$)

$$\langle \Phi(r, x) \, \Phi(s, y) \rangle_{\Gamma = \{\emptyset\}} = \int_s^\infty e^{-tm_0^2} \int P_{xy}^t \, (d\omega) dt$$

$$= \left(e^{-s(-\Delta + m_0^2)} (-\Delta + m_0^2)^{-1} \right) (x, y)$$

$$= \int \frac{d^3 p}{(2\pi)^3} \, e^{ip(x-y)} \, \frac{e^{-s(p^2 + m_0^2)}}{p^2 + m_0^2} \, .$$

Momenta $p^2 \gg \frac{1}{s}$ are essentially suppressed.

Now with these cutoffs we can use the cluster expansion to reduce estimates on

$$S_{\Lambda, t}^{(n)} (x_1 \ldots x_n) = \frac{\int \Phi(x_1) \, \ldots \, \Phi(x_n) \, e^{-V_{\Lambda, t}(\Phi)} \, d\mu_0}{\int e^{-V_{\Lambda, t}(\Phi)} \, d\mu_0}$$

to estimates on terms of the form

$$(3) \qquad \int F(\Phi, C_s) \, e^{-V_{\Lambda, t}} \, d\mu_{C_s} \, .$$

Estimates on expressions of this form are obtained through another expansion technique due to Glimm and Jaffe [7]: the <u>phase space cell expansion.</u>

The principal barrier to arriving at estimates on (3), uniform in t, is the presence of the exponential e^{-V}. Expanding it into a power series $\sum \frac{(-V)^n}{n!}$ just leads to the ordinary Feynman diagram expansion which is known to diverge. Instead we split phase space up in cells and write

$$V_{\Lambda, t} = \sum V_{\Lambda_i, T_1, \ldots T_4} + \text{other terms}$$

where

$$V_{\Lambda_i, T_1 \ldots T_4} = \int_{\Lambda_i} dx \int_{t_i \in T_i} : \prod_{i=1}^4 dt_i \, \psi(t_i, x): \, .$$

Here $\{\Lambda_i\}$ is a cover of the volume Λ and $\{T_j\}$ is a cover of the interval $[t, \infty)$.

Now we treat each phase cell $[\Lambda_i, T_j]$ separately. Notice that exponential clustering makes different cells nearly independent. If the t variables are restricted to high values, say larger than s, then we are dealing with a low momentum contribution and we can use the formal positivity of $:\Phi^4:$ to get

$$V_{\Lambda_i, s} \geq 0(1) \quad \text{and} \quad e^{-V_{\Lambda_i, s}} \leq 0(1)$$

if $\qquad s^{-1} \cdot \text{volume}(\Lambda_i) \leq 0(1).$

For small values of t, i.e. large momenta, we may write

$$e^{-V\cdots} \quad \text{as} \quad \sum_1^N \frac{(-V\cdots)^n}{n!} + \text{remainder.}$$

After we explicitly cancel renormalization terms the Feynman diagrams generated by this expansion give small contributions because they contain many lines that carry high momenta (small t's) only. Nevertheless, the expansion of $e^{-V\cdots}$ has to be truncated at some large N, because the number of diagrams grows like $(N!)^2$ and the high momenta only yield convergence factors proportional to $(t^\epsilon)^N$.

The high and low momentum steps have to be applied separately and the whole phase space cell expansion is defined inductively only, see [7] and [10] for details.

The combination of the cluster expansion and the phase space cell expansion leads to the estimates which are sufficient to establish convergence of the cutoff Euclidean Green's functions and to verify (E0) and (E4).

IV. <u>Outlook</u>

Our discussion has shown that many of the techniques developed in the study of $P(\Phi)_2$ models can be used for $(\Phi^4)_3$ as well. We conjecture that virtually all the results known for weakly coupled $P(\Phi)_2$ can be established for $(\Phi^4)_3$ by our methods. In particular we may ask:

- is there an upper mass gap and an n-particle cluster expansion?

- are there bound states?

- is there a critical point and a multiple phase region?

or more technically

- are correlation inequalities as useful as in 2 dimensions?

- is there a φ-bound? Is the field $\varphi(x)$ essentially self-adjoint?

There are other questions which have a trivial answer in $P(\Phi)_2$ models but which might be interesting to study in $(\Phi^4)_3$. For example $P(\Phi)_2$ models are known to be locally Fock ; $(\Phi^4)_3$ is not and that makes the analysis of the representation of the CCR and of the algebras of local observables an interesting topic.

The most challenging problem is left as an

<u>Exercise</u>: Generalize the construction described above to four space-time dimensions.

REFERENCES

1. J. Glimm, Commun. Math. Phys. <u>10</u>, 1 (1968).

2. K. Hepp, <u>Théorie de la Renormalization</u>, Lecture Notes in Physics, Vol. 2, Springer Verlag, Berlin 1969.

3. J. Fabrey, Commun. Math. Phys. <u>19</u>, 1 (1970).

4. J.-P. Eckmann, K. Osterwalder, Helv. Phys. Acta <u>44</u>, 864 (1971).

5. J.-P. Eckmann, Commun. Math. Phys. <u>25</u>, 1 (1972).

6. J. Fabrey, Jour. Math. Phys. <u>14</u>, 909 (1973).

7. J. Glimm, A. Jaffe, Fort. d. Physik <u>21</u>, 327 (1973).

8. J. Feldman, Commun. Math. Phys. <u>37</u>, 93 (1974).

9. Y. Park, Lattice Approximation of the $(\lambda\Phi^4-\mu\Phi)_3$ Field Theory in a Finite Volume, I.B.M. Preprint, and The $\lambda\Phi^4_3$ Euclidean Quantum Field Theory in a Periodic Box, Yonsei University preprint.

10. J. Feldman, K. Osterwalder, The Wightman axioms and the mass gap for weakly coupled $(\Phi^4)_3$ quantum field theories, Harvard preprint

11. J. Glimm, A. Jaffe, T. Spencer, Annals of Math. <u>100</u>, 585 (1974), and in G. Velo, A.S. Wightman (eds.), <u>Constructive Quantum Field Theory</u>, Lecture Notes in Physics, Vol. 25, Springer Verlag, Berlin 1973.

12. T. Spencer, Commun. Math. Phys. <u>39</u>, 77 (1974).

13. J. Dimock, Commun. Math. Phys. <u>35</u>, 347 (1974) and The $P(\Phi)_2$ Green's functions: smoothness in the coupling constant, preprint.

14. J. Fröhlich, Helv. Phys. Acta <u>47</u>, 265 (1974).

15. K. Osterwalder, R. Schrader, Commun. Math. Phys. <u>31</u>, 83 (1973).

16. K. Osterwalder, R. Schrader, Axioms for Euclidean Green's functions II, preprint, to appear in Commun. Math. Phys.

SOLUBLE MODELS AND THE MEANING OF NONRENORMALIZABILITY

John R. Klauder
Bell Laboratories
Murray Hill, New Jersey 07974

1. INTRODUCTION

At the present time there exists no satisfactory formulation
of the so-called nonrenormalizable quantum field theories, formally or
otherwise. Although the need for an infinite number of counterterms is
suggested by perturbation theory, this conclusion is often justified by
arguments based on the dimensionality of the coupling constant. Guided
by the solution of certain idealized models, we believe that the present
unsatisfactory situation can be understood and that an alternative picture suggests itself in which nonrenormalizable fields may effectively
become "renormalizable".[1] This alternative picture differs fundamen-
tally from the conventional one. Quite simply, in the language of
functional integration over fields, nonrenormalizable interactions act,
in part, formally like hard cores, so that once introduced their effects
can never be completely turned off, but instead as the coupling constant
tends to zero, the effects of the hard core invariably remain and give
rise to a (zero-coupling) pseudofree theory, which includes the hard
core, and which is generally very different than a conventional free
theory. If an interacting theory is expanded about the pseudofree
theory, rather than the free theory, then conceivably a better-behaved
perturbation expansion may arise. As will be indicated, certain soluble
models confirm this general view.

We begin with some heuristic considerations of covariant field
theory. These are followed by a variety of idealized models the solutions of which exhibit the effect in question. Finally, we return to
the covariant case to argue the plausibility of the effect in such cases.

Heuristic Remarks

Consider the functional integral formulation of a scalar field
as described in an n-dimensional Euclidean space time. Formally, the
generating functional of interest for a typical nonlinear interaction
reads

$$\mathfrak{n} \int e^{i \int h\Phi dx - \int \{ \frac{1}{2} [(\nabla\Phi)^2 + m^2\Phi^2] + \lambda\Phi^p \} dx} \, \mathcal{D}\Phi \, , \qquad (1)$$

where h(x) is a test function, $\Phi(x)$ a random field, $\mathfrak{n}$ a normalization
factor, and where $dx \equiv d^n x$, $\lambda > 0$, and p is even. Attempts to give
meaning to this expression generally involve cutoffs, normal ordering,

renormalization counterterms, etc. Some improvement, at least in form, comes from grouping the quadratic terms into a normal probability measure μ_F with covariance

$$\langle \Phi(x)\Phi(y)\rangle_F = (2\pi)^{-n} \int \frac{e^{ik(x-y)}}{k^2+m^2}\, dk \tag{2}$$

characteristic of the free (F) theory. Equation (1) then becomes

$$n' \int e^{i\int h\Phi dx - \lambda\int \Phi^p dx}\, d\mu_F(\Phi) \tag{3}$$

which still generally requires cutoffs, normal ordering, renormalization counterterms, limits, etc., for any kind of definition. The goal of these endeavors is to determine a proper probability measure μ incorporating the nonlinear interaction, and so that (1) or (3) becomes

$$\int e^{i\int h\Phi dx}\, d\mu(\Phi) \quad . \tag{4}$$

Consider, now, the dependence of the interacting measure μ on λ. We speak of a continuous perturbation whenever

$$\operatorname*{w-lim}_{\lambda\downarrow 0} \mu = \mu_F \quad , \tag{5a}$$

and of a discontinuous perturbation whenever

$$\operatorname*{w-lim}_{\lambda\downarrow 0} \mu = \mu_{PF} \neq \mu_F \quad , \tag{5b}$$

where the "pseudofree" (PF) measure μ_{PF} is generally orthogonal to μ_F. We believe that (5a) applies to renormalizable interactions while (5b) applies to nonrenormalizable interactions,[1] a point of view we shall attempt to make plausible. That the behavior in (5b) is at all possible may be heuristically understood by attributing the formal behavior of a hard core to the interaction which thus has the effect, after once being introduced, of creating a permanent change in the system.

2. EXAMPLES OF DISCONTINUOUS PERTURBATIONS

Perturbed Free Particle

Differential Equation Formulation[1,2]

The phenomena described above arises in simple quantum

mechanical problems as well. It is not difficult to show that the
differential operator

$$H = -\tfrac{1}{2} \frac{d^2}{dx^2} + \frac{\lambda}{|x-c|^\alpha} \tag{6}$$

is essentially self adjoint for $\alpha > 2$ while for $\alpha < 2$ or $\alpha = 2$ (and
$\lambda < 3/8$) this is not the case. In the range $\alpha \leq 2$, self-adjoint extensions may be chosen so that $H \to H_F = -\tfrac{1}{2}d^2/dx^2$, but for $\alpha > 2$,

$$H \to H_{PF} = -\tfrac{1}{2} \left.\frac{d^2}{dx^2}\right|_{B.C.} \tag{7}$$

where the boundary conditions (B.C.) require that all $\psi \in \mathcal{D}(H_{PF})$
satisfy $\psi(c) = 0$.

Path Integral Formulation[3]

Feynman-Kac techniques help transcribe these results to a
path space formulation. The interacting measure, related to the usual
Wiener measure μ_W by

$$d\nu(x) \equiv e^{-\lambda\int_o^T |x(t)-c|^{-\alpha}dt} \, d\mu_W(x) \;\;, \tag{8}$$

has the property for $\alpha > 2$ that

$$\underset{\lambda \downarrow 0}{\text{w-lim}} \; \nu = \mu'_{W,c} \tag{9}$$

the measure appropriate to an absorbing Wiener process defined as a
Wiener process in which all paths that reach $x = c$ are thenceforth neglected. The hard core picture is clearly applicable here.

For $1 \leq \alpha \leq 2$ it is possible to carefully choose "counterterms"
at $x = c$ leading to a measure that weakly converges to μ_W. In particular,
it suffices to define ν as the $\varepsilon \downarrow 0$ limit of measures in which the potential has the regularized form

$$(|x-c| + \varepsilon)^{-\alpha} - R_\varepsilon(\lambda,\alpha)\delta(x-c) \tag{10}$$

where [for simplicity we choose $(2-\alpha)^{-1}$ nonintegral]

163

$$R_\varepsilon(\lambda, \alpha) = \sum_{j=1}^{[(2-\alpha)^{-1}]} k_j \lambda^{j-1} \varepsilon^{(2-\alpha)j-1} \tag{11}$$

and the coefficients k_j are determined recursively from $k_1 \equiv (\alpha-1)^{-1}$ and

$$k_j \equiv -\frac{2}{[1-(2-\alpha)j]} \sum_{n=1}^{j-1} k_n k_{j-n} \quad , \quad j = 2, 3, \ldots \quad . \tag{12}$$

An interesting feature to observe is the essential λ dependence of the regularization term which can be likened to a need for counterterms expressed as a polynomial in the coupling constant familiar from renormalized perturbation theory.

Anharmonic Oscillator[1]

A variation on the foregoing example is based on the stationary Ornstein-Uhlenbeck process rather the Wiener process. Formally, we consider the path integration given by

$$\eta \int e^{i\int sx dt - \int\{\frac{1}{2}[\dot{x}^2 + x^2] + \lambda|x|^{-\alpha}\} dt}\, \mathcal{D}x \quad , \tag{13}$$

where $x(t)$ is a continuous sample path and $s(t)$ a test function, and we have chosen $c = 0$. The quadratic terms define the normal probability measure μ_F with covariance

$$\langle x(t)x(t')\rangle_F = (2\pi)^{-1} \int \frac{e^{i\omega(t-t')}}{\omega^2 + 1}\, d\omega = \tfrac{1}{2} e^{-|t-t'|} \tag{14}$$

so that (13) becomes

$$\eta' \int e^{i\int sx dt - \lambda\int|x|^{-\alpha} dt}\, d\mu_F(x) \quad . \tag{15}$$

Suppose the integrand here is defined through the regularization procedure described above (as $\varepsilon \downarrow 0$) to give the form

$$\int e^{i\int sx dt}\, d\mu(x) \quad . \tag{16}$$

If $\alpha \leq 2$, $\mu \to \mu_F$ as $\lambda \downarrow 0$, but if $\alpha > 2$ that is not the case. When $\alpha > 2$,

$$\operatorname*{w-lim}_{\lambda \downarrow 0} \mu = \mu_{PF} \quad , \tag{17}$$

where μ_{PF} is concentrated on paths $x(t)$ for which either $x(t) > 0$ or $x(t) < 0$. Such paths have zero μ_F measure, and so μ_{PF} and μ_F are mutually singular (orthogonal). This is undoubtedly the typical case, and clearly the hard core analogy is largely heuristic in such cases.

Independent-Value Quantum Field Theory[4]

The next example to be treated is a field theory, albeit an extremely unphysical one. Suppose we take Eq. (1) and discard the space-time gradient terms to yield a model heuristically represented by

$$\eta \int e^{i\int h\Phi dx - \int[\frac{1}{2}m^2\Phi^2 + \lambda\Phi^p]dx}\, \mathcal{D}\Phi \quad . \tag{18}$$

When $\lambda \equiv 0$ in the integrand we generate the free measure μ_F with characteristic functional

$$e^{-\frac{1}{2}m^{-2}\int h^2(x)dx} = \int e^{i\int h\Phi dx}\, d\mu_F(\Phi) \quad , \tag{19}$$

and it is consistent with conventional ideas that (18) may be rewritten as

$$\eta' \int e^{i\int h\Phi dx - \lambda\int\Phi^p dx}\, d\mu_F(\Phi) \quad . \tag{20}$$

This point of view is conceptually no different than in the covariant case except of course for the change in definition of μ_F. Ultimately, one expects to find a measure μ incorporating the interaction such that (18) reads

$$\int e^{i\int h\Phi dx}\, d\mu(\Phi) \quad . \tag{21}$$

If $p = 2$, and thus we deal with a mass perturbation, a consistent choice for μ is a normal distribution with variance parameter $(m^2 + 2\lambda)^{-1}$, and quite clearly w-lim $\mu = \mu_F$. However, if $p > 2$ the situation is completely changed.

Remarkably, when $p > 2$ a complete closed form solution to the formal expression (18) can be given,[4] namely

$$\exp\left(-\int dx \int \{1 - \cos[h(x)u]\} e^{-\frac{1}{2}m^2 u^2 - \lambda u^p}\, du/|u|\right) \quad , \tag{22}$$

and this expression (being a positive-definite functional) implicitly defines the interacting measure μ. The limit $\lambda \downarrow 0$ may be taken directly and leads to the characteristic functional

$$\exp(-\int dx \int \{1-\cos[h(x)u]\} e^{-\frac{1}{2}m^2 u^2} du/|u|) \quad , \tag{23}$$

which implicitly defines the measure μ_{PF}. Evidently

$$\text{w-lim } \mu = \mu_{PF} \neq \mu_F \quad , \tag{24}$$

and in fact $\mu_{PF} \perp \mu_F$.

Observe that every interacting model converges as $\lambda \downarrow 0$ to one and the same pseudofree model. Also it is rather clear that a perturbation development of the interacting theories--i.e., expansion of (22) as a power series in λ--leads to an asymptotic series, and moreover one for which Borel resummation techniques are successful. In other words, when expanded about the pseudofree model, these "nonrenormalizable" models are as well behaved as could be expected.

Ultralocal Quantum Field Theory[5]

Slightly more realistic than the last model, an ultralocal scalar field is formally given by the functional integral

$$\mathfrak{n} \int e^{i \int h\Phi dx - \int \{\frac{1}{2}\dot{\Phi}^2 + \frac{1}{2}m^2\Phi^2 + \lambda\Phi^p\} dx} \, \mathcal{D}\Phi \quad . \tag{25}$$

In the traditional view such an expression is formally equivalent to

$$\mathfrak{n}' \int e^{i \int h\Phi dx - \lambda \int \Phi^p dx} \, d\mu_F(\Phi) \tag{26}$$

where μ_F is a normal measure such that

$$e^{-\frac{1}{2}\int \frac{|\tilde{h}(k)|^2}{k_o^2 + m^2} dk} = \int e^{i \int h\Phi dx} \, d\mu_F(\Phi) \quad . \tag{27}$$

As before the aim is to find the interacting measure μ so that (25) is given by

$$\int e^{i \int h\Phi dx} \, d\mu(\Phi) \quad . \tag{28}$$

Clearly, if $p = 2$ it is consistent to choose μ as a normal measure (with shifted mass), and thus w-lim $\mu = \mu_F$. For $p > 2$ the situation changes completely.

The solution for $p > 2$ has been thoroughly discussed elsewhere in the framework of real-time quantum field theory.[5] A direct discussion of the solution in the Euclidean framework is complicated because the time-ordered vacuum expectation values do not exist. However, this is not too serious since time ordering is sensible for certain renormalized field powers, and with these a connection to the present discussion can be made. On the other hand, it is not necessary to make this connection explicit since many pseudofree characteristics can be exhibited in the real-time formulation.

We content ourselves with a discussion of the spectrum of the free, pseudofree and interacting theories. It should be no surprise that there is only a discrete spectrum when the spacial gradients are absent. The free theory is simply a collection of independent, equivalent harmonic oscillators, and the energy spectrum is simply

$$0, m, 2m, 3m, 4m, 5m, \ldots \quad .$$

The pseudofree energy spectrum (the derivation of which is not presented here) reads

$$0, 8m, 10m, 12m, 14m, 16m, \ldots \quad ,$$

which are the values appropriate to a four-dimensional space time. The generalization to an n-dimensional space time reads

$$0, 2nm, (2n + 2)m, (2n + 4)m, \ldots \quad .$$

In these spectra, the ground state has zero energy and is nondegenerate; all other energy levels are infinitely degenerate. When a nonlinear interaction is introduced, such simple numerical coefficients generally are lost, and analytic methods no longer suffice. We have calculated the first six energy levels numerically, based on equations given elsewhere,[5] for the quartic interaction where $p = 4$ in a four-dimensional space time. Although this is a renormalizable interaction for covariant theories, it is a nonrenormalizable one for ultralocal theories. In the figure the first six energy levels are plotted (in units of m) as a function of the coupling constant (here 2λ). These levels have some of the qualitative behavior of a quartic anharmonic oscillator, e.g., asymptotically behaving as $\lambda^{1/3}$ for large λ; but they differ significantly in that as $\lambda \downarrow 0$ the values $8, 10, 12, 14, \ldots$ emerge, as stated earlier. Although we have not verified the results explicitly we expect, from the type of equations involved, that a perturbation calculation of the energy levels about the pseudofree behavior leads to an asymptotic series and probably one summable by Padé approximants.

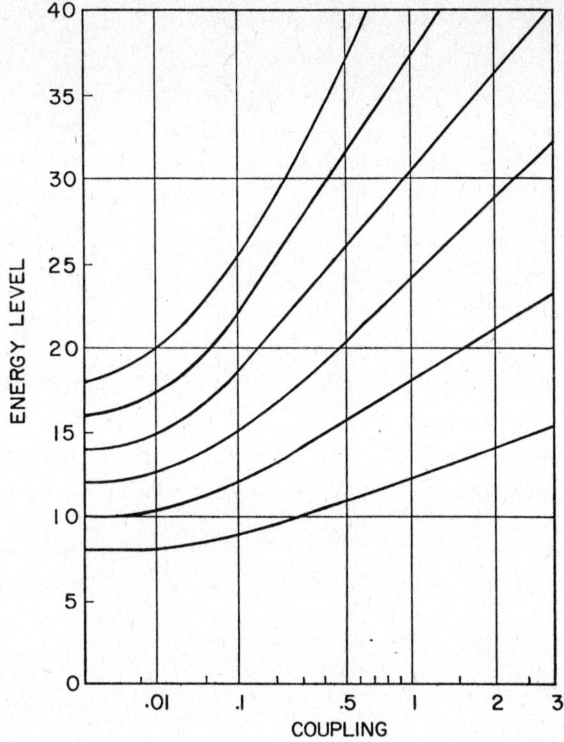

The interested reader may find further discussion of measures for ultralocal free and pseudofree theories in the references.

3. RELEVANCE FOR COVARIANT THEORIES

Discontinuous perturbations in quantum mechanics and in quantum field theory are a fact, and it is entirely possible that they may arise in covariant quantum field theory. No examples are known as yet, but certain plausible arguments suggest their existence and equivalence to conventional nonrenormalizable models. What are these arguments?

Consider the independent-value model, and focus on the two expressions $\int\phi^2 dx$ and $\int\phi^p dx$, which represent the free and interacting action, respectively. Formally, the L^2 functions control (in an unspecified way) the support of the free measure μ_F, while the $L^2 \cap L^p$ functions control the support of the interacting measure μ. If $p = 2$, there is no difference, but if $p > 2$, there is a profound difference. It is, heuristically at least, as if the presence of the interaction projects out ("hard core") all L^2-dictated elements not also present among those that are $L^2 \cap L^p$ dictated. This projection persists for all $\lambda > 0$, and formally determines the pseudofree theory as $\lambda \downarrow 0$.

Similar arguments can be advanced for the ultralocal case, and, suitably modified, even for the anharmonic oscillator discussed earlier. Let us ask the same type of question in the covariant case.

If we let $W_0 = \frac{1}{2}\int[(\nabla\Phi)^2 + m^2\Phi^2]dx$, and $W_I = \int|\Phi|^p dx$, where $dx \equiv d^n x$, then Sobolev-type inequalities[6,1] show that $W_I \leq KW_0^{p/2}$ where K is a Φ-independent, finite constant, provided $p \leq 2n/(n-2)$. On the other hand, if $p > 2n/(n-2)$ no such bound exists, and there are uncountably many Φ with $W_0 < \infty$ and $W_I = \infty$. Formally, at least, in Eq. (1) the term W_I projects out fields inconsistent with the combination of both W_0 and W_I whenever $p > 2n/(n-2)$, and this "hard core" effect remains even after $\lambda\downarrow 0$. For $p \leq 2n/(n-2)$ no such phenomena should arise since $W_0 < \infty$ implies $W_I < \infty$. To conclude the argument we need only point out that conventional renormalized perturbation theory asserts that the models with $p \leq 2n/(n-2)$ are renormalizable, while those with $p > 2n/(n-2)$ are nonrenormalizable. These arguments tend to suggest, therefore, that discontinuous perturbations and the associated pseudofree theories may have relevance in the general study of covariant quantum field theory, especially for the so-called nonrenormalizable theories.

Generalization

Let us consider the generalized class of models for which $W_0 = \frac{1}{2}\int(k^{2\xi} + 1)|\tilde{\Phi}(k)|^2 dk$ and $W_I = \int|\Phi(x)|^p dx$. If $\xi = 1$ these are the covariant models (with $m = 1$) discussed above. Generalized Sobolev-type inequalities[7] state that a Φ-independent, finite constant K exists such that $W_I \leq KW_0^{p/2}$ provided $p \leq 2n/(n-2\xi)$. No such bound exists when $p > 2n/(n-2\xi)$. Remarkably, only the ratio ξ/n enters these conditions, and this suggests[8] that in some sense we can simulate n-dimensional covariant quantum field theory in a <u>one</u>-dimensional space-time, i.e. time alone, if we simply set $\xi = 1/n$. What is simulated is the convergence or divergence of analogous graphs and the corresponding division into renormalizable and nonrenormalizable cases, which implies that these properties may also be studied in the context of one-dimensional stochastic processes, i.e., "noise". Finally we note that $\xi/n = 0$ for any n if $\xi = 0$. For $n = 1$, shot noise is a well-known δ-correlated process; for $n \geq 2$, the independent-value model discussed above is just the field-theory analogue of a special shot noise process.

REFERENCES

1. J. R. Klauder, Acta Physica Austriaca Suppl. XI, 341 (1973); Phys. Letters 47B, 523 (1973).

2. B. Simon, J. Funct. Anal. 14, 295 (1973); B. DeFacio and C. L. Hammer, J. Math. Phys. 15, 1071 (1974).

3. L. A. Shepp, J. R. Klauder and H. Ezawa, Ann. Inst. Fourier, Grenoble 24, 189 (1974); H. Ezawa, J. R. Klauder and L. A. Shepp, Ann. of Phys. 88 588 (1974); "Vestigial Effects of Singular Potentials in Diffusion Theory and Quantum Mechanics", J. Math. Phys., to be published.

4. J. R. Klauder, "On Model Fields with Independent Values at Every Space-Time Point", Acta Physica Austriaca, to be published.

5. J. R. Klauder, Lectures in Theoretical Physics, Vol. XIVB, W. Britten, ed. (Colorado Associated University Press, 1974), p. 329; Acta Physica Austriaca, Suppl. VIII, 227 (1971).

6. O. A. Ladyzenskaja, V. A. Solonnikov and N. N. Ural'ceva, Linear and Quasi-Linear Equations of Parabolic Type, Trans. of Math. Mono., Vol. 23 (American Math. Society, 1968).

7. V. P. Il'in and V. A. Solonnikov, Trudy MIAN SSSR 66, 205 (1962). We thank Prof. Solonnikov for helpful correspondence regarding this work.

8. J. R. Klauder, "Field Theory as Noise Theory", Physics Letters, to be published.

DISCUSSION

J. L. Lebowitz: In removing the regularization and then the coupling for the perturbed free particle, which links in the chain do not hold when the regularization is not the right one?

J. R. Klauder: Subject to minor restrictions, any regularization should at least have a convergent subsequence as $\varepsilon \downarrow 0$. However, in order to obtain Wiener measure as $\lambda \downarrow 0$ only a limited class of regularizations will work, which are effectively as indicated [Eqs. (10)-(12)] plus what might be termed finite renormalizations.

K. Symanzik: Is your perturbation expansion about the nonfree theory with $\lambda = 0$ in terms of integer powers of λ, without nonanalytic terms?

J. R. Klauder: For the model field theories I discussed (independent-value; ultralocal) the perturbation expansion in λ about the pseudofree theory is asymptotic so long as m > 0. However, this property fails if m = 0 [as is clear say from Eq. (22)]. Also it is clear that the perturbation expansion of the anharmonic oscillator [Eqs. (13)-(17)] about the pseudo-free theory is asymptotic for $2 < \alpha < 3$, while for $\alpha \geq 3$ nonanalytic terms arise.

A. Verbeure: Have you any idea how to find the measure for $\lambda = 0$ and for nonrenormalizable fields, without first solving the whole problem?

J. R. Klauder: Unfortunately, there are only a few general principles presently known that characterize pseudofree theories. For the special model field theories treated here, these principles plus the symmetry of the model determine the result, but for covariant models this is, at least so far, not the case.

BOUND STATES AND ASYMPTOTIC FIELDS
IN MODEL FIELD THEORIES

Yusuke Kato

Department of Engineering Mathematics, Utsunomiya University

Utsunomiya,321-31,JAPAN

In a recent paper with Sekine[1], I studied scattering and spectrum
of the translation-invariant Lee model, and proved, among other things,
the existence of the asymptotic fields for bound states, but under the
assumption of Galilei invariance. In this talk, I try to generalize the
results without the assumption of the Galilei invariance.

We first consider the total Hamiltonian H defined as the sum of the
free Hamiltonian H_0 and the interaction Hamiltonian H_I of the polynomial
type written in the momentum space as follows:

$$H_0 = \sum_{s=1}^{\gamma} \int w_s(p)a_s^*(p)a_s(p)dp, \tag{1}$$

$$H_I = \sum_{m,n}^{M,N} \int G_{mn}(p_1,\ldots,p_m;q_1,\ldots,q_n)\prod_i^m a^*(p_i)\prod_j^n a(q_j)(dp)_m(dq)_n. \tag{2}$$

The kernel G_{mn} should belong to such a function class that H_I be defined
on a dense domain in the Fock space F constructed by the vacuum vector
Ψ_0 and the creation and annihilation operators a_s^* and a_s, respectively.
In H_I we ignore the suffix s of a_s^*, a_s specifying the species of fields
and write each of them simply as a^* or a. When the interaction is translation-invariant, then

$$G_{mn} = \delta(\textstyle\sum p_i - \sum q_j)F_{mn}(p_1,\ldots,q_n). \tag{3}$$

We call the interaction term with m>n=0 (or n>m=0), with m≥n=1 (or n≥m
=1) or with m,n≥2 as zero-, one- or many-particle interaction term, respectively. It has been known[2] that the scattering problem of many-particle interaction, with certain restrictions on G_{mn}, is manageable by means
of the theory of wave operators quite analogously as in the quantum mechanics but that, in the case of zero- or one-particle interaction, the
direct application of the method is difficult due to the dressing or
polarization effect, respectively. For the study of the scattering and
spectral problem of these interactions, the concept of the asymptotic
fields defined as strong limits of adjusted physical one-particle operator has been revealed to be quite useful and the existence of the
strong limits has been established for a certain class of model field
theories[3,4], in which zero- and one-particle interactions are included
but H_I's are not translation-invariant, i.e., not momentum-conserving.
Hepp[5] studied the asymptotic fields in a model of identical particles
interacting via a 2-body local potential, i.e., a class of translation-

invariant many-particle interaction. Later, Høegh-Krohn[6] examined a translation-invariant model in which H_I contains general many-particle terms but no zero- or one-particle term, thus the Lee-type interaction being excluded. On this line of thought we study the Lee-type interaction with translational invariance as a typical example of one-particle interaction and examine whether the method of the asymptotic fields covers the all types of the translation-invariant interactions except the zero-particle one for which, as is well known, the Hamiltonian is difficult to be defined within the original Fock space F. The Lee-type Hamiltonian is given as:

$$H_0 = \int w_V(p)V^*(p)V(p)dp + \int w_N(q)N^*(q)N(q)dq + \int w_\theta(r)\,\theta^*(r)\,\theta(r)dr, \tag{4}$$

$$H_I = \lambda \int [F(p,r)V^*(p)N(q)\,\theta(r) + h.c.]\delta(p-q+r)dpdqdr, \tag{5}$$

V, N and θ obeying the boson commutation relations.

When the translation-invariant interaction Hamiltonian H_I given by Eq.(1) does not contain the zero-particle term, we define a physical one-particle operator as follows:

Definition. A vector of the form

$$O^*\{f\}\Psi_0 = \int O^*(p)f(p)dp\Psi_0 \in F$$

with its norm

$$\|O^*\{f\}\Psi_0\| = \|f\|,$$

is said to represent a physical one-particle state, whenever $O^*\{f\}$ is a linear operator having a dense domain in F and satisfies the condition

$$HO^*\{f\}\Psi_0 = O^*\{Ef\}\Psi_0 \tag{6}$$

for every $f \in L^2$ such that $Ef \in L^2$. The operator $O^*(p)$ is called the physical one-particle creation operator and $E(p)$ is called the energy function of the particle. The physical particle annihilation operator $O(p)$ is defined as the adjoint of $O^*(p)$.

We consider the cases where the physical particle operator $O^*(p)$ can be represented as

$$O^*(p) = Z^{1/2}(p) \sum_{n=\alpha}^{\beta} \delta(p-\textstyle\sum p_i)\,\phi_n(p_1,\ldots,p_n)\prod_{i=1}^{n} a^*(p_i) \tag{7}$$

with

$$Z(p) = \left[\sum_{n=\alpha}^{\beta} \int |\phi_n(p_1,\ldots,p_n)|^2 \delta(p-\textstyle\sum p_i)(dp)_n \right]^{-1}, \tag{8}$$

where $\beta \geqslant \alpha \geqslant 1$ and β being finite or infinite. We tentatively classify the physical one-particle operator into three categories:

1)$\alpha=\beta=1$. Physical one-particles identical with the corresponding free or bare particles, e.g., θ and N particles in the Lee-type model.

2)$\beta>\alpha=1$. Physical or dressed particle which is not identical with the free or bare particle but tends to it in the limit when the interaction vanishes, e.g., V particle in the Lee-type model.

3)$\beta\geqslant\alpha\geqslant2$. For a certain class of interaction kernels F_{mn}, bound states exist and the physical one-particle states describing them have no corresponding bare one-particle states; the corresponding bare state is a state describing a system of non-interacting particles, e.g., the bound state B in $V\theta-N\theta\theta$ sector in the translation-invariant Lee-type model[1].

Scattering and spectral problems comprising the bound states have been investigated by many authors with various methods. Here we note that the bound state in the quantum field theory has clear meaning only when the interaction is translation-invariant and we note also that, for the study of bound states, i.e., the physical one-particles of category 3), power series expansions of various quantities with respect to the coupling constant are expected to be not valid. Thus, in the study by the method of the asymptotic fields of the models including the Lee-type one, we give special emphasis on the translational invariance and bound states. We present some generalization of the previous results[1] in a summarized form for the case of Lee-type interaction. First we formulate conditions on the kernel F and the energy function E of the physical one-particle state.

Condition 1. For almost every p, $F(p,r)$ is a square-integrable function of r, and

$$\mathrm{ess\ sup}_{p}\ \left[\int |F(p,r)|^2 dr\right]^{1/2} \equiv M < \infty.$$

Condition 2. Let $\alpha \in C^{\infty}$ be such that $\beta\equiv1/\alpha$ is bounded and $F_1\equiv F\beta(r)$. There exist functions $c_n \in L^{\infty}$, $\xi_n \in L^2$ $(n = 1,2,\dots)$ such that

$$\lim_{N\to\infty}\mathrm{ess\ sup}_{p}\int \left|F_1(p,r) - \sum_{n}^{N} c_n(p)\xi_n(r)\right|^2 dr = 0.$$

Moreover, the Fourier transform $\tilde{\xi}_n$ is a function of $L^1\cap L^2$ and the following condition holds:

$$\sum \|c_n\|_{\infty}\ \|\tilde{\xi}_n\|_1 <\ \infty. \tag{9}$$

Condition 3. Put $f_t(p) = f(p)e^{-iE(p)t}$ for $f \in C_0^{\infty}$, where $E(p)$ is the energy function of a physical one-particle state and is bounded below. Then there exist positive constants A and μ such that

$$\|\tilde{f}_t\|_{\infty} < A|t|^{-\mu}\ \|\tilde{f}\|_1,\ \mu>1,\ \text{as } |t|\to\infty. \tag{10}$$

We also present a condition on the physical one-particle operator $O^*(p)$ given by Eq.(7).

Condition 4. For any $f \in C_0^\infty$, put

$$\phi_{f,n}(p;p_1,\ldots,p_{n-1}) \equiv f(p)\phi_n(p_1,\ldots,p_n)Z^{1/2}(p)\delta(p-\textstyle\sum p_i).$$

$$(n = \alpha,\ldots,\beta)$$

Then $\phi_{f,n}(p;\cdot)$ is a sufficiently smooth function of p so that its Fourier transform $\widetilde{\phi}_{f,n}(x;x_1,\ldots,x_{n-1})$ has the following property:

$$\int dx_1\ldots dx_{n-1}\Big|\int|\widetilde{\phi}_{f,n}(x;x_1,\ldots,x_{n-1})|dx\Big|^2 < \infty .$$

Remarks. We see that the inequality (9) in Condition 2 is quite similar to the one given by Høegh-Krohn[6] in connection with the many-particle interaction and that Condition 3 is valid for the Schrödinger or the Klein-Gordon kernel with $\mu=3/2$. We also see that Condition 4 holds for sufficiently smooth function $F(p,r)$ and smooth energy functions $E(p)$'s, which appear in the integral equation to determine ϕ_n. We saw that[1], in the case of Galilei-invariant Lee-type model, we can find the kernel F for which a bound state B exists in the $V\theta-N\theta\theta$ sector. Since in the Galilei-invariant case ϕ_n is independent of p, it is easily seen that all Conditions $1\sim4$ hold. For the more general Lee-type model which is translation-invariant but not Galilei-invariant, however, we present Conditions as those to be imposed on $F(p,r)$ and $E(p)$'s assuming that bound states exist in certain sectors.

Now the scattering and the spectral problems are solved.

Theorem 1.(asymptotic fields of θ and N). Under the assumption that Conditions 1,2 and 3 are satisfied the asymptotic fields for θ and N particles exist:

$$\theta_\pm^\#\{f\}\phi \equiv \operatorname*{s-\ell im}_{t\to\pm\infty}\theta_t^\#\{e^{-iEt}f\}\phi = \operatorname*{s-\ell im}_{t\to\pm\infty} e^{iHt}\theta^\#\{e^{-iEt}f\}e^{-iHt}\phi$$

$$\text{for any } f \in L^2 \text{ and } \phi \in D(H)\cap F(n,m),$$

where $F(n,m)$ denotes an arbitrary sector of F, i.e., a subspace invariant with respect to the total Hamiltonian H and $O^\#$ denotes O or O^*.

Theorem 2.(asymptotic fields of V_1,B and so on). Under the assumption that Conditions $1\sim4$ are satisfied for the physical one-particle operator O^*, the asymptotic fields of $O^\#$ exist:

$$O_\pm^\#\{f\}\phi \equiv \operatorname*{s-\ell im}_{t\to\pm\infty} O_t^\#\{e^{-iEt}f\}\phi,$$

$$\text{for any } f \in L^2 \text{ and } \phi \in D(H) \cap F(n,m).$$

Theorem 3.(commutation relations). Let $O_\pm^\#\{f\}\phi$ exist. Then

$$e^{iHt}O_\pm^\#\{e^{-iEt}f\}e^{-iHt}\phi = O_\pm^\#\{f\}\phi,$$

or

$$[\ H,\ O_\pm^*\{f\}\ ] = O_\pm^*\{Ef\},$$

and we have the commutation relations

$$[\ O_\pm\{f\},\ O_\pm^*\{g\}\ ]\phi = (f,g)\phi,\ \text{for}\ \phi \in F(n,m),$$

all other pairs giving zero. Let ϕ be an eigenstate of the total Hamiltonian H. Then

$$O_\pm\{f\}\phi = 0\ \text{ for all }\ f \in L^2.$$

Theorem 4.(decomposition of the Fock space). The Fock space F decomposes in two ways as a tensor product of two Hilbert spaces:

$$F = F_+ \otimes F'_+ = F_- \otimes F'_-,$$

where $F_\pm$ is the Fock representation of the $\pm$asymptotic fields constructed from an eigenstate of the total Hamiltonian H.

Theorem 5.(spectrum of the total Hamiltonian). Define, in $F_\pm$, the "free" Hamiltonian for the asymptotic fields $O_\pm^\#$:

$$H_{0\pm} \equiv \int E(p)O_\pm^*(p)O_\pm(p)dp.$$

Then for $\phi = \phi_\pm \otimes \phi'_\pm \in F_\pm \otimes F'_\pm = F$ we have the relation of the following form:

$$H\phi = H_{0\pm}\phi_\pm \otimes \phi'_\pm + \phi_\pm \otimes H'_{0\pm}\phi'_\pm.$$

References

1). Y.Kato and K.Sekine, Prog. Theor. Phys.52(1974),659.

2). K.O.Friedrichs, Perturbation of Spectra in Hilbert Space, AMS, Providence, R.I.(1965).

   L.Van Hove, Physica 24(1958),137.

   L.D.Faddeyev, Dokl. Akad Nauk SSSR.(1963),573.

3). Y.Kato and N.Mugibayashi, Prog. Theor. Phys.30(1963),103.

4). Y.Kato and N.Mugibayashi, Prog. Theor. Phys.45(1971),628.

5). K.Hepp, 'Axiomatic Field Theory' in 1965 Brandeis Lectures in Theoretical Physics, M.Chretien and S.Deser,eds.(Gordon and Breach, New York,1966),Vol.1,pp.135-246.

6). R.Høegh-Krohn, J.Math.Phys.(1969),639.

FEYNMAN PROPAGATORS ASSOCIATED WITH
THE VENEZIANO MODEL

Masatsugu Minami

RIMS, Kyoto University, Kyoto 606, Japan

It will be interesting to find out if Veneziano's dual-resonance model contains something which corresponds to the Feynman-Dyson perturbation theory and what it looks like if it really does. We have tried it purely within the functional-integral formalism and succeeded in showing that it is possible if the number of transverse degrees of space-time, δ, is 24 and the intercept of the leading Regge trajectory is unity.

To establish the perturbation theory, we first need to define the counterpart of the Feynman propagator. The propagator in the ordinary theory is associated with a one-dimensional arc $\{t \mid 0<t<\beta\}$. On the other hand, the medium in the Veneziano case should have a width of ℓ and should be associated with the two-dimensional rectangular domain $R_0^\beta = \{(s,t) \mid 0<t<\beta,\ 0<s<\ell\}$. The momentum distributions at $t=0$ and $t=\beta$ therefore should continuously depend on s and we denote them by $p_1(s)$ and $p_3(s)$ respectively. Then the representation of the <u>momentum-space</u> propagator our model should have is given by

$$\tilde{\Delta}_F[y_4,y_3(s),y_2,y_1(s)] = \int_0^\infty d\beta e^{\alpha_0 \pi \beta/\ell} \tilde{K}[y_4,y_3(s),y_2,y_1(s);\beta] \qquad (1)$$

where

$$\tilde{K}[y_4,y_3(s),y_2,y_1(s);\beta] = \int \mathcal{D}'y(s,t)\exp\frac{1}{4\pi\alpha'}\int_{R_0^\beta}dsdt[(\frac{\partial y}{\partial s})^2+(\frac{\partial y}{\partial t})^2], \qquad (2)$$

α_0 and α' being constant (to be identified with the intercept and Chew's slope respectively) and $y(s,t)$ being a continuous Minkowski vector on R_0^β. $\int \mathcal{D}'y(s,t)\cdots$ denotes the functional integration (restricted to a certain gauge to make the integral convergent) over all $y(s,t)$ satisfying the boundary condition: $y(s,t)=y_1$, y_2, y_3 and y_4 at $(s,t=0)$, $(s=\ell,t)$, $(s,t=\beta)$ and $(s=0,t)$ respectively. Here y_2 and y_4 are s-independent (we assume there is no outgoing momentum at the sides $s=0$ and $s=\ell$) and $y_1(s)$ and $y_3(s)$ are given by

$$\partial y_1(s)/\partial s = -2\pi\alpha'p_1(s), \quad \partial y_3(s)/\partial s = -2\pi\alpha'p_3(s) \qquad (3)$$

The kernel $\tilde{K}$, (2), can further be simplified by use of the <u>harmonic</u> $y^h(s,t)$ such that

$$\tilde{K} = g(\beta) \cdot \tilde{K}^h[y_4, y_3(s), y_2, y_1(s); \beta] \tag{4}$$

where

$$\tilde{K}^h = \exp \frac{1}{4\pi\alpha'} \int_{R_0^\beta} dsdt \left[\left(\frac{\partial y^h}{\partial s}\right)^2 + \left(\frac{\partial y^h}{\partial t}\right)^2 \right]. \tag{5}$$

Solving the Dirichlet-type boundary-value problem, $y^h(s,t)$ is explicitly given as

$$y^h(s,t) = y_4 + \frac{s}{\ell}(y_2 - y_4) + \frac{2}{\pi}\sum_{\nu=1}^\infty \frac{Y_{1\nu}}{\nu} \frac{\sin\nu(\pi/\ell)s \cdot \sinh\nu(\pi/\ell)(\beta - t)}{\sinh\nu(\pi/\ell)\beta}$$

$$+ \frac{2}{\pi}\sum_{\nu=1}^\infty \frac{Y_{3\nu}}{\nu} \cdot \frac{\sin\nu(\pi/\ell)s \cdot \sinh\nu(\pi/\ell)t}{\sinh\nu(\pi/\ell)\beta}, \tag{6}$$

where

$$Y_{1\nu} = \frac{2}{\pi}(y_1-y_4) + \frac{2}{\pi}(y_2-y_1)(-)^\nu + y_{1\nu}, \quad Y_{3\nu} = \frac{2}{\pi}(y_4-y_3) + \frac{2}{\pi}(y_3-y_2)(-)^\nu + y_{3\nu}, \tag{7}$$

$y_{1\nu}$ and $y_{3\nu}$ being given by the expansions of boundary values:

$$y_1(s) = y_1 + \sum_{\nu=1}^\infty (y_{1\nu}/\nu)\sin\nu(\pi/\ell)s, \quad y_3(s) = y_3 + \sum_{\nu=1}^\infty (y_{3\nu}/\nu)\sin\nu(\pi/\ell)s. \tag{8}$$

By virtue of (6), the rhs of (5) can further be rewritten as a (somewhat <u>formal</u>) series:

$$\tilde{K}^h = \exp\left\{ \frac{1}{4\pi\alpha'} \frac{\beta}{\ell}(y_2-y_4)^2 + \frac{1}{8\alpha'} \sum_{\nu=1}^\infty \frac{Y_{1\nu}^2 + Y_{3\nu}^2}{\nu} \coth\frac{\nu\pi\beta}{\ell} \right.$$

$$\left. + \frac{1}{8\alpha'} \sum_{\nu=1}^\infty \frac{2(Y_{1\nu}, Y_{3\nu})}{\nu} \operatorname{cosech} \frac{\nu\pi\beta}{\ell} \right\}. \tag{9}$$

To determine $g(\beta)$ in (4), one can use the "composition law" of the propagation kernels:

$$\int \tilde{K}[0, y_3(s), 0, y(s); \beta-\beta'] \mathcal{D} y(s) \tilde{K}[0, y(s), 0, y_1(s); \beta']$$

$$= \tilde{K}[0, y_3(s), 0, y_1(s); \beta]. \tag{10}$$

The result is:

$$g(\beta) \sim r_0^{\delta/24} (\vartheta_2^0 \vartheta_3^0 \vartheta_4^0)^{-\delta/6}, \tag{11}$$

where $r_0 = \exp(-\pi\beta/\ell)$, and $\vartheta_j^0 \equiv \vartheta_j(0 | \frac{\log r_0}{\pi i})$ denote Jacobi's theta functions.

As a special case, let us put $y_{1\nu}=0$ and $y_{3\nu}=0$ (in (8)) so that y_1 and y_3 become constant. Then we are left with the <u>four-leg case</u> where external momenta are given by difference vectors of y_j:

$$2\pi\alpha' p_1 = y_1 - y_4, \quad 2\pi\alpha' p_2 = y_2 - y_1, \quad 2\pi\alpha' p_3 = y_3 - y_2, \quad 2\pi\alpha' p_4 = y_4 - y_3, \tag{12}$$

where p_j are subjected to $\alpha' p_j^2 = -\alpha_0$. In this case $y^h(s,t)$ given by (6) takes the form

$$y^h(s,t) = -4\alpha' \sum_{j=1}^{4} p_j \ \mathrm{Im} \ \log \vartheta_j\left(-\frac{s+it}{2\ell} \ \Big| \ \frac{\log r_0}{\pi i}\right). \tag{13}$$

up to a constant. On the other hand, $\tilde{K}^h$ given by (9) needs some regularization since its exponent is to have such terms as $p_j^2 \log \vartheta_1^0$. To regularize, we put the requirement that $x^h(s,t)$ where $x^h(s,t)$ is given by the condition that $x^h(s,t)+iy^h(s,t)$ is <u>holomorphic</u> in R_0^β should reduce to the ordinary position vector when $\ell \longrightarrow 0$ (that is, $x^h(s,t) \xrightarrow[\ell\to 0]{} -(2\pi\alpha'/\ell)(p_1+p_2)t+\mathrm{const})$. This is assured by the replacement of ϑ_1^0 by $\vartheta_1'(0)/\pi$. By this replacement, $\tilde{K}^h$ takes the form

$$\tilde{K}^h = (\vartheta_2^0 \vartheta_4^0)^{4+4\alpha_0} (\vartheta_3^0)^{8\alpha_0-8}$$
$$\times \ (\vartheta_2^0/\vartheta_3^0)^{-4-4\alpha_0-4\alpha'(p_1+p_2)^2} (\vartheta_4^0/\vartheta_3^0)^{-4-4\alpha_0-4\alpha'(p_1+p_4)^2}, \tag{14}$$

which really reduces to the ordinary kernel $\exp[\beta(p_1+p_2)^2]$ in the limit $\ell \to 0$ (where $m\alpha'=\ell$). This in turn assures us that the amplitude gives rise to a tower of resonances in the $(p_1+p_2)^2$-channel.

Changing the integration variable β in (1) to λ defined by

$$\lambda = (\vartheta_2^0/\vartheta_3^0)^4 \tag{15}$$

we obtain

$$\tilde{\Delta}_F \longrightarrow \int_0^1 d\lambda \, r_0^{-\alpha_0+\delta/24} (\vartheta_3^0)^{-\delta/6-4+8\alpha_0} (\vartheta_2^0\vartheta_4^0)^{-\delta/6+4\alpha_0}$$
$$\times \ \lambda^{-1-\alpha_0-\alpha'(p_1+p_2)^2} (1-\lambda)^{-1-\alpha_0-\alpha'(p_1+p_4)^2} \tag{16}$$

This is just Veneziano's beta-function amplitude when $\alpha_0-(\delta/24)=0$ and $-\delta/6-4+8\alpha_0=0$, from which we obtain $\alpha_0=1$ and $\delta=24$. As easily seen, this condition is needed by the requirement that the amplitude should also yield a tower of resonances in the $(p_1+p_4)^2$-channel. Note that the formula is invariant under the interchange of β/ℓ by ℓ/β (λ becomes $1-\lambda$). This is in well accord with Veneziano's dualism. It should further be noted that $g(\beta)$ given by (11) has not been invariant under this (modular) transformation, however the deviation is so cancelled by the replacement $\vartheta_1(0)\to\vartheta_1'(0)/\pi$ that the output remains invariant.

Having successfully defined the propagators, we can give higher order "corrections" as in perturbation theory by treating the propagators as building blocks. For example, the quantity defined by

$$\int_0^\infty d\beta_1 d\beta_2\, e^{\pi\alpha_0(\beta_1+\beta_2)/\ell}\,\tilde{K}[y_{41},y_{42},y_3(s),y_{21},y_{22},y_1(s);\ \beta_1,\beta_2],\quad (17)$$

where

$$\tilde{K}=\int\tilde{K}[y_{42},y_3(s),y_{22},y(s);\ \beta_2]\,\mathfrak{D}\,y(s)\tilde{K}[y_{41},y(s),y_{21},y_1(s);\ \beta_1]\quad (18)$$

will reduce to the 5-point function. In this case, $(2\pi\alpha')^{-1}(y_{41}-y_{42})$ or $(2\pi\alpha')^{-1}(y_{22}-y_{21})$ gives the <u>fifth</u> external momentum. In general, iterating successively these procedures, we can exactly arrive at the multi-particle Bardaçki-Ruegg formula:

$$\int_0^1\prod_{j=1}^{n-1}d\lambda_j\,\lambda_j^{-\alpha_j-1}\prod_{0\le j\le k\le n}(1-\lambda_j\lambda_{j+1}\cdots\lambda_{k-1})^{-2\alpha'(p_j,p_k)}\quad (19)$$

where

$$\lambda_j=\frac{\vartheta_2^2(v_j)}{\vartheta_3^2(v_j)}\cdot\frac{\vartheta_3^2(v_{j+1})}{\vartheta_2^2(v_{j+1})}\ ,\quad v_j=\frac{i(\beta_1+\cdots+\beta_j)}{2\ell}\quad (20)$$

under the proviso $\delta=24$ and $\alpha_0=1$ again.

We must further note that, slightly generalizing the boundary conditions, we can also obtain the amplitudes relevant to the media which contain some slits in them (multi-loop amplitudes). In these cases, however, there appear some involved points concerning the <u>zeroth-modes</u> (this problem is well solvable in our framework).

In closing, I wish to give a few words on the continuous position vectors which we have used and also on a prototype of our propagator. It should be noted that it is not $y^h(s,t)$ but $x^h(s,t)$ that is a direct continuation of the position vector x appearing in the ordinary $\Delta_F(x)$. The vector $y^h(s,t)$ can rather be said a continuous version of the position-vectors of <u>dual-graph's</u> vertices (dual-position vectors). Therefore $y^h(s,t)$ corresponds to a <u>stream function</u> if one regards the original position $x^h(s,t)$ as a <u>potential function</u> (on a continuous version R_0^β of a linear graph). This is the reason why we were able to obtain $x^h(s,t)$ from $y^h(s,t)$ by the condition that $x^h(s,t)+iy^h(s,t)$ should be holomorphic in R_0^β (or x^h and y^h should satisfy the Cauchy-Riemann relations $\partial x^h/\partial s=\partial y^h/\partial t$, $\partial x^h/\partial t=-\partial y^h/\partial s$).

Finally, we should say that it is not quite unnatural also to write the ordinary Feynman propagator in terms of the dual-position y. Let us write

$$\tilde{\Delta}_F(p) = \frac{1}{p^2-m^2} = \int_0^\infty d\beta e^{-(m^2-p^2)\beta} \tag{21}$$

for the propagator in momentum space. Let us next introduce the difference vector of dual-position vectors by $y_1-y_2=\beta p$, and further $y^\ell(t)$ of the form

$$y^\ell(t) = \frac{y_2-y_1}{\beta}t + y_1. \tag{22}$$

Then we readily have

$$\tilde{\Delta}_F(p) = \int_0^\beta d\beta e^{-\beta m^2} \cdot \exp\int_0^\infty dt\left(\frac{dy^\ell(t)}{dt}\right)^2. \tag{23}$$

This is really a prototype of our propagator (1) in the light of (4) and (5).[We no longer rewrite the rhs of (23) although we could have written down the path-integral expression by using $1/\beta$ as the integration variable. This is the reason that we have not taken account of the zero-th mode in $\int \mathcal{D}'y(s,t)\cdots$ of (2), (10).] The <u>linear</u> vector of (22) becomes $y^h(s,t)$ which is <u>harmonic</u> in (s,t) in the two-dimensional generalization.

The author's contributions to this type of functional approach will be found in Prog. Theor. Phys. <u>45</u>(1971), 208; <u>46</u>(1971), 614; <u>48</u> (1972), 974, <u>48</u>(1972), 1308; <u>52</u>(1974), 1031; <u>52</u>(1974), 1890, <u>53</u>(1975), 237; Nuovo Cim. <u>19A</u>(1974), 13. Refer also to: Technical notes RIMS-141, 165, 166.

DUAL STRING MODELS AND QUANTUM GRAVITY

Tamiaki Yoneya

Department of Physics, Hokkaido University

Sapporo, Japan 060

Since it turned out that the closed-string sector of ghost-free
dual models contains a massless spin-2 state, possible connection between the dual models[1] and the theory of gravity has perhaps long been
anticipated by many people. As shown by the present author[2] and
SCHERK and SCHWARZ[3] some time ago, the massless spin-2 state of the
Virasoro-Shapiro model, which is a simplest closed string model, can
indeed be described by the Einstein theory in a certain low energy
limit. By this, we mean that the on-mass-shell S-matrix elements in
the Virasoro-Shapiro model, taking the massless spin-2 states as the
external lines, exactly coincide with the S-matrix elements obtained
from the Einstein Lagrangian $\frac{1}{\kappa^2}\sqrt{-g}R$ to all orders of the gravitational constant in the tree approximation, if we take the limit of zero
slope parameter in the Virasoro-Shapiro model with the quantity $g\sqrt{\alpha'}$
being fixed and identified with the gravitational constant κ . Here,
the quantization of gravity is done perturbatively following GUPTA,
FEYNMAN and DEWITT[4]. The constants g and α' are the dimensionless
coupling constant and the slope parameter of the model, respectively.
Since a massless scalar state also survives in the zero-slope limit,
the complete Lagrangian describing the Virasoro-Shapiro model in the
limit is that of a scalar-tensor theory of gravity[3]. It is also reasonable to expect that the connection also exists between the non-planar
version of the Neveu-Schwarz model and the gravity theory.

This connection indicates that the built-in gauge symmetry of the
dual models embodies in some sense the principle of general relativity.
However, it is of course hard to understand why the gauge symmetry of
the dual models that is a reflection of the geometrical nature of the
string dynamics has connection with general relativity, the equivalence
principle, and so on. To gain some insights in this problem it will be
interesting to make the theory of dual strings generally covariant.
For the string with only orbital degrees of freedom, this problem has
already been touched upon in reference[5]. In the following, we shall
consider the fermionic string of RAMOND[6]. Then our problem is to
construct the dual model counter-part of the theory of interacting

gravitational and spinor fields in usual local field theory. So we
shall closely follow the procedure of usual tetrad formalism[7].

We demand that the following three requirements be satisfied by
the generally relativistic theory of the fermionic string:
(i) The equations must be invariant under general coordinate transformation in space-time.
(ii) The principle of equivalence requires that the special relativity
should apply in each locally inertial frame. Thus the equations should
be invariant with respect to local Lorentz transformation defined at
each space-time position independently.
(iii) The super-gauge symmetry should not be violated.
It is not evident at all whether or not the requirement (iii) is compatible with the requirements (i) and (ii). Demonstrating just this
compatibility and an interesting interplay between them is our main
task in the following. For a more detailed account, see[8].

First, to implement the requirements (i) and (ii), we consider the
infinitesimal local transformations of the form

$$\delta x^{\mu}(\theta) \;=\; f^{r}(x(\theta)) \;, \tag{1}$$

$$\delta S^{\pm\,\alpha}(\theta) = \varrho^{\alpha\beta}(x(\theta))\, S_{\beta}^{\pm}(\theta)\,, \quad (\varrho^{\alpha\beta} = -\varrho^{\beta\alpha}) \tag{2}$$

where $x(\theta)$ is the four-coordinate of the string, parametrized by θ from
0 to π , and $S^{\pm}(\theta)$ are two component four-spinors defined on the string.
The super-gauge operators are given as

$$F_{n} = \frac{\sqrt{\alpha'}}{\pi} \int_{0}^{\pi} \left\{ \left(S^{+}(\theta)\cdot p(\theta) + S^{-}(\theta)\cdot x'(\theta) \right) \cos n\theta + i \left(S^{-}(\theta)\cdot p(\theta) + S^{+}(\theta)\cdot x'(\theta) \right) \sin n\theta \right\} d\theta \tag{3}$$

where $p(\theta)$ is the four-momentum canonically conjugate to the four-coordinate. The operators F_{n} can be made invariant under the transformation
by making the following replacement;

$$p_{\mu}(\theta) \;\longrightarrow\; \mathbb{P}_{\alpha}(\theta) \equiv b_{\alpha}^{\ r}(x(\theta))\left(p_{\mu}(\theta) + \tfrac{1}{2}i\, A^{\beta\gamma}_{\ \ r}(x(\theta))\, M_{\beta\gamma}(\theta)\right) \;,$$

$$x^{\mu}{}'(\theta) \;\longrightarrow\; \mathbb{V}^{\alpha}(\theta) \equiv h^{\alpha}_{\ \mu}(x(\theta))\, x^{\mu}{}'(\theta) \;,$$

where

$$M_{\alpha\beta}(\theta) \;=\; \tfrac{1}{4}\left[S_{\alpha}^{+}(\theta),\, S_{\beta}^{+}(\theta)\right] + \tfrac{1}{4}\left[S_{\alpha}^{-}(\theta),\, S_{\beta}^{-}(\theta)\right] \;,$$

and $h^{\alpha}_{\ \mu}$ is the usual tetrad field and $b_{\alpha}^{\ r}$ is its inverse:

$$g_{\mu\nu}(x) = h^{\alpha}_{\ \mu}(x) h_{\alpha\nu}(x) , \quad b_{\alpha}^{\ r}(x) h^{\alpha}_{\ \nu}(x) = \delta^{r}_{\ \nu} , \quad b_{\alpha}^{\ r}(x) h^{\beta}_{\ \mu}(x) = \delta_{\alpha}^{\ \beta} .$$

In terms of the tetrad and its inverse the "spin connexion" $A_{\alpha\beta\mu}$ is expressed as

$$A_{\alpha\beta\mu} = \tfrac{1}{2} h^{\delta}_{\ r} (C_{\delta\alpha\beta} - C_{\alpha\beta\delta} - C_{\beta\delta\alpha}),$$
$$C^{\tau}_{\ \alpha\beta} = (b_{\alpha}^{\ r} b_{\beta}^{\ \nu} - b_{\beta}^{\ r} b_{\alpha}^{\ \nu}) \partial_{\nu} h^{\tau}_{\ \mu} . \tag{4}$$

In the quantized theory of a string, we can consistently linearize the system of the equations because the seagull terms can always be set to zero for the properly regularized matrix elements in the Fock space of a string. We denote the gauge operators, obtained from the above replacement and linearized by setting $b_{\alpha}^{\ r} = \delta_{\alpha}^{\ r} + a_{\alpha}^{\ r}$ and only retaining the zeroth and first order terms in $a_{\alpha}^{\ r}$, by $\tilde{F}_n(a)$. The generally relativistic equations for the fermionic string are then given as

$$\left(\tilde{F}_n(a) - \sqrt{\alpha'} M \delta_{n,0} \right) |\psi\rangle = 0 , \quad (n \geq 0) . \tag{5}$$

From the invariance of $\tilde{F}_n$ under the transformations (1) and (2), we can derive the following gauge identity

$$[F_n , T[f^r] + M[e^{\alpha\beta}]] = \tilde{F}_n (\partial_{\alpha} f_r + e_{\alpha\mu}) - F_n \tag{6}$$

where $T[f]$ and $M[e]$ are the generators for the transformations (1) and (2), respectively. Using (6) we can show that (i) the anti-symmetric part of the $a_{\alpha\mu}$ does not couple with the S-matrix elements describing the interaction of the tetrad field and the fermionic string and (ii) gravitational gauge invariance is satisfied for the S-matrix elements.

If the field $a_{\alpha\mu}$ is massless, transversal and traceless, we can show that the gauge operators $\tilde{F}_n$ satisfy the closed algebra of the same form as that for free F_n :

$$\{ \tilde{F}_n , \tilde{F}_m \} = -2 \tilde{L}_{n+m} - \tfrac{1}{2} d m^2 \delta_{m,-n} ,$$
$$[\tilde{L}_m , \tilde{F}_n] = (\tfrac{1}{2} m - n) \tilde{F}_{n+m} , \tag{7}$$
$$[\tilde{L}_m , \tilde{L}_n] = (m - n) \tilde{L}_{m+n} + \tfrac{1}{8} d m^3 \delta_{m,-n} ,$$

where the first can serve as the definition of $\tilde{L}_n$. The requirement (iii) is thus compatible with (i) and (ii). Necessity of the restrictions on the field $a_{\alpha\mu}$ indicates that the super-gauge symmetry is only the symmetry of the on-mass-shell physical S-matrix elements. We stress that in obtaining the closed algebra it is very essential to require the invariance under the local Lorentz transformation and the identity (4) as well as the invariance under general coordinate transformation. We notice that these requirements are almost equivalent to Einstein's principle of equivalence. It is very interesting that such an interplay exists between the super-gauge symmetry, which is usually regarded as an internal symmetry of the fermionic string, and the external-geometric property of space-time, in which the string is embedded.

On the other hand, the topological property of the amplitude obtained from the wave equation (5) indicates that the external graviton state represented by the tetrad is nothing but the massless spin-2 state of the close-string sector of the Neveu-Schwarz-Ramond model; that is, by performing a suitable duality transformation we can extract from the amplitude the part representing the self-interaction among gravitons, and show that it is just the non-planar version[9]of the Neveu-Schwarz model. Possibility of consistent linearization owes to this duality property of the amplitude.

Incidentally, we have been able to construct a dual model for interacting fermionic and pomeronic strings in a manifestly Lorentz covariant way. The prodedure adopted here also teaches us that the super-gauge structure of the model indeed embodies some intrigueing connection with the principle of general relativity.

The author wishes to thank the Japan Society for the Promotion of Science for financial support.

REFERENCES

1   For reviews, see C. Rebbi, Phys. Reports 12C, 1 (1974);
     S. Mandelstam, Phys. Reports 13C, 260 (1974).
2   T. Yoneya, Prog. Theor. Phys. 51, 1907 (1974).
     See also, T. Yoneya, Lett. Nuovo Cim. 8, 951 (1973).
3   J. Scherk and J. H. Schwarz, Nucl. Phys. B81, 118 (1974).
4   S. N. Gupta, Proc. Phys. Soc. 65A, 161 (1952).
     R. P. Feynman, Acta Phys. Polon. 24, 697 (1963).
     B. S. DeWitt, Phys. Rev. 160, 1113 (1967).
5   M. Ademollo, A. D'Adda, R. D'Auria, E. Napolitano, P. Di Vecchia,
     F. Gliozzi and S. Sciuto, Nucl. Phys. B77, 189 (1974).
6   P. Ramond, Phys. Rev. D3, 2415 (1971).
7   See, e. g., S. Weinberg, "Gravitation and Cosmology" (New York,
     N. Y., 1972), p365;
     R. Utiyama, Phys. Rev. 101, 1597 (1956).
8   T. Yoneya, Hokkaido University preprint (1975), "Interacting
     Fermionic and Pomeronic Strings".
9   J. H. Schwarz, Nucl. Phys. B46, 61 (1972).

PHYSICAL STATES SATISFYING SUPERGAUGE CONDITIONS IN DUAL RESONANCE MODELS[1]

Gaku KONISI and Takesi SAITO*

Department of Physics, Kwansei Gakuin University
Nishinomiya 662

*Department of Physics, Osaka University
Toyonaka, Osaka 560

In the dual resonance models we give generating functions to calculate the number of the positive-norm on-shell states which are simultaneous eigenstates of the mass-squared operator R and the angular momentum J and satisfy the supergauge conditions.

In the dual pion model with the supergauge symmetry[2] the on-shell pysical states $|\psi, N\rangle$ satisfy the supergauge conditions

$$G_r |\psi, N\rangle = 0, \qquad r = 1/2, \; 3/2, \cdots \tag{1}$$

and

$$(L_o - \tfrac{1}{2})|\psi, N\rangle = (R - \alpha'p^2 - \tfrac{1}{2})|\psi, N\rangle = 0, \tag{2}$$

with

$$\alpha'p^2 = N - \tfrac{1}{2}, \tag{3}$$

where α' is the Regge slope, p^μ is the momentum of the state, and the ground state $|0, p\rangle$ corresponds to the "pion" with the mass $\alpha'p^2 = -1/2$. All the physical states satisfying (1) and (2) can be constructed by means of the spectrum generating operators B_r^i, B_r^+, A_n^i and A_n^+, which anticommute or commute with the supergauge operator G_r, and form the so-called spectrum generating algebra[3].

Let us denote the number of linearly independent states of the physical sector $\{R = N, J_3 = M\}$ by $T_{R,J_3}(N, M)$, where J_3 is the third component of the spin angular momentum of the string particle. Then the generating function for $T_{R,J_3}(N, M)$ is given by

$$G_{R,J_3}(t, z) \equiv \sum_{N,M} T_{R,J_3}(N, M)\, t^N z^M$$

$$= \frac{1-t}{1+\sqrt{t}} \prod_{r=1/2}^{\infty} [(1+zt^r)(1+t^r/z)(1+t^r)] \prod_{n=1}^{\infty} [(1-zt^n)(1-t^n/z)(1-t^n)]^{-1}. \tag{4}$$

By making use of the relation

$$T_{R,J}(N, M) = T_{R,J_3}(N, M) - T_{R,J_3}(N, M+1) + \delta_{N,1/2}\delta_{M,0}, \quad (5)$$

where $T_{R,J}(N, M)$ is the number of the $(2M+1)$-plets in the sector $\{R = N, J = M\}$, we get a generating function for $T_{R,J}(N, M)$:

$$G_{R,J}(t, z) \equiv \sum_{N,M} T_{R,J}(N, M)t^N z^M = (1-\frac{1}{z})G_{R,J_3}(t, z) + t^{1/2}. \quad (6)$$

The explicit values of the $T_{R,J}$'s for $R - J \leq 5$ are tabulated in Fig. 1. We can see that there occurs a saturation at $R = 2J - 3/2$ or $2J - 2$ along the line $R - J = $ const., and that there are a few missing states. Furthermore, we point out that the odd daughter states in the "ρ" family will be found experimentally in the $n\pi$-channel with $n \leq 4$, though not in the 2 -channel. On the other hand the three states $\varepsilon(J = 0, R = 1/2)$, $\rho'(1, 3/2)$ and $\varepsilon''(0, 5/2)$ in the "ρ" family are always missing.

1) G. Konisi and T. Saito, Prog. Theor. Phys. <u>52</u> (1974), 1902.
2) A. Neveu and J. H. Schwarz, Nucl. Phys. <u>B 31</u> (1971), 86.
3) R. C. Brower and K. A. Friedman, Phys. Rev. <u>D7</u> (1973), 535.

FIGURE CAPTIONS:

Fig. 1a The number of $(2J+1)$-plets for $R = 1/2, 3/2, 5/2, \cdots$(the family of "$\rho$" trajectories in the dual pion model). The points of saturation are indicated by squares.

Fig. 1b The number of $(2J+1)$-plets for $R = 0, 1, 2, \cdots$(the family of "π" trajectories in the dual pion model).

DISCUSSION:

Question (K. Kikkawa): Can you simplify your calculation if you use the light cone gauge?

Answer: Yes. We can get the same results using the light cone gauge. However the equivalence between the two formalisms is not obvious at first sight.

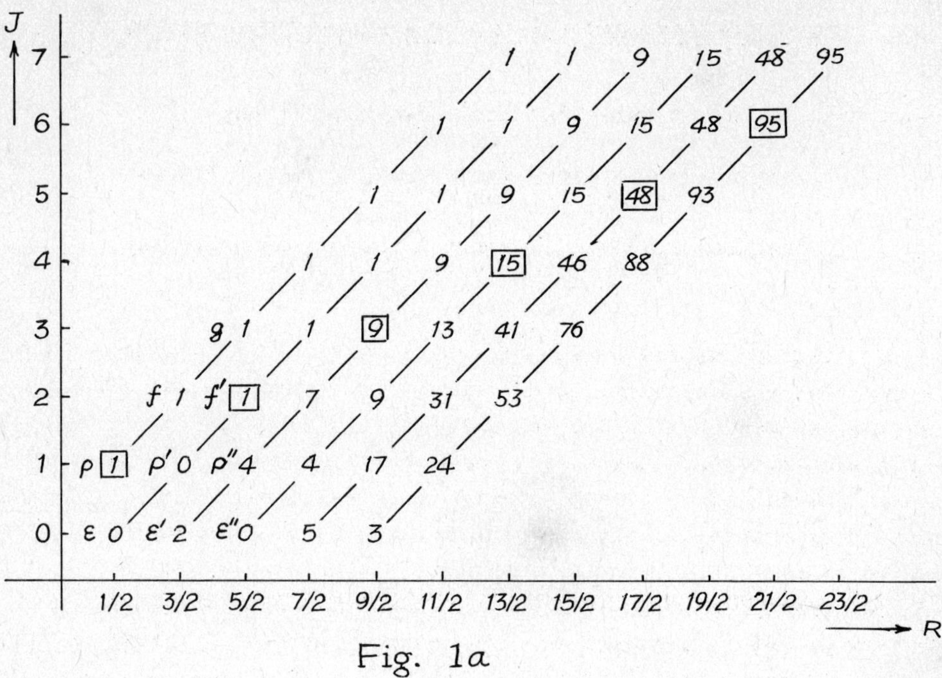

Fig. 1a

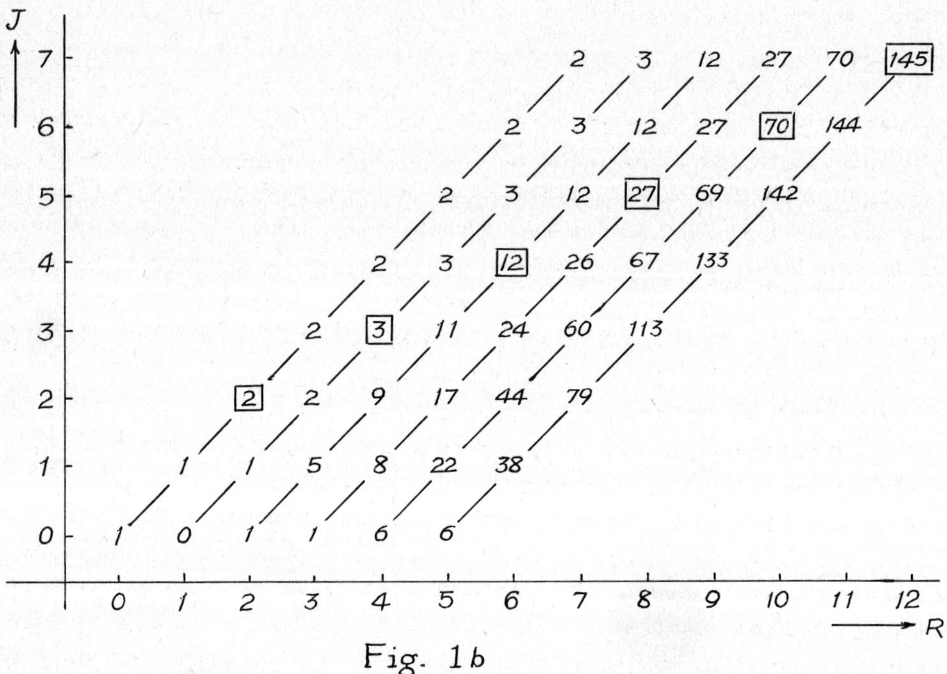

Fig. 1b

THE CANONICAL QUANTIZATION OF A RELATIVISTIC STRING

Yūichi CHIKASHIGE and Kiyoshi KAMIMURA[†]

Institute of Physics, University of Tokyo, Komaba
Meguro-ku, Tokyo 153, Japan

[†] Department of Physics, Tokyo Metropolitan University
Setagaya-ku, Tokyo 158, Japan

We describe a free motion of a relativistic string within the framework of the canonical formalism[1]. This problem was already discussed by Goddard, Goldstone, Rebbi and Thorn (G.G.R.T.)[2], but we believe it is worthwhile to reconsider this problem from a more general standpoint of view by quantizing a system with some constraints and to clarify the meaning of G.G.R.T.'s treatment.

We start with the famous Nambu-Gotō Lagrangian[3] which is defined in terms of two arbitrary parameters τ and σ on the world sheet swept out by a string in the Minkowski space. The Euler equations deduced from this Lagrangian, $\mathcal{L} = \{ (\dot{x} \cdot x')^2 - \dot{x}^2 x'^2 \}^{1/2}$, are not mutually independent because $\det(\partial^2 \mathcal{L} / \partial \dot{x}_\mu \partial \dot{x}_\nu) = 0$, where $x_\mu(\tau, \sigma)$ is a parametrization of the world sheet and $\dot{x}_\mu$ and x'_μ are the partial derivatives with respect to τ and σ. In passing from the Lagrangian formalism to the Hamiltonian formalism, we see that the Hamiltonian function, $p\dot{x} - \mathcal{L}$, vanishes identically and $\dot{x}_\mu$ cannot be represented as a function of x_μ and its conjugate momentum p_μ , since $\mathcal{L}$ depends on $\dot{x}_\mu$ linearly. Instead, there appear two constraints on x_μ and p_μ in this system such as $H(\sigma) \equiv p^2 + x'^2 = 0$ and $T(\sigma) \equiv p \cdot x' = 0$. Dirac proposed a way to construct the Hamiltonian formalism for a system with such a singular Lagrangian[4]. This generalized formalism has been applied to gauge theories and gravitational theory by Dirac himself and other authors. We apply this formalism to the string model.

Here, we introduced the extra condition to expand both of x_μ and p_μ in terms of cosine series in σ, which expresses one dimensional extension of a string. In addition to this, we can impose the coordinate condition $n \cdot x' = 0$ for a timelike or lightlike vector n_μ , the same as is introduced by G.G.R.T. As a result of these extra conditions, all constraints are classified into two groups, the first class and the second class constraints. In our case, $H^0 \equiv \int d\sigma H(\sigma)$ has zero Poisson brackets with other constraints so that $H^0 = 0$ is the first class and gives the generalized Hamiltonian when it is multiplied by an arbitrary function of σ, while remaining constraints have not always zero Poisson brackets with each other and so these form the second class constraints. They are used to eliminate some redundant variables from the theory. To this end, Dirac proposed that the Poisson brackets should be modified to the so-called Dirac brackets which become zero, when it is applied to all the second class ones. The classical mechanics can be constructed by means of Dirac brackets and we can easily check the validity of Poincaré invariance.

When one transits from a classical theory to a quantum theory, one applies the correspondence principle which dictates in our case that Dirac brackets should be replaced by commutators. Moreover, the first class constraint should not be considered as the operator equation but the condition for physical states ψ as $H^0\psi=0$. However, there is a serious difficulty, in this procedure, about ordering of products of operators appearing in Dirac brackets. But, for the lightlike vector n_μ, we can fortunately avoid this difficulty and in this case Dirac brackets for the transverse oscillation variables are just the same as the corresponding Poisson brackets. This is the lightlike gauge. In constructing quantum mechanics for this system, we fix the order of operators appearing in the Hamiltonian and the generators of Poincaré group in a suitable way. Physical states appear on the linear rising Regge trajectories and there are only the states with positive norm. The relativistic invariance must be confirmed again to construct Poincaré algebra explicitly, including ambiguities caused by the first class constraint. Our final expressions for those generators are found to be the same as those G.G.R.T. Consequently, as for the Poincaré algebra, we agree with G.G.R.T. who showed that the spacetime dimension is 26 and a tachyon state must be included to construct the relativistic quantum string model without ghost in the context of Dirac's generalized formalism.

References
(1) Y. Chikashige and K. Kamimura, UT-Komaba 74-17.
(2) P. Goddard, J. Goldstone, C. Rebbi and C.B. Thorn, Nucl. Phys. B56 (1973) 109.
(3) Y. Nambu, Lectures at the Copenhagen Summer Symposium (1970), T. Gotō, Prog. Theor. Phys. 46 (1971) 1560.
(4) P.A.M. Dirac, Canad. J. Math. 2 (1950) 129.

ALGEBRAIC METHODS AND QUARK STRUCTURE[*]

Feza Gürsey

Physics Dept., Yale University, New Haven, Conn. 06520

I. <u>Introduction</u>. The incorporation of charges in field theory has
brought new features such as gauge invariance for the electric charge,
spontaneous breakdown of symmetry (Goldstone-Nambu phenomenon) for
chiral charge and asymptotic freedom[1] (leading to scale invariance at
high energies) for a Yang-Mills type gauge theory of non abelian charges.
The quark puzzle has led to the introduction of color degrees of freedom
by tripling the quarks (Han-Nambu[2] and Gell-Mann[3]) and to the hypothesis
that only color singlets are observed while color non singlets are confined and violate the asymptotic condition of field theory, color being
linked with absolute conservation of a new $SU^c(3)$ group. Thus, the introduction of color could bring a new radical modification of field theory.
Our first question is: Why should $SU^c(3)$ play such a special role?

On the other hand, the success of the Weinberg-Salam gauge model[4]
with Higgs symmetry breaking mechanism in unifying weak interactions has
led to more ambitious schemes which also incorporate strong interactions.
Such unified field theories of hadrons and leptons necessitate the introduction of new quarks and new leptons in general[5]. The model should contain the observed strangeness conserving neutral currents but not the
strangeness changing one to the same order. Then a new charge (charm) is
needed with at least one additional colored quark[6]. Schemes with two[7]
and three[8] new quarks have also been proposed, especially in connection
with the discovery of new hadrons[9] that may be linked with charm[10].
Taking the most natural generalization of Weinberg's $SU_L(2)xU(1)$ model
as $SU_L(3)xSU_R(3)$[11], the phenomenological group suggested by experiment
seems to be $SU^c(3)xSU_L(3)xSU_R(3)xC$ where C is the charmed group that can
be $U(1)$, $SU(2)$ or $SU(3)$ depending on the number of additional quarks.
In a unified field theory, parameters like the Weinberg angle get determined if this group is imbedded in a simple or semisimple group G of
rank 7 or 8. Hence a satisfactory unified gauge field theory of leptons
and quarks must be based on a large group in a highly complex charge
space. Then we ask: Are there uniquely distinguished groups of rank 7 or
8 with some special property of their $SU^c(3)$ subgroup that makes it exact?

Let us now consider the space in which a given compact Lie group
acts. This can be a finite vector space (Hilbert space) or a module
space of hermitian or antihermitian matrices over the division algebras
$H(d)$ of dimension $d=1,2,4,8$ associated with real, complex numbers, quater-

nions and octonions respectively. While orthogonal, unitary and symplectic groups can be represented on vector spaces over $H(1)$, $H(2)$ and $H(4)$ or on modules over the same algebras, the exceptional groups cannot be represented over octonionic vectors but only over octonion modules. Thus a finite Hilbert space over octonions with octonions as scalars does not exist due to the non-associativity of $H(8)$. On the other hand, the module formulation generalizes to octonions. Can octonionic hermitian operators (exceptional observables) serve as a module space for exceptional groups?

Fourthly there seems to be a symmetry between local lepton and quark fields which is required for the cancellation of strangeness changing neutral currents. It is an unusual symmetry since leptons have integer charges while colored quarks have fractional charges. Where does such a symmetry arise naturally?

Finally we would like to know if a natural chain of symmetry breaking exists for our unifying gauge group such that it leads to a hierarchy of masses and coupling constants.

Here I would like to suggest that the exceptional groups, owing to their very special algebraic properties may well provide the basis for a theory where the questions we have asked could be answered in principle.

Indeed, all the exceptional groups are associated with octonions. The simplest G_2 is a subgroup of the others and is the automorphism group of the octonion algebra. All exceptional groups can be represented on modules consisting of special octonionic matrices. Since octonions have seven imaginary units, if we identify one of them with the imaginary unit of the Quantum Mechanical Hilbert space we have six additional imaginary units with automorphism group $SU(3)$. This is a uniquely distinguished exact group of our octonionic module space[12] so that we can identify it with the color group $SU_c(3)$. It is a common subgroup of all the exceptional groups G_2, F_4, E_6, E_7 and E_8 with ranks shown as subscripts. The ranks of the E series which has a unique place in Lie Groups are of the right order.

As also shown by Jordan et al.[13], observables associated with hermitian operators generalize to 3x3 octonionic hermitian matrices called exceptional observables which serve as a module space for the representations of exceptional groups.[14] Hence we can associate the charge space with the space of exceptional observables. It is interesting that there is no octonionic vector space in this charge space on which the exceptional groups can be represented linearly. This phenomenon may be linked with quark containment[15].

Further, the special role assigned to one octonionic unit, say e_7, leads to the splitting of the octonions[12] into a longitudinal part involving 1 and e_7 and a transverse part containing the six remaining units $e_1, \ldots, e_6$. Exceptional **observables** and modules will also split into longitudinal and transverse parts and the same electric charge operator assigns integer charges to longitudinal octonions and fractional charges to their transverse parts. This suggests the identification of these two parts of the basic octonionic module field **as leptons and quarks** respectively[15], which combine within a complete octonion.

Finally the automorphism groups of the subalgebras of the algebras[16] of octonionic modules for the exceptional groups provide a natural hierarchy of subgroups. These subgroups and their associated stability groups can be displayed in Freudenthal's magic square[17] and have geometrical interpretations in terms of the corresponding Tits geometries[18] that generalize projective geometries.

2. <u>The Hurwitz and Rozenfeld Algebras</u>. According to the Hurwitz theorem the algebras $H(d)$ with $d=1,2,4,8$ are the only normed division algebras. The elements of $H(8)$ are octonions $a = a_0 + e_n a_n$ $(n=1,\ldots,7)$ where e_n have multiplication table

$$e_m e_n = -\delta_{mn} + \phi_{mnr} e_r \qquad (2.1)$$

where the antisymmetrical tensor ϕ is equal to one for (n,m,r) given by (123) and its cyclic permutations following the order (1243657). The algebra defined by (2.1) is non associative and has G_2 with 14 parameters as its automorphism group. The associative subalgebras with the 3 imaginary units (e_m, e_n, e_r) are quaternion algebras which can be represented by the 2x2 Pauli matrices $i\sigma_j$, and have SU(2) as automorphism groups. Longitudinal octonions (complex numbers) $a_0 + e_7 a_7$ have no continuous automorphism groups. The octonion conjugate of a is $\bar{a} = a_0 - e_n a_n$. The norm $N(a)$ and the inverse are

$$N(a) = [a\bar{a}]^{1/2}, \quad a^{-1} = \bar{a}/(a\bar{a}) . \qquad (2.2)$$

The associator $[abc] = (ab)c - a(bc)$ is alternative, i.e.

$$[abc] = -[\bar{a}\bar{b}\bar{c}] = -[bac] = [bca] = [cab] . \qquad (2.3)$$

The Rozenfeld algebras are direct products defined by

$$H(d,d') = H(d) \otimes H(d') \qquad (2.4)$$

and have dimension dd'. Thus $H(2,4)$, $H(4,8)$ denote respectively the algebras of complex quaternions and quaternionic octonions. The infinitesimal action of G_2 on the octonion x is given by[14]

$$x' = x - \frac{1}{4}[[a,b],x] + \frac{3}{4}[a,b,x] . \qquad (2.5)$$

The corresponding automorphism groups are $g(d,d')$

$$g(i,j) = 1 \ (i,j=1,2), \ g(i,4) = SU(2), \ g(4,8) = SU(2) \times G_2 \ \text{etc.} \qquad (2.6)$$

3. <u>The Algebras $J(d,d')$ of Generalized Observables</u>. We consider 3x3 hermitian matrices J over $H(d,d')$ with

$$J = J^\dagger = \bar{J}^T \ \text{with} \ \bar{e}_n = -e_n, \ \bar{e}'_m = -e_m \ . \qquad (3.1)$$

$J(1,8)=J(8,1)$ are elements of the Jordan algebra of exceptional observables for which the product is

$$J_1 \cdot J_2 = \tfrac{1}{2} (J_1 J_2 + J_2 J_1) \ . \qquad (3.2)$$

It is a commutative non associative and power associative algebra. The determinant is defined by[14]

$$I \ \text{Det} \ J = J^{\cdot 3} - J^2 \text{Tr} J + \tfrac{1}{2} J\{(\text{Tr} J)^2 - \text{Tr} J^2\} \qquad (3.3)$$

where I is the 3x3 unit matrix. For non singular matrices the inverse of J is

$$J^{-1} = J^{\#} (\text{Det} \ J)^{-1}, \qquad (3.4)$$

$$J^{\#} = (J^2 - J \text{Tr} J) - \tfrac{1}{2} I \ \text{Tr}(J^2 - J \text{Tr} J). \qquad (3.5)$$

With the property

$$J^{-1} \cdot J = I, \quad J^{-1} \cdot J^2 = J \ . \qquad (3.6)$$

Using Gell-Mann's 3x3 λ matrices we can write

$$A^\dagger = A = I\alpha_0 + \lambda_3 \alpha_3 + \lambda_8 \alpha_8 + \tfrac{1}{2}(\lambda_1 + i\lambda_2)a_3 + \tfrac{1}{2}(\lambda_4 + i\lambda_5)\bar{a}_2 \qquad (3.7)$$

$$+ \tfrac{1}{2}(\lambda_6 + i\lambda_7)a_1 + \tfrac{1}{2}(\lambda_1 - i\lambda_2)\bar{a}_3 + \tfrac{1}{2}(\lambda_4 - i\lambda_5)a_2 + \tfrac{1}{2}(\lambda_6 - i\lambda_7)\bar{a}_1.$$

Here a_1, a_2, a_3 are octonions and $\alpha_0, \alpha_3, \alpha_8$ are real. If A is traceless $(\alpha_0=0)$ it is a module with 26 real parameters. The associator is

$$(A,J,B) = (A.J).B - A.(J.B) \qquad (3.8)$$

and has the properties

$$(A,J,A) = 0, \ (A,J,A^2) = 0, \ (A,J,A^{-1}) = 0. \qquad (3.9)$$

Now, if A,B are traceless and infinitesimal

$$J' = J + (A,J,B) \qquad (3.10)$$

is the 2x26=52 parameter representation of the Lie algebra F_4 over the octonionic J module.[14] We have

$$(A,J,B) = - \tfrac{1}{4} [[A,B],J] + \tfrac{1}{4} \sum_{A,B,J} [ABJ] \qquad (3.11)$$

where

$$[ABJ] = (AB)J - A(BJ) \qquad (3.12)$$

and the summation is over the six permutations of A,B,J. Thus F_4 is associated with the traceless antihermitian matrix $[A,B]$ with 38 para-

meters. The ordinary matrix multiplication associators [A B J] give a
diagonal G_2 action on J with 14 parameters completing the group F_4. This
procedure generalizes to J(d,d') over other Rozenfeld algebras. The dimensions of the corresponding automorphism groups are given by the dimension of the antihermitian matrices plus that of the automorphism of H(d,d')
so that we have

$$\text{dim. } G(d,d'): \quad \begin{array}{cccc} 3 & 8 & 18+3=21 & 38+14=52 \\ 8 & 16 & 32+3=35 & 64+14=78 \\ 21 & 35 & 60+6=66 & 116+17=133 \\ 52 & 78 & 133 & 220+28=248 \end{array} \qquad (3.13)$$

corresponding to the groups of the magic square

$$G(d,d'): \quad \begin{array}{cccc} SU(2) & SU(3) & Sp(6) & F_4 \\ SU(3) & SU(3){\times}SU(3) & SU(6) & E_6 \\ Sp(6) & SU(6) & SO(12) & E_7 \\ F_4 & E_6 & E_7 & E_8 \end{array} \qquad (3.14)$$

We also consider a reduction of these Lie algebras with respect to 2x2
Jordan algebras over H(d,d'). This provides the stability groups of the
magic square given by the table below. The symmetric coset spaces
G(d,d')/S(d,d') have dimension dim G-Dim S=2dd' corresponding to a pair
of elements H(d,d') and provide the framework for Tits geometries[17]

$$S(d,d'): \quad \begin{array}{cccc} U(1) & SU(2){\times}U(1) & SO(5){\times}SO(3) & SO(9) \\ SU(2){\times}U(1) & SO(4){\times}U(1){\times}U(1) & SU(4){\times}SU(2){\times}U(1) & SO(10){\times}SO(2) \\ SO(5){\times}SO(3) & SU(4){\times}SU(2){\times}U(1) & SO(8){\times}SO(4) & SO(12){\times}SO(3) \\ SO(9) & SO(10){\times}SO(2) & SO(12){\times}SO(3) & SO(16) \end{array}$$
$$(3.15)$$

For each group of the magic square we find 2 chains of symmetry
breaking associated with the tables (3.14) or (3.15). This generalizes
the results of Michel and Radicati[19] for SU(3)xSU(3) or G(2,2).

4. **Quark Structure of the Exceptional Groups.** The exceptional groups
G(8,d) all have subgroup $SU^c(3)$ which commutes with the automorphism
groups G(2,d') of complex hermitian matrices involving e_7. Thus we find
the maximal subgroups:

$$G(8,d') \supset SU^c(3) \times G(2,d') . \qquad (4.1)$$

Thus F_4, E_6, E_7, E_8 have subgroups $SU^c(3){\times}SU(3)$, $SU^c(3){\times}SU(3){\times}SU(3)$, $SU^c(3)$
${\times}SU(6)$ and $SU^c(3){\times}SU(3){\times}SU(3){\times}SU'(3)$. The last one corresponds to $SU^c(3)$
${\times}SU_L(3){\times}SU_R(3){\times}SU'(3)$, the SU'(3) being the group of 3 charmed quarks.
This is just the group of Barnett's six quark model.[8]

5. **Lepton-Quark Symmetry.** We can illustrate the lepton-quark symmetry
on E_6 which has as module a complex J (27 dimensional over H(2)). We

can write the left handed spinor field (longitudinal + transverse)

$$J^L = J^L_\ell + J^L_t = \begin{pmatrix} \alpha & \hat{\mu}_R & \hat{e}_R \\ \mu_L & \beta & \nu_\mu \\ e_L & \nu_e & \gamma \end{pmatrix} + \begin{pmatrix} 0 & \lambda'_L & \mathcal{N}'_L \\ \lambda_L & 0 & \mathcal{P}'_L \\ \mathcal{N}_L & \mathcal{P}_L & 0 \end{pmatrix}, \quad (\hat{e}_R = i\sigma_2 e^*_L), \quad (\alpha = \alpha_L = \hat{\alpha}_R). \qquad (5.1)$$

The fields α, β, γ correspond to additional neutral Majorana leptons. The right handed fields are given by $J^R = J^R_\ell + J^R_t$. Now if $q = \frac{1}{2}(\lambda_3 + \lambda_8/\sqrt{3})$ is the charge operator the charge gauge reads

$$J'^L = \exp(e_7 \theta q)\, J^L \exp(-e_7 \theta q) . \qquad (5.2)$$

This assigns the correct charges to leptons and fractional charges to the quarks $\lambda, \mathcal{N}, \mathcal{P}$ and the charmed quarks $\lambda', \mathcal{N}', \mathcal{P}'$. If J^L is used for a weak Lagrangian the resulting scheme is similar to Barnett's model[8].

Note that E_7 has a 56 dimensional module representation consisting of J^L, $(J^L)^*$ and a complex number that can be associated with a Dirac field. The smallest module representation of E_8 is its adjoint representation.

6. <u>Symmetry Breaking Chains</u>. Let us give an example of a nonlinear realization of the group with the stability group acting linearly as in spontaneous symmetry breaking. Starting from E_8 and isolating the color group we have $E_8 \supset SU(3) \times E_6$ from (4.1). Now we represent E_6 nonlinearly with its stability group $SO(10) \times SO(2)$ acting linearly by spontaneous breaking. This is just a model considered by Fritzsch and Minkowski[5]. Table (3.15) then gives the chain

$$SO(10) \times SO(2) \supset SU(4) \times SU(2) \times U(1) \supset SU(2) \times U(1) \times SU(2) \times U(1) \supset SU(2) \times U(1)$$
$$\qquad (6.1)$$

The last group can be identified with the Weinberg group and $SU(4)$ at an intermediate stage with the charmed $SU(4)$ which results from the breaking of the charmed $SU(6)$. Another example based on E_7 is

$$E_7 \supset SU^c(3) \times SU(6), \quad SU(6) \supset SU(4) \times SU(2) \times U(1). \qquad (6.2)$$

<u>Acknowledgement</u>. I am grateful to M. Günaydin, M. Gell-Mann, P. Ramond and R. Slansky for discussions, numerous contributions and clarifications of many ideas discussed here. My thanks are due to the National Science Foundation for the award of a travel grant.

*Research(Yale Report COO-3075-101) supported in part by the U.S. Atomic Energy Commission under grant no. AT(11-1)-3075.

<u>References</u>

1. D. Gross and F. Wilczek, Phys. Rev. Lett. <u>30</u>, 1343 (1973), H. Politzer, ibid, 30, 1346 (1973).

2. M.Y. Han and Y. Nambu, Phys. Rev. 139, B1006 (1965), Y. Nambu and
   M.Y. Han, ibid, D10, 674 (1974).
3. M. Gell-Mann, Acta Phys. Austr. Suppl. 9, 733 (1972), H. Fritzsch
   and M. Gell-Mann, in Proc. XVI Int. Conf. High Energy Physics, vol.
   2, p. 135 (NAL, Batavia, 1973).
4. S. Weinberg, Phys. Rev. Lett. 19, 1264 (1967), A. Salam, Proc. 8th
   Nobel Symp. (Wiley, New York 1968).
5. See for instance, B.W. Lee, in Proc. XVI Int. Conf. High Energy Physics,
   vol. 4, p. 249 (NAL, Batavia, 1973); H. Fritzsch and P. Minkowski,
   Unified Interactions of Leptons and Hadrons, Cal-Tech preprint 68-74
   (1974).
6. S.L. Glashow, J. Iliopoulos and L. Maiani, Phys. Rev. D2, 1285 (1970).
7. H. Georgi and S. Glashow, Phys. Rev. Lett. 32, 438 (1974).
8. R.M. Barnett, Phys. Rev. Lett. 34, 41 (1975).
9. J.J. Aubert et al., Phys. Rev. Lett. 33, 1404 (1974), J.-E. Augustin
   et al., ibid. 33, 1406 (1974).
10. A. de Rújula and S. Glashow, Phys. Rev. Lett. 34, 46 (1975).
11. S. Weinberg, Phys. Rev. D5, 1962 (1972).
12. M. Günaydin and F. Gürsey, Nuovo Cimento Lett. 6, 401 (1973), J. Math.
    Phys. 14, 1651 (1973), Phys. Rev. D9, 3387 (1974), M. Günaydin, Ph.D.
    Thesis, Yale Univ. 1973 (unpublished).
13. P. Jordan, J. von Neumann and E.P. Wigner, Ann. Math. 35, 29 (1934).
14. See for instance R.D. Schafer, Introduction to Non Associative Al-
    gebras (Acad. Press, 1966), N. Jacobson, Exceptional Lie Algebras
    (M. Dekker, 1971).
15. F. Gürsey, The Johns Hopkins Univ. Workshop on Current Problems in
    High Energy Particle Theory, p. 15, (Johns Hopkins Univ., 1974).
16. B.A. Rozenfeld, Proc. Colloq. Utrecht on Foundations of Geometry,
    p. 135 (1962).
17. H. Freudenthal, Proc. Colloq. Utrecht, p. 145 (1962).
18. J. Tits, Proc. Colloq. Utrecht, p. 175 (1962).
19. L. Michel and L.A. Radicati, Ann. Inst. H. Poincaré, 18, 185 (1973).

Discussion

TAKABAYASI Is your group decoupled from the Poincaré group?

GÜRSEY In the examples I have treated the charge groups with $SU(3)$ factorized are decoupled from the Poincaré group $\mathcal{P}$. e_7 is used however as
the imaginary unit for the representation of $\mathcal{P}$ in Hilbert space. There
are non compact forms of E_7 and E_8 that could unify $\mathcal{P}$ or the conformal
group with the charge group.

OHNUKI How do you explain the quark confinement in your model?

GÜRSEY According to current beliefs quark confinement is dynamical and
caused by infrared slavery in gauge theories with exact $SU(3)$. This
mechanism would work for any color group. If $SU(3)$ plays a special
role there may also be a non dynamical reason, associated with the fact
that exceptional observables can factorize when their determinant vanishes, but the factors do not form linear representations of the exceptional group. The longitudinal parts of these factors (larks) may obey
the asymptotic condition but not the transverse quark parts. Color
singlets will always be longitudinal and lie in a convention Hilbert
space and be observable in the usual sense.

<u>UNITARITY AND ASYMPTOTIC CONDITION IN A MODEL WITH DIPOLE GHOST</u>

K. Sekine

Department of Physics, Meisei University
Hino, Tokyo, Japan

The purpose of this talk is to clarify the mechanism of obtaining a unitary "physical" S-matrix in a simple model involving dipole ghost scattering.

The model is defined by the Hamiltonian $H = H_0 + H_1$, where

$$H_0 = \int d^3k\, \omega(k)\, \theta^*(k)\, \theta(k) + m_N N^* N +$$

$$+ m(A^* B + B^* A) + \lambda A^* A, \qquad \omega(k) = \sqrt{m_\theta^2 + k^2},$$

$$H_1 = K^* K, \qquad K = \theta\{g\} N + \theta\{g_1\} A + \theta\{g_2\} B.$$

Here $\theta\{g\}$ etc. are smeared out fields with normalizable functions, i.e.

$$\theta\{g\} = \int d^3k\, \bar{g}(k)\, \theta(k), \qquad g \in L^2.$$

The commutation relations are

$$[\theta\{g\}, \theta^*\{f\}] = (g, f)_{L^2},$$

$$[N, N^*] = [A, B^*] = [B, A^*] = 1,$$

all other pairs commuting. There are two constants of the motion :

$$\mathcal{N}_1 = N^* N + A^* B + B^* A, \qquad \mathcal{N}_2 = \int d^3k\, \theta^*(k)\theta(k),$$

as a result of which the state vector space $\mathcal{F}$ is decomposed into the direct sum of "sectors" : $\mathcal{F} = \oplus\, \mathcal{F}(n_1, n_2)$.

In the sector $\mathcal{F}(1,1)$, we can solve explicitly the off-shell Lippmann-Schwinger equation for an arbitrary inhomogeneous term.

Passing onto the energy shell, we obtain the "in" and "out" scattering states $|N, \varphi, \pm\rangle$, $|A, \varphi, \pm\rangle$ and $|B, \varphi, \pm\rangle$, for φ such that $\omega\varphi \in L^2$, satisfying

$$H\,|N,\varphi,\pm\rangle = |N, (m_N+\omega)\varphi, \pm\rangle\,,$$

$$H\,|A,\varphi,\pm\rangle = |A, (m+\omega)\varphi, \pm\rangle\,,$$

$$H\,|B,\varphi,\pm\rangle = |B, (m+\omega)\varphi, \pm\rangle + |A, \lambda\varphi, \pm\rangle,$$

and

$$\langle N,\varphi,\pm\,|\,N,\psi,\pm\rangle = (\varphi,\psi)_{L^2}\,,$$

$$\langle N,\varphi,\pm\,|\,A,\psi,\pm\rangle = 0\,,$$

$$\langle N,\varphi,\pm\,|\,B,\psi,\pm\rangle = 0\,,$$

$$\langle A,\varphi,\pm\,|\,A,\psi,\pm\rangle = 0\,,$$

$$\langle A,\varphi,\pm\,|\,B,\psi,\pm\rangle = (\varphi,\psi)_{L^2}\,,$$

$$\langle B,\varphi,\pm\,|\,B,\psi,\pm\rangle = 0\,.$$

On the basis of these relations we can establish the following propositions :

(1) $\quad |N,\varphi,\pm\rangle + |A,\varphi_1,\pm\rangle + |B,\varphi_2,\pm\rangle = 0,$

if and only if $\varphi = \varphi_1 = \varphi_2 = 0$.

(2) Let $m_\pm^N$ be the eigenspace spanned by $|N,\varphi,\pm\rangle$, and $m_\pm^A$ be the eigenspace spanned by $|A,\varphi,\pm\rangle$. Then, $m_\pm = m_\pm^N \oplus m_\pm^A$ is an invariant positive subspace, but not a positive definite one. In fact, for any $|\psi\rangle \equiv |N,\psi,\pm\rangle + |A,\psi_1,\pm\rangle \in m_\pm$, we find $\langle\psi|\psi\rangle = \|\psi\|^2 \geqslant 0$, which implies also that $m_\pm^A \subset m_\pm$ is the set of vectors of norm zero.

(3) Let $m = m_+ \cap m_-$. This, too, is an invariant positive subspace which we call "physical" subspace. For every $|\psi\rangle \in m$ we have

$$\langle A,\varphi,\pm\,|\,\psi\rangle = 0\,, \qquad \forall \varphi \in L^2.$$

(4) Let $m_\pm^{AB}$ be the invariant subspace spanned by $|A,\varphi,\pm\rangle$ as well as $|B,\varphi,\pm\rangle$. If we have the equality $m_+^N \oplus m_+^{AB} = m_-^N \oplus m_-^{AB} \equiv \mathcal{L}$, which we may call "asymptotic completeness", then the following two statements are equivalent :

    a) $|\psi\rangle \in \mathcal{L}$ and $\langle A,\varphi,\pm\,|\,\psi\rangle = 0,\ \forall \varphi \in L^2.$

    b) $|\psi\rangle \in m$.

(5) A vector $|\Psi> \in M$ can be written in two ways :

$$|\Psi> = |N, \psi, +> + |A, \psi_1, +> = |N, \psi', -> + |A, \psi_1', ->,$$

with

$$<\Psi|\Psi> = \|\psi\|^2 = \|\psi'\|^2.$$

Let $\{e_n\}$ be a complete orthonormal set of functions in L^2. Then,

$$\sum_n |<N, e_n, -|\Psi>|^2 = \sum_n |(e_n, \psi')|^2 = \|\psi'\|^2 = \|\psi\|^2,$$

and

$$\sum_n |<\Psi|N, e_n, +>|^2 = \sum_n |(\psi, e_n)|^2 = \|\psi\|^2 = \|\psi'\|^2.$$

These two equations express the conservation of probability, or "unitarity" as is usually called by physicists.

(6) A vector $|\Psi> \in M_+^A \cap M_-^A \subset M$, i.e. a vector of norm zero in our physical subspace, makes no contribution to the "physical" S-matrix :

$$<N, \varphi, -|\Psi> = <\Psi|N, \varphi, +> = 0 .$$

An explicit calculation reveals that actually $M_+ \neq M_-$. The "physical" S-matrix is nontrivial if $m > m_N$.

Let $\theta_\pm\{\varphi\}$ be the "in" and "out" fields satisfying

$$<N|\theta_\pm\{\varphi\} = <N, \varphi, \pm| .$$

Then we confirm the asymptotic condition

$$\lim_{t \to \mp \infty} <N|e^{iHt}\theta\{e^{-i\omega t}\varphi\}e^{-iHt}|\Psi> = <N|\theta_\pm\{\varphi\}|\Psi>.$$

for

$$|\Psi> = |N, \psi, \pm> + |A, \psi_1, \pm> \in M_\pm ,$$

provided that $g\bar{\varphi}, \bar{g}\psi, \bar{g}_2\psi_1 \in \mathcal{D}$. On the basis of this condition we can justify the use of the reduction formula such as

$$<N, \varphi, -|\Psi> = (\varphi, \psi) +$$

$$+ i \int_{-\infty}^{\infty} dt \int d^3k \, e^{i\omega t} \bar{\varphi}(k) \left(-i\frac{d}{dt} + \omega\right) <N|\theta(k, t)|\Psi>,$$

in order to evaluate our "physical" S-matrix, i.e. for $|\Psi> \in M$.

INVARIANT GAUGE FAMILIES INHERENT IN ABELIAN-GAUGE FIELD THEORY

Kan-ichi Yokoyama and Reijiro Kubo
Research Institute for Theoretical Physics
Hiroshima University, Takehara, Hiroshima 725

We present some universal features inherent in the whole neutral-vector field theory. Our problem essentially concerns an aspect on gauges. In Q.E.D. one-parameter covariant gauges for A_μ are specified by the form[1] of commutators of the asymptotic fields $A_\mu^{(in)}$ (or $A_\mu^{(out)}$)

$$[A_\mu^{(in)}(x), A_\nu^{(in)}(y)] = i\delta_{\mu\nu}D(x-y) - i\tau\partial_\mu\partial_\nu\tilde{D}(x-y), \qquad (1)$$

where $\tilde{D}(x) = -i(2\pi)^{-3}\int d^4k\varepsilon(k)\delta'(k^2)e^{ikx}$, $[\Box\tilde{D}(x) = D(x)]$. A given value of τ determines a gauge for A_μ. In view of the original concept on gauge, we should be able to have a certain transformation for A_μ which connects different values of τ with one another. The conventional Gupta-Bleuler theory is apparently incomplete in this respect, because it confines oneself only within the Feynman gauge ($\tau = 0$). The Nakanishi-Lautrup formalism[2] yields the commutators of the form (1), but in its framework, once one fixes τ at a certain value, one can no longer change the value in a manifestly covariant way. In order to ensure equivalence of all available gauges on the basis of manifest covariance, we should further enlarge the framework of the Nakanishi-Lautrup formalism by introducing a scalar dipole-ghost field B, as proposed recently.[3] We call B a <u>gaugeon</u> field.

In this way, we have presented Q.E.D. in general covariant gauges,[3] in which two kinds of dipole-ghost fields exist ; one corresponds to the longitudinal photon field, while the other is the gaugeon field. Our state-vector space is of indefinite metric. Then, physical states |phys> are subject to constraints ensuring that 1) none of the dipole ghosts are included in |phys> and 2) the Maxwell equations hold in the sense of expectation values for A_μ in |phys>. In this formalism there emerge two invariant gauge families in the sense that two arbitrary gauges belonging to the same family are connected with each other in a manifestly covariant way through a q-number gauge transformation for the relevant fields. The gaugeon field plays an essential role in this transformation. Here, we emphasize that such a gauge structure of theory is not only allowed for massless vector fields but also inherited by massive ones. In what follows, we shall introduce a unified description of this problem for the whole Abelian-gauge field.[4] We again emphasize that the description is well defined at the massless limit, so that it smoothly continues to that of our covariant-gauge Q.E.D.,[3] and it leads to a renormalizable theory. Our framework includes Nakanishi's Landau-gauge formalism[5] for neutral vector fields as a particular representation.

As the starting point of our theory, we take a free Lagrangian density, in which three kinds of auxiliary fields B, B_1 and B_2 are introduced in addition to a neutral vector field U_μ, in the form[5]

$$L_o = -\tfrac{1}{4} G_{\mu\nu} G_{\mu\nu} - \tfrac{1}{2} m^2 U_\mu U_\mu + B_1 \partial_\mu U_\mu - \partial_\mu B \partial_\mu B_2 - \tfrac{1}{2} \varepsilon \Lambda^2 , \tag{2}$$

where $G_{\mu\nu} = \partial_\mu U_\nu - \partial_\nu U_\mu$, $\Lambda = B_2 + \alpha B_1 + (\varepsilon/2)(1 + \varepsilon\alpha^2)m^2 B$, ε is a sign factor ($\varepsilon = \pm 1$) and α denotes our gauge parameter corresponding to $\tau = 1 + \varepsilon\alpha^2$.

Free-field equations follow from L_o in the form

$$\partial_\nu G_{\nu\mu} - m^2 U_\mu = \partial_\mu B_1 , \quad \partial_\mu U_\mu = \varepsilon\alpha\Lambda , \quad (\Box B_1 = -\varepsilon\alpha m^2 \Lambda)$$
$$\Box B = \varepsilon\Lambda , \quad \Box B_2 = \tfrac{1}{2}(1 + \varepsilon\alpha^2)m^2 \Lambda . \tag{3}$$

We further have $(\Box - m^2)\Lambda = 0$, from which we find

$$(\Box - m^2)\Box U_\mu = (\Box - m^2)\partial_\mu U_\mu = (\Box - m^2)\Box B = (\Box - m^2)\Box B_i = 0, \quad (i = 1,2). \tag{4}$$

Canonical quantization, together with the field equations, yields commutators at two arbitrary space-time points x and y in the form

$$[U_\mu(x), U_\nu(y)] = i\delta_{\mu\nu}\Delta(x-y) - i \frac{(1+\varepsilon\alpha^2)}{m^2} \partial_\mu \partial_\nu [\Delta(x-y) - D(x-y)] ,$$

$$[U_\mu(x), B(y)] = i \frac{\varepsilon\alpha}{m^2} \partial_\mu [\Delta(x-y) - D(x-y)] ,$$

$$[U_\mu(x), B_1(y)] = i(1 + \varepsilon\alpha^2)\partial_\mu D(x-y) - i\varepsilon\alpha^2 \partial_\mu \Delta(x-y) ,$$

$$[U_\mu(x), B_2(y)] = \tfrac{1}{2} \alpha(1 + \varepsilon\alpha^2)\partial_\mu [\Delta(x-y) - D(x-y)] ,$$

$$[B(x), B(y)] = i \frac{\varepsilon}{m^2} [\Delta(x-y) - D(x-y)] ,$$

$$[B(x), B_1(y)] = -i\varepsilon\alpha[\Delta(x-y) - D(x-y)] , \tag{5}$$

$$[B(x), B_2(y)] = \tfrac{1}{2}(1 + \varepsilon\alpha^2)\Delta(x-y) + \tfrac{1}{2}(1 - \varepsilon\alpha^2)D(x-y) ,$$

$$[B_1(x), B_1(y)] = -i(1 + \varepsilon\alpha^2)m^2 D(x-y) + i\varepsilon\alpha^2 m^2 \Delta(x-y) ,$$

$$[B_1(x), B_2(y)] = -\tfrac{1}{2}\alpha(1 + \varepsilon\alpha^2)m^2 [\Delta(x-y) - D(x-y)] ,$$

$$[B_2(x), B_2(y)] = \tfrac{1}{4}\varepsilon(1 + \varepsilon\alpha^2)^2 m^2 [\Delta(x-y) - D(x-y)] .$$

We note that all of the expressions obtained so far tend to those given in our previous work[3] at the limit $m \to 0$. The commutators in (5) suggest that our state-vector space is of indefinite metric. Hence, we define our physical states by the constraints

$$B_1^{(+)}(x)|\text{phys}> = B_2^{(+)}(x)|\text{phys}> = 0 . \qquad (6)$$

If $m \neq 0$, it follows $B^{(+)}(x)|\text{phys}> = 0$. Since (6) persist all the time, 1) our physical state-vector space includes no state of negative norm and 2) the Proca equations hold for expectation values of U_μ in $|\text{phys}>$.

Suppose that the following q-number gauge transformation is carried out for the relevant fields :

$$\left.\begin{array}{l} U_\mu \rightarrow \hat{U}_\mu = U_\mu + \lambda \partial_\mu B , \quad (B \rightarrow \hat{B} = B) \\[2mm] B_1 \rightarrow \hat{B}_1 = B_1 - \lambda m^2 B , \quad B_2 \rightarrow \hat{B}_2 = B_2 - \lambda B_1 + \frac{1}{2}\lambda^2 m^2 B \end{array}\right\} \qquad (7)$$

with an arbitrary parameter λ. Then, all the field equations in (3) and all the commutators in (5) are kept form-invariant under (7), and then the gauge parameter α is transformed into $\hat{\alpha}$ given by $\hat{\alpha} = \alpha + \lambda$ with ε unchanged. We can prove $L_0(\hat{A}_\mu , \hat{B} , \hat{B}_i ; \hat{\alpha} , \varepsilon) = L_0(A_\mu , B , B_i ; \alpha , \varepsilon)$.

Since ε remains unchanged under (7), there emerge two gauge families classified by $\tau = 1 + \alpha^2$ and $\tau = 1 - \alpha^2$ respectively. Any two gauges belonging to the same family are equivalent because they are connected with each other through (7). In this sense these two gauge families are invariant with respect to (7). Such a gauge structure of theory does not depend on whether $m \neq 0$ or $m = 0$, but it has a universal meaning over the whole Abelian-gauge field.

Although we have so far considered only the case of free fields, the essential point of gauge structure does not alter at all in the presence of interactions. In this case we have the field equation for U_μ

$$\partial_\nu G_{\nu\mu} - m^2 U_\mu = \partial_\mu B_1 - j_\mu . \qquad (8)$$

Assuming that j_μ is conserved as well as that it commutes with B and B_i, we can carry out a consistent renormalization.[4] The constraints (6) also persist due to $\partial_\mu j_\mu = 0$. Therefore, the unitarity of the S-matrix is guaranteed. Here, the renormalizability of the theory is manifest.

REFERENCES

1) L.D. Landau and I.M. Khalatnikov, Soviet Phys.-JETP 2 (1956), 69.
2) N. Nakanishi, Progr. Theor. Phys. 35 (1966), 1111 ; 38 (1967), 881 ;
   Progr. Theor. Phys. Suppl. No. 51 (1972), 1.
   B. Lautrup, K. Dan. Vidensk. Selsk. Mat.-fys. Medd. 35 (1967), No.11.
3) K. Yokoyama, Progr. Theor. Phys. 51 (1974), 1956.
   K. Yokoyama and R. Kubo, ibid. 52 (1974), 290.
4) K. Yokoyama, Progr. Theor. Phys. 52 (1974), 1669.
   R. Kubo and K. Yokoyama, ibid. 53 (1975), No. 3.
5) N. Nakanishi, Phys. Rev. D5 (1972), 1324.

<u>SPACE-TIME APPROACH TO ANOMALIES IN THE WARD-TAKAHASHI</u>

<u>IDENTITIES AND IMPLICATIONS IN PHYSICAL PROCESSES</u>

Nobuya   Nakazawa
Department of Physics
Kogakuin University

Tokyo 162, Japan

and

Masao   Yamada
Medical College of Miyazaki

Miyazaki 889, Japan

The anomalies present in the Ward-Takahashi (W-T) identities of the <u>a</u>xial type <u>n</u>-point <u>f</u>unctions (a-n-f) have been discussed by several authors using regulator methods[1] and ε-separation methods[2]. These methods, however, produce ambiguities in the renormalization procedure. In particular, these treatment do not give a clear answer, why there occur anomalies even in a-5-f, which is finite and well defined.

On the other hand, SCHREIER[3] discussed the a-3-f in configuration space under the assumption that the conformal invariance and the SU(3)✕ SU(3) symmetry become exact at the short distance. In his approach, divergent quantities never appear and instead of them, distributions are treated. He showed that the anomaly of the W-T identity of a-3-f occurs when, in taking the equal time limit of commutators, the third current gets pinched between the two in the commutator. We extend his analysis to the cases of a-4-f and a-5-f in detail and the anomalies in the W-T identities are obtained in a simple non-perturbative way[4]. It is shown that the pinching mechanism is the common origin of the anomalies.

First, we assume that there appear no anomalies in the absorptive part of the W-T identity of a-n-f. Therefore, the anomalies must be written in terms of delta functions and their derivatives. In other words, the anomalies are nothing but polynomials in momentum space. Then, we can prove that there occur no anomalies in the case of a-n-f, where n is more than 6.

Second, we notice that we need only to discuss the short distance behaviours of a-n-f's, because the anomalies occur only when all the space-time coordinates coincide. Using the short distance expansions of operator products[5], we can prove that we have only to examine a-n-f's, obtained by calculating simple loop diagrams with n-vertices connected by zero-mass fermion propagators.

If we introduce the external vector field $\mathcal{V}_k^\alpha(x)$ and the axial vector field $\mathcal{A}_k^\alpha(x)$ which are coupled to the vector and axial vector current with the charge g, respectively, the results can be described by a simple equation:

$$\frac{\partial}{\partial x^\mu}A_i^\mu(x) = -\frac{g^2}{4\pi}\frac{C}{4\pi}\,\varepsilon_{\alpha\beta\gamma\delta}\mathrm{Tr}[\frac{\lambda_i}{2}F_V^{\alpha\beta}(x)F_V^{\gamma\delta}(x) + \frac{\lambda_i}{2}F_A^{\alpha\beta}(x)F_A^{\gamma\delta}(x)], \qquad (1)$$

where $\quad F_V^{\alpha\beta}(x) = \partial^\alpha\mathcal{V}^\beta(x) - \partial^\beta\mathcal{V}^\alpha(x) - ig[\mathcal{V}^\alpha(x),\mathcal{V}^\beta(x)] - ig[\mathcal{A}^\alpha(x),\mathcal{A}^\beta(x)],$

$$F_A^{\alpha\beta}(x) = \partial^\alpha\mathcal{A}^\beta(x) - \partial^\beta\mathcal{A}^\alpha(x)\ ig\ [\mathcal{V}^\alpha(x),\mathcal{A}^\beta(x)] - ig[\mathcal{A}^\alpha(x),\mathcal{V}^\beta(x)],$$

$$\mathcal{V}^\alpha(x) = \frac{\lambda_k}{2}\mathcal{V}_k^\alpha(x)\ , \quad \mathcal{A}^\alpha(x) = \frac{\lambda_k}{2}\mathcal{A}_k^\alpha(x)$$

and λ_k's are SU(3) matrices.

In order to understand the origin of these anomalies deeply, we discuss the so-called commutator anomalies. We can show that if one thinks of the limiting procedure involved, say,

$$\mathop{\mathrm{Lim}}_{\varepsilon\to 0}\ T(A^0(y_0+\varepsilon,\vec{x})V^\nu(y)V^\rho(z)V^\sigma(u))$$

for the equal time commutator $[A^0(x),\ V^\nu(y)]$, then anomaly occurs when z^0 and/ or u^0 lie between y^0 and $y^0+\varepsilon$. Then, the anomalies in the equal time commutator can be described by a simple formula of an effective operator equations,

$$[A_i^0(x),\ V_j^\nu(y)]\delta(x^0-y^0) = if_{ijn}A_n^\nu(x)\delta^4(x-y)$$

$$+ \frac{iC}{16\pi^2}(\varepsilon^{\nu\underline{\alpha}\beta\gamma} + g_0^\nu\varepsilon^{0\underline{\alpha}\beta\gamma})[\mathrm{Tr}(\frac{\lambda_i}{2}\{\frac{\lambda_j}{2},\ F_{\beta\gamma}^V(x)\})\partial_{\underline{\alpha}}\delta^4(x-y)$$

$$+ \mathrm{Tr}(\frac{\lambda_i}{2}\{[\frac{\lambda_j}{2},\ -ig_{\underline{\alpha}}(x)],F_{\beta\gamma}^V(x)\}+\frac{\lambda_i}{2}\{[\frac{\lambda_j}{2},-ig_{\underline{\alpha}}(x)],F_{\beta\gamma}^A(x)\})\delta^4(x-y)] \quad (2)$$

where $\underline{\alpha}$ runs through 1,2,and 3, the space part of the indices only. This is a generalization of the commutator anomaly obtained by ADLER et al.[6] It is clear that the differences between Eqs.(1) and (2) should be attributed to the seagull terms. This completes our arguments about the origin of the anomalies in the W-T identities that they come from the common mechanism of "pinching" , apart from the effects of the seagull terms. We stress that our results are independent of any perturbation theories.

Finally, we examine the implications of these anomalies in the physical processes. Whenever we consider the following matrix element using the P.C.A.C. relation,

$$\int <A\pi|T(J_\mu(x)J_\nu(0))|B>\Delta^{\mu\nu}(x)d^4x$$

we must always take account of anomalies. We derive general formula for these anomaly terms in the above matrix element and discuss the following processes[7].

  1) $\eta \longrightarrow 3\pi$ decay   2) Non-leptonic decay of hyperon
  3) $K \longrightarrow 2\pi$ decay   4) $K \longrightarrow 3\pi$ decay

In the first processes, the presence of the photon propagator $D_{\mu\nu}(x)$ ($= \Delta_{\mu\nu}(x)$) enhances the singularity and the anomaly term is identified to be the electromagnetic origin of the u_3-tad-pole term. In the rest of the above processes, the weak boson propagator enhances the singularity and anomalies become observable ones. Fortunately, the effect of the anomaly term in the case of $K \longrightarrow 3\pi$ decay is negligible compared with the normal terms.

References;

  (1) R.W. Brown et al.   : Phys. Rev. $\underline{186}$ 1491 (1969)
  (2) W.A. Bardeen   : Phys. Rev. $\underline{184}$ 1848 (1969)
  (3) E.J. Schreier   : Phys. Rev. $\underline{D3}$ 980 (1971)
  (4) N. Nakazawa and M. Yamada : Ann. Phys. $\underline{87}$ 457 (1974)
  (5) R.J. Crewther   : Phys. Rev. Letters, $\underline{28}$ 1421 (1972)
  (6) S.L. Adler et al.   : Phys. Rev. $\underline{184}$ 1470 (1969)
  (7) M. Yamada and N. Nakazawa   : Medical college of Miyazaki
               Preprint MM-1 and MM-2

<u>DISPERSION APPROACH TO WARD-TAKAHASHI IDENTITIES</u>

K. Nishijima
Department of Physics
University of Tokyo
Tokyo, Japan

A dispersion formulation of field theories is presented and is applied to the derivation of various Ward-Takahashi identities.

1. Introduction

This is a brief summary of a series of works on the derivation of various W-T identities carried out in collaboration with R. Sasaki. The conventional field theoretical calculations often lead to ambiguities and also miss anomalies. The S matrix theory, on the other hand, is free from these objections, but we do not have a complete set of dispersion relations so that the theory represents an approximate dynamical scheme.

Thus we choose an approach standing midway between them. This approach is based on unitarity and dispersion relations, but the object is not the S matrix but the whole collection of Green's functions for which we have a complete set of dispersion relations. It provides us with a convenient basis for deriving various W-T identities unambiguously.

2. Unitarity and Dispersion Relations

We shall first introduce unitarity and dispersion relations for Green's functions. We consider the neutral scalar theory and define the following renormalized functions:

$$\tau(x_1,\ldots,x_n) = (-i)^n K_{x_1}\ldots K_{x_n} <0|T[\phi(x_1)\ldots\phi(x_n)]|0> \qquad (2.1)$$

The Fourier transform of the τ function represents the S matrix element when all the four-momenta are on the mass shell.

The unitarity condition for the τ functions follows from the LSZ reduction formalism.

$$\tau(x_1,\ldots,x_n) + \tau^*(x_1,\ldots,x_n)$$

$$+ \sum_{comb}' \sum_{\ell=0}^{\infty} \frac{i^{\ell}}{\ell!} \int (du)(dv)\tau(x_1',\ldots,x_k',u_1,\ldots,u_{\ell})$$

$$\times \Delta^{(+)}(u_1-v_1)\ldots\Delta^{(+)}(u_{\ell}-v_{\ell})\tau^*(x_{k+1}',\ldots,x_n',v_1,\ldots,v_{\ell})$$

$$= 0 , \tag{2.2}$$

where $i\Delta^{(+)}$ is the contraction function and the summation has to be taken over all possible divisions of $(x_1,\ldots,x_n)$ into two groups, one entering τ and the other entering τ^* excluding k=0 and n.

Next we consider a local operator A(x) and define

$$\tau_A(x;x_1,\ldots,x_n)=(-i)^{n+1}K_{x_1}\ldots K_{x_n}<0|T[A(x)\phi(x_1)\ldots\phi(x_n)]|0> \tag{2.3}$$

An arbitrary matrix element of A(x) is obtained from (2.3) by taking its Fourier transform with the help of the LSZ reduction formula. The unitarity condition for this set is given by

$$\tau_A(x;x_1,\ldots,x_n) + \tau_A^*(x;x_1,\ldots,x_n)$$

$$+ \sum_{comb} \sum_{\ell=0}^{\infty} \frac{i^{\ell}}{\ell!} \int (du)(dv)[\tau_A(x;x_1',\ldots,x_k',u_1,\ldots,u_{\ell})$$

$$\times \Delta^{(+)}(u_1-v_1)\ldots\Delta^{(+)}(u_{\ell}-v_{\ell})\tau^*(x_{k+1}',\ldots,x_n',v_1,\ldots,v_{\ell})$$

$$+ (\tau_A\to\tau, \ \tau^*\to\tau_A^*)]$$

$$= 0 \tag{2.4}$$

This unitarity condition is linear in τ_A and τ_A^* so that it will be referred to as the linear unitarity condition.

Next we proceed to dispersion relations. First, we shall denote the connected parts of τ and of τ_A by ρ and ρ_A, respectively, and shall introduce their Fourier transforms by

$$\rho(x_1,\ldots,x_n) = \frac{-i}{(2\pi)^{4(n-1)}}\int (dp)\delta^4(\sum_j p_j)\exp(i\sum_j p_j x_j)$$

$$\times \mathcal{G}(p_1,\ldots,p_n), \tag{2.5}$$

$$\rho_A(x;x_1,\ldots,x_n) = \frac{-i}{(2\pi)^{4n}}\int (dp)d^4q\,\delta^4(q+\sum_j p_j)\exp[i(qx+\sum_j p_j x_j)]$$

$$\times \mathcal{A}(q;p_1,\ldots,p_n) \tag{2.6}$$

These functions, $\mathcal{G}$ and $\mathcal{A}$, are functions of scalar products of four-momenta and may eventually be denoted by $\mathcal{G}(p_\alpha p_\beta)$ and $\mathcal{A}(p_\alpha p_\beta)$, respectively. They are known to satisfy the parametric dispersion relations.

$$\mathrm{Re}\begin{pmatrix}\mathcal{G}(p_\alpha p_\beta \cdot \xi)\\ \mathcal{A}(p_\alpha p_\beta \cdot \xi)\end{pmatrix} = \frac{P}{\pi}\int_\infty^\infty \frac{d\xi'}{\xi'-\xi}\epsilon(\xi')\mathrm{Im}\begin{pmatrix}\mathcal{G}(p_\alpha p_\beta \cdot \xi')\\ \mathcal{A}(p_\alpha p_\beta \cdot \xi')\end{pmatrix}, \tag{2.7}$$

where ξ is a common scaling parameter to be multiplied into all the scalar products of the form $p_\alpha p_\beta$. Now we have a complete set of dispersion relations in the sense that Eq.(2.7) is valid for any n except we need subtractions in some cases.

3. Subtractions and Power Counting

Subtractions in dispersion relations are so introduced as to reproduce the renormalized perturbation theory. This is our guiding principle to settle the subtraction conditions. As a simple example we shall consider the theory corresponding to the following Lagrangian:

$$\mathcal{L} = -\tfrac{1}{2}[(\partial_\mu\phi)^2 + m^2\phi^2] - \frac{\lambda}{24}\phi^4 \tag{3.1}$$

The subtraction conditions for the two-point function $\mathcal{G}^{(2)}(p)= \mathcal{G}^{(2)}(p,-p)$ follow from $\mathcal{G}^{(2)}(p)=-(p^2+m^2)^2\Delta'_F(p)$ and the Lehmann representation.

$$\mathcal{G}^{(2)}(p) = 0, \qquad \frac{\partial}{\partial p^2}\mathcal{G}^{(2)}(p) = -1 \quad \text{for } p^2 + m^2 = 0. \tag{3.2}$$

The four-point function, in the lowest non-vanishing order, is given by

$$\mathcal{G}^{(4)} = \lambda \tag{3.3}$$

suggesting one subtraction for the four-point function. Thus we assume

$$\mathrm{Re}\,\mathcal{G}^{(4)}(p_\alpha p_\beta \cdot \xi) = \mathcal{G}^{(4)}(0)+\frac{\xi}{\pi}\int \frac{d\xi'}{\xi'(\xi'-\xi)}\epsilon(\xi')\mathrm{Im}\,\mathcal{G}^{(4)}(p_\alpha p_\beta \cdot \xi') \tag{3.4}$$

When higher order corrections are taken into account we may fix the
subtraction condition as

$$\mathcal{G}^{(4)}(0) = \lambda \tag{3.5}$$

corresponding to the subtraction point $p_1=p_2=p_3=p_4=0$.

In order to study the subtraction problem in perturbation theory
we shall consider the ultraviolet asymptotic behavior of $\mathcal{G}(p_\alpha p_\beta \cdot \xi)$
for large values of ξ. We shall assume a power law:

$$\mathcal{G}^{(n)}(p_\alpha p_\beta \cdot \xi) \sim \xi^{c(n)/2} \quad \text{for } \xi \to \infty \tag{3.6}$$

except possibly   for logarithmic factors. In perturbation theory
the powers $c(n)$ are determined by combining unitarity with the
assumed dispersion relations. Namely, the power of the absorptive
part of a $\mathcal{G}$ function can never exceed that of the full $\mathcal{G}$ function
so that we obtain, by rewriting the absorptive part by means of the
unitarity condition, the following set of inequalities:

$$d(n) \geqq \underset{\substack{k+\ell>2 \\ n-k+\ell>2}}{\text{Max}}[d(k+\ell)+d(n-k+\ell)] \tag{3.7}$$

where $d(n)=c(n)+n-4$. From the subtraction conditions (3.2) and (3.5)
we find

$$c(2) \geqq 2, \qquad c(4) \geqq 0 \tag{3.8}$$

The only solution of (3.7) and (3.8) is given by

$$d(n) = 0, \text{ or } c(n) = 4 - n \tag{3.9}$$

This result can be extended to more general cases.

When we have both spinless and spinor fields the τ functions are
defined

$$\tau(x_1,\ldots,x_n;y_1,\ldots,y_\ell;z_1,\ldots,z_\ell)$$

$$= (-i)^n K_{x_1}\ldots K_{x_n} D_{y_1}\ldots D_{y_\ell} \vec{D}_{z_1}\ldots\vec{D}_{z_\ell}$$

$$\times <0|T[\phi(x_1)\ldots\phi(x_n)\psi(y_1)\ldots\psi(y_\ell)\bar{\psi}(z_1)\ldots\bar{\psi}(z_\ell)]|0>, \tag{3.10}$$

where $D=\gamma\partial+M$, $\tilde{D}=\gamma^T\partial-M$, and M is the fermion mass. In this case we introduce the power $c(n,m)$ with $m=2\ell$ and

$$d(n,m) = c(n,m) + n + \tfrac{3}{2}m - 4 \qquad (3.11)$$

Then we find that renormalizable theories are characterized by

$$d(n,m) = 0 \qquad (3.12)$$

For the function $\mathcal{A}^{(n,m)}$ we define the power $a(n,m)$ just as we introduced $c(n,m)$ for $\mathcal{G}^{(n,m)}$, and define

$$b(n,m) = a(n,m) + n + \tfrac{3}{2}m - 4 \qquad (3.13)$$

Then from the linear unitarity condition and (3.12) we find

$$b(n,m) = b_0, \text{ constant.} \qquad (3.14)$$

This parameter b_0 is referred to as the index of the set $\{\mathcal{A}\}$. As an example we shall consider

$$A(x) = \tfrac{1}{2}\phi^2(x). \qquad (3.15)$$

In the free field approximation we have

$$\mathcal{A}^{(2)}(q;p_1,p_2) = 1, \quad \text{all others} = 0 \qquad (3.16)$$

This suggests $a(2)=0$ so that $b_0=-2$. When higher order corrections are included we formulate the subtraction condition as

$$\mathcal{A}^{(2)}(0;p,-p) = 1 \quad \text{for } p^2 + m^2 = 0 \qquad (3.17)$$

The relation (3.15) is rather symbolic and our $\mathcal{A}$ functions are regularized through subtracted dispersion relations.

4. Linear Identities

In this section we study linear relationships among different sets of $\mathcal{A}$ functions. The W-T identities fall into this category. We present a basic theorem on the linear relationship in the scalar theory.

Theorem: Let $\{\mathcal{A}\}$, $\{\mathcal{B}\}$, $\{\mathcal{C}\}$ and $\{\mathcal{D}\}$ be sets of functions

of indices 0 or -2, transforming as scalar functions and satisfying
the linear unitarity condition, then they are linearly dependent.

As a consequence of this theorem we find a linear relationship
of the form:

$$a \, \mathcal{A}^{(n)} + b \, \mathcal{B}^{(n)} + c \, \mathcal{C}^{(n)} + d \, \mathcal{D}^{(n)} = 0 \qquad (4.1)$$

When all the subtraction conditions for the sets are given, these sets
of functions are uniquely determined at least in the sense of
perturbation theory. Now take a linear combination

$$\mathcal{S}^{(n)} = a \, \mathcal{A}^{(n)} + b \, \mathcal{B}^{(n)} + c \, \mathcal{C}^{(n)} + d \, \mathcal{D}^{(n)} \qquad (4.2)$$

The index of the set $\{\mathcal{S}\}$ is assumed to be equal to 0 without loss of
generality. The set $\{\mathcal{S}\}$ satisfies the linear unitarity condition
and is determined uniquely when the following three subtraction
constants are given:

$$\mathcal{S}^{(2)}(0;p,-p) \quad \text{and} \quad \frac{\partial}{\partial p^2} \, \mathcal{S}^{(2)}(0;p,-p) \quad \text{for } p^2 + m^2 = 0,$$
$$\mathcal{S}^{(4)}(0;0,0,0,0) \qquad (4.3)$$

These three constants are linear in a,b,c and d, so that we can
always make them vanish by choosing the four coefficients appropriately,
and then we have

$$\mathcal{S}^{(n)} = 0 \qquad \text{for all } n \qquad (4.4)$$

In this way the relationship (4.1) is obtained.

5. Ward-Takahashi Identities

The theorem given in the preceding section can trivially be
generalized and we shall give some examples.

First, we shall derive anomalous trace identities and the Callan-
Symanzik equations. For this purpose we introduce the energy-momentum
tensor $T_{\mu\nu}$ and the corresponding set of functions.

$$\{\mathcal{T}_{\mu\nu}(q;p_1,\ldots,p_n)\} \qquad (5.1)$$

This set satisfies the following W-T identities in the scalar case:

$$q_\mu \mathcal{T}_{\mu\nu}(q;p_1,\ldots,p_n) = \sum_j \frac{p_j^2+m^2}{(p_j+q)^2+m^2-i\varepsilon}(p_j+q)_\nu \mathcal{G}(p_1,\ldots,p_j+q,\ldots,p_n)$$

$$(5.2)$$

The set $\{\mathcal{T}_{\mu\nu}\}$ as the solution of the above equations is uniquely determined for q=0. Then we introduce a new set corresponding to $T=m^2\phi^2$:

$$\{\mathcal{T}(q;p_1,\ldots,p_n)\}$$

$$(5.3)$$

We further introduce two other sets:

$$\{\sum_j \frac{p_j^2+m^2}{(p_j+q)^2+m^2-i\varepsilon}\mathcal{G}(p_1,\ldots,p_j+q,\ldots,p_n)\},$$

$$(5.4)$$

$$\{\frac{\partial}{\partial\lambda}\mathcal{G}(p_1,\ldots,p_n)\}$$

$$(5.5)$$

The indices of these sets are equal to 0 except that the index of $\{\mathcal{T}\}$ is equal to -2. Application of the theorem leads to the relation

$$\mathcal{T}_{\mu\mu}(0;p_1,\ldots,p_n) - \mathcal{T}(0;p_1,\ldots,p_n) \quad nd_\phi(\lambda)\mathcal{G}(p_1,\ldots,p_n)$$

$$= \beta(\lambda)\frac{\partial}{\partial\lambda}\mathcal{G}(p_1,\ldots,p_n)$$

$$(5.6)$$

These are precisely the anomalous trace identities.

Now let us denote the Green's functions defined by dropping the Klein-Gordon operators in Eqs.(2.1) and (2.3) by printed letters instead of script letters, then (5.6) clearly holds also for the printed letters. Furthermore, we have for the latter

$$T_{\mu\mu}(0;p_1,\ldots,p_n) = (n - m\frac{\partial}{\partial m})G(p_1,\ldots,p_n)$$

$$(5.7)$$

Substituting (5.7) into the printed version of (5.6) we arrive at the Callan-Symanzik equations:

$$[m\frac{\partial}{\partial m} + \beta(\lambda)\frac{\partial}{\partial\lambda} + n\gamma_\phi(\lambda)]G(p_1,\ldots,p_n) = -T(0;p_1,\ldots,p_n),$$

$$(5.8)$$

where $\gamma_\phi=d_\phi-1$ is the anomalous dimension of the field ϕ and is defined as the following expansion coefficient:

$$\sigma_T(0;p,-p) = -2m^2 + 2\gamma_\phi(p^2 + m^2) + O((p^2 + m^2)^2) \qquad (5.9)$$

Next, we shall apply a similar argument to the derivation of the Adler anomaly for the axial vector current in QED.

$$A_\lambda = i\bar{\psi}\gamma_\lambda\gamma_5\psi, \quad P = i\bar{\psi}\gamma_5\psi, \quad C = EH = \tfrac{i}{8}\epsilon_{\alpha\beta\gamma\delta}F_{\alpha\beta}F_{\gamma\delta} \qquad (5.10)$$

Application of the theorem then leads to the following W-T identities:

$$iq_\lambda \mathcal{A}_\lambda = 2m\mathcal{P} + \mathcal{W} - \tfrac{2\alpha}{\pi}\mathcal{C} \qquad (5.11)$$

The last term is the so-called Adler anomaly, and the set of functions $\{\mathcal{W}\}$ is defined by

$$\mathcal{W}(q;k_1,\ldots,k_n;p_1,\ldots,p_\ell;\bar{p}_1,\ldots,\bar{p}_\ell)$$

$$= \sum_{j=1}^{\ell} [(ip_j\gamma + m)(-i(p_j + q)\gamma + m)^{-1}(i\gamma_5)_j$$

$$\times \; \mathcal{G}(k_1,\ldots,k_n,p_1,\ldots,p_j+q,\ldots p_\ell,\bar{p}_1,\ldots,\bar{p}_\ell)$$

$$+ \; \mathcal{G}(k_1,\ldots,k_n,p_1,\ldots,p_\ell,\bar{p}_1,\ldots,\bar{p}_j+q,\ldots,\bar{p}_\ell)$$

$$\times \; (i\gamma_5)_j(i(\bar{p}_j+q)\gamma+m)^{-1}(-i\bar{p}_j\gamma+m)]. \qquad (5.12)$$

It is characteristic of this set $\{\mathcal{W}\}$ that all members vanish identically when all the momenta $p_1,\ldots,p_\ell,\bar{p}_1,\ldots,\bar{p}_\ell$ are on the mass shell.

There are further applications of the theorem and we shall mention only the results.

(1) The W-T identities in quantum electrodynamics have no anomalous terms and the derivation is straightforward.

(2) It is possible to prove that the Schwinger term in spinor electrodynamics is a c number.

(3) R. Sasaki has succeeded in formulating the σ model entirely in terms of renormalized expressions alone.

References

1. K. Nishijima, Progr. Theor. Phys. 51, 1193 (1974)
2. K. Nishijima and R. Sasaki, Progr. Theor. Phys. 53, No.1 (1975)

Discussion

Slavnov (question): Can you make some comments concerning the
applicability of your procedure to the gauge theories. It seems that
your original arguments are not strictly applicable to these theories.
I mean the positivity of the scalar product.
Nishijima (answer): It is formally possible to incorporate the indefinite metric into our scheme as we are doing in the case of quantum
electrodynamics, but we need an extra proof that the S matrix is
unitary.

RENORMALIZATION OF SUPERSYMMETRIC GAUGE THEORIES

A. A. Slavnov

Steklov Mathematical Institute
Moscow   USSR

Supersymmetric generalization of gauge theories was given firstly by Wess and Zumino [1]. However these authors worked in a nonsupersymmetric gauge, so it was far from evident that the resulting renormalized theory was supersymmetric.

We propose a manifestly supersymmetric (ss) renormalization procedure for gauge theories. The difficulties that one encounters in the renormalization of ss gauge theories are due to the fact that in any gauge different from the non-invariant Wess-Zumino gauge the Lagrangian is essentially nonlinear. For the abelian case ss action is given by the expression:

$$S = \int [\tfrac{1}{8}(\overline{D}D)^2 \{\Psi_+^* e^{g\Psi}\Psi_+ + \Psi_-^* e^{-g\Psi}\Psi_-\} - \tfrac{M}{2} (\overline{D}D)\{\Psi_-^*\Psi_+ + \Psi_+^*\Psi_-\}]dx + S_0^\psi. \qquad (1)$$

Here the notations of Salam and Strathdee [2] are used. ψ is a superfield including electromagnetic field

$$\psi(x,\theta) = c(x) + \theta\,\bar{\gamma}_5\chi(x) +$$

$$\tfrac{1}{4}\{\bar{\theta}\theta M(x) + \bar{\theta}\gamma_5\theta N(x) + \bar{\theta}i\gamma_\nu\gamma_5\theta A_\nu(x) + \bar{\theta}\theta\bar{\theta}\gamma_5\lambda(x)\} + \tfrac{1}{32}(\bar{\theta}\theta)^2\Delta(x). \quad (2)$$

θ_α are x-independent anticommuting Majorana spinors; $c(x)$, $M(x)$, $\Delta(x)$ are pceudo-scalar fields, $N(x)$-scalar, $\chi(x)$, $\lambda(x)$ are spinors, and $A_\nu(x)$ is a vector field identified with the electromagnetic field. Matter fields are described by the chiral   superfields $\Psi_\pm(x,\theta)$:

$$\Psi_\pm(x,\theta) = \exp\{\mp \tfrac{1}{4}\bar{\theta}\hat{\partial}^1\gamma_5\theta\}[A_\pm(x) + \bar{\theta}\zeta_\pm(x) + \tfrac{1}{2}\bar{\theta}\,\frac{1\pm i\gamma_5}{2}\,\theta F_\pm(x)], \qquad (3)$$

where A and F are scalars, and ζ are chiral spinors. D is a covariant derivative

$$D_\alpha = \frac{\partial}{\partial\bar{\theta}^\alpha} - \frac{i}{2}(\gamma_\mu\theta)_\alpha\,\frac{\partial}{\partial x^\mu}\ . \qquad (4)$$

S_0^ψ is an action for the free field ψ, and in the abelian case it can be written explicitly in terms of the components

$$S_0^\psi = \int [-\tfrac{1}{4}(\partial_\mu A_\nu - \partial_\nu A_\mu)^2 - \tfrac{i}{2}\bar\lambda\hat\partial\lambda + \tfrac{1}{2}\Delta^2]dx. \tag{5}$$

This Lagrangian is invariant with respect to generalized gauge transformation

$$\psi_\pm^\Omega = \Omega_\pm \psi_\pm ; \qquad \exp\{g\psi^\Omega\} = \Omega_- \exp\{g\psi\}\Omega_+^{-1}, \tag{6}$$

where $\Omega_\pm$ is an arbitrary chiral superfield satisfying the relation $\Omega_+^* = \Omega_-^{-1}$. Due to this invariance the components c, M, N, χ and $\partial_\mu A_\mu$ do not correspond to physical degrees of freedom and in fact can be "gauged away". We shall call them "gauge" components.

S.s gauge fixing term can be chosen in the form

$$\int \frac{1}{8\beta} (D\bar D)^2 (\Box\psi_+\Box\psi_-)dx = \int \frac{1}{2\beta} \{[\partial_\mu(\Box c-\Delta)]^2 + [\partial_\mu(\partial_\nu A_\nu)]^2 + (\Box M)^2 + (\Box N)^2 -$$
$$- i(\Box\bar\chi - i\partial_\mu\bar\lambda\gamma_\mu)\hat\partial(\Box\chi - i\hat\partial\lambda)\}dx. \tag{7}$$

The Lagrangian (1) is essentially nonlinear, but due to anti-commutativity of spinors θ_α, only c-component may enter in an arbitrary degree. For our choice of the gauge condition the propagators of c-field decrease asymptoticaly as k^{-4}. Therefore the inclusion of internal c-lines does not spoil the convergence of a diagram. More precisely one can estimate the index of divergency of arbitrary diagram as follows

$$n \leq y - e_A - e_{A_\mu} - 2e_\Delta - 2e_F - 3/2e_\zeta - 3/2\,e_\lambda; \qquad A \equiv A, B; \ F \equiv F, G. \tag{8}$$

where e_I means a number of I-external lines.
There exist primitively divergent diagrams with an arbitrary number of "gauge" external lines, but the number of "physical" external lines, A_μ^{tr}, λ, Δ, is limited.

The diagrams with "gauge" external lines can be expressed in terms of the diagrams without such lines with the help of a generalized Ward identities. For the case under consideration these identities can be written in the form

$$\frac{\delta}{\delta\Omega_+} \int \exp\{iS+i\int[\frac{1}{8\beta}(\overline{D}D)^2(\square\psi_+^\Omega\square\psi_-^\Omega)+\Sigma J_{\psi_i}\psi_i^\Omega +\Sigma J_{\Psi_{+i}}\Psi_{+i}^\Omega]dx\}\Pi d\psi_i d\Psi_i=0. \tag{9}$$

where Ω_+ is the infinitesimal chiral superfield $\Omega_+ = \{U_+, V_+, W_+\}$,
and $\frac{\delta}{\delta\Omega_+}$ meams $\frac{\delta}{\delta U_+}$, $\frac{\delta}{\delta V_+}$, $\frac{\delta}{\delta W_+}$. J_i are the sources of the
corresponding fields. This identity can be written explicitly in terms
of the components. For example

$$\beta^{-1}\square^2 <TM(x)M(y)> = \delta(x-y) \tag{10}$$

etc.

It follows from (9) that if the diagrams without gauge external
lines are finite, all the diagrams should be automatically finite.
Only 2,3 and 4-leg diagrams without gauge external lines are primitively
divergent. The corresponding counterterms can be interpreted as wave
function, mass and charge renormalization. They satisfy also the
second system of Ward identities which follows from the supersymmetry
of the Lagrangian (1).

Finally the renormalized Lagrangian looks as follows

$$L_R = \frac{z}{8}(\overline{D}D)^2\{\Psi_+^*e^{g\psi}\Psi_+ +\Psi_-^*e^{-g\psi}\Psi_-\} - \frac{M+\delta M}{2}(\overline{D}D)\{\Psi_-^*\Psi_+ +\Psi_+^*\Psi_-\} -$$
$$- \frac{z_3}{2}\{\frac{1}{2}(\partial_\mu A_\nu -\partial_\nu A_\mu)^2 + i\overline{\lambda}\hat{\partial}\lambda - \Delta^2\}. \tag{11}$$

So only 3 independent counterterms are needed:common wave function
renormalization for the superfield ψ and wave function and mass
renormalization for the matter fields [3].

The analogous procedure is developed for the supersymmetric
nonabelian gauge theories [4]. It follows from our result that such
theories are asymptotically free (provided that the number of matter
supermultiplets is not too large). These theories give also an interesting example of the essentially nonlinear Lagrangian that can be renormalized by a finite number of counterterms.

In the derivation of these results we used the axiomatic version
of the path integral method, given in our paper [5]. The starting point
in this formalism is the definition of the quasigaussian path integral

$$\int \exp\{i \int \psi(x) \frac{K(x-y)}{2} \psi(y)dxdy + i \int \psi(x)\eta(x)dx\} \prod_x d\psi \stackrel{def}{=}$$

$$= (-i)^n \frac{\delta^n}{\delta\eta(x_1)\ldots\delta\eta(x_n)} \exp\{ - \frac{i}{2}\int \eta(x)K^{-1}(x-y)\eta(y)dxdy\}. \qquad (12)$$

It is shown that the object defined by this formula possesses all the algebraic properties of a usual integral: We can integrate by parts, change variables according to the well known rules, change the order of integration etc. These properties are sufficient to justify the manipulations with path integrals in the framework of the perturbation theory.

REFERENCES

1. J. Wess and B. Zumino, Nuclear Phys. B70, 39 (1974).
2. A. Salam and J. Strathdee,   Triest Preprint, IC/74/42.
3. A. A. Slavnov,   Preprint JINR E2-8308 (1974).
4. A. A. Slavnov,   Renormalization of Supersymmetric Gauge Theories. II Nonabelian Case. Submitted to Nuclear Physics.
5. A. A. Slavnov,   Theor. and Math. Phys. 22, 177 (1975).

Discussion

Klauder (Question): Can you clarify the charge renormalization in the renormalized Lagrangian for the Abelian theory ?

Slavnov (Answer): In complete analogy with the usual quantum electrodynamics there exists the relation $z_1 = z_2$. Due to this relation charge renormalization is given by $g \to z_3^{-1/2}g$. Lescaling the fields one can write the renormalized Lagrangian in the form (1) with the substitution $g \to z_3^{-1/2}g$.

STOCHASTIC CALCULUS

Kiyosi Itô
Research Institute for Mathematical Sciences
Kyoto University
Kyoto, Japan

1. Stochastic differentials.

Let (Ω, P) be a complete probability space and $\mathcal{F}(P)$ denote the σ-algebra of all P-measurable sets. Let $\{\mathcal{F}_t : 0 \leq t < \infty\}$ be a right continuous increasing family of sub-σ-algebras of $\mathcal{F}(P)$ such that $\mathcal{F}_0$ includes all P-null sets, and let Q denote the family of all local quasi-martingales ($\mathcal{F}_t$) with continuous sample functions. We denote by $\mathcal{M}$ and $\mathcal{A}$ respectively the local martingales ($\mathcal{F}_t$) in Q and the locally bounded variation processes in Q.

Q is a vector space over the reals $\mathbb{R}$ and both $\mathcal{M}$ and $\mathcal{A}$ are subspaces of Q. Q is also a commutative ring over $\mathbb{R}$ and $\mathcal{A}$ is a subring of Q.

The _differential_ dX of $X \in Q$ is defined to be the random interval function induced from X:

$$(dX)(I) = X(t) - X(s) \qquad \text{for} \qquad I = (s,t].$$

Let dQ, $d\mathcal{M}$ and $d\mathcal{A}$ denote

$$\{dX : X \in Q\}, \qquad \{dM : M \in \mathcal{M}\} \text{ and } \{dA : A \in \mathcal{A}\}$$

respectively. dQ is a vector space over $\mathbb{R}$ and both $d\mathcal{M}$ and $d\mathcal{A}$ are subspaces of dQ and

$$dQ = d\mathcal{M} \oplus d\mathcal{A}.$$

dQ is also a Q-module with respect to the _Q-multiplication_:

$$(Y\,dX)(I) = \int_I Y(t)\,dX(t) \qquad (\underline{\text{stochastic integral}} \ [1],[2],[3])$$

and both $d\mathcal{M}$ and $d\mathcal{A}$ are submodules of dQ. Introducing product in dQ by

$$dX\,dY = d(XY) - X\,dY - Y\,dX,$$

we can regard dQ as a commutative ring over Q. Then $d\mathcal{A}$ is a subring of dQ. It is easy to check that

$$dQ\,dQ \subset d\mathcal{A} \quad \text{and} \quad dQ\,d\mathcal{A} = = 0, \text{ so } (dQ)^3 = 0.$$

The <u>stochastic chain rule</u> is formulated as follows: If $F \in C^2(\mathbb{R}^n)$ and $X^i \in Q$ $(i=1,2,\ldots,n)$, then $F = F(X^1 \ldots X^n) \in Q$ and

$$\text{(C)} \quad dF = \sum_{i=1}^{n} \partial_i F \, dX^i + \tfrac{1}{2} \sum_{i,j=1}^{n} \partial_i \partial_j F \, dX^i dX^j.$$

The <u>symmetric Q-multiplication</u> $Y \circ dX$ is defined by

$$Y \circ dX = YdX + \tfrac{1}{2} dYdX.$$

dQ is also a Q-module with this multiplication. The stochastic integral $(Y \circ dX)(I)$ is often denoted by

$$s \cdot \int_I Y(t) dX(t)$$

and is called the <u>symmetric stochastic integral</u> that was introduced by Fisk [1] and Stratonovich [4]. They proved that the above chain rule (C) can be formulated in the same form as in the ordinary differential calculus:

$$\text{(C}_s) \quad dF(X) = \sum_{i=1}^{n} \partial_i F \circ dX^i, \quad \text{where} \quad F \in C^3(\mathbb{R}^n).$$

This can be derived immediately from (C) and is called the <u>symmetric stochastic chain rule</u>.

2. Stochastic differential equations.

Let $B(t) = (B^1(t), B^2(t), \ldots, B^n(t))$ be an n-dimensional Brownian motion and $\tilde{\mathcal{F}}_t$ denote the σ-algebra generated by $B^i(s)$ $(i=1,2,\ldots m, s \leq t)$ and all P-null sets. Then each $B^i(t)$ $(i=1,2,\ldots,m)$ is a martingale relative to the family $\mathcal{F}_t = \cap_{\varepsilon > 0} \mathcal{F}_{t+\varepsilon}$. Applying the results of Section 1 we have

$$dB^i \in d\mathcal{M}, \quad dt \in d\mathcal{A}, \quad dB^i \, dB^j = \delta_{ij} \, dt \quad \text{and} \quad dB^i dt = 0.$$

Consider the following two stochastic differential equations on X_α $(\alpha = 1,2,\ldots,m)$:

$$\text{(E)} \quad dX^\alpha = a^\alpha(t,X) \, dt + \sum_{i=1}^{n} \sigma_i^\alpha(t,X) dB^i$$

and

$$\text{(E}_s) \quad dX^\alpha = \bar{a}^\alpha(t,X) \circ dt + \sum_{i=1}^{n} \bar{\sigma}_i^\alpha(t,X) \circ dB^i,$$

where $a^\alpha(t,x)$, $\sigma_i^\alpha(t,x)$, $\bar{a}^\alpha(t,x)$ and $\bar{\sigma}_i^\alpha(t,x)$ belong to class C^3.

From now on the summation sign $\sum$ will be omitted as is usual in differential geometry.

Theorem 3.1. The equation (E) can be written in the form (E_s) with the coefficients

$$(1) \quad \bar{a}^\alpha = a^\alpha - \tfrac{1}{2}(\partial_\beta \sigma_i^\alpha) \sigma_i^\beta \quad \text{and} \quad \bar{\sigma}_i^\alpha = \sigma_i^\alpha,$$

and the equation (E_s) can be written in the form (E) with the co-efficients

$$(i) \quad a^\alpha = \bar{a}^\alpha + \tfrac{1}{2}(\partial_\beta \bar{\sigma}_i^\alpha) \bar{\sigma}_i^\beta \quad \text{and} \quad \sigma_i^\alpha = \bar{\sigma}_i^\alpha .$$

Proof. Using the results of Section 1 we have

$$
\begin{aligned}
& a^\alpha \, dt + \sigma_i^\alpha \, dB^i \\
= \;& a^\alpha \, dt + \sigma_i^\alpha \circ dB^i - \tfrac{1}{2} d\,\sigma_i^\alpha \, dB^i \\
= \;& a^\alpha \, dt + \sigma_i^\alpha \circ dB^i - \tfrac{1}{2}(\partial_\beta \sigma_i^\alpha \, dX^\beta) \, dB^i \\
= \;& a^\alpha \, dt + \sigma_i^\alpha \circ dB^i - \tfrac{1}{2} \partial_\beta \sigma_i^\alpha (\sigma_j^\beta \, dB^j) \, dB^i \\
= \;& (a^\alpha - \tfrac{1}{2} \sigma_j^\beta \partial_\beta \sigma_i^\alpha) \, dt + \sigma_i^\alpha \circ dB^i \\
= \;& \bar{a}^\alpha \circ dt + \bar{\sigma}_i^\alpha \circ dB^i ,
\end{aligned}
$$

proving the first assertion. Similarly for the second.

Theorem 3.2. The equation (E) determines a diffusion process with generator

$$\mathcal{J} = a^\alpha \partial_\alpha + \tfrac{1}{2} \sigma_i^\alpha \sigma_i^\beta \partial_\alpha \partial_\beta$$

and the equation (E_s) determines a diffusion process with generator

$$\mathcal{J}_s = \bar{a}^\alpha \partial_\alpha + \tfrac{1}{2}(\bar{\sigma}_i^\alpha \partial_\alpha)(\bar{\sigma}_i^\beta \partial_\beta).$$

Proof. Let $f \in C^2(\mathbb{R}^m)$. Using the same argument as in the proof of the last theorem we have

$$
\begin{aligned}
(3) \quad df(x) &= \partial_\alpha f \, dX^\alpha + \tfrac{1}{2} \partial_\alpha \partial_\beta f \, dX^\alpha \, dX^\beta \\
&= (a^\alpha \partial_\alpha f + \tfrac{1}{2} \sigma_i^\alpha \sigma_i^\beta \partial_\alpha \partial_\beta f) \, dt + \sigma_i^\alpha \partial_\alpha f \, dB^i .
\end{aligned}
$$

Since the last term $\sigma_i^\alpha \partial_\alpha f \, dB^i$ belongs to $d\mathcal{M}$, the first assertion follows at once. Similarly for the second.

An interesting example of (E_s) is the <u>Stroock equation</u> [8] determining a spherical Brownian motion:

$$(S) \quad dX^\alpha = \sum_{i=1}^{n} (\delta_{i\alpha} - X^i X^\alpha |X|^{-2}) \circ dB^i , \quad \alpha = 1, 2, \ldots, n,$$

where $|X|$ denotes the length of the vector $(X^1, X^2, \ldots, X^n)$.

$$d|X|^2 = 2 \sum_{\alpha=1}^{n} X^\alpha \circ dX^\alpha$$

$$= 2 \sum_{\alpha=1}^{n} \sum_{i=1}^{n} (X^\alpha \delta_{\alpha i} - \frac{X^i (X^\alpha)^2}{|X|^2}) \circ dB^i$$

$$= 2 \sum_{i=1}^{n} (X^i - X^i) \circ dB^i$$

$$= 0,$$

so $|X| = $ const. This shows that the solution of (S) is a diffusion process on a sphere Γ with center at the origin. Let θ be any rotation around the origin and let $\sigma(X)$ denote the coefficient matrix on the right hand side of (S). Then

$$\sigma(\theta X) = \theta \, \sigma(X) \, \theta^{-1},$$

so

$$d(\theta X) = \theta \, dX = \theta \, \sigma(X) \circ dB = \theta \, \sigma(X) \, \theta^{-1} \circ \theta \, dB = \sigma(\theta X) \circ d(\theta B).$$

Since θB is also an n-dimensional Brownian motion, the process θX has the same probability law as X. Thus $X(t)$ is a rotation invariant diffusion process on the sphere Γ, i.e. a spherical Brownian motion.

3. Stochastic parallel transport

The parallel transport in an affinely connected C^3 manifold $M = (M, \Gamma_{jk}^i)$ is given by the Levi-Civita equation:

$$du = \Gamma_k \, u \, dx^k$$

where u is a tensor of type (m,n). The stochastic analogue of this concept was introduced by Itô [5] for the special case of tensors of type $(0,n)$ in a Riemannian manifold and was extended by Dynkin [6] to the general case. We will discuss this in terms of stochastic calculus [7].

Let $C = C(t)$, $0 \le t < \infty$, be a random curve on M. It is said to belong to class Q if for every C^3 function $f: M \to R$ the process $f(C(t))$, $0 \le t < \infty$, belongs to Q. This definition can be phrased in terms of local coordinates: If for every local coordinate $(X^i(t))$ of C and for every pair of previsible stopping times $(\mathcal{F}_t)$ $T_1 \le T_2$ such that $X^i(t)$ is well-defined for $T_1 \le t \le T_2$ the process $X^i((t \vee T_1) \wedge T_2)$, $0 \le t < \infty$, belongs to Q, then C is said to belong to class Q.

The spherical Brownian motion of the last section is of class Q.

Now we want to define the parallel transport $U(t)$, $0 \le t < \infty$, of a tensor u of type (m,n) at a along a random curve C of class Q starting at a. Since the sample curve of C is not smooth in general, we cannot use the Levi-Civita equation in its original form, so we define $U(t)$ as follows. Take any division points of $[0, \infty)$:

$$\Delta = (0=t_0 < t_i < \ldots)$$

and construct a random curve $C_\Delta = C_\Delta(t, \omega)$ by connecting $C(t_i, \omega)$ with $C(t_{i+1}, \omega)$ by a geodesic curve for every i. Then C is piecewise smooth, so we can define the parallel transport $U(t)$ of u for each sample curve by the Levi-Civita equation:

$$d\, U_\Delta = \Gamma_k\, U_\Delta\, dx^k, \qquad U_\Delta(0) = u.$$

By a routine method we can prove the existence of the limit process U of U_Δ as

$$|\Delta| \equiv \sup_i |t_i - t_{i-1}| \to 0.$$

The limit process U is called the <u>stochastic parallel transport</u> of u along C.

Theorem 4.1. The above stochastic parallel transport U is determined by the stochastic differential equation:

$$dU = (\Gamma_k U)\circ dx^k, \qquad U(0) = u.$$

Proof. Routine. We should note that this differential equation has the same form as in differential geometry because of use of symmetric Q-multiplication.

We can use stochastic parallel transport to discuss diffusions of tensors [5], [6], [7] and rolling along Brownian motion on a Riemannian manifold [7].

BIBLIOGRAPHY

1. D.L. Fisk, Quasi-martingales and stochastic integrals, Technical Report No.1, Dept Mathe., Michigan State Univ., 1963.
2. P. Courrège; Intégral stochastiques et martingales de carré integrable, Sem. Th. Potential 7. Inst. Henri Poincare, Paris, 1963.
3. H. Kunita and S. Watanabe: On square integrable martingales, Nagoya Math. J. 30 (1967) 209-245.

4. R.L. Stratonovich: Conditional Markov processes and their applica-
    tion to optimal control, Elsevier, New York (1968).

5. K. Itô, The Brownian motion and tensor fields on Riemannian mani-
    fold, Proc. Int. Congress Math. 1962 (Stockholm), 536-539.

6. E.B. Dynkin, Diffusion of tensors, Soviet Math. Dokl. Vol.9 (1968),
    No.2, 532-535.

7. K. Itô, Stochastic parallel displacement, Report of Symposium on
    Stochastic Differential equations, Univ. Victoria, Canada, to appear
    in Springer Lecture Note Series.

8. D.W. Stroock: On the growth of stochastic integrals, Z. Wahr. 18
    (1971) 340-344.

QUESTIONS AND ANSWERS

Question (T. Hida). The speaker emphasizes the significance of the second definition of stochastic integral, that is the one due to Stratonovich. So far as differential (stochastic) calculus is concerned, I agree with the speaker's opinion. However as soon as we come to integral calculus such as multiple integrals with respect to martingales, the first defi- nition of the integral, namely the Itô integral, seems to be more impor- tant. How is the speaker's opinion on this?

Answer (Itô). The first definition is more basic, but in this lecture I emphasized that there are quite a few cases in which the second one is convenient.

Question (J.L. Lebowitz). Can you also conveniently describe the Ornstein- Uhlenbeck process on a sphere?

Answer (Itô). I hope so. There is one point to be noted. Since the Ornstein-Uhlenbeck process in $\mathbb{R}^n$ is a velocity process of a random motion in $\mathbb{R}^n$, the spherical analogue is not a process on a sphere but a process on the space of tangent bundles of the sphere.

ASYMPTOTIC PROPERTIES OF THE SPECTRAL DISTRIBUTIONS

OF DISORDERED SYSTEMS

Masatoshi Fukushima

Department of Mathematics
Osaka University
Toyonaka, Osaka, Japan

Let us first consider a discrete version of the Schrödinger operator with a random potential q :

$$(1) \quad (H^\omega u)(a) \ = \ -(H^O u)(a) \ + \ q(a,\omega)u(a), \qquad a \in Z^\nu.$$

Here u is a function on the ν-dimensional lattice space Z^ν. H^O is a second order difference operator defined by

$$(2) \quad (H^O u)(a) \ = \ \frac{\sigma^2}{2} \sum_{i=1}^{\nu} \{u(a_1,\cdots,a_i-1,\cdots,a_\nu) - 2u(a)$$
$$+u(a_1,\cdots,a_i+1,\cdots,a_\nu)\} , \qquad a \in Z^\nu ,$$

with a positive constant σ. We assume that $\{q(a,\omega) \ ; \ a \in Z^\nu\}$ is an ergodic stationary random field defined over some probability space $(\Omega, \mathcal{B}, P)$.

Denote by $L^2(Z^\nu)$ the space of all square summable functions on Z^ν with inner product $(u, v) = \sum_{a \in Z^\nu} u(a)v(a)$. For each $\omega \in \Omega$, the operator H^ω induces a self-adjoint operator A^ω on $L^2(Z^\nu)$ by

$$(3) \quad \begin{cases} \mathcal{D}(A^\omega) \ = \ \{u \in L^2(Z^\nu) \ ; \ H^\omega u \in L^2(Z^\nu)\} \\ A^\omega u \ = \ H^\omega u, \quad u \in \mathcal{D}(A^\omega). \end{cases}$$

Let $\{E_\lambda^\omega \ ; \ \lambda \in R^1\}$ be the resolution of the identity associated with A^ω : $A^\omega = \int_{-\infty}^{\infty} \lambda dE_\lambda^\omega$. We can then define the <u>spectral distribution function</u> $\rho(\lambda)$ of the ensemble of operators $\{H^\omega \ ; \ \omega \in \Omega\}$ by

$$(4) \quad \rho(\lambda) \ = \ <(E_\lambda^\omega I_O, \ I_O)>$$

where $I_O(a) = \delta_O(a)$,$a \in Z^\nu$, and $< \ >$ denotes the expectation in ω with respect to the probability measure P.

ρ can be identified with the so-called <u>integrated density of states</u> of a disordered physical system governed by $\{H^\omega; \ \omega \in \Omega\}$. Indeed we have the following ergodic theorem : Let Λ be a rectangle containing the origin with sides parallel to axes. Let

$\lambda_1^\omega \leqq \lambda_2^\omega \leqq \lambda_3^\omega \leqq \cdots \leqq \lambda_N^\omega$ be the eigenvalues of the problem :

$$(H^\omega u)(a) = \lambda u(a), \quad a \in \Lambda - \partial\Lambda ; \quad u(a) = 0, \quad a \in \partial\Lambda.$$

We put, for each λ, $\mathcal{H}^\omega(\lambda ; \Lambda) = \sum_{\substack{\lambda_i^\omega \leq \lambda \\ i=1}} 1$. Then there exists a

set $\Omega_0 \in \mathcal{B}$ with $P(\Omega_0) = 1$ such that, for each $\omega \in \Omega_0$,

$$(5) \qquad \lim_{\substack{L^{(i)}(\Lambda)\to\infty \\ i=1,2,\cdots,\nu}} \frac{\mathcal{H}^\omega(\lambda;\Lambda)}{|\Lambda|} = \rho(\lambda)$$

at every continuity point λ of $\rho(\lambda)$. Here $L^{(i)}(\Lambda)$ (resp. $|\Lambda|$) is the side length (resp. volume) of Λ.

Recent probabilistic investigations have revealed several asymptotic properties of the spectral distribution function near the end point of its spectrum ([1],[2],[3]). Here we state some of them.

Assume that $q(a)$, $a \in Z^\nu$, are non-negative valued independent and identically distributed random variables. In this case $\rho(\lambda)$ vanishes for negative λ and then increases from 0 exponentially :

Theorem 1. ([2])

(i) If $P(q(0) = 0) < 1$, then $\overline{\lim}_{\lambda\downarrow 0}\sqrt{\lambda}\log\rho(\lambda) < 0.$

(ii) If $P(q(0) = 0) > 0$ and $\nu = 1$, then $-\infty < \underline{\lim}_{\lambda\downarrow 0}\sqrt{\lambda}\log\rho(\lambda)$.

A similar phenomena occurs in the one dimensional Schrödinger operator with the potential q being the Poisson impulses :

$$(6) \quad H^\omega u(a) = -\frac{1}{2} u''(a) + q(a,\omega)u(a), \quad a \in R^1.$$

More precisely let $k(da,\omega)$ be the Poisson random measure with mean measure being the Lebesgue measure : $k(E,\omega)$ is distributed according to the Poisson law with mean $|E|$ and $k(E_1)$, $k(E_2),\cdots,k(E_n)$ are independent for disjoint E_j's. The operator H^ω should be understood in a generalized sense ([5]):

$$(6)' \quad H^\omega u(a) = \frac{-\frac{1}{2} du'(a) + k(da,\omega)u(a)}{da} .$$

Then we can associate with the ensemble of operators $\{H^\omega\}$ its spectral distribution function $\rho(\lambda)$, $\lambda \in R^1$, just in the same manner as in (5).

Theorem 2.   The spectral distribution function ρ of the random operator (6)' behaves as follows :

$$-\infty < \varliminf_{\lambda \downarrow 0} \sqrt{\lambda}\,\log \rho(\lambda) \; \leq \; \varlimsup_{\lambda \downarrow 0} \sqrt{\lambda}\,\log \rho(\lambda) \; < \; 0 .$$

This behaviour was first observed by I.M.Lifschitz ([4]), but the above theorem is due to S.Nakao ([3]).

Returning to the random difference operator (1), we mention some fine structure of the associated resolution of the identity $\{ E_\lambda^\omega, \lambda \in R^1 \}$. Suppose that $\nu = 1$ and that $q(a)$, $a \in Z^1$, are independent random variables with common distribution ϕ.

Theorem 3.   If the support of ϕ contains at least two points and $\int_{-\infty}^{\infty} |x|\,\phi(dx) < \infty$, then there exists a set $\Omega_0 \in \mathcal{B}$ with $P(\Omega_0) = 1$ such that, for each $\omega \in \Omega$, the spectrum of the self-adjoint operator A^ω defined by (3) admits no absolutely continuous part : the one dimensional measure $d_\lambda(E_\lambda^\omega u, u)$ is singular for any $u \in L^2(Z^1)$.

This theorem was discovered by A.Casher and J.L.Lebowitz ([6]) in the case of semi-infinite lattice $Z_+^1 = \{0, 1, 2, \cdots\}$ with the fixed end boundary condition at -1.   By extending an idea of K.Ishii ([7]), L.A.Pastur was recently able to get the result for the doubly infinite case Z^1.

The proof of Theorem 1 and Theorem 2 is carried out by making use of the Feynman-Kac expression of the Laplace transform of $\rho(\lambda)$ in terms of the associated auxiliary Markov process and then by invoking a Tauberian theorem of exponential type.

Theorem 3 is a consequence of an exponential growth character of the generalized eigenvector of the random operator H^ω.   Such a character has been derived by H.Matsuda and K.Ishii ([8]) and Y.Yoshioka ([9]) from the Furstenberg limit theorem ([10]) on the product of random unimodular matrices.

References

[1] L.A.Pastur, Spectra of random self-adjoint operator, Russian
    Mathematical surveys 28(1973), 1-67.

[2] M.Fukushima, On the spectral distribution of a disordered system
    and the range of a random walk, Osaka J. Math., 11(1974), 73-85.

[3] S.Nakao, On the spectral distribution of the Schrödinger operator
    with random potential, to appear.

[4] I.M.Lifschitz, Energy spectrum structure and quantum states of
    disordered condensed system, Uspehi Fiz.Nauk, 83(1964),617-663.

[5] K.Ito and H.P.McKean,Jr.,Diffusion processes and their sample
    paths, Springer-Verlag,1965.

[6] A.Casher and J.L.Lebowitz, Heat flow in regular and disordered
    harmonic chains, J.Math. Phys., 12(1971), 1701-1711.

[7]K.Ishii, Localization of eigenstates and transport phenomena in the
    one-dimensional disordered systems, Prog.Theor.Phys. Suppl. No53
    (1973),77-138.

[8] H.Matsuda and K.Ishii, Localization of normal modes and energy
    transport in the disordered harmonic chain, Prog.Theor.Phys.Suppl.
    No 45(1970),56-86.

[9] Y.Yoshioka, On the singularity of the spectral measures of a semi-
    infinite random system, Proc. Japan Acad., 49(1973), 665-668.

[10]H.Furstenberg, Noncommuting random products, Trans.Amer.Math.Soc.,
    108(1963), 377-428.

A REMARK TO HARRIS'S THEOREM ON PERCOLATION

M. Miyamoto

Yoshida College, Kyoto University

Let T be the set of bonds in the 2-dimensional square lattice Z^2. We adopt the following

NOTATIONS.

$$\Omega = \{0, 1\}^T ,$$

$$X_t(\omega) = \omega(t) \quad \text{for} \quad t \in T \quad \text{and} \quad \omega \in \Omega ,$$

$$X = \{X_t ; t \in T\} ,$$

$$X^{-1}(i) \equiv X^{-1}(i, \omega) = \{t \in T ; X_t(\omega) = i\} \quad \text{for} \quad i = 0, 1 ,$$

P is a probability measure on Ω .

Harris proved the following

__THEOREM.__ If a random field $(\Omega, P ; X)$ is independent at each $t \in T$, and if $P\{X_t = 1\} = P\{X_t = 0\} = \frac{1}{2}$ for every t , then neither $X^{-1}(0)$ nor $X^{-1}(1)$ has infinite connected components a. s. (P) .

His method can be applied to generalize the above to a dependent case. For $V \subset T$, let $\mathcal{B}_V$ be the σ-algebra generated by $\{X_t ; t \in V\}$, and let $\mathcal{B}_\infty = \cap \mathcal{B}_{Vc}$, where V runs

over the set of all finite subsets of T . Let ∂V be the set

of bonds which touch bonds in V .

THEOREM.   We assume that a random field $(\Omega, P ; X)$ satisfies

the following conditions.

1) (Spatial symmetry)   P is invariant under shift, rotation by

a right angle and reflection in the axis in T .

2) (Symmetry of configurations)   P is invariant under inter-

change of 0 and 1 .

3) P is everywhere dense.

4) $\mathcal{B}_\infty$ is trivial, if it is measured by P .

5) (The FKG inequality)   If $f(\omega)$ and $g(\omega)$ are non-decreas-

ing functions of $\omega \in \Omega$, then

$$\int f(\omega) g(\omega) dP(\omega) \geq \int f(\omega) dP(\omega) \cdot \int g(\omega) dP(\omega) .$$

6) (Markovian property)   For each $\Lambda \in \mathcal{B}_V$,

$$P(A | \mathcal{B}_{Vc}) = P(A | \mathcal{B}_{\partial V}) .$$

Then neither $X^{-1}(0)$ nor $X^{-1}(1)$ has infinite connected com-

ponents a.s. (P) .

Reference

T. E. Harris;   A lower bound for the critical probability in a
certain percolation process. Proc. Cambridge
Phil. Soc., 56(1960),13-20.

THE ERGODICITY OF THE MOTION OF A PARTICLE IN A POTENTIAL FIELD

Izumi KUBO

Department of Mathematics
Nagoya University
Nagoya, Japan

1. Introduction

The ergodicity of classical dynamical systems which appear actually
in the statistical mechanics was discussed by Ya. G. Sinai [11].   He
announced the ergodicity of the dynamical system of particles with
central potentials of a special type which are enclosed in a rectangular
box.   However no proofs have been published yet, except the simplest
one-particle model which is called a Sinai billiard system.   The
ergodicity of the system and the K-property are shown in [13], [8], and
the Bernoulli property is shown in [5].   The purpose of this report
is to show the ergodicity and the Bernoulli property of the motion of
a particle in a potential field.   The proofs for a special case of a
compound central potential field are separately presented in [9], [10].

2. Observation

Let T be a two dimensional torus and let $U_\iota(q)$, $\iota = 1,2,\ldots,I$,
be several potential functions with finite ranges $\overline{Q}_\iota$'s defined on T.
Suppose that the potential ranges do not overlap and that the boundary
∂Q_ι of the range $\overline{Q}_\iota$ is a closed curve of C^3-class.   Assume that
every $U_\iota(q)$ is continuous in the torus T and is continuously
differentiable in $\overline{Q}_\iota$.

Observe the motion of a particle with mass m and energy E in
the potential field.   Denote by $\{S_t\}$ the flow induced from the
dynamical system ; that is, for each (q, p) the point $S_t(q, p)$ means
the state of the particle at time t whose initial state is (q, p),
where $q = (q^{(1)}, q^{(2)}$ and $p = (p^{(1)}, p^{(2)})$ are the position and the
momentum.   As usual the flow $\{S_t\}$ can be restricted to the energy

surface M_E, $\quad M_E \equiv \{(q, p) \; ; \; (p^{(1)})^2 + (p^{(2)})^2 = 2m(E-U(q)), \quad q \in Q_E\}$,

where $Q_E \equiv \{q \; ; \; U(q) \leq E\}$. Moreover the measure

$$d\mu_E \equiv \text{const. } d\omega dq^{(1)} dq^{(2)}$$

on M_E is invariant under $\{S_t\}$, where $(p^{(1)}, p^{(2)}) = (\kappa\cos\omega, \kappa\sin\omega)$ with $\kappa^2 = 2m(E-U(q))$. Let π be the natural projection from M_E to the configuration space Q_E ; $\pi(q, p) = q$. Put $Q \equiv T - \overset{I}{\underset{\imath=1}{\cup}} \overline{Q}_\imath$ and $M_0 \equiv \pi^{-1}(Q)$. It is easily seen that almost every orbit of the particle crosses the boundary $\partial Q = \underset{\imath}{\cup}\partial Q_\imath$ of Q. A point q of the boundary can be parametrized by $(\imath, r)$, where $\imath$ is the number of the curve $\partial Q_\imath$ which contains q, and r is the arclength between the point q and a fixed origin of $\partial Q_\imath$ measured along the curve $\partial Q_\imath$ clockwise. Further, a point (q, p) in the set $\pi^{-1}(\partial Q)$ can be parametrized by $(\imath, r, \varphi)$, where φ is the angle between p and the inward normal of $\partial Q_\imath$ at q. Put

$$M \equiv \{(\imath, r, \varphi) \in \pi^{-1}(\partial Q) \; ; \; \tfrac{1}{2}\pi \leq \varphi \leq \tfrac{3}{2}\pi\},$$

namely M is the set of all incident vectors at ∂Q. Then a transformation T_* of M is defined by that $T_*^{-1}(\imath, r, \varphi) = (\imath_1, r_1, \varphi_1)$ shows the next incident position and angle after starting from $(\imath, r, \varphi)$. Then it is seen that T_* preserves the measure ν of M defined by

$$d\nu = -\nu_0 \cos\varphi d\varphi dr d\imath$$

with normalized constant ν_0, where $d\imath$ means the unit mass on each $\imath$.

Since it is true that $\{S_t\}$ is ergodic if and only if T_* is ergodic, it is important to investigate the transformation T_*. Then the transformation T_* can be resolved into the product

$$T_* = T_1 T,$$

where T_1 is the transformation such that for $(\imath, r', \varphi') \equiv T_1^{-1}(\imath, r, \varphi)$, the point $(\imath, r', \pi-\varphi)$ shows the position and the momentum at the instant when the particle goes out from $\overline{Q}_\iota$, and T is *the Sinai billiard transformation* which maps $T_*^{-1}(\iota, r, \varphi) = (\iota_1, r_1, \varphi_1)$ to (ι, r', φ').

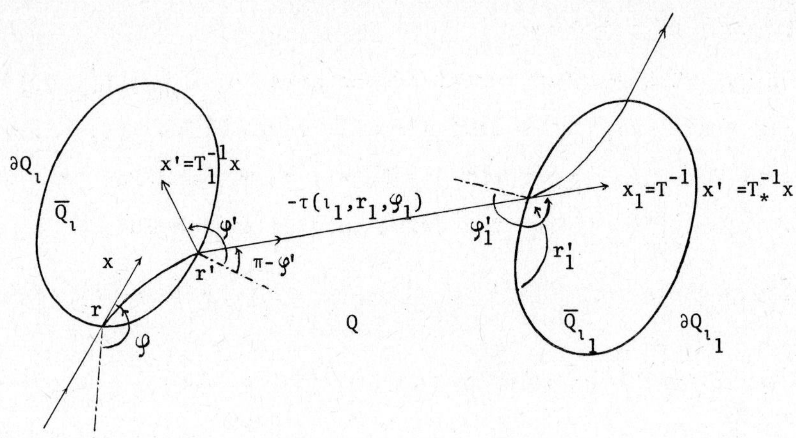

Generally, let us call a transformation T_* *a perturbed billiard transformation* if T_* is expressed in the form

$$T_* = T_1 T$$

with the Sinai billiard transformation T defined for Q and with a C^2-diffeomorphism T_1 of M which preserves the measure ν and each set $\pi^{-1}(\partial Q_\iota)$, $\iota = 1, 2, \cdots, I$.

Obviously, the transformation T_* induced from $\{S_t\}$ is a perturbed billiard transformation. If the perturbation of T by T_1 is small, then the ergodic theoretical properties of T should be preserved under the perturbation.

3. Theorems

In this section, assume the following assumptions (H-1) $\sim$ (H-3) :

(H-1) Every $\overline{Q}_\iota$ is a strictly convex domain with the boundary ∂Q_ι of C^3-class.

(H-2) T_1 is a C^2-diffeomorphism of M which preserves the measure ν and satisfies

$$T_1(\iota, r, \tfrac{1}{2}\pi) = (\iota, r, \tfrac{1}{2}\pi) \quad \text{and} \quad T_1(\iota, r, \tfrac{3}{2}\pi) = (\iota, r, \tfrac{3}{2}\pi).$$

Denote by $k(\iota, r)$ the curvature of Q_ι at r and by $-\tau(\iota_1, r_1, \varphi_1)$ the distance between (ι_1, r_1) and (ι, r') measured along the orbit with $(\iota, r', \varphi') = T(\iota_1, r_1, \varphi_1)$. Put $A \equiv \dfrac{\partial r_1}{\partial r}$, $B \equiv \dfrac{\partial r_1}{\partial \varphi}$, $C \equiv \dfrac{\partial \varphi_1}{\partial r}$ and $D \equiv \dfrac{\partial \varphi_1}{\partial \varphi}$.

(H-3)   There exist positive constants η, L_{max}, $L_{min}^+(\iota)$ and $L_{min}^-(\iota)$, $1 \le i \le I$, such that for (ι, r, φ) with $(\iota, r', \varphi') = T_1^{-1}(\iota, r, \varphi)$ and $(\iota_1, r_1, \varphi_1) = T_*^{-1}(\iota, r, \varphi)$ and for $s \ge L_{min}^+(\iota)$ (resp. $s \ge L_{min}^-(\iota_1)$) the following inequalities hold

$$-Cs-D \ge 1+\eta \qquad\qquad \left(\text{resp. } -\frac{\cos\varphi_1}{\cos\varphi}(Cs+A) \ge 1+\eta\right),$$

$$L_{max} \ge \frac{As+B}{Cs+D} \ge L_{min}^+(\iota_1) \qquad\qquad \left(\text{resp. } L_{max} \ge \frac{Ds+B}{Cs+A} \ge L_{min}^-(\iota)\right),$$

$$\frac{\cos\varphi_1}{\cos\varphi}(A+\tfrac{1}{s}B)(Cs+D) \ge 1+\eta \qquad\qquad \left(\text{resp. } \frac{\cos\varphi_1}{\cos\varphi}(D+\tfrac{1}{s}B)(Cs+A) \ge 1+\eta\right).$$

The assumption (H-3) may look rather complicated.   However the assumption is satisfied if the perturbation by T_1 is small.   For example, the following proposition gives sufficient conditions under which the assumption (H-3) is satisfied.   For a function f, put $[f]^+ \equiv \max\{f, 0\}$.

Proposition 1.

(i)   If there exists a positive constant η such that

$$-D \ge 1+\eta, \qquad -\frac{\cos\varphi_1}{\cos\varphi} A \ge 1+\eta$$

and if $-B \ge 0$, $-C \ge 0$, then the assumption (H-3) is fulfilled.

(ii)   If

$$\frac{1}{k_{min}}\left[-\frac{\partial\varphi'}{\partial r}\right]^+ + \left(1 + \frac{1}{k_{min}|\tau|_{min}}\right)\left[-\frac{\partial\varphi'}{\partial\varphi} + 1\right]^+ < 1,$$

$$k_{max}\left(1 + \frac{2}{k_{min}|\tau|_{min}}\right)\left[\frac{\partial\varphi}{\partial r'}\right]^+ + \left(1 + \frac{1}{k_{min}|\tau|_{min}}\right)\left[-\frac{\partial\varphi}{\partial\varphi'} + 1\right]^+ < 1,$$

then the assumption (H-3) is fulfilled, where $k_{min} \equiv \min_{\iota,r} k(\iota, r)$,

$$k_{max} \equiv \max_{\imath, r} k(\imath, r) \quad \text{and} \quad |\tau|_{min} \equiv \min_{\imath, r, \mathcal{S}} |\tau(\imath, r, \mathcal{S})|.$$

Theorem 2.   Under the assumptions (H-1), (H-2) and (H-3),   the transformation $T_* = T_1 T$ is Bernoullian,   and hence it is ergodic and a K-system.

Theorem 3.   Let $\{S_t\}$ be the flow given in §1 which is governed by potential functions $\{U_\imath\}$.   If $\{U_\imath\}$ and the energy E of the particle admit the assumptions (H-1), (H-2) and (H-3),   then $\{S_t\}$ is ergodic.   Moreover, if $\{S_t\}$ has no point spectrum, then $\{S_t\}$ is a K-system and a Bernoulli flow.

If $U_\imath$'s are central potentials,   then one can state the sufficient condition for the ergodicity in terms of $\{U_\imath\}$.   A central potential $V(s)$ is called *bell-shaped* if

(BS-1) $V(s)$ is continuous for $s > 0$ and $V(s) = 0$ for $s > R$ with some R,

(BS-2) $V(s)$ belongs to C^2-class in $(0, R)$ and there exist left derivatives $V'(R-0)$ and $V''(R-0)$,

(BS-3) $-sV'(s)$ is monotone decreasing and $V'(R-0) < 0$.

Theorem 4.   If every $U_\imath$ is bell-shaped and if the energy E satisfies the condition

$$0 < E < \frac{1}{4} \min_{\imath}\{- \frac{R_\imath |\tau|_{min}}{R_\imath + |\tau|_{min}} U'(R_\imath - 0)\},$$

then $\{S_t\}$ is ergodic.   Moreover if $\{S_t\}$ has no point spectrum, then $\{S_t\}$ is a K-system and a Bernoulli flow.

REFERENCES

[1]   Бунимович, Л. А. : Об ергодических свойствах некотрых
        бильярдов.   Функц. анализ. 8(1974), 73-74.

[2]   Бунимович, Л. А. : Центральная предельная теорема для одного
        класса бильярдов.   Теор. вероят. и ее примен. 19
        (1974), 63-83.

[3] Бунимович, Л. А. : О бильярдах, близких к рассеивающим.
     Матем. сб. 94(1974), 49-73.

[4] Bunimovich, L.A. = Sinai, Ya.G. : On a fundamental theorem in the
        theory of dispersing billiards, Math. USSR-Sb. 90 (1973),
        407-423.

[5] Galavotti, G.=Ornstein, D.S. : Billiards and Bernoulli schemes
        (preprint).

[6] Kubo, I. : Ergodicity of the dynamical system of a particle on a
        domain with irregular walls, Proc. 2nd Japan-USSR Symp. Prob.
        Theory.   Lecture Notes in Math. 330, Springer (1973),
        287-295.

[7] Kubo, I. : The motion of a particle in a central field,
        数理解析研究所講究録 204, RIMS (1974), 100-109.

[8] Kubo, I. : Billiard systems, Lecture Notes (preparing).

[9] Kubo, I. : Perturbed billiard systems I.   The ergodicity of the
        motion of a particle in a compound central field.   (preprint).

[10] Kubo, I. = Murata, H. : Perturbed billiard systems II.   The
        Bernoulli property.   (preparing).

[11] Sinai, Ya. G. : On the foundations of the ergodic hypothesis for a
        dynamical system of statistical mechanics, Sov. Mat. Dokl. 4
        No.6 (1963), 1818-1822.

[12] Sinai, Ya. G. : Classical dynamical systems with countably multiple
        Lebesgue spectra II, A.M.S. Transl. (2) 68 (1968), 34-68.

[13] Sinai, Ya. G. : Dynamical systems with elastic reflections,
        Russian Math. Surveys 25 (1970), 137-189.

<u>ERGODICITY OF SOME SIMPLE MODEL SYSTEMS OF INFINITELY</u>

<u>MANY PARTICLES</u>

Toshio Niwa
Department of Mathematics, Kyoto Univ.
Kyoto, Japan

1. Introduction

The aim of our talk is to propose some simple model system of infinitely many classical particles and to investigate its ergodic properties.

By many authors such as Sinai and Lanford dynamical systems with infinitely many degrees of freedom have been constructed [3],[4].   But their time evolutions are so complicated.   It is hardly possible to study their ergodic properties except the cases of ideal gases in arbitrary dimension and the system of hard rods in one-dimensional space [1],[2].

Recently Hardy et al have proposed some simple 2-dimensional system of classical particles [5].   Simplicity of its dynamics allows to study some of its ergodic properties.   Nevertheless it is still very difficult problem.   They have studied it by "linearizing" the time evolution and could get some results.

Here we study the ergodic properties of its time evolution without "linearizing" but we assume the domain where collisions do occur is bounded.

We prove that the system so obtained is Bernoulli, so is mixing, and the time correlation functions are decreasing exponentially.

Unfortunately our system has no interactions between particles except those of which are in some bounded domain, so the system is to be considered as "perturbed ideal gas".   It is also possible to consider it as a finite system surrounded by ideal gas.

2. Description of the model.

Now we describe the system more precisely [5].   Let Z^2 be 2-dimensional lattice.

The phase space $\mathcal{X}$ of allowed configurations of particles is
$$\mathcal{X} = \{ X ; X : Z^2 \times P \rightarrow \{ 0, 1 \} \}$$
where $P = \{ V = (v_x, v_y) \in Z^2 ; |v_x| + |v_y| = 1 \}$ is a momentum space.

Let $\mathcal{X}_a$ be the space of configurations of particles on a site $a \in Z^2$:
$$\mathcal{X}_a = \{ X_a ; X_a : \{ a \} \times P \rightarrow \{ 0, 1 \} \} .$$
Then naturally
$$\mathcal{X} = \prod_{a \in Z^2} \otimes \mathcal{X}_a .$$

The time evolution of the system is defined as follows.

Let T_0 be a translation of $\mathcal{X}$, that is,
$$T_0 X (a, v) = X(a - v, v) \text{ for } \forall X \in \mathcal{X} , \forall a \in Z^2 , \text{ and } \forall v \in P.$$

Collision   C   is defined by the interaction operator φ_a on each lattice site a, that is a bijection mapping of $\mathcal{X}_a$ and satisfies the following conditions.

1) Preservation of the total momentum:
$$\sum_{v : \varphi_a X_a (v) = 1} v = \sum_{v : X_a (v) = 1} v ,$$

2) Preservation of the total number of particles:
$$\sum_{v \in P} \varphi_a X_a (v) = \sum_{v \in P} X_a (v).$$

Under these conditions φ_a is trivial, that is,
$$X_a = \varphi_a X_a \text{ for } \forall X_a \in \mathcal{X}_a$$

or given by

$$(\varphi_a X_a)(v) = X_a(v) + (-1)^{v_x}[X_a(v^1)X_a(v^3)\{1 - X_a(v^2)\}\{1 - X_a(v^4)\}$$
$$- \{1 - X_a(v^1)\}\{1 - X_a(v^3)\}X_a(v^2)X_a(v^4)]$$

for $\forall X_a \in \mathcal{X}_a$, $\forall v \in P$, where $v^1 = (1, 0)$, $v^2 = (0, 1)$, $v^3 = (-1, 0)$ and $v^4 = (0, -1)$.

This is the one considered in [5].

C is defined by

$$CX = \prod_{a \in Z^2}^{\otimes} \varphi_a X_a \quad \text{for} \quad \forall X = \prod_{a \in Z^2}^{\otimes} X_a \in \mathcal{X} = \prod_{a \in Z^2}^{\otimes} \mathcal{X}_a.$$

The time evolution T of $\mathcal{X}$ is defined by

$$T = CT_o. \qquad (\text{or} \quad T = T_o CT_o).$$

Under some conditions there exists a unique equilibrium state ρ on $\mathcal{X}$, that is, T-invariant probability measure, which has no correlations between particles with different velocities and lattice sites [5].
Henceforth we fix such state ρ.

Now we assume that the interaction operators φ_a on $\mathcal{X}_a$ are trivial if $a \notin V$ for some bounded domain V of Z^2.

3. Results and some remarks.

We investigate the ergodic properties of the dynamical system $(\mathcal{X}, T, \rho)$.
We get

Theorem. $(\mathcal{X}, T, \rho)$ is a Bernoulli System, therefore, has a mixing property, and the time correlation functions are decreasing exponentially, that is, for any cylinder sets A and B (i.e. subsets of $\mathcal{X}$ defined on the sites that are in some bounded domain of Z^2) we have

$$|\rho(T^n(A) \cap B) - \rho(A)\rho(B)| \leq \text{const.} \ r^n \quad \text{for} \quad \forall n \geq 1$$

where const. and r are constants depending only on the supports of A and B and $0 < r < 1$.

<u>Remark</u> These results can be extended to more general one.

Firstly dimension of the lattice is not necessary to have two.
Secondly interactions between different sites can be permitted, in this case "dissipative" character of the interactions is enough.

References

1. Ya. G. Sinai: Ergodic properties of a gas of one dimensional hard rods with an infinite number of degrees of freedom. Func. Anal. Appl. 6-1, 41-50, (1972)
2. ——, K. L. Volkovysskii: Ergodic properties of an ideal gas with an infinite number of degrees of freedom, Func. Anal. Appl. 5-4, 19-21, (1971)
3. ——, The construction of cluster dynamics for dynamical systems of statistical mechanics, Vestinik Moscow Univ. 1, 152-158, (1974)
4. O. E. Lanford III: The classical mechanics of one-dimensional systems of infinitely many particles. I, II. Comm. Math. Phys. 9, 176-191, (1968), II, 257-292, (1969)
5. J. Hardy et al: Time evolution of a two-dimensional model system I. Jour. Math. Phys. 14-12, 1746-1759, (1973)

NEW EXACT SOLUTIONS OF THE CLASSICAL
YANG-MILLS FIELD EQUATION

Akio Hosoya and Jun Ishida
Department of Physics
Osaka University

Toyonaka, Japan

Nowadays many particle physisists guess that the hadrons are perhaps something like vortex-lines or "strings" in the ether.[1] It will be interesting for gauge theoretists if there exist classical solutions for gauge fields which exhibit vortex-lines.[2][3] In this talk we wish to present a method to solve the 0(3) gauge field equation in the case of magnetostatics. We shall obtain a new wide class of classical solutions of the equation:

$$\partial_i F^a_{ij} + g\,\epsilon^{abc} W^b_i F^c_{ij} = 0 \ , \tag{1}$$

where F^a_{ij} is the field strength of the gauge field W^a_i. We seek for such solutions of eq. (1) that the magnetic field $H^a_i = (1/2)\epsilon_{ijk} F^a_{jk}$ has the following form:

$$(1/2)\,\epsilon_{ijk} F^a_{jk} \equiv H^a_i = D^{ab}_i A^b \ , \tag{2}$$

where A^a is a scalar triplet field and D^{ab}_i is the covariant derivative. Eq. (2) is motivated by the Abelian magnetic field in the magnetostatics: $H_i = \partial_i \Omega$, Ω being a magnetic potential. Using the field equation (1) we can show that the ansatz on the magnetic field (2) severely restricts the form of the vector potential to be

$$W^a_i = \frac{\epsilon^{abc}(\partial_i A^b)A^c}{g\,|A|^2} + \text{terms proportional to } A^a \ . \tag{3}$$

To get the explicit form of the magnetic potential A^a we insert eq. (3) back into the left-hand side of eq. (2). We obtain the condition for A^a as follows.

$$\partial_i |A| = \epsilon_{ijk}\,\epsilon^{abc}(\partial_j A^a)A^b(\partial_k A^c)\,/\,g\,|A|^2 \ . \tag{4}$$

Explicit solutions of eq. (4) can be written in terms of the polar coordinates $|A|$, θ and ϕ in the isotopic space as follows.

$$|A| = \int d\sigma \; \frac{d\, D(\vec{x}-\vec{X}(\sigma))}{d\sigma} \; G(\vec{X}(\sigma)) \quad ,$$

$$\cos\Theta\, d\Phi = \int d\sigma \; g \, \epsilon_{ijk} \frac{d\vec{X}_j(\sigma)}{d\sigma} \, \partial_k^x D(\vec{x}-\vec{X}(\sigma)) \, G(\vec{X}(\sigma)) \quad , \tag{5}$$

where $\vec{X}(\sigma)$ is the position of the string-like singularity and $G(\vec{x})$ can
be interpreted as magnetic dipole densities at $\vec{X}(\sigma)$.   When we specifically put $G = {}^*g = $ constant we have an analog of Dirac's monopole
theory for the group $0(3)$.   Indeed in this case we observe that the
energy density $E = 1/2 \cdot (H_i^a)^2 = 1/2 \cdot (\partial_i A)^2$ is independent of the location of the string $\vec{X}(\sigma)$.   It is also remarked that eq. (4) leads to
the charge quantization condition of two units:

$$ {}^*g\, g / 4\pi = 2 \times (n/2) \quad , \quad (n: \text{integer}) \quad . \tag{6}$$

When the Higgs fields are properly introduced, it is expected that our
gauge fields are confined near the string $\vec{x} = \vec{x}(\sigma)$ in the Higgs ether
to form vorteces.

"Discussion"

Ukawa (question): The charge quantization condition of Schwinger's
version can be obtained by considering the finite string in the
Abelian case.   What is the reason why you can obtain the integral
valued charge quantization condition considering the halfly infinite
string in the non-Abelian case?

Hosoya (answer): The coefficient $1/4\pi$ in the condition $g{}^*g/4\pi = $ integer comes from the total solid angle 4π in the $0(3)$ case.   The corresponding quantity will be 2π in the Abelian case.

Ukawa (question): Is it possible to generalize your theory to larger
groups?

Hosoyo (answer): I think so.   But unfortunately we have not succeeded up to now because of the complexities.

"References"

(1) H. B. Nielsen and P. Olesen, Nucl. Phys. B61 45 (1973).

(2) Y. Nambu, Phys. Rev. D15, 4262 (1974).

(3) A. Hosoya and J. Ishida, "Non-Abelian Magnetic Charges" Osaka University preprint (1974).

TOPOLOGY OF HIGGS FIELDS

J. Arafune*, P.G.O. Freund[+] and C.J. Goebel[§]

*National Laboratory for High Energy Physics
Oho-Machi, Tsukuba-Gun, Ibaraki 300-32, Japan

[+]Enrico Fermi Institute and Department of Physics
University of Chicago, Chicago, Ill 60637, U.S.A.

[§]Department of Physics
University of Wisconsin, Madison, Wis 53706, U.S.A.

Since this article will be published in the Journal of Mathematical Physics, only the abstract and the discussions are given here.

Abstract   It is shown that the conserved magnetic charge discovered by 't Hooft in nonabelian gauge theories with spontaneous breaking is not associated with the invariance of the action under a symmetry group. Rather, it is a topological characteristic of an isotriplet of Higgs fields in a three-dimensional space: the Brouwer degree of the mapping from a large sphere in the configuration space to the unit sphere in the field space provided by the normalized Higgs fields $\hat{\phi}^a = \phi^a(\phi^b\phi^b)^{-1/2}$. The use of topological methods in determining magnetic charge configuration is outlined. A peculiar interplay between Dirac strings and zeros of the Higgs fields under the gauge transformations is pointed out. The monopole-antimonopole system is studied.   (to be published in Journal of Mathematical Physics)

DISCUSSIONS

(1) Question by B.Schroer: Are the conservation laws a property of
a particular solution, a property of a class of solutions or an
intrinsic property of the dynamics?
Answer: They are due to the dimension of the space and the special
number of the Higgs fields. They are true for any reasonable
solutions.

(2) Question by B.Schroer: Do you expect this classical solution
to be useful in a future quantum field theory?
Answer: Yes, I think so. But it is not clear to me whether the
conservation law has a positive meaning in the quantized theory.

(3) Question by N.Nakanishi: I am afraid that $"\dot{Q}_{mag} = 0"$ does not
necessarily follow from $\partial_\mu j^\mu_{mag} = 0$, because in this derivation
one has to assume the vanishing of j^μ_{mag} at very distant points.
This assumption may not be satisfied in this case, because the
normalized field $\hat{\phi}^a = \phi^a / |\vec{\phi}|$, ($|\hat{\vec{\phi}}| = 1$), is used in the definitions
of j^μ_{mag} . Furthermore, I think that Q_{mag} may change discontin-
uously if all the components of $\vec{\phi}$ vanish at a particular *time*
$x^0 = a^0$.
Answer: Fortunately we are dealing with fields with spontaneous
breaking.   In order that the energy of a given system is to be
finite $|\vec{\phi}|$ must approach a non-vanishing constant at distant
points. Therefore vanishing of $\vec{\phi}$ for all the space at a time a^0
is not allowed; for it gives infinite energy at that time. For
simplicity we examined only a case where there are finite number
of zeros of Higgs fields. Since $j^0_{mag} = 0$ for non-vanishing $|\vec{\phi}|$,
the boundary condition is satisfied to obtain $\dot{Q}_{mag} = 0$ in such a
case.

SPONTANEOUS BREAKING OF CHIRAL SYMMETRY IN

A VECTOR-GLUON MODEL

Toshihide Maskawa and Hideo Nakajima
Department of Physics, Kyoto University

Kyoto, Japan

Solutions of the self-consistent equation for the fermion propagator in a vector-gluon model are fully examined in the light of spontaneous breaking of chiral symmetry.[1] In other words, the question is whether the fermion mass is spontaneously generated with the axial-vector current kept conserved.[2] In our analysis, we choose a suitable gauge ("Landau-like" gauge) for the vector-gluon and make for simplicity an pole-approximation to the massive vector-gluon propagator[*] and no radiative correction to the vertex function in the Schwinger-Dyson equation for the fermion self-energy part. It can be shown that these approximations together with the ladder approximations to an axial-vector vertex function and to a peudoscalar vertex function, reproduce consistently the Ward-Takahashi identity for the axial-vector current. Therefore we can safely talk about the axial-vector current conservation within these approximations. Setting $i\gamma p+m_0+\sum(p)\equiv i\alpha(p^2)$, we obtain nonlinear equations for the function α and β, characterized by a set of parameters, i.e., the coupling constant g, the bare mass of the fermion m_0, and a cutoff Λ which is introduced as a regulator mass when necessary. The results of our analysis are as follows: (1) It can be proved that the nonlinear equations without tht cutoff have solutions only in the case of $m_0=0$ and that the number of the solutions is infinity of continuum, i.e., there is a free continuous parameter specifying the solution, if $g^2/4\pi < (16/33)^2\pi$. Apparently this situation does not come from the freedom of fixing the mass scale as is thought in massless QED since the gluon is massive. Then we argue that this situation reflects the fact that if a cutoff Λ is introduced and led to infinity, $m_0(m,\Lambda)$ tends to zero, while the value of $2m_0 j_p(=-i\partial_\mu j_{5\mu})$ at the renormalization point tends not necessarily to zero, but to a value suitably fixed by m since the renormalization constant z_p^{-1} tends to infinity.[3] In other words, we obtain the solutions ($\beta\neq 0$) with $m_0=0$, but these are not necessarily chiral invariant; or an infinitesimal bare mass can slip into the solutions, in the cutoff version.

---

[*] We do not deal with spontaneous breakdown of gauge symmetry itself here, and assume that we already have massive vector-gluon somehow.

(2) We investigate the self-consistent equations with sufficiently large but finite cutoff Λ in order to make the above argument more definite. In this case the equation $m_0=0$ certainly means chiral invariance since no infinity appears in the formalism. And we obtain one and only one solution for any fixed m_0 if $g^2/4\pi \lesssim \pi/4$. Therefore when we set $m_0=0$, we are left with the only solution $\beta(p^2)\equiv 0$, which means that no "super-conducting" solution exists for such a small value of g^2 irrespective of the value of Λ, and no Nambu-Goldstone boson appears. (3) For rather strong coupling, however, more specifically, for $g^2/4\pi \gtrsim \pi/4$, we can demonstrate that the solution for the equation with the cutoff is not unique and there does exist another solution $\beta(p^2) > 0$ ("super-conducting" solution) than the solution $\beta(p^2)\equiv 0$ ("normal-state" solution) when $m_0=0$. Further the existence of many "super-conducting" solutions, is inferred. (4) It is also found that in the region $g^2/4\pi> 8\pi$, the "normal-state" solution for the equation without the cutoff, if any, should necessarily have an unphysical singularity, i.e., the propagator should have a singurality in the space-like momentum region. This fact implies that the "normal-state" solution becomes unstable for a sufficiently large value of g^2. To the question whether the spontaneous breaking of chiral symmetry occurs in a vector-gluon model or not, we have thus answered "no for the weak coupling, and yes for the strong coupling" as far as the lowest order approximations for the vector vertex and for the fermion-antifermion scattering kernel are concerned. The latter half of our answer would be valid also for the full theory, provided that the higher order corrections do not spoil badly the applicability of our method to the proofs. For the proofs of the results stated above, we fully utilized fixed point theorems in consideration of nonlinear integral equations.

In connection with our analyses we would like to comment on some attempts of the dynamical Higgs mechanism of the references 4 and 5. In the dynamical Higgs mechanism, i.e., the Higgs mechanism without canonical Higgs' scalar, a composite Nambu-Goldstone boson (massless excitation) which supersedes the canonical Higgs scalar is necessary. Generally it is utilized the fact that if one really has spontaneous symmetry-breaking mass generation of the fermion, there appears successfully the massless excitation. However, at the point where one chooses a required solution $(\beta\neq 0)$, one should be very careful not to take a solution resulting from nonconservation of the current (in the sense stated in the part (1)), but to take a solution which makes the current conserve. Despite of careful reasoning on this point in the reference 4, solutions essentially same as the solution resulting from nonconservation of the current were finally taken in the both papers.

We have proved in the part (2) that no spontaneously symmetry-breaking solution exists if the coupling is weak. Therefore, although their formalisms are promising the solutions that they took are not suitable for their formalisms or the solutions cause internal inconsistency in the formalisms.[*)] Recently, there have appeared similar comments on this point in the reference 6.

REFERENCES

1) T. Maskawa and H. Nakajima, Kyoto university preprint, KUNS 311.

   T. Maskawa and H. Nakajima, Prog. Theor. Phys. <u>52</u> (1974), 1326.

2) R. Haag and A.J. Maris, Phys. Rev. <u>132</u> (1963), 2325.

   K. Johnson, M. Baker and R. Willey, Phys. Rev. <u>136</u> (1964), B1111.

   Th.A.J. Maris, V.E. Herscovitz and G. Jacob, Phys. Rev. Letters 12 (1964), 313.

   H. Pagels, Phys. Rev. Letters <u>28</u> (1972), 1482: Phys. Rev. <u>D7</u> (1973), 3689.

   P. Langacher and H. Pagels, phys. Rev. <u>D9</u> (1974), 3413.

3) M. Baker and K. Johnson, Phys. Rev. <u>D3</u> (1971), 2516.

   K. Johnson, M. Baker and R. Willey, Phys. Rev. <u>136</u> (1964), B1111.

4) R. Jackiw and K. Johnson, Phys. Rev. <u>D8</u> (1973), 2386.

5) J.M. Cornwall and R.E. Norton, Phys. Rev. <u>D8</u> (1973), 3338.

6) J. Smit, Phys. Rev. <u>D10</u> (1974), 2473.

DISCUSSION

<u>Question</u> (B. Schroer) Is your axial current the gauge invariant current (which in renormalized perturbation theory developes anomalies) or the chiral symmetry current?

<u>Answer</u> (Nakajima) In our approximation scheme, there appears no Adler-anomaly term.

---

*) In the paper of the reference 4, the equation $\partial_\mu j_{5\mu}=0$ is derived from the equation of motion $\partial_\nu F_{\mu\nu}=g'j_{5\mu}$ and the antisymmetry of $F_{\mu\nu}$, while the solution ($\beta\neq0$) of the Schwinger-Dyson equation for weak coupling requires an infinitesimal bare mass in the cutoff version and therefore the operator equation, $\partial_\mu j_{5\mu}=2m_0 i\bar\Psi\gamma_5\Psi$ of which the r.h.s. is clearly not vanishing, follows from the equation of motion of the fermion field. On the other hand $\partial_\nu F_{\mu\nu}=g'j_{5\mu}$ is still valid even in the presence of m_0. In these situation that naive manipulation of a given Lagrangian may cause the internal inconsistency, one cannot conclude the axial-vector current conservation only because of that antisymmetry of $F_{\mu\nu}$. In other words, in order to make the original formalism successful, one should take another solution ($\beta\neq0$) which requires no bare mass even in the cutoff version. The situation is exactly the same also in the reference 5.

QUANTUM FIELD-THEORETICAL APPROACH TO
SPONTANEOUSLY BROKEN GAUGE INVARIANCE

Noboru Nakanishi

Research Institute for Mathematical Sciences
Kyoto University
Kyoto 606, Japan

I review my previous work[1] on the Higgs mechanism, adding some new results, and mention its application to the Schwinger model done mainly by K. R. ITO[2].

As is well known, when phase symmetry is spontaneously broken in gauge theory, the gauge field acquires a non-zero mass, while the Goldstone boson disappears according to the Higgs mechanism[3]. It is impossible, however, to invalidiate the Goldstone theorem in the usual framework of local field theory. It is important, therefore, to explain why and how the Higgs mechanism is compatible with the Goldstone theorem in the manifestly covariant quantum field theory.

I consider the Higgs model as an example. Its Lagrangian in the covariant formalism is given by

$$\mathcal{L} = -\frac{1}{4}F^{\mu\nu}F_{\mu\nu} + B\partial^{\mu}A_{\mu} + \frac{1}{2}\alpha B^2$$
$$+ (\partial^{\mu}+ieA^{\mu})\phi^{\dagger}\cdot(\partial_{\mu}-ieA_{\mu})\phi + \frac{1}{2}\mu_0^2\phi^{\dagger}\phi - \frac{1}{4}\lambda(\phi^{\dagger}\phi)^2, \qquad (1)$$

where B is an auxiliary scalar filed, α being a gauge parameter. Field equations for the gauge field A_{μ} are

$$\Box A_{\mu} - (1-\alpha)\partial_{\mu}B = j_{\mu}, \qquad \partial^{\mu}j_{\mu} = 0,$$
$$\partial^{\mu}A_{\mu} + \alpha B = 0, \qquad \Box B = 0. \qquad (2)$$

Canonical quantization is carried out. Then one can compute all the four-dimensional commutation relations involving B; especially,

$$[B(x), \phi(y)] = e\phi(y)D(x-y). \qquad (3)$$

The subsidiary condition

$$B^{(+)}(x) \,|\,phys> = 0 \qquad (4)$$

is imposed upon physical states so as to avoid negative probability.

Now, spontaneous symmetry breaking is introduced by setting

$$\sqrt{2}\phi(x) = v + \varphi(x) + i\chi(x),$$

$$v = v^* \neq 0, \qquad <0|\varphi(x)|0> = <0|\chi(x)|0> = 0 \tag{5}$$

as usual. Then the commutation relation (3) implies

$$[B(x), \chi(y)] = -ie[v + \varphi(y)]D(x-y). \tag{6}$$

Therefore,

$$<0|[B(x), \chi(y)]|0> = -iM \cdot D(x-y), \tag{7}$$

where $M = ev$ stands for the zeroth approximation to the mass of the gauge particle.

Let A_μ^{in}, B^{in}, φ^{in}, and χ^{in} be the asymptotic fields of A_μ, B, φ, and X, respectively. Since $B^{in} = B$, the subsidiary condition (4) holds also for in-states. From (7) one obtains

$$[B^{in}(x), \chi^{in}(y)] = -iM \cdot D(x-y). \tag{8}$$

This is a very important relation, which yields the following conclusions.

1. χ^{in} is massless, that is, it should be the Goldstone field. Thus, the Goldstone theorem holds.

2. The Goldstone bosons are unphysical because they do not satisfy the subsidiary condition (4). Hence the Higgs mechanism works for physical states. Here it can be shown that physical fields are $U_\mu^{in} = A_\mu^{in} - Z_\chi M^{-2}\partial_\mu B^{in} - M^{-1}\partial_\mu \chi^{in}$, B^{in}, and φ^{in}.

Thus, <u>the consistency between the Goldstone theorem and the Higgs mechanism is established in a manifestly covariant way.</u>

The Goldstone field χ^{in} can be shown to satisfy

$$[\chi^{in}(x), \chi^{in}(y)] = iZ_\chi D(x-y) + i\alpha M^2 E(x-y), \tag{9}$$

where $E(x) \equiv -(\partial/\partial m^2)\Delta(x,m^2)|_{m=0}$,   and

$$[Q, \chi^{in}(x)] = -iM, \tag{10}$$

where Q denotes the generator of the gauge transformation. Furthermore, the equation for χ^{in} is

$$\square \chi^{in} = -\alpha M \cdot B^{in}. \tag{11}$$

Thus the Goldstone bosons are dipole ghosts for $\alpha \neq 0$. This may be the first example of dipole-ghost Goldstone bosons. [A similar conclusion has been reached by YOKOYAMA (private communication) independently.]

The formalism developed above has been applied to the Schwinger model[4] (the two-dimensional massless QED) by ITO.[2] This model is known to be exactly solvable, and the gauge field acquires a non-zero mass. According to KOGUT and SUSSKIND,[5] the so-called Schwinger mechanism is different from the Higgs mechanism because there is no spontaneous breakdown of symmetry. An opposite conclusion, however, is deduced on this point: <u>the Schwinger mechanism is nothing but a special case of the Higgs mechanism.</u>

Field equations are again given by (2), but in the Schwinger model, one has $j_\mu = J_\mu - m^2 A_\mu$, where $m^2 = e^2/\pi$ and J_μ is a current composed by free fermion fields. According to KLAIBER,[6] one can construct an "associated boson" field $X(x)$ such that

$$J_\mu = \partial_\mu X. \tag{12}$$

Therefore A_μ satisfies

$$(\square+m^2)A_\mu - \partial_\mu[(1-\alpha)B + X] = 0. \tag{13}$$

Hence all two-dimensional commutation relations are easily calculated, and one finds that

$$[X(x),\ X(y)] = im^2 D(x-y) + i\alpha m^4 E(x-y), \tag{14}$$

$$\langle 0|[Q,\ X(x)]|0\rangle = [Q,\ X(x)] = -im^2 \neq 0, \tag{15}$$

$$\square X = -\alpha m^2 B. \tag{16}$$

Eq. (15) shows that there is indeed spontaneous breakdown of symmetry and that X is the Goldstone field. Furtheremore, I emphasize that <u>there is complete parallelism between the Schwinger model and the Higgs model</u> as is seen by comparing (14)-(16) with (9)-(11), that is, $m^{-1}X$ corresponds to χ^{in}. (Detailed accounts will appear elsewhere.)

REFERENCES

1. N. Nakanishi, Progr. Theoret. Phys. <u>49</u>, 640(1973); <u>50</u>,1388(1973).

2. K. R. Ito, Progr. Theor. Phys. to be published.

3. P. Higgs, Phys. Rev. <u>145</u>, 1156(1966).

4. J. Schwinger, Phys. Rev. <u>128</u>, 2425(1962).

5. J. Kogut and L. Susskind, Phys. Rev. <u>9</u>, 3501(1974).

6. B. Klaiber, in <u>Boulder Lectures in Theoretical Physics</u>, Vol. XA (1967), p.141.

DISCUSSIONS

ITO (comment): I comment that the Schwinger model can be solved more rigorously by the generalized Bogolubov transformation. I have found that its bare Hamiltonian cannot be positive without renormalization counterterms and that photons become massive by those counterterms.

SWIECA (question): Is it true that the problem of whether there is or not spontaneous breakdown in the Schwinger model depends on the gauge used?

NAKANISHI (answer): Gauge dependence might be possible; I am not sure. The complete parallelism between the Schwinger model and the Higgs model exists at least in the covariant gauges.

SEKINE (question): By what equations have you defined the in-fields for dipole ghosts? Isn't there any trouble?

NAKANISHI (answer): They can be defined consistently even for dipole-ghost fields by using the Yang-Feldman formalism.

REVIEWS ON AXIOMATIC STUDY OF SYMMETRY BREAKING

Helmut Reeh

Institut für Theoretische Physik der Universität Göttingen

The work on broken symmetry within the axiomatic frame is by now rather extended and it is impossible to give a complete review here. To restore some balance I give a few references [1] concerning points not touched upon in the following. I shall limit myself on the notion of spontaneously broken symmetry. Most of the results are not very new. Earlier and more complete   reviews   are given in [2].

We use the Wightman framework of relativistic quantum field theory, at the end the framework has to be enlarged slightly to accommodate for supergauge transformations. $\mathcal{R}, \mathcal{H}, \Omega, \mathcal{U}(\Lambda, a)$ denote the algebra of unbounded strictly local fields, the Hilbert space, unique vacuum vector and unitary representation of the Poincaré group respectively.

I

A symmetry transformation is given by a group $\mathcal{O}\ni g$ and a representation by automorphisms α_g of $\mathcal{R}$. Here automorphism denotes a 1-1 map of $\mathcal{R}$ onto $\mathcal{R}$ preserving the algebraic structure, with $(\alpha(A))^* = \alpha(A^*)$ and being unitarily implemented on every fixed local subring of $\mathcal{R}$ (replacing the norm preservation in case of an algebra of bounded operators). Furthermore, it is assumed that α_g commutes with the 4 translations,

$$\alpha_g(A(x)) = \alpha_g(A)(x), \qquad A \in \mathcal{R} \tag{1}$$

where $A(x) = U(1,x) A U^{-1}(1,x)$. A field theory is specified by its vacuum functional $A \to \varphi_0(A) := (\Omega | A \Omega)$. Due to (1), $\varphi_g(A) = (\Omega | \alpha_g(A)\Omega)$ is again a translationally invariant functional on $\mathcal{R}$ and one has the two possibilities

(i) $\varphi_g(A) = \varphi_0(A)$ for all $A \in \mathcal{R}$: α_g is a conserved symmetry.

(ii) $\varphi_g(A) \neq \varphi_0(A)$ for some $A \in \mathcal{R}$: α_g is a spontaneously broken symmetry.

The symmetry transformation in both cases can be presented on $\mathcal{H}$ by a linear invertable operator V_g representing $\mathcal{O}$, commuting with $U(1,x)$, and with $\alpha_g(A) = VAV^{-1}$. V_g is defined by

$$A\Omega \to \alpha_g(A)\Omega := V_g A \Omega, \qquad \text{domain } \vartheta_{V_g} = \mathcal{R}\Omega.$$

<u>Lemma 1</u>: In case (i) V_g is unitary, in case (ii) V_g is not closable (i.e. the adjoint is not densely defined), hence not unitary.

Case (ii) seems to allow for the possibility that two fields are linked by an automorphism such that, e.g., the 4 point functions are different, hence possibly the scattering amplitudes for the corresponding two kinds of particles. Up to now there is no concrete rigorous model theory exhibiting such features. Heuristic models like the σ-model show certain draw backs: One of the two kinds of particles is unstable [3].

II

For most conclusions one needs more specific structure: Guided by heuristic q.f.th. one assumes that $\mathcal{G}$ is composed of one parameter subgroups continuously connected to the identity, the infinitesimal transformations defined by integrals over densities

$$A \to \lim_{r \to \infty} \; i\,[Q_r, A] \quad , \quad A \in \mathcal{R}$$

$$Q_r = \int j_o(x)\, \vartheta_r(\vec{x})\, \eta\,(x^o)\; d^4x \; .$$

Here $j_\mu(x)$ is a real local field, $\mu = 0, 1, 2, 3$, $\partial^\mu j_\mu(x) = 0$. For the following j_μ is assumed to be invariant under translations.$^{+)}$ The two cases then read

(i) $\lim\limits_{r \to \infty} (\Omega | [Q_r, A]\, \Omega) = 0$ for all $A \in \mathcal{R}$:   conserved symmetry.

(ii) $\lim\limits_{r \to \infty} (\Omega | [Q_r, A]\Omega) \neq 0$ for some $A \in \mathcal{R}$:   spontaneously broken symmetry.

One may now define a linear operator Q on $\mathcal{H}$ by $A\Omega \to \lim\limits_{r \to \infty} [Q_r, A]\Omega = QA\Omega$, domain $\vartheta_Q = \mathcal{R}\Omega$.

<u>Lemma 1'</u> : In case (i) Q is hermitean, in case (ii) Q is not closable.$^{*)}$

For further properties of Q in particular concerning selfadjointness see [4]. Famous result is the Goldstone theorem.

<u>Thm 2</u> : Denote by E_o the projection on the mass zero vectors in $\mathcal{H}$. Then

$$\lim_{r \to \infty} (\Omega | [Q_r, A]\,\Omega) = \lim_{r \to \infty} (\Omega | Q_r E_o A\Omega) - (\Omega | A E_o Q_r \Omega) \tag{2}$$

$$= 2 \lim_{r \to \infty} (\Omega | Q_r E_o A\Omega) . \tag{3}$$

A clean proof of (2) was first presented in [5], of (3) in [6].

Useful for the proof of this and other assertions is

<u>Lemma 3</u>: If $\| Q_r \Omega \| \leq c < \infty$ or $\| E_o Q_r \Omega \| \leq c < \infty$ for $r \to \infty$, then $\lim\limits_{r \to \infty} (\Omega | [Q_r, A]\Omega) = 0$.

To our knowledge, this conclusion is not invertable, not even if $\eta (x^o)$ is chosen dependent on r [8].

Vectors from $E_o \mathcal{H}$ contributing in (2) are called Goldstone states. Their quantum numbers depend on Q_r. In case j_μ is a covariant vector field they have helicity 0. In case of space time dimension 4 and 3 one knows that this situation indeed may occur [2]. In case of dimension 2 it was first pointed out in [9] that the situation is different.

<u>Thm 4</u>: In case of space time dimension 2 and j_μ a covariant vector field, $\| Q_r \Omega \|$ is bounded, hence $\lim\limits_{r \to \infty} (\Omega | [Q_r, A] \Omega) = 0$ by lemma 3.

The only zero mass particles observed in nature are the photon and the neutrinos with helicity 1 and 1/2. Therefore one has to investigate currents of more complicate transformation properties. Let $j_\mu(x) = t_{\mu m}(x)$ where m stands for a collection of vector and spinor indices and

$$\mathcal{U}(\Lambda,a)\, t_{\mu m}(x)\, \mathcal{U}^{-1}(\Lambda,a) \;=\; \Lambda^{-1}{}_\mu{}^{\mu'} D_m{}^{m'}(\Lambda^{-1})\, t_{\mu'm'}(\Lambda x + a)$$

with a finite representation of the homogeneous Lorentz group (h. L. g.).

Consider first the case that D is a one valued representation. Then we have [10]

<u>Thm 5</u>: Decompose $t_{\mu m}$ to get irreducible representations (decomposition of $\Lambda \otimes D$)
If no covariant vector current occurs, then $\lim\limits_{r \to \infty} (\Omega | [Q_{rm}, A] \Omega) = 0$ for all $A \in \mathcal{R}$.

Hence there are no Goldstone states with helicity $\neq 0$. However, it has to be noted that the Wightman framework is assumed for the proof. In case of an indefinite metric the situation may be different: In the Gupta Bleuler frame the photon states may be interpreted as Goldstone states of certain gauge transformation [10, 11].

In case D is a two valued representation slightly less can be proved [10].

<u>Thm 6</u>: Decompose $t_{\mu m}$ into irreducible parts (decomposition of $\Lambda \otimes D$). If one gets only representations of type (b+h, b) with $2b+h \geq 2$, then
$\lim\limits_{r \to \infty} (\Omega | [Q_r, A] \Omega) = 0$ for all A.

The remaining cases $(\tfrac{1}{2}, 0)$, $(\tfrac{3}{2}, 0)$ occur for a conserved current transforming according to $\Lambda \otimes$ (spin 1/2-representation) as expected for spin 1/2 Goldstone states.

There are, of course, problems arising if one deals with spin 1/2 currents which should be local relative to $\mathcal{R}$ when $\mathcal{R}$ contains the currents themselves.

In the framework of super gauge transformations [12] these difficulties can be avoided and within this scheme indeed Goldstone states with spin 1/2 may occur. However, the frame of local quantum field theory has to be enlarged by adding a certain underworld involving elements of a Graßmann algebra and in consequence an indefinite metric. I would like to demonstrate this by presenting a simple example.

III

Consider a free Dirac spinor field $\psi(x)$ of mass zero on a Fock space $\mathcal{H}$ and a Graßmann algebra $\Lambda(E_4)$ with elements e_B over a 4 dimensional vector space E_4. B denotes a subset of $\{1, 2, 3, 4\}$. The elements of $\Lambda(E_4)$ are used besides of the complex numbers to multiply operators and states. This we do by a direct product. On $\mathcal{H}$ there is an operator I being 1 on integer spin states, -1 on half integer spin states. Consider the conserved currents

$$j^{\mu}_{\beta}(x) = e_{(\beta)} \otimes I \, \gamma^{\circ}_{(\beta)\sigma} \, \gamma^{\mu}_{\sigma\varrho} \, \psi_{\varrho} + \hat{\psi}_{\sigma} \gamma^{\mu}_{\sigma(\beta)} \, I \otimes e_{(\beta)}$$

$$J^{\mu}_{\beta}(x) = \frac{1}{i} \left[e_{(\beta)} \otimes I \, \gamma^{\circ}_{(\beta)\sigma} \, \gamma^{\mu}_{\sigma\varrho} \, \psi_{\varrho} - \quad h.c. \right]$$

with summation over repeated indices unless in brackets ($\beta = 1, 2, 3, 4$). The I is to get relative locality with respect to ψ, the e_{β} accomplish j^{μ}_{β}, J^{μ}_{β} to obey local commutativity, the latter being needed for iterated commutators and finite transformations. With $q_{r\beta} = \int j^{\circ}_{\beta}(x) \, \vartheta_{r}(\bar{x}) \, \eta(x^{\circ}) \, d^4x$, $Q_{r\beta} = \int J^{\circ}_{\beta}(x) \, \vartheta_{r}(\bar{x}) \, \eta(x^{\circ}) \, d^4x$ one easily computes

$$[q_{r\beta}, \psi_{\lambda}(y)] = - e_{(\beta)} I \, \delta_{\lambda(\beta)}$$

$$[q_{r\beta}, \hat{\psi}_{\lambda}(y)] = e_{(\beta)} I \, \gamma^{\circ}_{(\beta)\lambda}$$

$$[Q_{r\beta}, \psi_{\lambda}(y)] = - i \, e_{(\beta)} I \, \delta_{\lambda(\beta)}$$

$$[Q_{r\beta}, \hat{\psi}_{\lambda}(y)] = - i \, e_{(\beta)} I \, \gamma^{\circ}_{(\beta)\lambda}$$

for sufficiently large r. Since all iterated commutators vanish for large r one gets immediately the finite transformations

$$\Psi_\lambda \to \Psi_\lambda - i\tau\, \delta_{\lambda(\beta)}\, e_{(\beta)}\, I \qquad\qquad \text{for } q_{\tau\beta}$$

$$\Psi_\lambda \to \Psi_\lambda + \sigma\, \delta_{\lambda(\beta)}\, e_\beta\, I \qquad\qquad \text{for } Q_{\tau\beta}$$

($\tau,\delta \in \mathbb{R}'$) defining automorphisms on $\mathcal{A} = \Lambda(E_4)\otimes(\mathcal{R}, I)$, which replaces $\mathcal{R}$ in the enlarged framework. It can be shown that these transformations cannot be unitarily implemented. Therefore one has the situation of a spontaneously broken (super gauge) symmetry. This is also true in case of two dimensional space time.

More details of the example can be found in [14]. In particular it turns out that the enlarged state space has to carry an indefinite metric if one wants an involution to be defined on $\Lambda(E_4)$ which coincides with the adjoint operation with respect to the scalar product. (In theorem 6 only one simple commutator is considered. For that one could also replace the $e_{(\beta)}$ by c numbers thus avoiding additional substructure and indefinite metric).

References

[1] Concerning general wisdom and constructive quantum field theory: A. S. Wightman: Coral Gables lecture 1972, Part D. Iverson, Perlmutter, Mintz, eds., Plenum Press 1973

J. Glimm, A. Jaffe, Th. Spencer: Erice lecture 1973. Velo, Wightman eds., Springer Verlag 1973

Concerning perturbation theory:

C. Becchi, A. Rouet, R. Stora: Basko Polje lecture 1974

General questions and breaking of boost transformations:

J. H. Borchers, R. N. Sen: "Relativity Groups in the Presence of Matter", preprint Göttingen 1974

[2] D. Kastler: Rochester Conference 1967

H. Reeh: Fortschritte d. Physik 16, 687 (1968)

J. A. Swieca: Cargese lecture 1969

C. Orzalesi: Rev. Mod Phys. 42, 381 (1970)

H. Reeh: Haifa lecture 1971. R. Sen, C. Weil eds. Jerusalem 1972, Keter Publishing

[3] B. Lee: Nuclear Phys. B 9, 649 (1969)

J. D. Becker: Thesis, Munich 1974

[4] D. Maison: Commun. Math. Phys. 14, 56 (1969) and preprint Munich 1970

J. A. Swieca: Cargese lecture 1969

K. Kraus, L. J. Landau: Commun. Math. Phys. 24, 243 (1972)

A. H. Völkel: "On self adjoint extensions of global charge operators", preprint, Rio de Janeiro April 1972

[5] H. Ezawa, J. A. Swieca: Commun. Math. Phys. 5, 330 (1967)

[6] D. Maison: Nuovo Cim. 11A, 389 (1972)

[7] H. Reeh: Fortschr. d. Physik 16, 687 (1968)

[8] M. Requardt: preprint, Göttingen Jan. 1975

[9] S. Coleman: Commun. Math. Phys. 31, 259 (1973)

E. Gal Ezer: forthcoming preprint, Tel Aviv University

[10] D. Maison, H. Reeh: Commun. Math. Phys. 24, 67 (1971)

[11] R. Ferrari, L. E. Picasso: Nucl. Phys. B31, 316 (1971)

[12] J. Wess, B. Zumino: Nucl. Phys. B 70, 39 (1974)

[14] W. D. Garber, H. Reeh: "Supergauges and Goldstone Spinors", preprint Göttingen Nov. 1974

Footnotes:

+) ϑ, η are real test functions, $\vartheta_r(\vec{x}) = \vartheta(\frac{|\vec{x}|}{r})$, $\vartheta(s) = 1$ for $s < 1$, 0 for $s \geq 2$. η has compact support and $\int (x^o)dx^o = 1$.

*) Case (ii) does not correspond to a not closable derivation as considered in the talk by D. Robinson.

Discussion

Swieca: Do you know the appropriate axiomatic setting to cover the case of gauge invariance of the second kind?

Reeh: I should say no. There are papers in which parts of the results are generalized (Y. Dothan, E. Gal-Ezer, Nuovo Cim. 12A, 465 (1972); R. Ferrari, Nuovo Cim. 14A, 386 (1973)).

Swieca: In the case of supersymmetries, do you have automorphisms or generalizations of them?

Reeh: Yes, the transformations are locally unitarily implemented. Details can be found in [14].

Schroer: There are cases which were not discussed in your spontaneous symmetry frame work: Tensor currents and spinor currents which have an explicit x-dependence, i.e. formally $[P_{\mu}, Q] \neq 0$. Can you comment on the possibility of having spontaneously broken symmetries of such currents.

Reeh: I do not have an answer ready for the case of spinor currents. I would conjecture that the general situation should be essentially the same, i.e. there should be a Goldstone-theorem too and examples of a spontaneous break down, just as one knows, e.g., in the case of dilatation and conformal currents (compare, e.g. my Haifa lecture, ref. [2]).

STRUCTURE OF RENORMALIZED FEYNMAN AMPLITUDES

O.I. Zavialov
Steklov Mathematical Institute
Moscow

Let $\langle \mathcal{V}, \mathcal{L} \rangle$ be a Feynman graph, $\mathcal{V}$ standing for the set of all vertices, $\mathcal{L}$ - for the set of all lines. The Bogolubov-Parasiuk R-operation [1] has then the following structure

$$R = \sum_{\mathcal{V} = \mathcal{V}_1 + \ldots + \mathcal{V}_m} \Lambda(\mathcal{V}_1) \ldots \Lambda(\mathcal{V}_m),$$

where the sum goes over all partitions of $\mathcal{V}$ into non-empty non-overlapping subsets $\mathcal{V}_1, \ldots, \mathcal{V}_m$. The Λ-operation is defined recursively by

$$\Lambda(\mathcal{V}_i) = 1 \quad \text{if} \quad |\mathcal{V}_i| = 1, \quad |\mathcal{V}_i| - \text{the total number of vertices in } \mathcal{V}_i;$$

$$\Lambda(\mathcal{V}_i) = -M(\mathcal{V}_i) \sum_{\mathcal{V}_i = \mathcal{V}_{i_1} + \ldots + \mathcal{V}_{i_m}} \Lambda(\mathcal{V}_{i_1}) \ldots \Lambda(\mathcal{V}_{i_m}) + T(\mathcal{V}_i)$$

in all other cases. The sum goes over all partitions of the set $\mathcal{V}_i$ into non-empty non-overlapping subsets $\mathcal{V}_{i_1}, \ldots, \mathcal{V}_{i_m}$ none of which coinsides with $\mathcal{V}_i$. Here $M(\mathcal{V}_i)$ acts non-trivially only on proper graphs and is then the substraction operator of order $a \geq \omega(\mathcal{V}_i, \mathcal{L}_i)$ where $\omega(\mathcal{V}_i, \mathcal{L}_i)$ is the superficial divergence of $\langle \mathcal{V}_i, \mathcal{L}_i \rangle$. $T(\mathcal{V}_i)$ is a finite renormalization operator transforming the coefficient function of $\langle \mathcal{V}_i, \mathcal{L}_i \rangle$ into a given polynomial of order a. In the case of zero finite renormalization (all $T=0$) the R-operation is known to reduce to a sum over all forests of the diagram of the appropriate products of M [2,3].

Next follows the explicit solution of the above recursive relations in the case of arbitrary finite renormalization. Let $\mathcal{V}_1, \ldots, \mathcal{V}_k$ be the set of all divergent subgraphs of the diagram. Then

$$R = \; \vdots \; \prod_{i=1}^{k} (1 - M(\mathcal{V}_i) + T(\mathcal{V}_i)) \; \vdots \tag{1}$$

Here the three-point-product is defined as follows. Let $A(\mathcal{V}_i)$ stand either for $M(\mathcal{V}_i)$ or for $T(\mathcal{V}_i)$. Then

$$\vdots \; A(\mathcal{V}_{i_1}) \ldots A(\mathcal{V}_{i_j}) \; \vdots \; = 0$$

if among $\mathcal{V}_{i_1}, \ldots, \mathcal{V}_{i_j}$ there is at least one pair of overlapping subgraphs or if there is at least one pair $\mathcal{V}_{i_\ell}, \mathcal{V}_{i_m}$ such that $\mathcal{V}_{i_\ell} \subset \mathcal{V}_{i_m}$ and $A(\mathcal{V}_{i_m}) = T(\mathcal{V}_{i_m})$. In all other cases the three-point-product is an ordinary product taken in the "natural" order (that is if $\mathcal{V}_{i_\ell} \subset \mathcal{V}_{i_m}$ then $A(\mathcal{V}_{i_\ell})$ stands to the right of $A(\mathcal{V}_{i_m})$).

Another structure relation which is of some use in calculations reads.   Let two R-operations correspond to finite-renormalization but differ in number of substractions a and b (b<a). Then

$$R_b \equiv \; \vdots \; \prod_i (1 - M_b(\mathcal{V}_i)) \; \vdots \; =$$

$$= \; \vdots \; \prod_i (1 - M_a(\mathcal{V}_i) + T_{ab}(\mathcal{V}_i)) \; \vdots \tag{2}$$

where

$$T_{ab}(\mathcal{V}_i) = (M_a(\mathcal{V}_i) - M_b(\mathcal{V}_i)) \vdots \prod_{\mathcal{V}_{i_j} \subset \mathcal{V}_i} (1 - M_b(\mathcal{V}_{i_j})) \vdots \tag{3}$$

Being just a redefinition of T-product, R-operation can be treated on the other hand as a result of the presence of counterterms in the Lagrangian, the T-product being unchanged (if some intermediate regularization is introduced). Since it is often convenient to perform calculations just on the level of intermediate regularization, the following question arises. Given an R-operation, to which counterterms does it correspond?

As far as the S-matrix is concerned the answer is the following

$$S \equiv RT \exp\{ig \int \mathcal{L}_0(x)dx\} =$$

$$= T \exp\{\Lambda T[e^{ig \int \mathcal{L}_0(x)dx} - 1]\} \tag{4}$$

where Λ is the constituent of R-operation defined above. The relation (4) means that the renormalized Lagrangian is

$$\mathcal{L}_{ren}(x) = \Lambda T\{\mathcal{L}_0(x) \frac{\exp[ig \int \mathcal{L}_0(x')dx'] - 1}{ig \int \mathcal{L}_0(x')dx'}\} . \tag{5}$$

One can easily see that all counterterms are hermitian, so almost trivial additional arguments prove the unitarity of the renormalized S-matrix.

The above identities can be readily generalized. For example,

$$RT\{:\varphi^n:(x)e^{ig \int \mathcal{L}_0(x')dx}\} =$$

$$= T\{[\Lambda T:\varphi^n(x): e^{ig \int \mathcal{L}_0(x')dx'}] e^{ig \int \mathcal{L}_{ren}(x')dx'}\} \tag{6}$$

Such equalities together with structure relation (3) can be used to give a simple proof (and some generalizations) of the Zimmermann identities between composite operators [3].

The following generalization of α-parametric representation for the case of divergent diagrams is known [2, 4, 5]. Put into correspondence to every line "ℓ" the Feynman parameter α_ℓ, and to every divergent subgraph $\mathcal{U}_i$ of the diagram the parameter κ_i. Introduce the variables β:

$\beta_\ell = \alpha_\ell$ if the line "ℓ" does not enter any of the divergent subgraphs;

$\beta_\ell = \kappa_{i_\ell} \ldots \kappa_{j_\ell} \alpha_\ell$ if the line "ℓ" enters the divergent subgraphs

$$\mathcal{U}_{i_\ell} \ldots \mathcal{U}_{j_\ell} .$$

Let $p_1, p_2, \ldots$ be the external momenta of the diagram. Then

$$G_{ren}(p_1, p_2, \ldots) \cong \int_0^\infty d\alpha_1 \ldots d\alpha_L \, e^{-i\Sigma m_\ell^2 \alpha_\ell} \int_0^1 d\kappa_1 \ldots d\kappa_k R_{\kappa_1} \ldots R_{\kappa_k} \cdot$$

$$\frac{1}{D^2(\beta)} \exp\{i \frac{A_{ij}(\beta)}{D(\beta)} p_i p_j\} Q(\kappa,\alpha) .$$

Here

$$R_{\kappa_i} = \frac{(1-\kappa_i)^{\omega_i/2}}{(\omega_i/2)!} \left[\frac{\partial}{\partial \kappa_i}\right]^{\omega_i/2 + 1} \kappa_i^{2n_i}$$

is the generalized Taylor operator; $D(\beta)$ and $A_{ij}(\beta)$ are the standard homogeneous Feynman forms, but constructed of β rather than of α.
The function Q is supposed to be an identity in the quoted literature.
In this case the representation (7) corresponds to zero substraction
point that is zero finite renormalization. Since the physical restrictions demand that the substraction point should not be zero at least
for two-point functions we are going to specify $Q(\kappa, \alpha)$ in such a way
that the representation (7) should correspond to an arbitrary substraction point.

Introduce another set of variables corresponding to each of the
divergent subgraphs $\mathcal{V}_q$

$$\beta_\ell^{(q)} = \alpha_\ell \quad \text{if the line "}\ell\text{" does not enter any of the divergent}$$
$$\text{subgraphs contained in } \mathcal{V}_q$$

$$\beta_\ell^{(q)} = \kappa_{i_\ell}^{(q)} \ldots \kappa_{j_\ell}^{(q)} \alpha_\ell \quad \text{if the line "}\ell\text{" enters divergent subgraphs}$$
$$\mathcal{V}_{i_\ell}^{(q)}, \ldots, \mathcal{V}_{j_\ell}^{(q)} \quad \text{which are contained in } \mathcal{V}_q .$$

Let the substraction point for $\mathcal{V}_q$ be specified by

$$p_i^{(q)} p_j^{(q)} = u_{ij}^{(q)} . \tag{8}$$

Then define

$$Q(\kappa, \alpha) = \prod_{q=1}^{k} \exp\{i(1-\kappa_q)u_{ij}^{(q)} \frac{A_{ij}^{(q)}(\beta^{(q)})}{D^{(q)}(\beta^{(q)})}\} \tag{9}$$

where $A_{ij}^{(q)}$ and $D^{(q)}$ are the corresponding Feynman forms for the
subgraph $\mathcal{V}_q$ (constructed of $\beta(q)$ rather than of α or β).

One can check that if there is no overlapping the representation (7) with Q given by (9) corresponds to R-operation with substraction points $u_{ij}^{(q)}$. We claim that it remains valid for diagrams with overlapping provided that the substraction points are non-zero only for two-point subgraphs and that the theory is renormalizable.

The representation (7) can be used to obtain a simple proof of convergence [6,7] and the asymptotic expansions of the renormalized amplitudes (the appropriate technique has been developed in [8, 2]).

The detailed discussion of the related topics is going to appear in Theoretical and Mathematical Physics.

Literature

1. N.N. Bogolubov, O.S. Parasiuk., Acta Math., $\underline{97}$, 227 (1957)

2. O.I. Zavialov, B.M. Stepanov., Soviet Nuclear Phys. $\underline{1}$, 922 (1965)

3. W. Zimmermann., Brandeis Lectures 1970, MIT 1971

4. T. Appelquist., Annals of Physics, $\underline{54}$, 27 (1969)

5. M. Bergere, J. Zuber., Saclay Preprint D.Ph T/73-31 (1973)

6. S.A. Anikin, M.K. Polivanov, O.I. Zavialov., Dubna preprint E2-7433 (1973)

7. S.A. Anikin, O.I. Zavialov, M.K. Polivanov., Theor. Math. Phys. $\underline{17}$ 189 (1973)

8. O.I. Zavialov., Soviet Phys. JETP $\underline{47}$ 1099 (1964)

ON THE REGULARIZATION IN THE CALLAN-SYMANZIK EQUATION

Yasunori Fujii

Institute of Physics, University of Tokyo-Komaba, Tokyo 153, Japan

and

Yasushi Takahashi

Department of Physics, The University of Alberta, Edmonton, Canada

There is a number of different approaches to the Callan-Symanzik (CS) equation, as roughly classified to the following three categories:

(I)   Method of response equation[1], renormalization group approach.
(II) Dispersion theory[2].
(III) Canonical theory of broken scale invariance (CTBSI)[3].

In the approaches of the third category, regularization procedure is indispensable, either by Pauli-Villars method or by the dimensional regularization technique. What we want to do here is to push further the most conservative approach of CTBSI with the Pauli-Villars regulators. The reason is that this old-fashioned but well-defined method still seems to provide us with the solid basis of the much intuitive understanding of the various anomalies, yet has never been fully examined in the literatures. We confirm that the CS equation is derived in this approach in a completely general manner.

From CTBSI follows a "dimensional equation" for the unrenormalized Green's function Γ; the right-hand side is another Green's function with an insertion of the trace of the energy-momentum tensor. If the naive dimensional analysis (NDA) was justified, this equation would be put into the form

$$m_0 \frac{\partial}{\partial m_0} \Gamma = -\Delta\Gamma \ , \tag{1}$$

from which the CS equation follows immediately. Without having regulator fields, however, NDA obviously fails due to the presence of divergent integrals. As a consequence a close tie between the CS equation and CTBSI is lost. One may start with (1) which can be obtained as a formal relation among the divergent Feynman amplitudes without any reference to CTBSI. We want, however, to restore the above mentioned tie by including the regulators.

Incidentally we can use the CS equation to derive the Pais-Epstein formula[4] which expresses the unwanted "self-stress" S by the amount of failure of NDA for the self-mass δm;

$$S = \frac{1}{3}(1 - m_0 \frac{\partial}{\partial m_0}) \, \delta m \,.$$

The "unregularized" δm results in a nonvanishing S; an immediate contradiction with the Lorentz covariance of a one-particle state. This is perhaps the most unambiguous argument to show that the regularization is necessary. The same argument shows also that the insertion in the right-hand side of the CS equation is not the trace of the true energy-momentum tensor.

If we introduce the regulators in the Lagrangian, NDA holds ture. Consequently we have (m; the renormalized mass)

$$m\frac{\partial}{\partial m} \, \Gamma_r = -\Delta\Gamma_r \,, \tag{2}$$

for the renormalized Green's functions. This equation looks similar to the one which was dismissed by Coleman[5] as being false. In the right-hand side of (2), however, we have an insertion of the trace of the true energy-momentum tensor that includes the contribution from the regulators as well as the one from the physical fields. To the former part one cannot apply Weinberg's theorem on the asymptotic behavior. To separate the part to which one can apply Weinberg's theorem, we pick up the term of the physical fields multiplied by a divergent coefficient so that the result is finite. We subtract this term from the right-hand side of (2). The result is then transferred to the left-hand side. This yields exactly the anomalous terms given by the CS equation. Every step of the calculation is well-defined.

We can extend the method easily to the more complicated cases where there are two (or more) massive fields. In separating the finite part expressed only in terms of the physical fields, we make full use of the BPH theorem. The procedure is essentially the generalization of the previous lowest-order calculation[6]. Our derivation is completely general and is independent of the detailed properties of the regulator fields.

Part of this work was done through the collaboration with Ken-ichi Shizuya.

DISCUSSION

Schroer: Do you have an "intuitive picture" for the method of dimensional regularization?

Fujii: Yes, there are almost one-to-one correspondences between the calculations with the two regularization methods. For example, to our regulator contribution in $\Delta\Gamma_r$ corresponds a term proportional to n-4, which is cancelled by a denominator 1/(n-4) to result in the anomalies.

Symanzik: I should like to give some arguments in favor of the response equation interpretation of the PDE discussed by Fujii. If one renormalizes a family of Lagrange theories such that different members of the family correspond to different values of the parameters (masses, renormalized coupling constants, etc.) then adding to the Lagrangian a finite local scalar operator such that hereby only a change of the Lagrangian within the considered family of Lagrangians is effected, then the infinitesimal operation to this insertion leads to a response equation, and the finiteness of the parametric function occuring therein is a direct consequence of the finiteness of the insertion. One is indeed led to the whole family of such insertions. They have been given for ϕ^4 theory by Lowenstein, who calls them generalized vertex operations, and are equivalently obtained by Nishijima's use of homogeneous generalized unitarity relations reported here. That some insertions may have a (treacherous!) similarity to the trace of the energy-momentum tensor at zero momentum transfer is accidental and plays no role whatsoever for the derivation of the response equation, nor does Poincaré invariance.

REFERENCES

1. K. Symanzik, Comm. Math. Phys. <u>18</u>, 227 (1970); <u>23</u>, 49 (1971).

2. K. Nishijima, Invited talk of this Conference.

3. C. Callan, Phys. Rev. <u>2D</u>, 1541 (1970).

4. A. Pais and S.T. Epstein, Rev. Mod. Phys. <u>2</u>, 445 (1949).

5. S. Coleman, Dilatation, 1971 International Summer School of Physics, Ettore Majorana.

6. S. Coleman and R. Jackiw, Ann. Phys. (N.Y.) <u>67</u>, 552 (1971).

THE STATISTICS OF PARTICLES
IN LOCAL QUANTUM THEORIES

S. Doplicher
Istituto Matematico, Università
di Roma, Italy

Introduction

This is a review talk on joint work with R.Haag and J.E. Roberts
(see [1,2,3] and references given there to previous works).

Let us start with the question: why is there an alternative between
the Bose and the Fermi type of particle statistics?

This brings in succession two other questions: what are the
possible statistics compatible with general principles; and: how is
formulated the concept of statistics in terms of general principles.

The usual description in terms of field operators cannot be entirely
satisfactory since fields are not observable in general, (e.g. as soon
as they do not commute with one another at spacelike distances) and
also because you have to introduce from the outset into the formalism
the type of statistics appearing in your theory, by assuming specific
commutation relations at spacelike distances.

These comments apply also to the superselection structure of a
theory. For the sake of strong interaction physics with short range
forces, that structure is customarily embodied into the field formalism
requiring that the exact internal symmetries of the theory are described
by a compact group, the gauge group of the theory. The choice of this
group is in practice suggested by the empirical patterns of elementary
particles and resonances. This group acts locally on fields but leaves
the observables uneffected since they are by definition gauge invariant.

Now first principles ought to be formulated precisely in terms of
observables — if you start from there, no special commutation property
nor gauge invariance is built in explicitly.

Our input is the algebra of all local observable $\mathcal{O}$ acting on the
Hilbert space $\mathcal{H}_0$ they generate on the vacuum state vector Ω: i.e.
the vacuum superselection sector alone is given.

Equivalently you might think as given the abstract C*-algebra $\mathcal{O}$
and a pure state ω_0 on $\mathcal{O}$, the vacuum expectation functional.

This means: $\mathcal{O}$ is an irreducible C*-algebra acting on $\mathcal{H}_0$.

The main postulate is <u>locality</u>. To each nice bounded region $\mathcal{O}$ in space-time (double cones) a subalgebra $\mathcal{A}(\mathcal{O})$ of $\mathcal{A}$ is assigned, in a way which preserves inclusions; by definition $\mathcal{A}(\mathcal{O})$ is generated by the local observables which can be measured within the space-time limitations of $\mathcal{O}$. The closure of the union of all $\mathcal{A}(\mathcal{O})$'s is $\mathcal{A}$.

Einstein causality together with Quantum Mechanics say that observables affiliated with spacelike separated double cones commute. This locality postulate in turn defines what is meant by "local observables". So it is natural to strengthen it by the so called duality requirement, so that <u>the</u> postulate becomes:

(i) $\mathcal{A}(\mathcal{O})$ is the set of all bounded operators on $\mathcal{H}_0$ commuting
      with every observable spacelike located to the double cone $\mathcal{O}$.

In other words: $B \in \mathcal{A}(\mathcal{O})$ iff for each double cone in the spacelike complement $\mathcal{O}'$ of $\mathcal{O}$ and each $A \in \mathcal{A}(\mathcal{O}_1)$ we have

$$AB = BA \ .$$

By recent work of Bisognano and Wichmann the duality requirement appears necessary for the existence of a suitable underlying Wightman field theory.

Another requirement of this nature, which is actually used in a very limited way, is the additivity assumption saying that any small region in space time contains sufficiently many local observables to build up any local algebra: if $\mathcal{O}_1 \cup \ldots \cup \mathcal{O}_n \supset \mathcal{O}$ (all double cones) then

(i') $\{ \mathcal{A}(\mathcal{O}_1) \cup \ldots \cup \mathcal{A}(\mathcal{O}_n) \}'' \supset \mathcal{A}(\mathcal{O})$.

A detailed analysis of the possible superselection structures can be done on the basis of postulate (i). The statistics can be defined as a property of each superselection sector and classified in a simple way.

To relate this to properties of one particle states and scattering states one needs in full the general assumptions of relativistic quantum theories, namely

(ii)   relativistic covariance of the local algebras
(iii)   the vacuum is Poincaré invariant and is a ground state in $\mathcal{H}_0$.

As mentioned this is not needed for most of the general analysis; to be quite careful, an algebraic consequence of (i'), (ii), (iii) is freely used as a technical assumption, when needed [1].

Superselection Sectors and Statistics.

Let $\mathcal{S}_0$ be the set of vector states on $\mathcal{A}$ coming from $\mathcal{H}_0$ and $\mathcal{S}(\mathcal{A})$ the set of all states on $\mathcal{A}$. Since $\mathcal{A}$ is irreducible on $\mathcal{H}_0$, $\mathcal{S}_0$ is a collection of pure states among which the superposition principle holds, and forms the <u>vacuum superselection sector</u>.

How does the structure of $\mathcal{A}$ determine the collection of all superselection sectors?

We want a universal recipee to select from $\mathcal{S}(\mathcal{A})$ the subset of "elementary" states which in the end will carry each a finite number of elementary "charges".

The main criterion is that $\omega \in \mathcal{S}_r$ should describe a deviation from the vacuum state which becomes negligible in far away regions of space. In general:

$$\mathcal{S}_0 \subsetneq \mathcal{S}_r \subsetneq \mathcal{S}(\mathcal{A}) \ .$$

The superselection sectors are then the coherence classes of pure states in $\mathcal{S}_r$:

$$\text{superselection sectors} = \mathcal{S}_r \cap \text{Pure states of } \mathcal{A}/\underset{\approx}{\sim}$$

where for pure states ω_1, ω_2 on $\mathcal{A}$, $\omega_1 \cong \omega_2$ means equivalently a. or b.:

a. $\pi_{\omega_1} \cong \pi_{\omega_2}$;

b. there is $B_{12} \in \mathcal{A}$ s.t., for all $A \in \mathcal{A}$,

$$\omega_2(A) = \omega_1(B_{12}^* A B_{12})$$

i.e. ω_2 results from ω_1 by a physical operation B_{12}, which typically creates pairs but no single charge; a. and b. being equivalent by the theorem of Kadison.

As you see the criterion to select $\mathcal{S}_r$ is of central importance. By making the foregoing precise, one can prove that, in theories describing only short range forces, $\mathcal{S}_r$ is the vector states of the representations of a special type, the localized morphisms. [1].

These are * isomorphisms ρ of $\mathcal{A}$ into $\mathcal{A}$ each localized in some double cone $\mathcal{O}_\rho$, i.e.:

$$\rho(A) = A \quad \text{if} \quad A \in \mathcal{O}(\vartheta) \quad \text{and} \quad \mathcal{O} \text{ is a double cone}$$
$$\text{spacelike to} \quad \mathcal{O}_\rho \ .$$

Let Δ be the collection of those localized morphisms which up to a unitary equivalence can be localized everywhere (a less stringent condition suffices, see [1]). This initial result implies

1. each sector contains a <u>strictly localized state</u> ω_ρ: here $\rho \in \Delta$ and if $A \in \mathcal{O}$

$$\omega_\rho(A) = \omega_0 \circ \rho(A) \equiv (\Omega, \rho(A)\Omega)$$

so that $\omega_\rho = \omega_0$ on observables spacelike to $\mathcal{O}_\rho$ (in electrodynamics this would imply absence of electric charge by the law of Gauss).

2. if ω_1, ω_2 are such states, $\omega_i = \omega_0 \circ \rho_i$, localized in mutually spacelike double cones, we can define an <u>exact product state</u>

$$(1) \qquad \omega_1 \times \omega_2 = \omega_0 \circ \rho_1 \rho_2$$

i.e. $\omega_1 \times \omega_2 = \omega_2 \times \omega_1$ and

$$\omega_1 \times \omega_2(A) = \omega_1(A)$$

if A is an observable localized in a double cone spacelike to the localization region $\mathcal{O}_{\rho_2}$ of ω_2, and conversely.

Then we have the Theorem: if ω_1, ω_2 are strictly localized states in the same sector, and $\rho \in \Delta$ is any morphism in the class of that sector,

$$(2) \qquad \omega_1 \times \omega_2 \text{ is } \underline{\text{pure}} \text{ iff } \rho(\mathcal{O}) = \mathcal{O}.$$

Now the product (1) composes the "charges" of the two sectors; such "charges" are labelled by the equivalence class of irreducible $\rho \in \Delta$. Allowing mixtures, as seems appropriate by (2), we can see that the equivalence classes of our representations are just the elements of the quotient

$$(3) \qquad \Delta / \mathcal{Y}$$

of the <u>semigroup</u> Δ modulo <u>inner</u> localized automorphisms. The generalized charges (3) form a <u>commutative</u> semigroup; those corresponding to the localized <u>automorphisms</u> Γ(i.e. $\rho \in \Gamma$ if $\rho \in \Delta$ and $\rho(\mathcal{O}) = \mathcal{O}$) form an abelian <u>group</u> $\Gamma / \mathcal{Y}$ in $\Delta / \mathcal{Y}$, the group of "simple" charges.

The generalized charges $\Delta/\mathcal{J}$ of a theory are found so far by classifying a subset of the states $\mathcal{J}(\mathcal{A})$ of $\mathcal{A}$. It is however possible to construct them at least in principle from the vacuum sector: if ω_ξ is a strictly localized state with charge ξ and $\mathcal{O}_n$ is a sequence of double cones moving off to infinity in a spacelike direction, there is a sequence $\omega_n \in \mathcal{J}_0$ of states with no overall charge s.t.

$$(4) \qquad \begin{array}{l} \omega_n(A) = \omega_\xi(A) \quad \text{if} \quad A \quad \text{is an observable} \\ \text{associated to any double cone specelike to} \quad \mathcal{O}_n; \\[1ex] \omega_n(A) \to \omega_\xi(A) \;, \; A \in \mathcal{A}. \end{array}$$

Thus ω_ξ is the limit of states with the same charge ξ in the localization region of ω_ξ and with a compensating charge in $\mathcal{O}_n$. If ξ is a simple charge i.e. $\xi \in \Gamma/\mathcal{J}$, the compensating charge is ξ^{-1}. If all charges are simple, the dual G of the discrete abelian group $\Gamma/\mathcal{J}$ is the gauge group of the theory and is compact <u>abelian</u>. In theories with non abelian gauge groups there are conversely non simple charges as seen by the physical example of single nucleon states in a fully SU_2 invariant theory: if ω_1, ω_2 are such states the product state $\omega_1 \times \omega_2$ is a <u>mixture</u> of singlet and triplet (compare (2)).

Simple sectors can also characterized by the fact that the duality condition holds in the associated representations of $\mathcal{A}$.

A deeper characterization is: the simple sectors obey an ordinary (Bose or Fermi) statistics; the other sectors obey a parastatistics.

This brings us to our second question: what is the statistics of a sector?

The product operation (1) allows us to define quite generally such a notion without referring to one particle states.

Let ξ be a sector and $\rho \in \Delta$ a representative.

Let $\omega_1,\dots,\omega_n$ be mutually spacelike localized states of the same charge ξ . The product state

$$(5) \qquad \omega_1 \times \omega_2 \times \cdots \times \omega_n$$

is symmetric; it is a vector state in the representation ρ^n. Now $\rho(\mathcal{A})$ is irreducible but $\rho^n(\mathcal{A})$ is reducible for $n \geq 2$ unless ξ is simple. In that case there are several linearly independent state vectors for (5) and they can change from one another under permutation of the factors. Indeed let $\Psi_1,\cdots,\Psi_n$ be state vectors for $\omega_1,\dots,\omega_n$ resp. in the representation ρ; Ψ_i is unique up to a phase and given by

$$\Psi_i = U_i^* \Omega$$

where U_i is a local unitary in $\mathcal{O}$. To the product (5) corresponds a product

$$(6) \qquad \Psi_1 \times \Psi_2 \times \ldots \times \Psi_n = (U_1^* \times U_2^* \times \ldots \times U_n^*)\Omega$$

where $U_1 \times U_2 = U_1 \rho(U_2)$ is an associative composition law between all pairs of intertwining operators (see [1]).

One gets easily from this the permutation behaviour

$$(7) \qquad \Psi_{p^{-1}(1)} \times \ldots \times \Psi_{p^{-1}(n)} = \varepsilon_\rho^{(n)}(p)\, \Psi_1 \times \ldots \times \Psi_n$$

where $\varepsilon_\rho^{(n)}(p)$ is a unitary in $\mathcal{O}$ commuting with $\rho^n(\mathcal{O})$ and p is an element of the permutation group $\mathbb{P}(n)$. The $\varepsilon_\rho^{(n)}(p)$ could depend upon the choice of $\omega_1, \ldots, \omega_n$; however:

<u>Theorem</u>. $\rho \in \mathbb{P}(n) \to \varepsilon_\rho^{(n)}(p)$ is a representation of $\mathbb{P}(n)$ depending upon ρ alone; its equivalence class $\varepsilon_\xi^{(n)}$ depends upon the sector ξ alone.

This result makes it possible and natural to define the <u>statistics of the sector</u> ξ to be the sequence $\{\varepsilon_\xi^{(n)}; n=1,2,\ldots\}$. Each $\varepsilon_\xi^{(n)}$ is specified by a set of irreducible representations each with infinite multiplicity.

Which are the possible statistics? A priori, continuously many. However

<u>Theorem</u>. To each sector ξ is associated a number $\lambda(\xi)$, the statistics parameter of ξ, with values inverse integer or zero, which determines entirely the statistics of ξ as follows:

if $\lambda(\xi) = 0$, $\varepsilon_\xi^{(n)}$ contains all representations of $\mathbb{P}(n)$;

if $|\lambda(\xi)| = 1/d(\xi)$, $\varepsilon_\xi^{(n)}$ contains all representations

of $\mathbb{P}(n)$ with at most $d(\xi)$ antisimmetrizations resp.

symmetrizations for $\lambda(\xi) > 0$ resp. $\lambda(\xi) < 0$.

We see here that ξ obeys the ordinary statistics if ξ is a simple sector; the converse holds too i.e. ξ is simple iff $d(\xi) = 1$.

We say that ξ is a finite sector if $\lambda(\xi) \neq 0$ and call $|\lambda(\xi)|^{-1} = d(\xi)$ the order of the parastatistics; ξ is paraBose resp. paraFermi if $\lambda(\xi) > 0$ resp. $\lambda(\xi) < 0$.

Let Σ be the smallest set in $\Delta/\mathcal{Y}$ containing the finite sectors, their products and their subrepresentations (but not all direct sums!). Then:

<u>Theorem</u>. Each $\xi \in \Sigma$ is a finite sum of irreducible elements of Σ; to each $\xi \in \Sigma$ there is a statistics parameter $\lambda(\xi)$ as above and

$$\xi \notin \Sigma \ \to \lambda(\xi)^{-1} \in \mathbb{Z}$$

is a homomorphism of the commutative semigroup Σ into the multiplicative semigroup $\mathbb{Z}$. If $\xi \in \Sigma$ and $\xi = \xi_1 \oplus \xi_2$ then

$$d(\xi) = d(\xi_1) + d(\xi_2)$$

(8)

$$\text{sign } \lambda(\xi) = \text{sign } \lambda(\xi_i) \ .$$

It was mentioned before that each state in a sector $\xi \in \Delta/y$ is the limit of bilocalized states carrying the charge ξ and a compensating charge which makes them neutral (i.e. in the vacuum sector). If ξ is finite this compensating charge can be found by the the opposite process on such states, of removing ξ far away to spacelike infinity.

The resulting state belongs to the <u>conjugate sector</u> characterized by:

<u>Theorem</u>. Let ξ be a finite sector; there is one and only one sector $\bar{\xi}$ such that

(9) $$\xi \cdot \bar{\xi} \ \simeq \ \text{vacuum representation} \ \oplus \cdots \ ;$$

the sectors $\xi, \bar{\xi}$ have then the same statistics

$$\lambda(\xi) = \lambda(\bar{\xi})$$

and the vacuum charge appears only once in (9).

Note that this <u>charge conjugation of superselection quantum numbers</u> exists, as the statistics of the sectors and its classification, solely on the grounds of the locality postulate (i), without use of space-time covariance principles.

Note also the analogy between the structure of Σ and that of the dual object of a compact gauge group.

We call also attention on the fact that, in theories of short range forces, by the results of this section superselection charges appearing in compounds must also appear isolated, with a compatible statistics in the sense of the preceding theorems.

Gauge groups and parastatistics

Statistics has been analyzed irrespectively of field commutation

properties, actually in a formulation where no field at all is given
to transfer charge quantum numbers.

If however you do deal with a theory specified by a field algebra
$\mathscr{F}$ and a gauge group G acting on it so that the observables are
precisely the fixed points

$$(10) \qquad \mathscr{O} = \mathscr{F} \cap \mathscr{U}(G)'$$

then you can relate the superselection structure of $\mathscr{O}$ and the statistics intrinsically defined by it to properties of G and of the field
algebra $\mathscr{F}$.

Firstly there is a one to one correspondence from the set of all
classes of irreducible representations of G into the irreducible
elements of Σ; let

$$(11) \qquad u \in \hat{G} \to \xi_u \in \Sigma$$

be this correspondence; we have the theorem

$$(12) \qquad \dim u = d(\xi_u)$$

that is <u>parastatistics appear necessarily</u> as soon as the exact symmetry
group is <u>not abelian</u>.

Actually the relations (11) and (12) are particular cases of a full
correspondence between the dual structure of the gauge group G and
the intertwining operators between the representations in Σ (see [2]).

This indicates that a compact gauge group should always be associated with finite sectors, although the solution of this problem is not
compelling for the physical interpretation of a theory: the "dual gauge
structure" is determined by the observables and gives all the informations
usually derived from the gauge group itself: reduction of products,
"Clebsh-Gordan coefficients" etc. (see [2,3]).

About the theorem expressed by (12) we stress the fact that
parastatistics might appear also if all fields obey ordinary commutation
or anticommutation properties; the only relevant hypothesis is that
fields should commute <u>with observables</u> at spacelike distances. In turn
the statistics of the sector ξ_u determines partially the commutation
properties of fields transforming like $\bar{u}$ (i.e. carrying the charge
ξ_u): if ψ, ψ' are irreducible tensor field operators of type $u, \bar{u}$
respectively which are spacelike located, then necessarily

$$\int_G \alpha_g (\psi\psi' \mp \psi'\psi) dg = 0$$

where $\mp = -\text{sign}(\lambda(\xi_u))$.

Statistics and Particles

If you assume full Lorentz invariance, axioms (i), (ii), (iii) above, we have that:

a. the spectrum condition $P_\mu P^\mu \geq 0$ follows from convariance in all finite sectors.

b. if there is a one particle state $[m,j,\xi]$ in the finite sector ξ, s.t. there are finitely many descrete irreducible representations of the Poincaré group with mass m in the sector ξ, then there are <u>antiparticle</u> states $[m,j,\bar{\xi}]$ and the multiplicities are the same:

$$\nu([m,j,\xi]) = \nu([m,j,\bar{\xi}])$$

c. in the previous hypothesis,

$$(-1)^{2j} = \text{sign}(\lambda(\xi)).$$

This expresses the <u>generalized connection of spin and statistics</u>; note that here none of these concepts is mediated by algebraic or covariance properties of fields [2].

If particles $[m,j,\xi]$ appear isolated from the continuum in the sector ξ, by additivity of the spectrum [2] there is necessarily a mass gap in the vacuum sector and scattering theory can be developed. Indeed the exact product operation we defined between strictly localized state vectors can be used to deduce, by the usual limiting procedure, an asymptotic product between one particle state vectors [2]. The resulting scattering states $\Psi_1 \overset{ex}{\times} \cdots \overset{ex}{\times} \Psi_n$ are state vectors in a representation given by a localized morphism.

Since scattering states describe assembly of asymptotically free particles we can ask what is the statistics of these particles. For this sake we study n identical particles of "charge" ξ each in a state specified by a vector Ψ_i and a common reference morphism ρ. Then $\Psi_1 \overset{ex}{\times} \cdots \overset{ex}{\times} \Psi_n$ is a state vector of ρ^n and we find:

$$(13) \qquad \Psi_{p^{-1}(1)} \overset{ex}{\times} \cdots \overset{ex}{\times} \Psi_{p^{-1}(n)} = \varepsilon_\rho^{(n)}(p)\, \Psi_1 \overset{ex}{\times} \cdots \overset{ex}{\times} \Psi_n$$

namely the statistics of the sector ξ defined in terms of strictly localized states coincides with that of the asymptotically free particles of charge ξ.

The state vectors (13) have natural tensor product metric properties
and can be used to calculate transition probabilities for scattering
processes

$$W_{\alpha\beta} = \langle \Psi^{in}_{\alpha_1} \times \cdots \times \Psi^{in}_{\alpha_n} , E^{out}_{\beta} \Psi^{in}_{\alpha_1} \times \cdots \times \Psi^{in}_{\alpha_n} \rangle$$

where E^{out}_{β} is the appropriate support projection for the outgoing
state vector $\Psi^{out}_{\beta_1} \times \cdots \times \Psi^{out}_{\beta_m}$ in the weak closure of the algebra $\mathcal{OL}$
of quasilocal observables, describing the possible final configurations
contained in the (generally non pure) outgoing state.

Problems
========

Of the many open problems related to the subject, I mention only
three. The following statements should become theorems under same
natural additional postulate besides (i) to (iii) above.
  1. Each localized isomorphism of $\mathcal{OL}$ into itself is in Δ.
  2. Each $\rho \in \Delta$ is covariant under Poincaré transformations (from
      the covering group).
  3. Each irreducible element in $\Delta/\mathcal{J}$ has finite statistics;
i.e. no sector with infinite statistics occurs; see also [2, Appendix].
I hope that the relevance of these problems is clear from the content
of this talk.

References
==========

[1] S. Doplicher, R. Haag, J.E. Roberts: "Local Observables and Particle
        Statistics" I,Commun. Math. Phys. <u>23</u>, 199 (1971).
[2] S. Doplicher, R. Haag, J.E. Roberts: "Local Observables and Particle
        Statistics" II,Commun. Math. Phys. <u>35</u>, 49 (1974).
[3] S. Doplicher, J.E. Roberts: "Fields, Statistics and non Abelian
        Gauge Groups" Commun. Math. Phys. <u>28</u>, 331 (1972).

Discussion
==========

Kamefuchi (Question): Have you got anything to say about the CPT
theorem within your formalism?

Doplicher (Answer): Starting with local algebras generated by locally
commuting or anticommuting bounded field operators, Henri Epstein proved
in 1967 that, if you assume asymptotic completeness, the S-matrix is
CPT invariant.

You can formulate asymptotic completeness in our scattering theory,
and this assumption should imply quite in the same way CPT invariance of
the S matrix.

RELAXATION AND FLUCTUATION OF MACROVARIABLES

Ryogo Kubo

Department of Physics, University of Tokyo

Bunkyo-ku, Tokyo

Abstract

An extensive macrovariable X of a system of a large size Ω may have an extensive property of its probability distribution; namely the time-dependent probability function has the form

$$P(x,t) = C \exp \Omega\phi(x,t), \quad x = X/\Omega.$$

This ansatz has been proved by assuming a Markoffian process with transition probability satisfying a homogeneity condition. The function $\phi(x,t)$ can be identified with the action integral and the problem can be formulated by the Hamilton-Jacobi method, which is naturally related to a path-integral formalism. In normal cases, the distribution is Gaussian corresponding to a central limit theorem. Evolution of the mean value and the variance is determined by simple equations which contains the first and second moments of the basic transition probability.

Let X be an extensive macrovariable or a set of such variables in a large system with a size Ω, and x the corresponding density defined by

$$x = X/\Omega = \varepsilon X, \qquad \varepsilon = \Omega^{-1} \tag{1}$$

Examples are the numbers of molecules of certain species in a reaction vessel, the total spin in a magnetic (Ising spin) system, the population of certain class of people in a city and so on. The variable x is considered to make a stochastic process for which a time-dependent probability function $P(x,t)$ is defined. Some time ago, the author conjectured that this has an extensive property in the sense that it has the asymptotic form

$$P(x,t) = C \exp \Omega\phi(x,t) \tag{2}$$

for a large Ω[1]. This Ansatz was proved to be true under the assumption that the process X(t) is Markoffian and its transition probability satisfies a certain condition of homogeneity to be given later. Recently,

Suzuki has proved the Ansatz to hold under more general conditions[2].
If the Ansatz (2) is valid, the deterministic path $x = y(t)$ will be
determined by

$$\phi(x,t) = \text{max.} \quad \text{for } x = y(t) \tag{3}$$

and the fluctuation

$$z = x - y(t)$$

is governed by the function ϕ. In <u>normal</u> cases, where ϕ is regular at
$x = y$, the distribution is nearly Gaussian with the variance

$$<(x - y(t))^2> = -\varepsilon\phi''(y)^{-1} \equiv \varepsilon\sigma(t). \tag{4}$$

This corresponds to a central limit theorem for such a extensive macro-
varialbe X.

Limiting ourselves to Markoffian cases, we assume that the process
is described by the Chapman-Kolmogorov equation

$$P(X,t) = -\int dr\, W(X \to X+r, t)\, P(X,t)$$
$$+\int dr\, W(X-r \to X, t)\, P(X-r, t), \tag{5}$$

where the transition probability W is assumed to be of the form

$$W(X \to X+r, t) = \Omega\, w(x, r, t); \tag{6}$$

in other words, the elementary jump of the state is essentially dependent on the density x and the magnitude of jump, the size Ω appearing
only as the proportionality factor. This is a reasonable assumption
realized in a great many cases of birth and death processes in physical
and non-physical problems. Equation (5) can be written as

$$P(x,t) = - H(x, \varepsilon\frac{\partial}{\partial x}, t)\, P(x,t) \tag{7}$$

with the operator H defined by

$$H(x, p, t) = \int dr\, (1 - e^{-rp})\, w(x,r,t) \tag{8}$$

or

$$H(x, p, t) = \sum_{n=1}^{\infty} \frac{(-)^{n-1}p^n}{n!}\, c_n(x,t) \tag{9}$$

where

$$c_n(x,t) = \int dr\, r^n\, w(x,r,t) \tag{10}$$

is the n-th moment of the transition probability.

We immediately notice that Eq. (7) is similar to a Schrödinger equation so that the asymptotic properties of the solution $P(x,t)$ for $\varepsilon \to 0$ may be treated in an analogous way. Referring to our paper[3] published about a year ago for the details, we summarize in the following some of the main points.

Propagation of the extensive property

The characteristic function $Q(\xi,t)$ is defined by

$$Q(\xi, t) = \int P(x,t)\, e^{ix\xi}\, dx . \tag{11}$$

It is shown that $Q(\xi,t)$ keeps the form

$$Q(\xi, t)\, \exp\, [\tfrac{1}{\varepsilon}\psi(i\varepsilon\xi,\, \varepsilon,\, t)] \tag{12}$$

if it is of this form at an initial time t_0.

By using the steepest descent evaluation, we have $P(x,t)$ in the form, Eq. (2). In particular, the extensivity holds for the initial condition $P(x,t_0) = \delta(x-x_0)$, so that the transition probability $P(x_0 t_0 | x,t)$ is extensive.

The proof assumes the convergence of cumulants of all orders and the analyticity of ψ. The propagation of extensivity may break down if these assumptions cease to be valid at some t.

Evolution equations

In normal cases, $P(x,t)$ is approximated by a Gaussian distrubution

$$P(x, t) \sim C\, \exp[\, -\frac{1}{2\varepsilon\sigma(t)}\, (x - y(t))^2] . \tag{13}$$

It is easily proved that $y(t)$ and $\sigma(t)$ obey the evolution equations,

$$\dot{y}(t) = c_1(y,\, t), \tag{14}$$

$$\dot{\sigma}(t) = 2\frac{\partial c_1}{\partial y}\,\sigma + c_2(y,\,t), \tag{15}$$

which have been obtained by van Kampen from a different point of view.[4] These equations can easily be generalized to a many variable case;

$$\dot{y}_k(t) = c_{1k}(y,t), \tag{16}$$

$$\dot{\sigma}_{jk}(t) = \sum_{\ell}(\sigma_{j\ell}\frac{\partial c_{1k}}{\partial y_\ell} + \frac{\partial c_{1j}}{\partial y_\ell}\,\sigma_{\ell k}) + c_{2jk}. \tag{17}$$

These equations can be applied to a number of problems and lead to some important consequences. The standard Brownian motion corresponds to the assumption,

$$c_1(y) = -\,\gamma y\,, \qquad c_2 = \text{const.} \tag{18}$$

The evolution equations then give immediately the well-known basic formulae for a Brownian motion. If $\gamma < 0$ in Eq. (18), $y = 0$ is unstable; the variance σ increases exponentially as y grows. An anomalous enhancement of fluctuation is seen to be a general phenomenon when a system departs from an unstable situation to reach a new stable equilibrium.

In a more than two variable case, there may arise a limit cycle for the motion y(t). By a change of a parameter of the system, this may appear as a kind of phase transition.[5]

Hamilton-Jacobi formalism

Equation (7) gives

$$\frac{\partial}{\partial t}\phi = -\,H(x,\,\frac{\partial\phi}{\partial x},\,t) \tag{19}$$

to determine the function ϕ. The equations for the characteristics are then

$$\frac{dx}{dt} = \frac{\partial H}{\partial p}\,, \qquad \frac{dp}{dt} = -\,\frac{\partial H}{\partial x}\,, \tag{20}$$

$$\frac{dJ}{dt} = -H + p\frac{\partial H}{\partial p} \equiv L\,, \qquad \frac{dq}{dt} = -\,\frac{\partial H}{\partial t}\,. \tag{21}$$

To solve the Cauchy problem of Eq. (19) for a given initial function

$$\phi(x,\,t_0) = f(x),$$

we impose the initial condition

$$x(t_0) = \xi\,, \qquad p(t_0) = f'(\xi),\quad J(t_0) = f(\xi)$$

$$q(t_0) = -\,H(\xi,\,f'(\xi),\,t_0) \tag{22}$$

and find the solutions

$$x = x(t,\xi), \quad p = p(t,\xi), \quad J = J(t,\xi), \quad q = q(t,\xi).$$

This gives a parametric representation of $\phi(x,t)$ as

$$\phi(x,t) = J(t,\xi), \quad x = x(t,\xi). \tag{23}$$

This method is more general than using the evolution equations (14), (15); it can be used even when the variance σ does not exist.

Path-integral formulation

The asymptotic solution of Eq. (7) may also be represented by a path integral

$$P(x_0 t_0 | x\ t) = \int d\vartheta(x(t))\ \exp[\frac{1}{\varepsilon}\int_{t_0}^{t} ds\ L(x,\dot{x},s)], \tag{24}$$

where the Lagrangian L is that defined by Eq. (21).
The action integral

$$\phi = \int_{t_0}^{t} L(x(s),\ \dot{x}(s),\ s)\ ds \tag{25}$$

is maximized for the path which satisfies the Hamilton equation of motion, (20).

The Gaussian form (13) corresponds to the approximation

$$L(x,\ \dot{x},\ t) = -\ \frac{1}{2c_2(x,t)}\ \{\dot{x}(t) - c_1(x,t)\}^2. \tag{26}$$

If c_2 is a constant, the process is easily seen to be that described by the Langevin equation

$$\dot{x} = c_1(x,t) + R(t) \tag{27}$$

with a Gaussian white random noise

$$\langle R(t)\ R(t')\rangle = c_2\ \delta(t - t'). \tag{28}$$

The stochastic equation (27) is interpreted in Itô's sense, which is in accord with the Eq. (24) with the Lagrangian as given by (26).

Generalization to a field variable

So far we have confined ourselves to a single or a finite number of macrovariables. This is allowed for uniform systems. In non-uniform systems, we have to consider a field of macrovariables. The generalization of the asymptotic evaluation of fluctuations to such a case is not so direct. A scaling theory has been proposed by Mori in this connection.[6]

Here we only remark that it is possible to extend our formalism to a field function $\psi(\vec{r},t)$ which follows the Langevin equation

$$\frac{\partial}{\partial t}\psi(\vec{r},t) = c_1(\psi,\vec{r},t) + R(\psi,\vec{r},t), \tag{29}$$

where R is a white noise. If R is characterized by its cumulants,

$$\int \cdots \int <R(\psi,\vec{r}_1 t_1) \cdots R(\psi,\vec{r}_n t_n)>_c d\vec{r}_1 dt_1 \cdots d\vec{r}_n dt_n$$

$$r < r_j < r+\Delta r, \qquad t < t_j < t+\Delta t \tag{30}$$

$$= c_n(\psi,\vec{r},t)\Delta\vec{r}\,\Delta t, \qquad\qquad n > 2 ,$$

the characteristic function of $\partial\psi/\partial t$ is given by

$$< \exp[-\int_{t_0}^{t_0+t} dt \int_\Omega d r\, \pi(\vec{r},t)\, \frac{\partial\psi}{\partial t}\,]>$$

$$= \exp\,[-\int_{t_0}^{t} dt \int_\Omega dr\, \mathcal{H}(\pi,\psi,\vec{r},t)\,] \tag{31}$$

with the "Hamiltonian"

$$\mathcal{H} = \sum_{n=1}^{\infty} \frac{(-)^{n-1}}{n!}\, \pi^n\, c_n(\psi,\vec{r},t). \tag{32}$$

If the volume Ω is large, an asymptotic evaluation can be made to obtain the distribution function of the field $\psi(\vec{r},t)$ in a path-integral form with the Lagrangian function

$$\mathcal{L}(\psi,\dot{\psi},\vec{r},t) = -\mathcal{H}(\pi,\psi,\vec{r},t) + \pi(\vec{r},t)\dot{\psi} . \tag{33}$$

This approach may be useful to treat non-uniform systems, but it has not been fully developed as yet.

Relaxation Spectra

Writing Eq. (7) as

$$\dot{P} = -\Gamma P \, , \tag{34}$$

we consider the eigenvalue problem

$$\Gamma \Psi_\alpha = \lambda_\alpha \Psi_\alpha . \tag{35}$$

We ask the asymptotic behavior of the eigenvalue spectrum for $\varepsilon \to 0$. It is seen that the eigenmodes are classified into two types. Normally, the first type of eigenmodes have eigenvalues independent of ε and corresponds to fluctuations around an equilibrium. The second type of eigenmodes have eigenvalues of the order of ε^{-1} and describe the decay of large deviations from equilibrium.

If the equilibrium is critical or marginal, the relaxation equation (14) becomes

$$\dot{y} = -\gamma y^k , \qquad k > 1 . \tag{36}$$

In such a case, there occurs an accumulation of eigenvalues at $\lambda = 0$, which corresponds to the phenomena of critical slowing down.

If Eq. (5) is a difference equation with the symmetry of detailed balance condition, the spectral density can be easily obtained with the use of the method of large perturbation introduced by Bethe many years ago. This treatment has also been described briefly in our previous paper.

References

1) R. Kubo, in Synergetics, H. Haken ed., Teubner, Stuttgart, 1973.

2) M. Suzuki, Physic Letters 50A (1974) 47, also a paper at this Conference.

3) R. Kubo, K. Matsuo and K. Kitahara, J. Stat. Phys. 9 (1973) 51.

4) N.G. van Kampen, Can. J. Phys. 39 (1961) 551; in Fundamental problems in Statistical Mechanics, E.G.D. Cohen ed., North Holland, Amsterdam (1962) 173.

5) K. Tomita, Prog. Theor. Phys. 51 (1974) 1731.

6) H. Mori, Prog. Theor. Phys. 52 (1974) 433, also a paper in this Conference.

HOLOMORPHIC FUNCTIONS WITH A NONNEGATIVE IMAGINARY PART IN THE FUTURE TUBE

V.S. Vladimirov
Steklov Institute of Mathematics
Moscow

1. The future tube $\tau^+ = \mathbb{R}^4 + iV^+$ is holomorphycally equivalent to "the generalized unit circle" τ_0 - a set of complex 2×2-matrices W satisfying $WW^* < I$. The skeleton of τ_0 is the group $U(2)$. The manifold

$$S = [X: \det(I + X) = 0, X \in U(2)]$$

corresponds to infinite points of the space $\mathbb{R}^4$. It is homeomorphic to the 3-dimensional unit sphere in which the north pole $(1,\vec{0})$ and south pole $(-1,\vec{0})$ are identified.

2. A function $f(W)$ is holomorphic and $\operatorname{Im} f \geq 0$ in τ_0 if and only if it is (uniquely) represented by a sum $f(W) = f_1(W) + f_0(W)$, where

$$f_1(W) = 2i \int\limits_{U(2)\backslash S} \frac{\sigma(dX)}{\det^2(I-WX^*)} + \overline{f}_1(0), \quad f_0(W) = 2i \int\limits_{S} \frac{\nu_0(dX)}{\det^2(I-WX^*)} + \overline{f}_0(0),$$

where the nonegative mesures σ and ν_0 are orthogonal on $U(2)\backslash S$ and on S resp. to the spherical functions $\Delta^{jm}_{q_1 q_2}(X)$ with the neutral indices $2j=2,3,\ldots,m=-2j+1,\ldots,-1,-j\leq q_1,q_2\leq j$. Moreover $\operatorname{Im} f_1(W) \geq 0$ and $\operatorname{Im} f_0(W) \geq 0$ in τ_0.

3. A function $f(z)$, $z=(z_0,\vec{z})$, is holomorphic and $\text{Im } f \geq 0$ in τ^+ if and only if it is (uniquely) represented by

$$f(z) = \frac{i}{\pi^3}[(z+\mathbf{i})^2]^2 \int_{R^4} \frac{\nu(dx')}{[(z-x')^2]^2[(x'-\mathbf{i})^2]^2} - i(\frac{2}{\pi})^3 \int_{R^4} \frac{\nu(dx)}{|(x+\mathbf{i})^2|^4} + (a,z)+q,$$

where $\mathbf{i} = (i,\vec{0})$, q is a real number, ν is a nonnegative measure satisfying condition

$$\text{Im } f(x + i0) = \nu$$

and a is a (real) vector from cone $\bar{V}^+$ such that the linear function (a,y) is the best linear minoranta of the (concave) function

$$h(y) = \lim_{t\to+\infty} \frac{\text{Im } f(ity)}{t}.$$

4. Applications are in a theory of convolution operators $Z*$ (Z is a $N \times N$-matrix with elements Z_{kj} from the space $\mathcal{D}'(\mathbb{R}^4)$) which are passive with respect to the future light cone $\bar{V}^+$:

$$\text{Re} \int_{-V^+} <Z*\varphi,\varphi> dx \geq 0, \quad \varphi=(\varphi_1,\ldots,\varphi_N) \in \mathcal{D}^{\times N}.$$

(extension of singularities, many-dimensional dispersion relations, inverse operators, a generalized Cauchy problem.) Many operators of mathematical physics are passive with respect to the cone $\bar{V}^+$, including eqs of Dirac, Maxwell, magnetic hydrodynamics, acoustics, electrical networks,...

Literature

V. S. Vladimirov, Matem. sb., 33(1974), 3-17; 34(1974), 499-515; ТМФ, 1(1969), 67-94;
Continuum Mechanics and Related Problems of Analysis, Moscow, 1972, pp.121-134.

Discussion

Borchers (question): In Wightman field theory the 2n-point functions have rather properties of kernel functions. Can you also derive estimates for such kernels?

Vladimilov (answer): I shall try.

ALGEBRAIC ASPECTS OF WIGHTMAN QUANTUM FIELD THEORY

H. J. Borchers

Institut für Theoretische Physik Göttingen

Since the existence of quantum mechanics there are two schemes for the
description of physical systems each with its own advantage.

Method one uses bounded operators and has the advantage of using mathematically nice objects so that a great technical machinery is at hand. Fundamental physical objects, however, do not belong to this class. By rewriting a physical
theory in terms of bounded operators some detailed information is usually getting
lost and one ends up with a scheme of quite general nature. This setting is extremely helpful for answering questions of general nature, but, I believe, that it is useless for constructing models.

The second method uses unbounded operators and has the advantage that
we can formulate the physical principles better than in the first scheme. This
leads to the fact that the objects of interest are usually explicitly defined. The
disadvantage lies purely on the mathematical side since one runs into all kinds of
problems and pathologies associated with unbounded operators.

In relativistic field theory the first scheme is known as Araki[1] - Haag-
Kastler[2] theory of local rings. The second scheme exists in two versions, the
LSZ formalism[3] and the Wightman field theory[4]. Since 1962 it is known[5]
that a Wightman field theory is nothing else than representation of a well defined
algebra, the so called test-function algebra. So the "only" problem which remains
is the study of the representation theory of this algebra. In my lecture I will try
to give a report on our knowledge of this algebra.

Let $\mathcal{S}(\mathbb{R}^4)$ be the Schwartz space of strongly decreasing C^∞ functions
then $\underline{\mathcal{S}}$ will denote the tensor algebra with identity generated by $\mathcal{S}(\mathbb{R}^4)$, i.e.
$\underline{\mathcal{S}}$ consists of terminating sequences of functions $\left\{f_0, f_1, \ldots f_n \ldots\right\}$ with $f_0 \in \mathbb{C}$
and $f_n \in \mathcal{S}(\mathbb{R}^{4n})$. $\underline{\mathcal{S}}$ is a topological $*$-algebra furnished with the direct
sum topology. For details see e.g. W. Wyss[6], G. Lassner and A. Uhlmann[7],
H. J. Borchers[8]. A Wightman field theory can be identified with a cyclic representation of this algebra given by a state W which is invariant and which anihi-

lates the twosided locality-ideal and the spectral-left-ideal.

The topology on $\mathscr{L}$ is a locally convex vector-space topology so that addition and multiplication by scalars are continuous operations. The product, however, is only separately continuous.

When we talk about a representation we will assume that we have a dense subspace $\mathcal{D}$ in a Hilbert-space $\mathcal{H}$ and in general unbounded operators π (f); $f \in \mathscr{L}$, which are defined on $\mathcal{D}$ having the property π (f) $\mathcal{D} \subset \mathcal{D}$ and $(\gamma, T(f)\varphi) = (\pi(f^*)\gamma, \varphi)$. This seems to be the most general notation of *- representations. One can define more restricted representations as it is done for instance by R. Powers [9], [10] which leads to results which are closer to results similar to those of rings of bounded operators, see also Y. Itagaki [11] . But since we want to stay with Wightman's axioms we have to stick to our definition of representations.

One further requirement is the assumption that the representation is weakly continuous. This means the mapping

$$ f \longrightarrow (\gamma, \pi(f)\varphi) \quad ; \quad \gamma, \varphi \in \mathcal{D} $$

shall be continuous. If the *- algebra is barrelled, as it is the case for $\mathscr{L}$ and the representation is weakly continuous then it is also strongly continuous i. e.

$$ f \longrightarrow \| \pi(f)\varphi \| \quad ; \quad \varphi \in \mathcal{D} $$

is continuous [12] thm 4.1 or [13] thm 3.7. A more detailed discussion of the whole problematic of continuity of representations can be found by G. Lassner [13] . Before discussing the representations of $\mathscr{L}$ more thoroughly let us first look at the algebra itself.

Let us first look at the purely algebraic properties. One knows:
a) $\mathscr{L}$ contains an identity
b) is free of divisors of zero
c) has a trivial center
d) only multiples of the identity are invertible
e) the identity is the only idempotent element
f) has a trivial radical.

Furthermore A. Uhlmann showed [14]
g) In $\mathscr{L}$ exists an euclidian algorithm.

This allows to define prime elements in $\mathscr{S}$ and to disentangle the structure of left ideals which are generated by a finite number of elements.

If we denote by $\mathscr{S}_h$ the real subspace of hermitian elements, then one defines the positive cone by

$$\mathscr{S}^+ = \left\{ \sum_i f_i^* f_i \qquad ; \text{ the sum is convergent} \right\}$$

Since $\mathscr{S}$ contains the identity follows that $\mathscr{S}^+ - \mathscr{S}^+ = \mathscr{S}_h$. Furthermore $\mathscr{S}^+$ is a proper cone this means $\mathscr{S}^+ \cap - \mathscr{S}^+ = \{0\}$. W. Wyss [6] observed that $\mathscr{S}^+$ is a strict b-cone that is for every bounded set $B \subset \mathscr{S}_h$ exists a bounded set $B' \subset \mathscr{S}^+$ such that $B \subset B' - B'$. H. J. Borchers [15] proved that $\mathscr{S}^+$ is a closed cone. J. Yngvason [16] showed that $\mathscr{S}^+$ is a cone with base. F. Brauer [17] and H. J. Borchers [18] showed that the positive cone $\mathscr{S}^+$ is generated by its extremal rays, more precisely every element $f^* f$ can be written in the form

$$f^* f = \sum_i g_i^* g_i$$

where $g_i^* g_i$ lies on an extremal ray and the sum converges in $\mathscr{S}$. Finally one has to mention that the positive cone $\mathscr{S}^+$ has no topological (and algebraical) interior point. This fact has the consequence that almost all extension problems of states are unsolvable.

Extremely little is known about the structure of closed left ideals. In particular one would like to have an algebraic characterization of those left ideals which are the intersection of left kernels of states. This property is automatic for closed left ideals in C^*-algebras, but not for $\mathscr{S}$ since one can construct examples of left-ideals which do not have this property. But the two interesting ideals have the property that they are intersections of left-kernels. This can be derived using the explicit structure of these ideals.

Next we are turning to the dual-space $\mathscr{S}'$. We will denote by $\mathscr{S}'^+$ the set of $\omega \in \mathscr{S}'$ with the property $\omega (f^* f) \geq 0$ for all $f \in \mathscr{S}$ and by $\mathscr{S}'_h$ the set of real functionals.

Using the fact that $\mathscr{S}^+$ is a proper cone it follows directly that $\mathscr{S}'^+ - \mathscr{S}'^+$ is a dense subspace of $\mathscr{S}'_h$. But J. Yngvason [17] gave an example of a continuous linear functional which cannot be decomposed into positive ones. In the same paper he derived a necessary and sufficient condition that a functional can be decomposed into positive functionals. We know already that the product is only separately continuous. Let $\mathcal{T}$ be the original topology and $\mathcal{N}$ be the final topology such that the product map

$$\underline{\mathcal{S}}(\mathcal{T}) \times \underline{\mathcal{S}}(\mathcal{T}) \longrightarrow \mathcal{S}(\mathcal{N})$$

is simultaneous continuous, then a functional is decomposable iff it is continuous in the $\mathcal{N}$ topology. This result is in agreement with the earlier result that for any state ω the function $f \to \omega (f^* f)^{1/2}$ is continuous.

That there exist linear functionals on $\underline{\mathcal{S}}$ which are not linear combinations of states follows from the fact that $\underline{\mathcal{S}}$ admits representations by unbounded operators. Namely if π is such a representation and $\gamma \in \mathcal{D}$ but φ is arbitrary in $\mathcal{H}$ then the matrixelement

$$f \to (\varphi, \pi(f)\gamma)$$

is continuous on $\underline{\mathcal{S}}$ but if it is not continuous in γ then one cannot make a decomposition into states. One further result of Yngvason [17] is that every continuous linear functional is of this form showing that there exist no strange continuous functionals on $\underline{\mathcal{S}}$.

Since $\underline{\mathcal{S}}$ is an algebra with identity follows that the set of states (normalized positive functionals) form a base of the cone $\underline{\mathcal{S}}'^+$. One also can show that this cone contains no topological interior point. More interesting is the fact that the cone $\underline{\mathcal{S}}'^+$ is generated by its extremal rays. This has been proved by Borchers and Yngvason [12] and I think I should make some remarks about the results of this paper.

If ω is a state and π_ω the representation given by the G.N.S. construction and if ω can be decomposed this means

$$\omega = \lambda \omega_1 + (1-\lambda) \omega_2 \quad ; \quad 0 < \lambda < 1$$

then exists a bounded positive operator C on $\mathcal{H}_\omega$ such that $\omega_1 (f^* \cdot g) = (\pi_\omega(f)\Omega, C\, \pi_\omega(g)\,\Omega)$. Since the operators $\pi_\omega(f)$ are generally unbounded follows that the expression $\pi_\omega(f)\, C\,\Omega$ must not necessarily be defined. With other words the operator C belongs only to the weak commutant of π_ω . As in the case of bounded operators the weak commutant is invariant under taking the adjoint of an operator and it is also closed in the weak operator topology. But in contrast to the bounded case it is not an algebra in general. This has the consequence that one has to alter the decomposition theory in such a way that it is also applicable to rings of unbounded operators. The main clue to this program is following observation: If π_1 and π_2 are two cyclic representations and if π_1 and π_2 are strange then the vector $\sqrt{\lambda}\,\Omega_1 + \sqrt{1-\lambda}\,\Omega_2$ need not necessarily be a cyclic vector for $\pi_1 \oplus \pi_2$, with other words the cyclic subspace generated by the vector $\sqrt{\lambda}\,\Omega_1 + \sqrt{1-\lambda}\,\Omega_2$ need not be dense in $\mathcal{H}_1 \oplus \mathcal{H}_2$. Therefore if one wants to decompose the restriction of $\pi_1 \oplus \pi_2$ to the cyclic subspace one first has to recover the full Hilbert-space $\mathcal{H}_1 \oplus \mathcal{H}_2$. That this can be done is essential part of the paper [12] . The outcome is the following:

Theorem: Let π be a representation of $\mathcal{G}$ defined on a dense **nuclear domain** $\mathcal{D}$ of a separable Hilbert space $\mathcal{H}$. Then exists a standard measure space (Λ, μ) a separable Hilbert space

$$\mathcal{H}' = \int_\Lambda^{\oplus} \mathcal{H}_\lambda \, d\mu(\lambda)$$

such that there exist dense domains $\mathcal{D}_\lambda \subset \mathcal{H}_\lambda$, representations π_λ defined on $\mathcal{D}_\lambda$ with :

a) if $\lambda \to \varphi_\lambda \in \mathcal{D}_\lambda$ is μ -measurable then also

$$\lambda \to \pi_\lambda(f)\,\varphi_\lambda \qquad \text{is } \mu \text{ -measurable}$$

and fulfills some continuity property .

b) $\mathcal{H}$ is a closed subspace of $\mathcal{H}'$ such that the projection of $\mathcal{D}$ into $\mathcal{H}_\lambda$ is $\mathcal{D}_\lambda$.

c) The representations π_λ have a trivial weak commutant.
(all these properties hold μ a.e.)

That one can also decompose unbounded representations is probably only true when $\mathcal{D}$ is a nuclear space which is the case for cyclic representations of nuclear algebras. Since this is the case for $\mathcal{G}$ we can decompose every state on $\mathcal{G}$ into extremal ones :

$$\omega = \int_\Lambda \omega_\lambda \, d\mu(\lambda)$$

where ω_λ are μ - a.e. extremal states that is the representation π_{ω_λ} all have a trivial weak commutant.

If the state ω is a Wightman functional then all the states ω_λ in the above decomposition are also Wightman functionals. This shows that the set of Wightman states form a face in the set of all states. If in a cyclic Wightman theory with cyclic vector Ω this Ω is also the only invariant vector in the representation Hilbert space then this representation has a trivial weak commutant, which means that the corresponding Wightman state is extremal. As one can show by examples the converse conclusion is not correct this means there exists extremal Wightman functionals such that the corresponding representation has an infinite dimensional subspace as invariant subspace. This effect does not occur in the theory of local rings, and is due to pathologies in the theory of commuting unbounded operators.

Symanzik [19] introduced euclidean fields as some kind of "analytic continuation" of Wightman fields. It is natural to apply the above decomposition theory also to such fields. This has been done by Borchers and Yngvason [20]. The question here is not the decomposition of a state on an abelian algebra into extremal states but rather the question when can a state of an abelian algebra be decomposed into characters. This is not always the case, but necessary and sufficient for this is that the state must be strictly positive, that is, if $P(x_1 \ldots x_n) \geqq 0$ is a positive polynomial then $\omega\,(P(f_1, \ldots f_n))$ has to be non negative for all $f_i \in \underline{\mathscr{S}}_h$. If the Schwinger functional is strict positive then it can be represented by a cylinder-measure over $\mathscr{S}'(R^+)$.

If one starts from a Wightman state then we can analytically continuate to the Schwinger points. This, however, defines a continuous linear functional only on some subspace of the symmetric part of $\underline{\mathscr{S}}$. As Osterwalder and Schrader [22] remarked this can be extended to a continuous linear functional to the whole symmetric part of $\underline{\mathscr{S}}$. Since the positive cone in $\underline{\mathscr{S}}$ has now interior points it seems to me a hopeless problem to show (except in concrete examples) that there exists an extension as positive functional. Since one does not know explicitly the structure of all positive polynomials, it is even much more hopeless to the extension as a strict positive functional. The beauty of strict positive functionals is that they can be represented by cylinder measures. In order to get a wider class one can study functionals which are representable by not necessarily positive cylinder measures, i.e. functionals which can be decomposed into strict positive ones. This program is under investigation in Göttingen and we could already say that such functionals can be characterized by pure continuity property. The question whether every analytic continuation of a Wightman functional has this continuity property is unsettled.

One group of problems which is still completely in the dark is the extension of positive functionals given on subspaces or subalgebras to a positive functional on all of $\underline{\mathcal{Y}}$. For Wightman functionals W. Wyss [22] could give some conditions. They are still in such a form that it is practically impossible to use them. But nevertheless Lassner and Hofmann [23] could use these conditions and give an abstract proof that there exist Wightman fields different from the generalized free fields.

It is known that for Wightman fields only unbounded representations of are of interest. But, nevertheless, it is known from [8] and [15] that $\underline{\mathcal{Y}}$ admits a large number of representations by bounded operators. This means we have on $\mathcal{Y}$ a great family of continuous norms and semi-norms which are also C^{*} -norms. M. Dubois-Violette [24] is trying to explore this fact. His present investigations are in the direction of the infinite dimensional moment problem. But I think that here is a new technique which might add to the understanding of the algebra $\underline{\mathcal{Y}}$ and which has to be explored in the future.

As final subject I will talk on combination of states. This is a method of combining two states to a new one in a nonlinear fashion. These combinations are mainly possible because $\underline{\mathcal{Y}}$ is a graded algebra. The first subject is the s-product which is well known in statistical mechanics to obtain the Meyer expansion from correlation function. In field theory it is used for obtaining the truncated functional ω^{t} from ω . The relation is

$$\omega = \exp \omega^{t} |_{s}$$

where we have to take the powers in the s-product. If now ω_1 and ω_2 are two states on $\mathcal{Y}$ then

$$\omega = \omega_1 \, s \, \omega_2$$

defines a new state on $\underline{\mathcal{Y}}$. This operation amounts in the language of operators to the following: Let $A(x)$ and $B(x)$ be the field obtained from ω_1 and ω_2 and $C(x)$ the field obtained from ω then we have:

$$C(x) = \text{cyclic part of} \left(A(x) \, 1_2 + 1_1 \, B(x) \right).$$

(For the proofs see [8]). Of interest is now the inverse process namely the decomposition of states into s-products. The reason for this is an observation made by

Symanzik, namely, if A(x) is a field which has complete asymptotic fields and if W is the Wightman functional belonging to A(x) then the field B(x) constructed from the functional W s W has in general no longer complete asymptotic fields. This makes the interest in the s-product evident. Some results in this direction have been obtained by Hegerfeldt [25], but, I need some notations before I can describe them. A state ω on $\mathcal{S}$ is called a prime state if it cannot be written as the s-product of two other states. On the other hand a state is called infinitely divisible if it is the n-th s-power of another state for every n = 1, 2, ... Hegerfeldt's result is now the following: Every state ω on $\mathcal{S}$ can be decomposed into the s-product of at most a countable number of prime states and a rest where the rest is either trivial or infinitely divisible.

His proof is a pure existence proof. What would be of great interest, any characterisation of prime states, is still missing. But nevertheless one can prove the existence of prime states since one can characterize the infinitely divisible states and show that the Wick polynomials do not belong to this class. The characterization of this class is by means of the truncated functional ω^t. The result is: A state ω is infinitely divisible if and only if ω^t is positive on the ideal of element with vanishing zeroth component. The generalized free fields belong to this class.

There is still another combination of states which works only for Wightman functionals. But since nothing has been done with it I will refrain from explaining it. There is also another group of results which I will not bring here. These are results obtained in connection with constructive field theory. Most of them start from additional assumptions suggested by models and not from a general axiomatic.

References

[1]   Araki, H.: "Einführung in die axiomatische Quantenfeldtheorie" Lecture Notes Zürich (1961/62)

[2]   Haag, R. and Kastler, D.: "An Algebraic Approach to Quantum Field Theory" J.Math. Phys., 16, 158 (1964)

[3]   Lehmann, H., Symanzik, K. and Zimmermann, W.: "Zur Formulierung quantisierter Feldtheorien" Nuovo Cimento, 1, 205 (1955)

[4]   Wightman, A.S.: "Quantum Field Theory in Terms of Vacuum Expectation Values" Phys. Rev. 101, 860 (1956)

[5]   Borchers, H.J.: "On Structure of the Algebra of Field Operators" Nuovo Cimento, 24, 214 (1962)

[6]   Wyss, W.: "On Wightman's Theory of Quantized Fields" Boulder Lecture Notes (1968)

[7]   Lassner, G. and Uhlmann, A.: "On Positive Functionals on Algebras of Test Functions for Quantum Fields" Comm. Math. Phys. $\underline{7}$, 152, (1968)

[8]   Borchers, H.J.: "Algebraic Aspects of Wightman Field Theory" Haifa Summer School (1971)

[9]   Powers, R.: "Self-Adjoint Algebras of Unbounded Operators I" Comm. Math. Phys., $\underline{21}$, 85 (1971)

[10] Powers, R.: "Self-Adjoint Algebras of Unbounded Operators II" Transactions A.M.S. $\underline{181}$, 261 (1974)

[11] Itagaki, Y.: "Self-Adjoint Algebras of Operators on a Rigged Hilbert Space" Notes and Abstracts of the Japan-U.S. Seminar on C*-Algebras and Applications to Physics, Kyoto (1974)

[12] Borchers, H.J. and Yngvason, J.: "On the Algebra of Field Operators. The Weak Commutant and Integral Decomposition of States" Comm. Math. Phys. in print

[13] Lassner, G.: "Topological Algebras of Operators" Rep. Math. Phys. $\underline{3}$, 279 (1972)

[14] Uhlmann, A.: "Algebraic Properties of the Test Function Algebra" Lectures given in Göttingen

[15] Borchers, H.J.: "On the Algebra of Test Functions" Prepublications de la RCP no 25, Vol. 15, Strassbourg 1973

[16] Yngvason, J.: "On the Algebra of Test Functions for Field Operators. Decomposition of Linear Functionals into Positive Ones" Comm. Math. Phys. $\underline{34}$, 315 (1973)

[17] Brauer, F.: "Über die Struktur des positiven Kegels in der Feldalgebra" Diplomarbeit, Göttingen (1973)

[18] Borchers, H.J.: "On the Positive Cone in the Algebra of Test Function" unpublished manuscript

[19] Symanzik, K.: "Euclidean Quantum Field Theory" Proc. of the Varenna Summer School XLV, ed. R. Jost, Academic Press 1969

[20] Borchers, H.J. and Yngvason, J.: "Integral Representations for Schwinger Functionals and the Moment Problem over Nuclear Spaces" Preprint, submitted to the Comm. Math. Phys.

[21] Osterwalder, K. and Schrader, R.: "Axioms for Euclidean Green's Functions" Comm. Math. Phys. $\underline{31}$, 83 (1973)

[22] Wyss, W.: "The Field Algebra and its Positive Linear Functionals" Comm. Math. Phys. $\underline{27}$, 223 (1972)

[23] Lassner, G. and Hofmann, G.: "Existence Proofs for Wightman Type Functionals" Preprint Dubna E 2 - 7536 (1973)

[24] Dubois-Violette, M.: "A Generalization of Classical Moment Problem
      on *-Algebras with Application to Relativistic Quantum Theory"
      Preprint (1974)
[25] Hegerfeldt, G.C.: "Relativistic and Euclidean Prime Fields" in
      preparation.

Discussion

Question by R. Arens:

On the algebra $\mathcal{P}$, is the map $x \to xx^*$ perhaps continuous?

Answer:

No, since the algebra contains an identity, continuity of the
map $x \to xx^*$ and joint continuity of the product are the same.

Question by M. Winnink:

Can the decompositions you have been discussing be related to
a simplicial structure of the set of states of interest?

Answer:

For the decomposition treated here the set of states do not form
a simplex even not for the set of Wightman states. I expect that no
face of the set of states, which might appear in physics, will form a
simplex. The reason for this is that the moment problem has a unique
solution only under very restrictive conditions. And I cannot imagine
that such condition will appear naturally.

THE INFRARED PROBLEM AND NON-FOCK REPRESENTATIONS
OF THE CANONICAL COMMUTATION RELATIONS

G. Reents

Physikalisches Institut, Universität Würzburg
Würzburg, Federal Republic of Germany

Some important aspects of the infrared problem, as for instance the creation of infinitely many particles or as the non-unitarity of the scattering operator in the usual Fock space, can already be discussed on the basis of the simple model where a quantized electromagnetic field interacts with a prescribed external c-number current. In this case Maxwell's equations can be solved explicitly. The solution may be expressed as an integral containing a Green's function and the current, plus incoming, or, respectively, outgoing free field. Adopting the radiation gauge one obtains the following relation between the creation (annihilation) operators of the respective fields [1]

$$\underset{\sim}{a}_{out}(\underset{\sim}{k}) = \underset{\sim}{a}_{in}(\underset{\sim}{k}) + i\underset{\sim}{j}(\underset{\sim}{k}) \ . \tag{1}$$

$\underset{\sim}{j}(\underset{\sim}{k})$ denotes the transverse components of the Fourier transform of the current on the mass shell of the photon. The $\underset{\sim}{a}(\underset{\sim}{k})$, $\underset{\sim}{a}^{*}(\underset{\sim}{k})$ satisfy the canonical commutation relations (CCRs)

$$[a_{n}(\underset{\sim}{k}), a^{*}_{m}(\underset{\sim}{k}')] = 2\omega\delta_{nm}\delta(\underset{\sim}{k}-\underset{\sim}{k}'), \quad \omega(\underset{\sim}{k}) = |\underset{\sim}{k}| \ . \tag{2}$$

If we assume the vacuum to be the incoming state the number of created particles is given by

$$\|\underset{\sim}{j}\|^{2} = \int d\mu(\underset{\sim}{k})\underset{\sim}{j}(\underset{\sim}{k})\underset{\sim}{j}^{*}(\underset{\sim}{k}), \quad d\mu(\underset{\sim}{k}) = \frac{d^{3}\underset{\sim}{k}}{2\omega} \ . \tag{3}$$

For realistic current densities, however, $|\underset{\sim}{j}(\underset{\sim}{k})|^{2} = \underset{\sim}{j}(\underset{\sim}{k})\underset{\sim}{j}^{*}(\underset{\sim}{k})$ behaves like ω^{-2} for small ω so that the integral (3) diverges logarithmically at $\omega = 0$. A transformation $\underset{\sim}{a}(\underset{\sim}{k}) \to \underset{\sim}{a}(\underset{\sim}{k}) + \underset{\sim}{f}(\underset{\sim}{k})$ like (1) is established by the operator

$$W(\underset{\sim}{f}) = \exp \int d\mu(\underset{\sim}{k})\{\underset{\sim}{a}^{*}(\underset{\sim}{k})\underset{\sim}{f}(\underset{\sim}{k}) - \underset{\sim}{a}(\underset{\sim}{k})\underset{\sim}{f}^{*}(\underset{\sim}{k})\} \ . \tag{4}$$

If we insist on a unitary scattering operator we have to look for representations of the CCRs which allow the automorphism (1) to be unitarily implemented. It is a well known fact, however, that $W(\underset{\sim}{f})$ is a unitary operator in Fock space if and only if $\|\underset{\sim}{f}\|^2 < \infty$ [2]. This shows that the Fock representation is not an appropriate candidate to solve our problem.

Nevertheless we want to start with the Fock space in order to construct new representations of the CCRs. Let Ω be the Fock vacuum and $W(\underset{\sim}{f})\Omega$ the coherent state generated by $W(\underset{\sim}{f})$ for $\underset{\sim}{f} \in L^2_\mu$. The inner product of two such states is given by [3]

$$(W(\underset{\sim}{g})\Omega,\ W(\underset{\sim}{f})\Omega) = \exp \tfrac{1}{2} \int d\mu \{ \underset{\sim}{g}^*\underset{\sim}{f} - \underset{\sim}{f}^*\underset{\sim}{g} - |\underset{\sim}{f}-\underset{\sim}{g}|^2 \}. \tag{5}$$

When $\underset{\sim}{f}$ and $\underset{\sim}{g}$ are also continuous this can be written as

$$(W(\underset{\sim}{g})\Omega,\ W(\underset{\sim}{f})\Omega) = \exp \int d\mu \log (W_{\underset{\sim}{k}}(\underset{\sim}{g})\Omega_{\underset{\sim}{k}},\ W_{\underset{\sim}{k}}(\underset{\sim}{f})\Omega_{\underset{\sim}{k}})_{\underset{\sim}{k}}; \tag{6}$$

$W_{\underset{\sim}{k}}(\underset{\sim}{f})$ means the usual Schrödinger representations of 2 degrees of freedom, i.e., $W_{\underset{\sim}{k}}(\underset{\sim}{f}) = \exp\{a^*_{\underset{\sim}{k}}\underset{\sim}{f}(\underset{\sim}{k}) - a_{\underset{\sim}{k}}\underset{\sim}{f}^*(\underset{\sim}{k})\}$ where $a^*_{\underset{\sim}{k}} = (\underset{\sim}{Q}-i\underset{\sim}{P})/\sqrt{2}$, $a_{\underset{\sim}{k}} = (\underset{\sim}{Q}+i\underset{\sim}{P})/\sqrt{2}$ and $\underset{\sim}{Q}$, $\underset{\sim}{P}$ are position and momentum operator in the Hilbert space $\mathcal{H}_{\underset{\sim}{k}} = L^2(\mathbb{R}^2)$. Of course, $\Omega_{\underset{\sim}{k}}$ has to be the ground state of the harmonic oscillator: $\Omega_{\underset{\sim}{k}}(\underset{\sim}{x}) = \pi^{-1/2} \exp(-\tfrac{1}{2}\underset{\sim}{x}^2)$. Then $W_{\underset{\sim}{k}}(\underset{\sim}{f})\Omega_{\underset{\sim}{k}}$ is a coherent state in $\mathcal{H}_{\underset{\sim}{k}}$ and the scalar product between two such vectors yields

$$(W_{\underset{\sim}{k}}(\underset{\sim}{g})\Omega_{\underset{\sim}{k}},\ W_{\underset{\sim}{k}}(\underset{\sim}{f})\Omega_{\underset{\sim}{k}})_{\underset{\sim}{k}} =$$
$$= \exp \tfrac{1}{2}\{ \underset{\sim}{g}^*(\underset{\sim}{k})\underset{\sim}{f}(\underset{\sim}{k}) - \underset{\sim}{f}^*(\underset{\sim}{k})\underset{\sim}{g}(\underset{\sim}{k}) - |\underset{\sim}{f}(\underset{\sim}{k})-\underset{\sim}{g}(\underset{\sim}{k})|^2 \}$$

so that equation (6) becomes clear. The coherent states $W(\underset{\sim}{f})\Omega$, however, with $\underset{\sim}{f} \in L^2_\mu \cap C_o$ form a total set in the Fock space. Formula (6) then says that the Fock space may be considered as a continuous tensor product of the Hilbert spaces $\mathcal{H}_{\underset{\sim}{k}}$ built up by continuous products of coherent states $\underset{\underset{\sim}{k}}{\otimes} W_{\underset{\sim}{k}}(\underset{\sim}{f})\Omega_{\underset{\sim}{k}}$ [4].

The vacuum state is clearly represented as the continuous product of the harmonic oscillator ground states, i.e., $\Omega = \underset{\underset{\sim}{k}}{\otimes} \Omega_{\underset{\sim}{k}}$. Since $\mathcal{H}_{\underset{\sim}{k}} = L^2(\mathbb{R}^2)$ does not depend on $\underset{\sim}{k}$ we shall drop this subscript whenever it is possible without danger of confusion.

At this point the following generalization presents itself.
Let $\phi_{\underset{\sim}{k}} \in \mathcal{H}_{\underset{\sim}{k}}$ be given by

$$\phi_{\underset{\sim}{k}}(\underset{\sim}{x}) = (b(\underset{\sim}{k})/\pi)^{1/2} \exp\{-i\alpha_1(\underset{\sim}{k})a^*_{\underset{\sim}{k}1}a_{\underset{\sim}{k}1} - i\alpha_2(\underset{\sim}{k})a^*_{\underset{\sim}{k}2}a_{\underset{\sim}{k}2}\}\exp\{-\tfrac{1}{2}b(\underset{\sim}{k})\underset{\sim}{x}^2\}$$

where $b(\underset{\sim}{k})$ is bounded and positive and $\alpha_1(\underset{\sim}{k})$, $\alpha_2(\underset{\sim}{k})$ are real functions.
After some elementary calculations one arrives at

$$(W_{\underset{\sim}{k}}(\underset{\sim}{g})\phi_{\underset{\sim}{k}}, \ W_{\underset{\sim}{k}}(\underset{\sim}{f})\phi_{\underset{\sim}{k}}) =$$

$$= \exp \tfrac{1}{2}\{\underset{\sim}{g}^*\underset{\sim}{f} - \underset{\sim}{f}^*\underset{\sim}{g} - b[\mathrm{Re}\ \underset{\sim}{u}_{\underset{\sim}{\alpha}}(\underset{\sim}{f}-\underset{\sim}{g})]^2 - \tfrac{1}{b}[\mathrm{Im}\ \underset{\sim}{u}_{\underset{\sim}{\alpha}}(\underset{\sim}{f}-\underset{\sim}{g})]^2\}$$

$$\equiv \exp d_{\underset{\sim}{g},\underset{\sim}{f}}(\underset{\sim}{k}), \quad (\underset{\sim}{u}_{\underset{\sim}{\alpha}}(\underset{\sim}{f}))_n = e^{i\alpha_n} f_n \ . \tag{7}$$

This suggests that one could define a (pre-) Hilbert space $\underset{k}{\otimes}(\mathcal{H}_{\underset{\sim}{k}}, \phi_{\underset{\sim}{k}})$ containing the vectors $W(\underset{\sim}{f})\Phi = \underset{k}{\otimes} W_{\underset{\sim}{k}}(\underset{\sim}{f})\phi_{\underset{\sim}{k}}$ with inner product $(W(\underset{\sim}{g})\Phi, \ W(\underset{\sim}{f})\Phi) = \exp \int d\mu(\underset{\sim}{k})d_{\underset{\sim}{g},\underset{\sim}{f}}(\underset{\sim}{k})$. One has to check that this bilinear form actually defines a scalar product, i.e., that $A_{ij} \equiv (W(\underset{\sim}{f}_i)\Phi, W(\underset{\sim}{f}_j)\Phi) = \exp \int d\mu(\underset{\sim}{k})d_{ij}(\underset{\sim}{k})$ is a positive matrix. Now, for all real λ, $\exp\lambda^2 d_{ij} = (W_{\underset{\sim}{k}}(\lambda\underset{\sim}{f}_i)\phi_{\underset{\sim}{k}}, W_{\underset{\sim}{k}}(\lambda\underset{\sim}{f}_j)\phi_{\underset{\sim}{k}})$ is obviously positive and so is A_{ij} because it may be considered as a limit of the coefficient-wise product of positive matrices [5]. The expectation functional of this representation is given by

$$E(\underset{\sim}{f}) = (\Phi, \ W(\underset{\sim}{f})\Phi)$$

$$= \exp - \tfrac{1}{2} \int d\mu \left(b(\underset{\sim}{k})[\mathrm{Re}\ \underset{\sim}{u}_{\underset{\sim}{\alpha}}(\underset{\sim}{f}(\underset{\sim}{k}))]^2 + \tfrac{1}{b(\underset{\sim}{k})}[\mathrm{Im}\ \underset{\sim}{u}_{\underset{\sim}{\alpha}}(\underset{\sim}{f}(\underset{\sim}{k}))]^2 \right) .$$

As is known, this functional determines completely, (up to unitary equivalence), the representation of the CCRs and also the space V of admissible test functions $\underset{\sim}{f}$ [6,7,8]. In our example the test function $\underset{\sim}{f}:\mathbb{R}^2 \to \mathbb{C}^2$ belongs to V if and only if $\sqrt{b}\,\mathrm{Re}(\underset{\sim}{u}_{\underset{\sim}{\alpha}}(\underset{\sim}{f})) \in L^2_\mu$ and $\tfrac{1}{\sqrt{b}}\mathrm{Im}(\underset{\sim}{u}_{\underset{\sim}{\alpha}}(\underset{\sim}{f})) \in L^2_\mu$. V is a real linear space and $E(\lambda\underset{\sim}{f})$ is continuous in the real parameter λ for $\underset{\sim}{f} \in V$. Among other things this implies that $W(\underset{\sim}{f})$ is unitary for $\underset{\sim}{f} \in V$. Two different functions $b(\underset{\sim}{k})$, $b'(\underset{\sim}{k})$ lead to the same expectation functional if $b=b'$ μ-almost everywhere. So, the restriction $b(\underset{\sim}{k})>0$ may be changed into $b(\underset{\sim}{k})>0$ μ-almost everywhere. Returning to the problem stated at the beginning we define $\alpha_n(\underset{\sim}{k})$ so that $\mathrm{Im}\left(\underset{\sim}{u}_{\underset{\sim}{\alpha}}(\underset{\sim}{ij}(\underset{\sim}{k}))\right) \equiv 0$. Then by a suitable choice of $b(\underset{\sim}{k})$ we achieve that $\sqrt{b}\,\mathrm{Re}(\underset{\sim}{u}_{\underset{\sim}{\alpha}}(\underset{\sim}{ij}))$ is square-integrable in the measure μ, i.e., $W(\underset{\sim}{ij})$ is unitary and so the automorphism (1) is unitarily implemented in this representation of the CCRs. The representation is cyclic by construction and since the $W(\underset{\sim}{f})$ are unitary with $W^*(\underset{\sim}{f}) = W(-\underset{\sim}{f})$ it is even irreducible.

It is unitarily inequivalent to the Fock representation which can be
seen, for instance, by the fact that it generates a different topology
on the test function space.

<u>Acknowledgement:</u> I would like to thank K. Kraus and L. Polley for
helpful comments made during many discussions. I would also like to
express my gratitude to M. Everitt for preventing me from making too
many mistakes in my English.

DISCUSSION

ZAVIALOV: What about Lorentz-invariance in your scheme? The generating
functional you have written seems to be not invariant.

REENTS: The functional is not invariant and this has to be expected.
Talking about non-Fock representations one has to take into account
that there is no vacuum state in the representation space and, therefore, there is no state which is invariant under all Lorentz transformations.

ZAVIALOV: What is the connection of your scheme with that by KULISH
and FADDEEV [9] ?

REENTS: In order to obtain the S-matrix related to equation (1) as a
limit of scattering operators at finite times one has to modify the
asymptotic dynamics as it has been done by FADDEEV and KULISH. Our representations, however, are not (generalized) coherent state representations. They are unitarily inequivalent to those used by FADDEEV
and KULISH.

REFERENCES

1.   Reents, G.: J. Math. Phys. $\underline{15}$, 31 (1974)
2.   Araki, H., Woods, E.J.: J. Math. Phys. $\underline{4}$, 637 (1963)
3.   Kibble, T.W.B.: J. Math. Phys. $\underline{9}$, 315 (1968)
4.   Guichardet, A., Wulfsohn, A.: J. Funct. Anal. $\underline{2}$, 371 (1968)
5.   Gelfand, I. M., Vilenkin, N.J.: Generalized Functions, Vol. IV
6.   Araki, H.: J. Math. Phys. $\underline{1}$, 492 (1960)
7.   Hegerfeldt, G. C., Klauder, J.R.: Commun. Math. Phys. $\underline{16}$, 329 (1970)
8.   Araki, H., Woods, E.J.: Rep. Math. Phys. $\underline{4}$, 227 (1973)
9.   Faddeev, L., Kulish, P.: Theoret. and Math. Phys. $\underline{4}$, 745 (1970)

CONTINUOUS REPRESENTATIONS OF THE TEST FUNCTION ALGEBRA
AND THE EXISTENCE PROBLEM FOR QUANTUM FIELDS

Gerd Lassner

Mathematics Department

Karl-Marx-University

Leipzig, GDR

1. Wightman functionals with certain properties

__1.1__ A quantum field $\varphi(x)$ (neutral and scalar for the sake of
technical simplicity) is described by a normed positive functional,
the Wightman functional

$$W(f) = \sum_{n \geqslant 0} \int W_n(x_1,\ldots,x_n)\, f_n(x_1,\ldots,x_n)\, dx$$

on the complete topological tensor $*$-algebra $\mathfrak{R}[\tau] = \mathcal{S}_0 \oplus \mathcal{S}_1 \oplus \mathcal{S}_2 \oplus \cdots$
over the Schwartz space $\mathcal{S}_1 = \mathcal{S}(R^4)$, where τ is the direct sum to-
pology. The elements of $\mathfrak{R}$ have the form $f = (f_0, f_1(x_1), f_2(x_1,x_2),\ldots$
$\ldots, f_n(x_1,\ldots,x_n),\ldots)$, where all but a finite number of $f_i \in \mathcal{S}_i$ are
equal to zero. From the Wightman functional W we come back to the
field $\varphi(x)$ by the GNS-representation of the test function algebra
$\mathfrak{R}$ /1/,/2/. The physical properties (locality, covariance, spectrality) are equivalent to $W(f) = 0$ for all $f \in J$, where J is a
certain linear subspace of $\mathfrak{R}$. $W(f)$ is uniquely determined by its
values on the hermitian part $\mathcal{U} = \{ f;\ f \in \mathfrak{R},\ f^* = f \},\ \mathfrak{R} = \mathcal{U} + i\mathcal{U}.$
Further, one has $J = L + iL$ and $\mathcal{U} \supset \mathfrak{R}_+ = \text{conv}\ \{ f^*f;\ f \in \mathfrak{R} \}$. $\mathfrak{R}_+$
is the cone of positive elements.

__1.2__ In a concentrated formulation a Wightman functional W is
a real linear functional on $\mathcal{U} = \mathcal{U}_0 \oplus \mathcal{U}_1 \oplus \mathcal{U}_2 \oplus \cdots$ with the properties

__1.__ $W(1) = 1$, __2.__ $W(f) \geqslant 0$ for all $f \in \mathfrak{R}_+$, __3.__ $W(f) = 0$ for all $f \in L$.

Because of the normalization __1.__ the Wightman functional is uniquely
determined by the hyperplane $E_W = \ker W$, whereby $\mathfrak{R}_+$ is lying on
one side of E_W (__2.__) and $L \subset E_W$ (__3.__). Therefore the problem
to find out "all" quantum fields is equivalent to the problem to determine all hyperplanes E_W with the properties described above (Fig. 1).

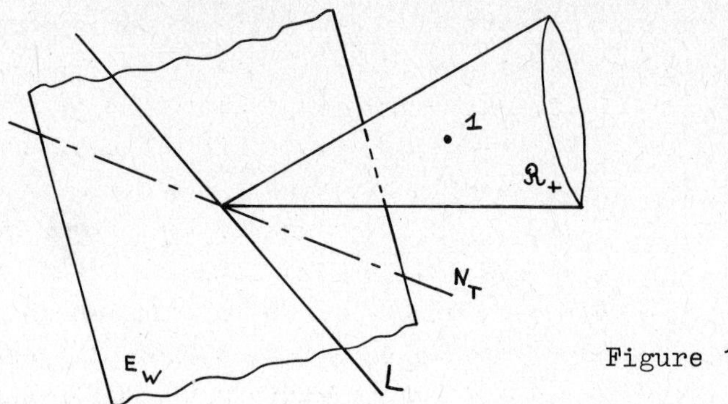

Figure 1

<u>1.3</u> The problem to determine a Wightman functional (field) with 'certain physical properties' is in some sense related to the problem to find such a hyperplane E_W , which contains yet a certain subspace N_T . E.g. if we ask for fields with given 2n-point function $W_{2n}(x_1,\ldots \ldots,x_{2n})$, then we have to seek a Wightman functional W on $\mathcal{A}$ which is the extension of the linear functional $T(f) = \int W_{2n}(x_1,\ldots,x_{2n}) \times$ $\times f_{2n}(x_1,\ldots,x_{2n})\, dx$ on $M = \mathcal{A}_{2n}$. This is a special case of a general situation.

We start with a linear subspace M of $\mathcal{A}$ (resp. $\mathcal{R}$) and a linear functional T on M . Then we ask for a Wightman functional on $\mathcal{A}$ (resp. $\mathcal{R}$) which is an extension of T . This is equivalent to the geometric condition $E_W \supset N_T = \ker T$ (see Fig. 1).

<u>1.4</u> To handle the extension problem formulated in the foregoing section one can employ functional analytical methods like the separation theorems for convex sets in topological spaces, as it was first done in /3/,/4/,/5/,/6/.

<u>Theorem.</u>(Wyss /7/)
There exists an extension of T to a Wightman functional W if and only if one of the following equivalent conditions is satisfied.

$$\text{i)} \quad \mathcal{A} \neq \overline{\mathcal{R}_+ + L + N_T}$$
$$\text{ii)} \quad -1 \notin \overline{\mathcal{R}_+ + L + N_T} \quad ,$$

where the closure is taken with respect to the direct sum topology $\mathcal{T}$.

<u>1.5</u> With the help of the foregoing theorem in /8/ a method could be demonstrated   to prove the existence of 'Wightman type' functio-

nals different from those of the free or generalized free fields. To
do that it was taken a certain four-particle function $g_4 = g_4(x_1,..,x_4)$
$\in \mathcal{O}_4$ such that $W(g_4) = 0$ for all Wightman functionals of the free
or generalized free fields. By $T(f) = \lambda + \mu\beta$ it was defined a real
linear functional on the two-dimensional space $M = \{f = \lambda 1 + \mu g_4\}$
$\subset \mathcal{O}$, where $\beta = T(g_4) > 0$. It could be proved that $-1 \notin \mathcal{R}_+ + L +$
$+ N_T$, where here L is only generated by the covariance and locality
conditions. Therefore it exists an extension W of T , which is of
'Wightman type' (without spectrality) and different from all generalized free fields, since $W(g_4) > 0$. With quite analogous methods in /9/
it was proved the existence of positive functionals satisfying almost
all conditions for an Euclidian field and different from a Gaussian
one (free field).

2. C*-like topologies on the test function algebra

<u>2.1</u> In a B*-algebra R the problem of extending a positive func-
tional from a subspace $M \subset R$ to the full algebra R has a simple
positive answer, since the cone R_+ of positive elements has interior points, e.g. 1 is an interior point. In the test function algebra
$\mathcal{R}$ the cone of positive elements has no interior points at all. Yng-
vason /10/ has shown that on the closed $*$-subalgebra $\mathcal{R}_0 = \mathcal{C} \oplus \mathcal{L}_1 \oplus \mathcal{L}_2 \oplus \cdots$
of $\mathcal{R}$, generated by the subspace $\mathcal{L}_1 = \{f \in \mathcal{S}_1; f^{(n)}(0) = 0, n=0,1,..\}$
of $\mathcal{S}_1$, positive functionals exist, which cannot extended to a posi-
tive functional on $\mathcal{R}$. But this algebra $\mathcal{R}_0$ is not a "C*-like subalgebra" in the sense defined in what follows.

<u>2.2</u>Now we are going to introduce the concepts of "C*-like" to-
pologies and "C*-like" subalgebras in a $*$-algebra R and explain
in the special case of the test function algebra the connection of these
concepts with the existence problem for "non-trivial" quantum fields.

Let $\mathcal{A}$ be an Op*-algebra of (unbounded) operators on a pre-Hil-
bert space $\mathcal{D}$ /11/. The uniform topology $\mathcal{T}_\mathcal{D}$ on $\mathcal{A}$ is defined by
the system of all seminorms

$$\| A \|_\mathcal{M} = \sup_{\phi, \psi \in \mathcal{M}} |\langle \phi, A \psi \rangle|$$

where $\mathcal{M}$ runs over all subsets $\mathcal{M} \subset \mathcal{D}$ such that $\| A \|_\mathcal{M} < +\infty$ for all
$A \in \mathcal{A}$. $\mathcal{A}$ is a locally convex topological $*$-algebra with respect to
the topology $\mathcal{T}_\mathcal{D}$.

Definition.

An Op*-algebra $\mathcal{A}[\mathfrak{T}_\mathfrak{D}]$, equipped with the uniform topology $\mathfrak{T}_\mathfrak{D}$ we call $\tilde{O}^*$-algebra. If it is complete, we call it O*-algebra. A topological algebra R, which is algebraically and topologically isomorphic to an $\tilde{O}^*$-algebra resp. O*-algebra we call $A\tilde{O}^*$-algebra resp. AO*-algebra. The topology $\mathfrak{T}$ on R is then called an $\tilde{O}^*$-topology resp. O*-topology ("C*-like" topology).

The O*-algebras are generalizations of the C*-algebras and the AO*-algebras are generalizations of the B*-algebras. It is therefore important also for AO*-algebras to have an abstract characterization like the property $\|a^*a\| = \|a\|^2$ of a B*-algebra. For barrelled $*$-algebras such a property is the normality of the cone of positive elements.

Theorem. (Schmüdgen /12/)

i) In an $A\tilde{O}^*$-algebra R the cone $R_+ = \mathrm{conv}\{a^*a;\ a \in R\}$ is normal.

ii) If in a barrelled $*$-algebra R with a unity e the cone R_+ is normal, then R is an $A\tilde{O}^*$-algebra.

2.3 On a B*-algebra R the O*-topology given by the norm $\|\cdot\|$ is uniquely determined. There is not another $\tilde{O}^*$-topology on R. Quite other is the situation for general topological algebras. E.g. on the test function algebra $\mathcal{R}$ different O*-topologies exist, as we shall see. First we remark yet the following

Theorem. /13/

The direct sum topology $\mathfrak{T}$ on $\mathcal{R}$ is not an O*-topology.

With the help of the foregoing Theorem that follows from the fact that the cone $\mathcal{R}_+$ is not normal with respect to the direct sum topology .

Now we define on $\mathcal{R}$ two important topologies, which are weaker than $\mathfrak{T}$. The first topology $\mathfrak{T}_\infty$ is defined by all seminorms

$$\mathfrak{T}_\infty : \qquad \|f\|_{(\gamma_n),k} = \sum_{n \geq 0} \gamma_n \|f_n\|_k$$

where (γ_n) runs over all sequences of positive numbers and k over all nonnegative integers. $\|f_n\|_k = \sup\limits_{x;\,i,j \leq k} |(1+|x_1|)^i \ldots (1+|x_n|)^i D_x^j \ldots D_x^j f_n|$ is the system of seminorms defining the topology of the Schwartz space $\mathcal{S}_n$. In difference to the direct sum topology $\mathfrak{T}$ the degree k of the norm $\|f_n\|_k$ does not depend on the degree n of the component.

The second important topology on $\mathfrak{R}$ is $\mathscr{N}$ /10/, the strongest locally convex topology on $\mathfrak{R}$ such that the multiplication on $\mathfrak{R}$ is a jointly continuous bilinear mapping $m: \mathfrak{R}[\mathfrak{I}] \times \mathfrak{R}[\mathfrak{I}] \to \mathfrak{R}[\mathscr{N}]$.

2.4 Theorem. /14,15/

Both topologies $\mathfrak{I}_\infty$ and $\mathscr{N}$ are 0^*-topologies on $\mathfrak{R}$. It is $\mathfrak{I}_\infty \lesssim \mathscr{N} \lesssim \mathfrak{I}$. $\mathscr{N}$ is the strongest 0^*-topology on $\mathfrak{R}$ which is weaker than $\mathfrak{I}$.

The two different 0^*-topologies $\mathfrak{I}_\infty, \mathscr{N}$ on $\mathfrak{R}$ are in a remarkable relation to the existence problem of the quantum field theory. Namely, let $\mathcal{B}$ be (the set of all Wightman functionals of) the Borchers class of the free field and $W(f) = \sum_n W_n(f_n)$ a Wightman functional of $\mathcal{B}$. Then there exists a nonnegative integer k such that $|W_n(f_n)| \le \mathfrak{I}_n \| f_n \|_k$, whereby k is independent of n . This can be easy shown. Therefore any W of $\mathcal{B}$ is already continuous with respect to $\mathfrak{I}_\infty$. Since the 0^*-topology $\mathscr{N}$ is stronger than $\mathfrak{I}_\infty$, on $\mathfrak{R}$ exist positive functionals $W_{\mathscr{N}}$, which are continuous with respect to $\mathscr{N}$ but uncontinuous with respect to $\mathfrak{I}_\infty$. If such a $W_{\mathscr{N}}$ annulled the Wightman kernel L, it would be a Wightman functional of a field outside of the Borchers class. We are sure that this idea can be improved to a rigorous and short proof of the existence of fields, different from those of the Borchers class.

2.5 Let $R[\mathfrak{I}]$ be a topological $*$-algebra having a faithful weakly continuous representation (semisimple). By $\mathfrak{I}_u$ we denote the strongest $\tilde{0}^*$-topology on R weaker than $\mathfrak{I}$. A closed $*$-subalgebra R_0 of $R[\mathfrak{I}]$ is called <u>$\tilde{0}^*$-subalgebra</u> (C^*-like subalgebra) if $\mathfrak{I}_u|_{R_0} = (\mathfrak{I}|_{R_0})_u$. The subalgebra $\mathfrak{R}_0$ (sec. 2.1) of the test function algebra $\mathfrak{R}$ is not an $\tilde{0}^*$-subalgebra (see /10/, proof of Theorem 9). It is yet an open problem, wether any positive functional on an $\tilde{0}^*$-subalgebra can be extended to a positive functional on the full algebra. Also from the following Theorem one can see that there is some connection between the $\tilde{0}^*$-subalgebras and the possibility to extend positive functionals.

Let $\mathfrak{R}_0$ as in <u>2.1</u> be a $*$-subalgebra of $\mathfrak{R}$ generated by an arbitrary closed linear subspace $\mathcal{L}_1$ of $\mathcal{S}_1$, then we have (cf. /10/, Theorem 7 and /14/, proof of Theorem 5.5)

Theorem.

If $\mathcal{L}_1$ has a topological complement in $\mathcal{S}_1$ then $\mathfrak{R}_0$ is an $\tilde{0}^*$-subalgebra and every positive functional on $\mathfrak{R}_0$ has an extension in $\mathfrak{R}$. Further $\mathcal{N}_{\mathfrak{R}_0} = \mathcal{N}|_{\mathfrak{R}_0}$. The topology $\mathcal{N}_{\mathfrak{R}_0}$ on $\mathfrak{R}_0$ is quite analogous defined as $\mathcal{N}$ (see <u>2.3</u>).

References

1. Borchers, H.J. On structure of algebra of field operators. Nuovo Cim. 24 (1962) 214
2. Uhlmann, A. Über die Definition der Quantenfelder nach Wightman und Haag. Wiss.Z.Karl-Marx-Univ. Leipzig, Math.-Naturw.R.11(1962)213
3. Lassner, G. Uhlmann, A. On positive functionals on algebras of test functions for quantum fields. Comm.math.Phys. 7 (1968) 152
4. Lassner, G. Uhlmann, A. On positive functionals on the algebra of test functions. Acta Univ. Wratislaviensis 99 Vo. I B (1968)
5. Karwowski, W. Sznajder, N. On the existence of the Wightman-type functionals. Prepr. Wroclaw No 220 (1970)
6. Karwowski, W. On translation invariance of the positive functionals of the algebra of test functions. Prepr. Wroslaw No. 232 (1971)
7. Wyss,W. The field algebra and its positive linear functionals. Comm.math.Phys. 27 (1972) 223
8. Hofmann, G. Lassner, G. Existence proofs for Wightman type functionals. Prepr. Dubna E2 - 7536 (1973)
9. Karwowski, W. Euclidian fields and positive functionals. Prepr. Wroclaw No. 305 (1974)
10. Yngvason, J. On the algebra of test functions for field operators. Comm.math.Phys. 34 (1973) 315
11. Lassner, G. Topological algebras of operators. Rep.Math.Phys.3 (1972)
12. Schmüdgen, K. The order structure of topological $*$-algebras of unbounded operators $\perp$. Rep.Math.Phys. to appear
13. Lassner, G. Über die Realisierbarkeit topologischer Tensoralgebren. Math.Nachr. 62 (1974) 89
14. Lassner, G. Onthe structure of the test function algebra. Prepr. Dubna E2-5254 (1970)
15. Lassner, G. O*-topologies on the test function algebra. Publ.Dep. Math. Univ. Lyon. to appear

UNBOUNDED DERIVATIONS OF C* ALGEBRAS

Derek W. Robinson[*]
Department of Physics
University of California
Berkeley, California 94720

1. Introduction

Symmetry groups play an important role in physical theories. In quantummechanical theories of a finite number of particles symmetry groups are traditionally given by unitary group representations on Hilbert space. These representations
are usually continuous and the notion of infinitesimal generator can be introduced.
The infinitesimal generators themselves have direct physical significance; the
generator of space translations is the momentum operator, the generator of time
translations is the Hamiltonian or energy operator, and the angular momentum
operator generates rotations. The question of when an operator is an infinitesimal
generator of a unitary group often arises and in particular one often asks whether
certain operators are suitable as Hamiltonians. The answer to this kind of question
is well-known. An operator H on a Hilbert space $\mathcal{H}$ generates a strongly continuous
one-parameter group of unitary operators on $\mathcal{H}$ if, and only if, the operator is self-
adjoint. Various criteria for self-adjointness have been given in terms of
deficiency spaces, sets of analytic vectors, positivity etc., and these criteria
have played a useful role in such contexts as scattering theory and statistical
mechanics.

In theories of infinite systems it appears both useful and necessary to
interpret symmetries in a more general fashion. The basic observables of the theory
can be taken to form a C* algebra $\mathcal{O}$ and the symmetries enter as groups of automorphism of $\mathcal{O}$. If the automorphism group is continuous in a suitable sense one
can again introduce the notion of an infinitesimal generator and such generators
will be symmetric derivations of $\mathcal{O}$, i.e operators δ defined on a dense *subalgebra
$D(\delta)\subset\mathcal{O}$ with the properties

*On leave from Univ. d'Aix-Marseille II, Luminy, Marseille

1. $\delta(AB) = \delta(A)B + A\delta(B)$ $\qquad$ $A,B \in D(\delta)$.

2. $\delta(A^*) = -\delta(A)^*$ $\qquad$ $A \in D(\delta)$.

3. If $\mathcal{O}\!\ell$ contains an identity element $\mathbb{1}$ then $\mathbb{1} \in D(\delta)$ and $\delta(\mathbb{1}) = 0$.

In general δ will be an unbounded operator on $\mathcal{O}\!\ell$ with a precise physical interpretation. It is known that δ is bounded, i.e

$$\|\delta(A)\| \leq \text{constant}\|A\|, \quad A \in D(\delta),$$

if, and only if, $D(\delta) = \mathcal{O}\!\ell$; bounded derivations have been extensively studied (for
a review see, for example, [7] Chapter 4). The analysis of unbounded derivations
is at a much more embryonic stage and it is only in the last year that a significant number of results concerning such derivations have appeared (see Bratteli and
Robinson [1], Powers and Sakai [3][4], and Sakai [8]; earlier results were derived
in Robinson [5], Sinai and Helemskii[2]) Naturally one of the important questions
concerning unbounded derivations is the analogue of the Hilbert space problem
previously mentioned; under what conditions does a derivation generate a strongly
continuous one-parameter group of *automorphisms of $\mathcal{O}\!\ell$. In this talk we announce
and describe various new results which characterize infinitesimal generators [6]
and review some of the general results given in [1] [3] [4] [8]. Although these
results have not as yet had any striking application to physical theories we are
hopeful that this theory will eventually play the same useful role that the Hilbert
space theory plays.

2. <u>Infinitesimal Generators</u>

Let $\mathcal{O}\!\ell$ denote a C* algebra and

$$\tau ; A \in \mathcal{O}\!\ell \mapsto \tau_t(A) \in \mathcal{O}\!\ell, \; t \in \mathbb{R}$$

a one-parameter group of * automorphisms of the C*-algebra satisfying the strong
continuity condition

$$\lim_{t \to 0} \|\tau_t(A) - A\| = 0, \; A \in \mathcal{O}\!\ell.$$

Next define δ by

$$\delta(A) = \lim_{t \to 0} [\tau_t(A) - A]/it$$

for the set $D(\delta)$ of $A \in \mathcal{O}\!\ell$ such that the limit exists. It is easily checked that δ

is a derivation of $\mathcal{O}$, e.g. the automorphic property

$$\tau_t(AB) = \tau_t(A) \, \tau_t(B)$$

leads to the first derivation property listed in the previous section,

$$\tau_t(A)^* = \tau_t(A^*)$$

the second, and

$$\tau_t(\mathbb{1}) = \mathbb{1}$$

the third. The density of $D(\delta)$ follows from consideration of certain 'regularized' elements of $\mathcal{O}$ and is a standard part of semi-group theory.

A derivation arising in the above manner will be called the infinitesimal generator of the group τ. It is of primary interest to characterize those derivations which generate groups. This is a problem analogous to the characterization of the symmetric operators on Hilbert space which are actually self-adjoint. The following result gives a characterization simlar to the Stone-von Neumann self-adjointness criterion.

<u>Theorem 1[1]</u> <u>let δ be a derivation of a C^* algebra $\mathcal{O}$. The following conditions are equivalent</u>

<u>1. δ is the infinitesimal generator of a strongly continuous one-parameter group of $*$-automorphisms of $\mathcal{O}$</u>

<u>2. δ is closed, $R(\delta \pm i) = \mathcal{O}$, and</u>

$$\| \delta(A) + zA \| \geq |\text{Imz}| \, \| A \| \tag{1}$$

In the foregoing statement $R(\delta \pm i)$ is the range of $\delta \pm i$, i.e.

$$R(\delta \pm i) = \{B; \ B = \delta(A) \pm iA, \ A \varepsilon D(\delta)\},$$

and the assumption that δ is closed means that if $\| A_n \| \to 0$ and $\| \delta(A_n) - B \| \to 0$ then B must be identically zero.

Theorem 1 should be compared to the Stone-von Neumann criterion; a symmetric operator H on a Hilbert space $\mathcal{H}$ is the infinitesimal generator of a strongly continuous one-parameter group of unitary operators on $\mathcal{H}$ if, and only if, H is closed and $R(H \pm i) = \mathcal{H}$, where now we have

$$R(H \pm i) = \{\psi \varepsilon \mathcal{H}; \ \psi = (H \pm i)\phi, \ \phi \varepsilon D(H)\}$$

Thus the two results differ principally because of the extra lower, bound assumption.

In the Hilbert space case the symmetry of H allows one to immediately conclude that

$$\| (H \pm i) \phi \|^2 \ \geq \ \| \phi \|^2$$

This inequality together with the assumption that $R(H \pm i) = \mathcal{H}$ proves that the resolvent operators

$$R(\pm i) \ = \ \frac{1}{H \pm i}$$

are everywhere defined and have norm smaller than one. Exploitation of this fact by the Hille-Yosida theory of semigroups allows the construction of a group of unitaries with H as infinitesimal generator. In the algebraic case this estimate is not necessarily true as the following example shows.

Example (Bratteli) Let $\mathcal{A} = C([0,1])$ the C*algebra of continuous functions over the interval [0,1] and define the derivation δ by

$$\delta(f)(x) \ = \ i \frac{df}{dx}(x)$$

where $D(\delta)$ is the set of absolutely continuous functions over [0.1].

It follows that δ is closed, $R(\delta \pm i) = \mathcal{A}$ but

$$(\delta + z)(e^{izx}) \ = \ 0$$

and consequently δ is not an infinitesimal generator.

The similarity of Theorem 1 and the Stone-von Neumann theorem suggests that other theorems concerning symmetric operators on Hilbert space might lift to theorems about derivations on C* algebras. A typical example would be Nelson's theorem on analytic vectors.

Let δ be a derivation. It is natural to define an analytic (entire) element of δ as an element $A \epsilon D(\delta^n)$, n=1, 2, 3, $\cdots$. Such that the function

$$z \epsilon \mathbb{C} \ \longmapsto \ e_z(A) \ = \ \sum_{n \geq 0} \frac{z^n}{n!} \delta^n(A) \ \epsilon \ \mathcal{A}$$

exists and is analytic in some neighbourhood of the origin (is entire). An analogue of Nelson's theorem would state that δ is an infinitesimal generator if, and only if δ is closed δ possesses a dense set of analytic (entire) elements, and estimate (1) is valid. It is unclear whether this theorem is true but a weakened form of it may be established in terms of geometric elements of δ. A geometric element of δ is defined to be an element $A \varepsilon D(\delta^n)$, n=1, 2,...., such that

$$\| \delta^n (A) \|^{1/n} \leq C_A$$

where C_A is independent of n. This is equivalent to demanding that

$$e^{-\lambda |t|} \sum_{n \geq 0} \frac{|t|^n}{n!} \| \delta^n (A) \| \xrightarrow[|t|=\infty]{} 0$$

for λ sufficiently large, i.e. A is required to be an entire element of δ with a certain restriction on the growth properties of $e_z(A)$. The following is now true.

<u>Theorem 2[6]</u> Let δ be a derivation of a C* algebra $\mathcal{A}$. The following conditions are equivalent

1. δ is the infinitesimal generator of a strongly continuous one-parameter group of * automorphisms of $\mathcal{A}$.

2. δ is closed, δ possesses a dense set of geometric elements.

$$\| \delta (A) + zA \| \geq |1mz| \| A \|$$

Although this result is weaker than the analytic element conjecture it does have at least one interesting consequence.

<u>Theorem 3[6]</u> Let δ be a derivation of a C* algebra $\mathcal{A}$ and suppose that δ is the infinitesimal generator of a strongly continuous one parameter group of *-automorphisms τ of $\mathcal{A}$.

If $D \subset D(\delta)$ is a dense *-subalgebra of $\mathcal{A}$ with the property that

$$\tau_t (D) \subset D, \qquad t \varepsilon \mathbf{R},$$

then it follows that D is a core for δ, i.e. the closure $\overline{\delta]}_D$ of the restriction of δ to D satisfies

$$\overline{\delta]}_D = \delta.$$

3 Closed Derivations

One of the basic properties that a derivation must have to qualify as an infinitesimal generator is the property of being closed. A symmetric operator H on a Hilbert space $\mathcal{H}$ always has the property of being closeable (H* and H** are closed extensions of H); the analogy between symmetric operators on $\mathcal{H}$ and derivations δ on a C* algebra $\mathcal{O}$ automatically leads to the conjecture that all derivations are closeable. This conjecture is, however, false by the following result of Bratteli and Robinson [1].

Theorem 4 Let $\mathcal{O}$ be the CAR algebra and B_n an increasing sequence of $2^n \times 2^n$ full matrix algebras generating $\mathcal{O}$.

There exists a non-zero derivation δ of $\mathcal{O}$ such that

1. Every B_n is in the domain $D(\delta)$ of δ

2. δ restricted to each B_n is zero.

Hence δ is not closeable.

The existence of abelian algebras with non-closeable derivations is established as a by-product of the construction used to prove Theorem 5.

The foregoing result proves that the property of closeability of derivations is a real restriction in contrast to the situation with symmetric operators. We next consider the problem of characterizing closeable derivations. This is an algebraic problem and the first criterion for closeability is given by a functional analytic property of the domain of the derivation

Theorem 5 Let δ be a derivation of a C* algebra

If δ is such that $A^{1/2} \varepsilon D(\delta)$ whenever $0 \leq A \varepsilon D(\delta)$ then δ is closeable.

Conversely if δ is closed and $A \varepsilon D(\delta)$ is positive and invertible than $A^{1/2} \varepsilon D(\delta)$.

The first statement of the theorem is given by Powers and Sakai [4] the second statement occurs in Bratteli and Robinson [1]. In fact the latter authors develop a more detailed functional-analytic description of the domains of closed derivations. The essential point is that if δ is closed and $A = A^* \varepsilon D(\delta)$ then the resolvent $(\lambda - A)^{-1}$ is also in $D(\delta)$ whenever λ is not in the spectrum $\sigma(A)$ of A.

Note that the foregoing result is not a good characterization of closed derivations because the converse statement places the extra requirement of invertibility.

In general the domain of a derivation is not closed under the square root operation. If $\mathcal{O}\mathcal{L} = C_o(\mathbb{R})$ and δ is the infinitesimal generator of translations then

$$f(x) = |x|^{3/2} e^{-x^2}$$

is such that $f\varepsilon D(\delta)$ but $f^{1/2} \notin D(\delta)$.

It remains an open question whether a derivation δ whose domain $D(\delta)$ is invariant under the formation of resolvents, i.e. $A\varepsilon D(\delta)$ implies $(\lambda -A)^{-1}\varepsilon D(\delta)$ for $\lambda\varepsilon\sigma(A)$, is automatically closeable.

A second criterion for closeability can be given in terms of an invariance condition. Assume for the moment that δ is the infinitesimal generator of an automorphism group τ and that ω is a state over $\mathcal{O}\mathcal{L}$ which is invariant under τ, i.e.

$$\omega(\tau_t(A)) = \omega(A)$$

for all $A\varepsilon\mathcal{O}\mathcal{L}$ and $t\varepsilon\mathbb{R}$. This invariance condition is equivalent to the following condition expressed in terms of δ

$$\omega(\delta(A)) = 0$$

for all $A\varepsilon D(\delta)$. The following result considers derivations with faithful invariant states.

Theorem 6 Let δ be a derivation of a C* algebra $\mathcal{O}\mathcal{L}$.

Assume that $\mathcal{O}\mathcal{L}$ possesses a state ω which generates a faithful cyclic representation $(\mathcal{H}_\omega, \pi_\omega, \Omega_\omega)$ and also satisfies the invariance condition

$$\omega(\delta(A)) = 0$$

for all $A\varepsilon D(\delta)$.

It follows that

1. δ is closeable

2. There exists a symmetric operator H_δ on $\mathcal{H}_\omega$ such that

$$D(H_\delta) = \{\psi;\ \psi=\pi_\omega(A)\Omega_\omega,\ A\varepsilon D(\delta)\}$$

$$\pi_\omega(\delta(A))\psi = [H_\delta,\ \pi_\omega(A)]\psi$$

for all $A\varepsilon D(\delta)$ and $\psi\varepsilon D(H_\delta)$.

The theorem as stated occurs in Bratteli and Robinson [1]; Powers and Sakai [4] give a special version of the theorem for UHF algebras.

It remains unclear whether Theorem 6 has a converse. Is it true that for each closed derivation of a C*-algebra $\mathfrak{A}$ ($\mathfrak{A} \ni \mathbb{1}$) there exists a state ω such that

$$\omega(\delta(A)) = 0$$

for all $A \in D(\delta)$? This result has been established in [1] for special algebras, C* algebras acting on a Hilbert space $\mathcal{H}$ and containing the C* algebra $\mathcal{LC}(\mathcal{H})$ of compact operators on $\mathcal{H}$ as subalgebra. It is also true if $\mathfrak{A} \ni \mathbb{1}$ and δ is an infinitesimal generator by a simple compactness and fixed point argument.

ACKNOWLEDGEMENTS The author's interest in the foregoing subject developed through collaboration with O. Bratteli. Part of the material of Section 2 was developed while the author was a guest of the Physics Dept. and the Maths. Dept. at the University of California, Berkeley. I am indebted to E. Wichmann and O.E. Lanford for their hospitality during this period.
Research support inpart by the National Science Foundation, under Grant # GP-42249X.

References

1. Bratteli, O. and D.W. Robinson; Marseille Preprint (1974).

2. Helemski, A. and Ya. Sinai; Functional Anal. Appl. **6** 343 (1973).

3. Powers, R. and S. Sakai; Univ. of Pennsylvania Preprint (1974)
        (to be published Commun. Math. Phys.)

4. Powers, R. and S. Sakai; Univ. of Pennsylvania Preprint (1974)
        (to be published in Jour. Funct. An.)

5. Robinson, D.W.; Commun. Math. Phys. **7** 337 (1968).

6. Robinson, D.W.; seminar, Univ. of California, Berkeley (1975)

7. Sakai, S.; C* algebras and W* algebras, Springer-Verlag, Berlin, (1970).

8. Sakai, S.; Univ. of Pennsylvania Preprint (1974), (to be published in Amer.
        Jour. Math.)

Discussion

Doplicher (Comment): It seems that the difference between the Hilbert space situation and the derivation situation is related to the different definition of adjoints: the adjoint of a Hilbert space linear operator can be also made to correspond to the transpose of a linear operator between Banach spaces; if the transpose is densely defined the initial operator is closable. For the sake of * automorphism groups you use rather the "skew adjointness" $\delta(A*)* = -\delta(A)$, which as you say does not force δ to be closable.

A METHOD FOR MAKING RELATIVISTIC NON-QUANTUM SYSTEMS BY CONSTRUCTING
THE HAMILTONIAN AND THE OTHER NINE GENERATING FUNCTIONS

Richard Arens
Department of Mathematics
University of California
Los Angeles, California 90024

For the construction of a relativistic completely Hamiltonian dynamical system
("RCHDS"), the ten generating functions $(H, P_i$, etc.) must satisfy commutation relations non-trivial even when invariance is not demanded [1, 5.2.3]. (In classical
mechanics, the space-time group is one-dimensional, and commutators do not arise.
Hence such systems can be obtained by selecting a Hamiltonian almost arbitrarily.)

We show here that an RCHDS can be obtained by selecting ten functions subject
only to mild restrictions not involving explicitly the structure constants of the Lie
algebra, and then solving algebraic equations for the actual generating functions. If
the latter should turn out not to involve the group parameters, then the system will
be invariant.

Mathematically, the problem is as follows. Let $T_1(N)$ be the cotangent bundle
for a manifold N. Let G be the space-time group, and let Q be the proposed configuration space. We want to obtain a symplectic realization of G in $T_1(G \times Q)$
such that for functions depending only on the coordinates in $T_1(G)$, the action is
the standard one of G in $T_1(G)$. To explain our method of doing so, we must introduce notation.

Let N have dimension n, and let $c^1, \ldots, c^n$ be covector fields linearly independent at each point (cf. the classical dq^i). Their conjugate momenta are functions $\zeta_1, \ldots, \zeta_n$ on $T_1(M)$ which make the sum $\zeta_i c^i$ equal to the canonic 1-form
(cf. $p_i dq^i$). Let $Z_1, \ldots, Z_n$ be the vector fields on N dual to the forms $c^1, \ldots, c^n$:
$\langle Z_i, c^j \rangle = \delta_i^j$. Then the Poisson bracket of two functions φ and ψ on $T_1(M)$ has
the form

$$\{\varphi, \psi\} = \frac{\partial \varphi}{\partial \zeta_k} Z_k \psi - \frac{\partial \psi}{\partial \zeta_k} Z_k \varphi + \rho^k_{ij} \zeta_k \frac{\partial \varphi}{\partial \zeta_i} \frac{\partial \psi}{\partial \zeta_j}$$

where $\rho^k_{ij} Z_k = [Z_i, Z_j]$ (and $dc^k = -\frac{1}{2} \rho^k_{ij} c^i \wedge c^j$). This is independent of linear
transformation of the ζ's. When $N = G$, our ζ's shall be the right-invariant
Maurer-Cartan forms and denoted by η's. The Z's are thus the right-invariant
vector fields $Y_1, \ldots, Y_g$ and the ρ's are the structure constants r^k_{ij}. When $N = Q$,
it suffices to treat only the case where the vector fields $U_1, \ldots, U_f$ on Q can be

chosen to commute and hence those ρ's are 0 (as classically). Finally, for $N = G \times Q$ we let $Z_1, \ldots, Z_n$ be $Y_1, \ldots, Y_g$, $U_1, \ldots, U_f$ and define the ζ's accordingly.

Let $\mathcal{D}$ be a submanifold of $T_1(G \times Q)$ with the property: at each point P of $\mathcal{D}$ there are coordinates $\varphi^1, \ldots, \varphi^{2n}$ in $T_1(G \times Q)$ such that $\mathcal{D}$ is defined by $\varphi^1 = \cdots = \varphi^g = 0$, and $\{\varphi^a, \varphi^b\} = 0$ for $a, b \leq g$ on $\mathcal{D}$. Such a $\mathcal{D}$ we call a regular Lie submanifold (RLSM). Such a $\mathcal{D}$ shall be called faithful if the projection of $\mathcal{D}$ on the factor manifold $G \times T_1(Q)$ is $1:1$ and regular. A realization (or action) of G on such a faithful RLSM produces a realization of G in $G \times T_1(M)$, i.e. a (possibly non-invariant) dynamical system with configuration space Q. The main theorem is that faithful RLSM does define an infinitesimal realization of G on $\mathcal{D}$, and that the dynamics is completely Hamiltonian.

By 'infinitesimal' we mean that the differential equations for the motions can be consistently set up, but (as in classical mechanics) there is no promise of more than local solutions. It means exactly that there is a realization of the Lie algebra of G by vector fields in $\mathcal{D}$ and on $G \times T_1(Q)$ and that the dynamorphisms Δ_x^y in $T_1(Q)$ are symplectic. It means exactly that $[1, 5.2.3]$ hold.

The problem is thus reduced to finding an effortless way (i.e. not requiring the observance of commutation relations or structure constants in G) of making faithful RLSM. This can often be done as follows. Reduce the canonical 1 form C to normal form $\pi_1 d\psi^1 + \cdots + \pi_n d\psi^n$ $(n = g + f)$. Here each coordinate commutes with all but one of the others. Select a set of g commuting ones as $\varphi^1, \ldots, \varphi^g$. With a little care, $\varphi^1 = \cdots = \varphi^g = 0$ will be smoothly solvable for the group momenta $\eta_1, \ldots, \eta_g$.

We now prove some of these statements, beginning with the Poisson bracket formula. Choose any coordinates $x^1, \ldots, x^n$ in Q and define the p_i in the usual way, so $p_i dx^i = C = C^j \zeta_j$. Contracting with Z_k gives $\zeta_k = p_i Z_k(x^i) = p_i Z_k^i$. The standard formula gives $\{x^i, \zeta_j\} = Z_j^i$ and

$$\{\zeta_i, \zeta_j\} = Z_i^m p_k \frac{\partial Z_j^k}{\partial x^m} - Z_j^m p_k \frac{\partial Z_i^k}{\partial x^m} = p_k [Z_i, Z_j]^k = p_k (\rho_{ij}^m Z_m)^k = \rho_{ij}^m \zeta_m.$$

The proposed formula also gives these results. Since $x^1, \ldots, x^n$, $\zeta_1, \ldots, \zeta_n$ form coordinates in N, it follows that the proposed $\{\varphi, \psi\}$ is equal to the standard one for all φ and ψ.

We require a theorem incorrectly stated by Whittaker $[2, p. 322]$ and attributed to S. Lie. Let $\underline{u^1, \ldots, u^g}$ <u>be an independent set of functions on a symplectic manifold N. Let $\mathcal{D}$ be the set of common zeros, and let $\mathcal{E}$ be the set of common zeros of the $\{u^i, u^j\}$. If now φ, ψ vanish on $\mathcal{D}$, then $\{\varphi, \psi\}$ vanishes on $\mathcal{D} \cap \mathcal{E}$.</u> For a proof let P belong to $\mathcal{D} \cap \mathcal{E}$ and extend the u^i to a coordinate system $u^1, \ldots, u^n$ at P. Surely $\{\varphi, \psi\} = \Sigma_{i,j=1,\ldots,n} \varphi_i \psi_j \{u^i, u^j\}$ where $\varphi_i = \partial \varphi / \partial u^i$, $\psi_j = \partial \psi / \partial u^j$. On $\mathcal{D}$, each φ_i, ψ_i is 0 for $i > g$ and the remaining $\{u^i, u^j\}$ are

0 on $\mathcal{E}$.

Now consider manifolds G and Q of dimensions g and f respectively and let $Y_1,\ldots,Y_g$ be a set of vector fields on G, linearly independent at each point, and let $U_1,\ldots,U_f$ be similar vector fields for Q. Let $A^1,\ldots,A^g$ be the forms dual to $Y_1,\ldots,Y_g$; and $B^1,\ldots,B^f$ - those dual to the U's. Let the canonic forms be $\Sigma_{a=1}^{g}\,\eta_a\,A^a$ and $\Sigma_{j=1}^{f}\,\theta_j B^j$ for G and Q. Then $\eta_1,\ldots,\eta_g$, $\theta_1,\ldots,\theta_f$ are the momenta conjugate to $A^1,\ldots,A^g$, $B^1,\ldots,B^f$ in $T_1(G\times Q)$.

Now let $\mathcal{N}$ be a faithful RLSM. Then evidently $\mathcal{N}$ can be described as the set of common zeros of functions $\eta_a + H_a$ where the H_a are defined on $G\times T_1(Q)$. By the Lie theorem, $\{\eta_a + H_a,\ \eta_b + H_b\} = 0$ on $\mathcal{N}$, and so does $\{\eta_a + H_a,\ \eta_b + H_b\} - r_{ab}^c(\eta_c + H_b)$ where these functions r_{ab}^c come from $[Y_a,Y_b] = r_{ab}^c Y_c$. We may demand $[U_i,U_j] = 0$, since a local treatment in Q suffices for our purposes. Therefore $\rho_{ab}^k = 0$ for $k > 0$, and $\rho_{ab}^c = r_{ab}^c$ for $c \leq g$. Thus from the $\{\varphi,\psi\}$-formula, $\{\eta_a,\eta_b\} = r_{ab}^c \eta_c$ and $\{\eta_a,H\} = Y_a H$ for a function not depending on the η's. Therefore $\{H_a,H_b\} + Y_a H_b - Y_b H_a - r_{ab}^c H_c = 0$ on $\mathcal{N}$. This is equivalent to $[1, 5.2.3]$ because for functions not depending on the η's, $\{H_a,H_b\}$ is their Poisson bracket <u>as defined on</u> $T_1(Q)$.

References

[1] Richard Arens, Hamiltonian formalism for non-invariant dynamics, Commun. Math. Phys. 34, 91-110 (1973).

[2] E. T. Whittaker, Analytical Dynamics, Cambridge University Press (1937).

AN ALGEBRAIC APPROACH TO THE THEORY OF K-FLOWS AND K-ENTROPY

Gérard G. Emch
Departments of Mathematics and of Physics
The University of Rochester
Rochester, N. Y. 14627
USA

An algebraic approach is proposed to the theory of Kolmogorov and Sinai. It includes as particular cases both the classical theory and a generalization to non-commutative dynamical flows. The connection between the latter case and the statistical mechanics of a quantum diffusion process is presented.

In this paper a general, conservative "dynamical flow" is understood to be a triple $\{\mathfrak{N}, \varphi, \alpha\}$ formed by a von Neumann algebra $\mathfrak{N}$ acting on a separable Hilbert space $\mathfrak{H}$; a faithful normal state φ on $\mathfrak{N}$; and a weakly-continuous map $\alpha: R \to \mathrm{Aut}(\mathfrak{N}, \varphi)$. Without loss of generality we can assume that there exists a vector Φ in $\mathfrak{H}$, of norm 1, cyclic and separating for $\mathfrak{N}$, such that $\langle \varphi; N \rangle = (N\Phi, \Phi)$ for all N in $\mathfrak{N}$; and that there exists a one-parameter, continuous group of unitary operators on $\mathfrak{H}$ such that for all t in R and all N in $\mathfrak{N}$;

$$\alpha(t)[N] = U(t) \, N \, U(-t) \quad \text{and} \quad U(t)\Phi = \Phi.$$

If we were to restrict our attention to the particular case where $\mathfrak{N}$ is abelian (and hence maximal abelian) our scheme would reduce to the Koopman-formalism description of a classical flow $\{\Omega, \mu, T\}$; $\mathfrak{N}$ is the natural representation of $\mathcal{L}^{\infty}(\Omega, \mu)$ as a von Neumann algebra of operators acting on $\mathfrak{H} = \mathcal{L}^{2}(\Omega, \mu)$; Φ is the vector $\Phi(\omega) = 1$; and $\alpha(t)[N](\omega) = N(T(t)[\omega])$.

In the general theory, as in the classical one, the definition and interpretation of the K-entropy depend on the notion of conditional entropy: the latter is used to obtain a measure of the information gained in performing a filtering measurement repeatedly in the course of time. Since we want to concentrate on the stochastic effects of the time-evolution itself rather than the randomness possibly introduced by the measuring process, we consider as "admissible" a partition $\mathfrak{F}$ the measurement of which does not perturb φ, i.e. a partition of the identity in mutually orthogonal projectors F_i in $\mathfrak{N}$, which satisfies: $\varphi = \Sigma_i \lambda_i \varphi_i$ with $\lambda_i = \langle \varphi ; F_i \rangle$ and $\langle \varphi_i ; \cdot \rangle = \langle \varphi ; F_i \rangle^{-1} \langle \varphi ; F_i \cdot F_i \rangle$.

In the general case it can be shown that a partition $\mathfrak{F}$ in $\mathfrak{N}$ is admissible if and only if it is contained in the centralizer $\mathfrak{N}_\varphi$ of $\mathfrak{N}$, the latter being identical with the algebra of fixed points of $\mathfrak{N}$ under the action of $\alpha_\varphi : \mathbb{R} \to \mathrm{Aut}(\mathfrak{N}, \varphi)$, the modular group of automorphisms of $\mathfrak{N}$ associated to φ. Notice that the "true" time-evolution is given here by $\alpha(\mathbb{R})$, not $\alpha_\varphi(\mathbb{R})$. In the classical theory $\alpha_\varphi(t) = \mathrm{id}$ for all t in $\mathbb{R}$; hence $\mathfrak{N}_\varphi = \mathfrak{N}$, and every partition of the identity in $\mathfrak{N}$ is admissible. Notice, in the general case, that $\mathfrak{F}$ is admissible if and only if $\alpha(t)[\mathfrak{F}]$ is admissible for every t in $\mathbb{R}$; hence the von Neumann algebra $C_n[\mathfrak{F}]$ generated by $\{\alpha(-1)[\mathfrak{F}], \alpha(-2)[\mathfrak{F}], \ldots, \alpha(-n)[\mathfrak{F}]\}$ with $\mathfrak{F}$ admissible, is contained in $\mathfrak{N}_\varphi$.

For every von Neumann subalgebra C of $\mathfrak{N}_\varphi$, the conditional expectation $\mathcal{E}_\varphi(\cdot | C) : \mathfrak{N} \to C$ w.r.t. φ exists, and we can therefore write $\hat{H}_\varphi(\mathfrak{F} | C) = \inf \{H_\varphi(\mathfrak{F} | \mathfrak{B}) | \mathfrak{B} \subseteq C; \mathfrak{B} \text{ abelian}\}$ where $H_\varphi(\mathfrak{F} | \mathfrak{B}) = \Sigma_i \langle \varphi ; h[\mathcal{E}_\varphi(F_i | \mathfrak{B})] \rangle$ with $h : x \in [0,1] \to -x \log x$. With $\hat{\mathfrak{F}}$ denoting the family of all admissible partitions, and $\hat{C}$ the family of all von Neumann subalgebras of $\mathfrak{N}_\varphi$, the map $\hat{H}_\varphi : (\mathfrak{F}, C) \in \hat{\mathfrak{F}} \times \hat{C} \to \hat{H}_\varphi(\mathfrak{F} | C) \in \mathbb{R}^+$ satisfies the following properties: (i) $\hat{H}_\varphi(\mathfrak{F} | C) = 0$ if and only if $\mathfrak{F} \subset C$; (ii) $\hat{H}_\varphi$ is increasing (resp. decreasing) in its first (resp. second) argument. We therefore call $\hat{H}_\varphi(\mathfrak{F} | C)$ the entropy of $\mathfrak{F}$, conditioned by C, w.r.t. φ. We then can define the K-entropy $\hat{H}_\varphi(\alpha)$ of the general dynamical flow $\{\mathfrak{N}, \varphi, \alpha\}$ as $\hat{H}_\varphi(\alpha) = \sup \{\hat{H}_\varphi(\mathfrak{F} | \alpha) | \mathfrak{F} \in \hat{\mathfrak{F}}\}$ where $\hat{H}_\varphi(\mathfrak{F} | \alpha) = \lim_{n \to \infty} \hat{H}_\varphi(\mathfrak{F} | C_n[\mathfrak{F}])$. Clearly, $\hat{H}_\varphi(\alpha)$ reduces to the Kolmogorov-Sinai dynamical entropy in the classical case where $\mathfrak{N}$ is abelian.

We now say that a dynamical flow $\{\mathfrak{N}, \varphi, \alpha\}$ is a "generalized K-flow" if there exists a von Neumann subalgebra G of $\mathfrak{N}$ with the following properties: (i) $\alpha(t)[G] \supseteq G$ for all t in $\mathbb{R}^+$; (ii) $\bigvee_{t \in \mathbb{R}} \alpha(t)[G] = \mathfrak{N}$; (iii) $\bigcap_{t \in \mathbb{R}} [\alpha(t)[G] \Phi] = \mathbb{C}\Phi$; (iv) G is stable under $\alpha_\varphi(\mathbb{R})$; (v) every maximal abelian von Neumann subalgebra $\mathfrak{X}$ of $\mathfrak{N}_\varphi$ is also maximal abelian in $\mathfrak{N}$.

This generalizes to $\mathfrak{N}$ non-abelian the concept of K-flows studied by Kolmogorov and Sinai. In particular, this generalization carries over the fact that the K-entropy defined above is strictly positive on generalized K-flows.

Any general dynamical system satisfying the above conditions (i)-(iv), and in addition the conditions: (v,a) φ is not a trace on $\mathfrak{N}$; and (v,b) for all t in R and all Z in $\mathfrak{N}\cap\mathfrak{N}'$, $\alpha(t)[Z] = Z$; satisfies then also (v) and is thus a K-flow, which we term "special non-abelian K-flow". These particular K-flows have a very special structure indeed. In addition of being ergodic, mixing and have Lebesgue spectrum with countably infinite multiplicity, they are such that φ is homogeneous and periodic in the sense of Takesaki (which is how (v) is proven); and moreover $\mathfrak{N}$ is a type III factor. Furthermore, the restriction of the flow to $\mathfrak{N}_\varphi$ gives again a generalized K-flow, where the algebra is now a type II_1 factor.

Special non-abelian K-flows happen to appear naturally in the statistical mechanics of certain quantum diffusion processes. Specifically, let $\{\mathfrak{N}_o,\varphi_o,\gamma_o\}$ be the triple constituted by: the von Neumann algebra generated by the representation of the one-degree-of-freedom CCR associated to the canonical equilibrium state φ_o, at temperature β, for the harmonic oscillator $H_o = (p^2+\omega^2 q^2)/2$; and γ_o is defined, for all t in $\mathbb{R}^+$ as the linear, continuous extension to $\mathfrak{N}_o$ of

$$\gamma_o(t)[W_o(z)] = W_o(e^{-\lambda t}z)\,\exp\{-\Theta|z|^2(1-e^{-2\lambda L})/4\}\text{with } \Theta = \coth(\beta\omega/2)$$

It is easy to check that for every z_o in $\mathbb{C}$, $\gamma_o(\mathbb{R}^+)$ restricted to the von Neumann algebra $W_o(\mathbb{R}z_o)''$ is the solution of a classical diffusion equation in the appropriate quadratic potential. In particular, for any normal state ψ_o on $\mathfrak{N}_o$: $\lim_{t\to\infty}\gamma_o(t)^*\psi_o = \varphi_o$.

One can then show that there exist: a von Neumann algebra $\mathfrak{N}_1$; a faithful normal state φ_1 on $\mathfrak{N}_1$; and a map $\alpha : \mathbb{R} \to \mathrm{Aut}(\mathfrak{N}=\mathfrak{N}_o\otimes\mathfrak{N}_1,\varphi = \varphi_o\otimes\varphi_1)$ such that $\mathcal{E}_o\alpha(t)\mathcal{E}_o = \gamma_o(|t|)\mathcal{E}_o$ for all t in $\mathbb{R}$, where $\mathcal{E}_o$ is the conditional expectation from $\mathfrak{N}$ onto $\mathfrak{N}_o$ w.r.t. φ; we thus have for every normal state ψ_o on $\mathfrak{N}_o$, every t in $\mathbb{R}^+$, and every N_o in $\mathfrak{N}_o$: $\langle\psi_o\otimes\varphi_1;\alpha(t)[N_o\otimes I]\rangle = \langle\psi_o;\gamma(t)[N_o]\rangle$. Hence $\{\mathfrak{N}_1,\varphi_1\}$ can be interpreted as a thermal bath for the quantum diffusion system $\{\mathfrak{N}_o,\varphi_o,\gamma_o\}$, with $\alpha(\mathbb{R})$ describing the interaction responsible for the approach to equilibrium in $\{\mathfrak{N}_o,\varphi_o,\gamma_o\}$. Actually $\{\mathfrak{N}=\mathfrak{N}_o\otimes\mathfrak{N}_1, \varphi=\varphi_o\otimes\varphi_1,\alpha\}$ can be minimally realized as the structure of the Ford-Kac-Mazur model of an infinite chain of coupled quantum harmonic oscillators in the van Hove limit.

In connection with the general structures proposed in this paper, it must be noticed that the dynamical system $\{\mathfrak{N}, \varphi, \alpha\}$ just described is a special non-abelian K-flow.

Bibliography

Positivity of the K-Entropy on Non-Abelian K-flows,
    Z. Wahrscheinlichkeitstheorie verw. Gebiete 29 (1974), 241-252.
Non-Abelian Special K-Flows,
    Journ. Funct. Analysis 18 (1975), April issue.
The Minimal K-Flow Associated To A Quantum Diffusion Process,
    in: Physical Reality and Mathematical Description, C. P. Enz
    and J. Mehra Eds., D. Reidel Publ., Dordrecht-Holland (1974),
    477-493.

SYMMETRY RESTORATION AT FINITE TEMPERATURE

R. Jackiw

Laboratory for Nuclear Science and Department of Physics
Massachusetts Institute of Technology
Cambridge, Massachusetts 02139, U.S.A.

Some field theories, possessing a symmetry, admit only solutions which do not respect the symmetry. This is the familiar Goldstone-Nambu phenomenon. However, when the theory is examined in an external environment, the hidden symmetry may be restored. We show how spontaneously broken global and local symmetries are restored in a temperature environment. The methods for studying this question are explained, and the critical temperature for symmetry restoration is computed in several models.

The formalism which we shall use is that of the generalized effective potential $V(\phi,G)$. That object, an extension of the familiar effective potential $V(\phi)$, is defined as follows, for a theory described by a Lagrangian $\mathcal{L}(\Phi)$[1]. Consider the vacuum persistence amplitude $Z(J,K)$ in the presence of space-time varying sources $J(x)$ and $K(x,y)$; $J(x)$ couples to $\Phi(x)$ and $K(x,y)$ to $\frac{1}{2}\Phi(x)\Phi(y)$. The generalized effective action $\Gamma(\phi,G)$ is defined by a double Legendre transform of $-i\ln Z(J,K) = W(J,K)$.

$$\frac{\delta W(J,K)}{\delta J(x)} = \phi(x), \quad \frac{\delta W(J,K)}{\delta K(x,y)} = \frac{1}{2}[\phi(x)\phi(y) + G(x,y)] \tag{1}$$

$$\Gamma(\phi,G) = W(J,K) - \int dx\,\phi(x)J(x)$$
$$-\frac{1}{2}\int dxdy[\phi(x)\phi(y) + G(x,y)]K(x,y) \tag{2}$$

In the physical theory the sources are absent, hence $\Gamma(\phi,G)$ satisfies

$$\frac{\delta\Gamma(\phi,G)}{\delta\phi(x)} = -J(x) - \int dy K(x,y)\phi(y) = 0$$

$$\frac{\delta\Gamma(\phi,G)}{\delta G(x,y)} = -\frac{1}{2}K(x,y) = 0 \tag{3}$$

It is clear from (1) that $\phi(x)$ is the expectation of the quantum field and $G(x,y)$ is the propagator of the theory. However translation invariance, which we do not expect to be spontaneously broken, implies that the field expectation is constant: $\phi(x) = \phi$, and that the propagator depends only on the coordinate difference: $G(x,y) = G(x-y)$. The generalized effective potential $V(\phi,G)$ is defined from $\Gamma(\phi,G)$ by using these translation invariant forms.

$$\Gamma(\phi,G)\bigg|_{\substack{\text{translation} \\ \text{invariant}}} = -V(\phi,G) \int d^4x \tag{4}$$

It is a function of ϕ and a functional of $G(x-y)$; for later convenience G is expressed in the momentum representation: $G(x-y) = \int \frac{d^4k}{(2\pi)^4} e^{-ik(x-y)} G(k)$. The stability equations (3) now become

$$\frac{\partial V(\phi,G)}{\partial \phi} = 0$$

$$\frac{\delta V(\phi,G)}{\delta G(k)} = 0 \tag{5}$$

For a definite example we chose a theory described by

$$\mathcal{L}(\Phi) = \tfrac{1}{2}\partial_\mu \Phi \partial^\mu \Phi - U(\Phi)$$

$$U(\Phi) = \frac{\lambda_o}{4!}\left(\frac{6\mu_o^2}{\lambda_o} - \Phi^2\right)^2 \tag{6}$$

and compute $V(\phi,G)$ in the Hartree-Fock approximation.[1]

$$V(\phi,G) = U(\phi) + \frac{1}{2}\int \frac{d^4k}{(2\pi)^4}\ln G(k)$$
$$-\frac{i}{2}\int \frac{d^4k}{(2\pi)^4}G(k)\left[k^2 + \mu_o^2 - \frac{\lambda_o\phi^2}{2}\right] + \frac{\lambda_o}{8}\left[\int \frac{d^4k}{(2\pi)^4} G(k)\right]^2 \tag{7}$$

The first term on the right hand side of (7) is the no-loop tree approximation, the next two terms represent the one-loop quantum correction, the last term arises from a two-loop graph. From (5), we find

$$\left(-\mu_o^2 + \frac{\lambda_o}{2}\int \frac{d^4k}{(2\pi)^4} G(k) + \frac{\lambda_o\phi^2}{6}\right)\phi = 0 \tag{8a}$$

$$iG^{-1}(k) = k^2 + \mu_o^2 - \frac{\lambda_o}{2} \int \frac{d^4k}{(2\pi)^4} G(k) - \frac{\lambda_o \phi^2}{2} \qquad (8b)$$

Upon defining the renormalized mass parameter

$$-m^2 = -\mu_o^2 + \frac{\lambda_o}{2} \int \frac{d^4k}{(2\pi)^4} G(k)$$

the above equation becomes

$$(-m^2 + \frac{\lambda_o \phi^2}{6}) \phi = 0$$

$$iG^{-1}(k) = k^2 + m^2 - \frac{\lambda_o \phi^2}{2} \qquad (9)$$

and the only consistent solution is the one which breakes spontaneously the $\Phi \leftrightarrow -\Phi$ symmetry of (6).

$$\phi^2 = \frac{6m^2}{\lambda_o}$$

$$iG^{-1}(k) = k^2 - 2m^2 \qquad (10)$$

[The solution $\phi = 0$ is unacceptable, since it leads to a propagator with imaginary mass.] Similar calculations can be performed in an O(N) invariant Φ^4 theory, [for which the Hartree-Fock approximation dominates when N is large] in different dimensions.[2] The behavior of spontaneous symmetry breaking with varying dimensionality is thereby exposed: a continuous symmetry can be broken only when the dimension of spacetime is greater than 2.

Another interesting question is whether a spontaneously broken symmetry can be restored at finite temperature. Qualitative arguments indicating that this should happen were given by Kirzhnits and Linde,[3] and subsequent detailed computations have established the phenomenon.[4] This topic can be readily analyzed by a straight forward extension of the formalism.

A field theory at a finite temperature, proportional to $1/\beta$, is conveniently described by its finite-temperature Green's functions, which are defined by a statistical average: $\text{tr } e^{-\beta H} T\Phi(x_1)\ldots\Phi(x_n)/\text{tr } e^{-\beta H}$. These Green's functions satisfy the same differential equations as the corresponding ones at zero temperature. However, the boundary conditions are different: in the complex time interval $[0,-i\beta]$ there are periodicity requirements. It follows that in a theory with interactions, the formulas for temperature Green's functions in terms of the elementary free-field propagator and vertices are the same as at zero temperature, except that the free-field propagators have a "momentum" representation consistent with the periodicity conditions.

$$D_\beta(x) = \int_k e^{-ikx} \frac{i}{k^2-m^2} \tag{11}$$

where $\int_k$ stands for $\frac{1}{-i\beta} \sum_n \int \frac{d^3k}{(2\pi)^3}$, $n = 0, \pm1\ldots$; $k^2 = k_o^2 - \underset{\sim}{k}^2$; and k_o equals $\frac{2\pi n}{-i\beta}$. Consequently symmetry behavior at finite temperature can be studied in terms of solution to eqs. (8) which are also valid at finite temperature.[5]

We are thus led to the system of equations

$$[-\mu_o^2 + \frac{\lambda_o}{2} \int_k G_\beta(k) + \frac{\lambda_o \phi_\beta^2}{6}] \, \phi_\beta = 0 \tag{12a}$$

$$iG_\beta^{-1}(k) = k^2 + \mu_o^2 - \frac{\lambda_o}{2} \int_k G_\beta(k) - \frac{\lambda_o \phi_\beta^2}{2} \tag{12b}$$

$$\phi_\beta = tr \, e^{-\beta H} \, \Phi(x)/tr \, e^{-\beta H}$$

$$\phi_\beta^2 + \int_k e^{-ik(x-y)} G_\beta(k) = tr \, e^{-\beta H} T\Phi(x)\Phi(y)/tr \, e^{-\beta H}$$

The symmetric solution $\phi_\beta = 0$ will be acceptable if the mass parameter in the propagator is positive.

Eq. (12b) is solved by $G_\beta^{-1}(k) = i[-k^2 + m_\beta^2]$ where the temperature dependent mass m_β satisfies [at $\phi_\beta = 0$] the gap equation.

$$m_\beta^2 = -\mu_o^2 + \frac{\lambda_o}{2} \int_k \frac{i}{k^2-m_\beta^2}$$

$$= -\mu_o^2 + \frac{\lambda_o}{2\beta} \sum_n \int \frac{d\underset{\sim}{k}}{(2\pi)^3} \frac{1}{\frac{(2\pi n)^2}{\beta^2} + \underset{\sim}{k}^2 + m_\beta^2}$$

$$= -\mu_o^2 + \frac{\lambda_o}{2} \int \frac{d\underset{\sim}{k}}{(2\pi)^3} \; \frac{1}{2\sqrt{\underset{\sim}{k}^2 + m_\beta^2}} + \frac{\lambda_o}{\beta^2} \; f(m_\beta^2 \beta^2) \tag{13a}$$

$$f(a^2) = \frac{1}{4\pi^2} \int_o^\infty \frac{x^2 dx}{\sqrt{x^2 + a^2}} \; [e^{\sqrt{x^2 + a^2}} - 1]^{-1} \tag{13b}$$

The integral occurring in (13a) is divergent; the gap equation must be renormalized. It is important that renormalization does not involve any counterterms beyond those of the zero temperature theory. To carry out the renormalization we evaluate the integral in (13a) with a cutoff, and rewrite that expression.

$$m_\beta^2 = -m^2 + \frac{\lambda}{\beta^2} \; f(m_\beta^2 \beta^2) + \frac{\lambda}{32\pi^2} \; m_\beta^2 \ln \frac{m_\beta^2}{m^2} \tag{13c}$$

Here m^2 is the renormalized mass parameter, and λ is the renormalized coupling constant; they are given in terms of the bare parameters and cutoff Λ.

$$-m^2 = -\mu_o^2 + \frac{\lambda_o}{32\pi^2} \; [\frac{\Lambda^2}{2} + m^2 \ln \frac{\Lambda^2}{m^2} - m^2]$$

$$\frac{1}{\lambda} = \frac{1}{\lambda_o} + \frac{1}{32\pi^2} [\ln \frac{\Lambda^2}{m^2} - 1]$$

Eq. (13c) is the renormalized gap equation for m_β^2 and we seek solutions for positive m_β^2. At low temperature $[\beta \to \infty]$ there is symmetry breaking and no positive solution exists. At high temperature $[\beta \to 0]$ one can satisfy (13c) with $m_\beta^2 > 0$. The critical temperature β_c^{-1} is that value of the temperature for which m_β^2 vanishes. Therefore from (13b) and (13c) we have

$$0 = -m^2 + \frac{\lambda}{\beta_c^2} \; f(0)$$

$$\frac{1}{\beta_c^2} = \frac{24m^2}{\lambda} \tag{14}$$

For $\beta^{-1} > \beta_c^{-1}$ symmetry is manifest, for $\beta^{-1} < \beta_c^{-1}$ the symmetry is broken.

In this calculation we have ignored higher loop corrections, which are proportional to higher powers of the coupling constant. Hence the entire approach is correct only if the coupling is small, in which case β_c^{-1}, given by (14), is indeed large. One can estimate that for small λ, the corrections are $0(\lambda^{1/2})$ i.e. $\beta_c^{-2} = \frac{24m^2}{\lambda}[1+0(\lambda^{1/2})]$. Eq. (13c) can be solved for m_β when $\beta \approx \beta_c$.

$$m_\beta = \frac{2\pi}{3}(\beta^{-1}-\beta_c^{-1}) \tag{15}$$

The computation may be extended to an $O(N)$ invariant Φ^4. [The Hartree-Fock approximation then dominates for large N.] Once again (14) and (15) are found, except that λ is replaced by $\lambda \frac{(N+2)}{3}$. Also if gauge fields are included, so that the theory is locally $O(N)$ invariant, one gets

$$\frac{1}{\beta_c^2} = \frac{m^2}{\lambda\frac{(N+2)}{72} + e^2 \frac{(N-1)}{4}} \tag{16}$$

where e is the gauge coupling strength.[6]

What is the significance of phase transitions in field theory which restore a spontaneously broken symmetry? One answer addresses itself to questions of principle. It may be thought that a theory with a hidden, spontaneously broken symmetry is equivalent to a theory without any symmetry at all, and that the various relationships that exist between masses, coupling constants etc. are merely consequences of perturbative unitary or renormalizability. However, if the physical environment can be arranged so that the symmetry becomes manifest, one can not doubt the existence of the symmetry. Practical application of this phenomenon must be confined to speculations about the early universe, since only in that environment were there temperatures sufficiently high to effect a phase transition. [The order of magnitude of β_c^{-1} may be estimated as follows. For global symmetries, we identify the vacuum expectation value of the field with f_π, the pion decay constant. Hence $f_\pi^2 \approx \frac{6m^2}{\lambda}$ and, $\beta_c^{-1}= 0(f_\pi) \approx 100$ MeV. For local symmetries, recall that the spontaneously generated vector meson mass is $\sqrt{2}ef_\pi$ and the mass of the Higgs particle is $\sqrt{2m^2}$. Hence (16) is also given by $\beta_c^{-2} = \left[\frac{e^2}{m_W^2} \frac{N+2}{6} + \frac{e^2}{m_H^2} \frac{N-1}{2}\right]^{-1}$. With $e^2 \approx 10^{-2}$ and $m_H < m_W$, we get $\beta_c^{-1} \approx 0(10m_H)$.] Thus a temperature environment for field theoretic phase transitions is not readily available. However,

One may study other environments which effect phase transitions described by critical parameters that have more immediate experimental consequences.[7]

REFERENCES

1. J. Cornwall, R. Jackiw and E. Tomboulis, Phys. Rev. D $\underline{10}$, 2428 (1974).
2. L. Dolan and R. Jackiw, Phys. Rev. D 9, 3320 (1974).
    H. Schnitzer, Phys. Rev. D $\underline{10}$, 1800 ($\overline{1}$974).
    S. Coleman, R. Jackiw and H. Politzer, Phys. Rev.D $\underline{10}$, 2491 (1974).
3. D. Kirzhnits and A. Linde, Phys. Lett. $\underline{42B}$, 471 (19$\overline{72}$).
4. S. Weinberg, Phys. Rev. D $\underline{9}$, 3357 (1974).
    L. Dolan and R. Jackiw, Phys. Rev. D $\underline{9}$, 3320 (1974).
    D. Kirzhnits and A. Linde, Lebedev Institute preprint.
5. This brief synopsis of field theory at finite temperature is of
    course an inadequate summary of the beautiful work of Martin, Schwinger
    and others. For a fuller account see for example L. P. Kadanoff
    and G. Baym Quantum Statistical Mechanics, (W. A. Benjamin, Menlo
    Park, 1962). Discussions of this topic which focus on the present
    application are also given in Refs.4.
6. The approximation and results of this section are familiar in sta-
    tistical mechanics, where they are known as "spherical model", "mean-
    field theory" etc. For a review see E. Stanley, Introduction to Phase
    Transitions and Critical Phenomena (Oxford Univ.Press, New York, 1971).
7. That phase transitions can be induced by high matter density has been
    shown by T.D.Lee and G.C.Wick, Phys.Rev. D $\underline{9}$, 2291 (1974). A.Salam
    and J. Strathdee, Nature $\underline{252}$, 569 (1974) and ICTP preprint, have
    studied field theoretic phase transitions in electromagnetic envi-
    ronments.

DISCUSSION

K. Symanzik: Could you comment on the gauge dependence or independence of your results for a gauge theory?
The critical temperature was calculated in a variety of renormalizable gauges parametrized by gauge constants (covariant gauges, R_ξ gauges, etc.) and the answer comes out independent of these constants, hence gauge invariant within that class. For non-renormalizable gauges (the unitary gauge and others) a well defined answer does not emerge. We attribute this to a failure of the perturbation theory.

J. Zittartz: Could you not get your answers just by dimensional arguments?
We give a numerical value for the critical temperature. Dimensional arguments, which were used initially by Kirzhnits and Linde, will yield the dependence of β_c^2 on m^2 and λ, but will not provide the numerical coefficient.

K.R. Ito: You consider an equation for the temperature dependent propagator. On the other hand, we can define phase transitions by the order parameter. Are these equivalent?
Yes; indeed Dolan has rederived our results by calculating the order parameter.

T. Kugo: What happens in massless scalar electrodynamics, where Coleman and Weinberg find symmetry breaking at zero temperature?
We have not done calculations for this massless theory, since our perturbation theory requires a mass parameter. However, I would speculate that finite temperature again restores the symmetry, and that a mass is generated for the charged particle, but not for the photon.

CRITICAL BEHAVIOUR OF STATIONARY
RANDOM FIELDS

G. Jona-Lasinio
Istituto di Fisica dell'Università-Roma
Gruppo GNSM - Padova

In recent years considerable progress has been made towards a
theoretical understanding of critical phenomena. The concept of Renormalization Group has provided the key ideas on which a very interesting qualitative, and to a certain extent quantitative, picture of critical behaviour has been built. From the mathematical stand point,
the techniques employed are entirely heuristic and at first look,
rather esoteric. However, it has been recently recognized that the
renormalization group approach is closely related to well-known problems in probability theory and that in fact leads to a generalization
of those problems[1]. All this opens the way to the possibility of
characterizing critical behaviour directly in terms of properties of
the stationary process underlying the microscopic description of
thermodynamic phenomena.

In this note, we briefly review the probabilistic interpretation
of critical behaviour, outlined in the two papers of Ref. (1).
We also point out the relevance of some ergodic properties of stationary processes in connection with this problem.

STABLE RANDOM FIELDS (S.R.F.)

The idea of <u>stable random field</u> generalizes to the case of multi-
dimensional processes the concept of <u>stable distribution</u>[2] well-known
in the theory of independent random variables.

To visualize the problem in a simple way, consider a one-dimensional lattice at each point of which is associated a real random variable X_i . This system will be described by a measure μ in the space
$$K = \Pi_{i \in Z} R \; , \; R \text{ is the real line.}$$
Construct now a new lattice by dividing the original system into
blocks of length L and associating with each block the random variable

$$\zeta_k = \frac{M_K - \langle M_K \rangle}{L^{p/2}}$$

where $M_K = \sum_{i \in \Lambda_K} X_i$

To the new lattice will be associated a new measure μ_L^ρ which can be written symbolically :

$$\mu_L^\rho = H_L^\rho \mu$$

H_L^ρ represents the transformation we have just described. If ρ is properly chosen, it may happen that by repeating the same operation an arbitrary number of times, we obtain convergence to a limit measure

$$\mu_\infty^\rho = \lim_{L \to \infty} H_L^\rho \mu$$

Clearly :

(1) $$\mu_\infty^\rho = H_L^\rho \mu_\infty^\rho$$

These equations are to be interpreted in the sense of weak convergence.

A measure satisfying Eq. (1) will be called a <u>Stable Random Field</u>. It is easy to verify that if μ_∞^ρ reduces to a product of identical factors, i.e. it describes a set of independent variables, Eq. (1) reduces to the usual definition of a stable distribution.

The interesting case is when μ_∞^ρ is not a product. In such a case μ_∞^ρ corresponds to a scaling invariant system and this provides the reason why S.R.F. are relevant for critical phenomena.

For independent variables, there is a beautiful theory which provides a complete classification of stable distributions. Since these are the only possible limit distributions for sums of identically distributed variables, a whole series of limit theorems which generalize the central limit theorem, is obtained.

The ideal of a theory of S.R.F. would be to develop along similar lines. As explained elsewhere[1], such a theory would provide a sound basis to the universality of critical phenomena.

AN EXAMPLE

A non trivial example of S.R.F. can be obtained from a stationary sequence well-known to probabilists[3]. One starts from a sequence of independent variables :

$$\ldots\ldots \xi_{-1},\ \xi_o,\ \xi_1,\ldots\ldots$$

normally distributed. One then constructs the Gaussian process :

$$Y_j = \sum_{-\infty}^{-1} |K|^{-a}\, \xi_{j+K}$$

characterized by the correlation function :

$$R_{ij} = <\ Y_i\ Y_j\ >$$

The next step consists in considering the process :

$$(2) \qquad X_j = Y_j^2 - <\ Y_j^2\ >$$

It can be shown that the normalized sums

$$\sum_1^n \frac{X_j}{[Var(\sum X_j)]^{\frac{1}{2}}}$$

for $\frac{1}{2} < a < \frac{3}{4}$ do not satisfy the central limit theorem as $n \to \infty$.

It is also easy to compute the limit correlation functions for block variables as they are all expressible in terms of R_{ij}.

The value $a = \frac{3}{4}$ corresponds to the onset of long range correlations for the process X_j. This can be seen by calculating its spectral function which for $\lambda \to 0$ has the form :

$$f(\lambda) \underset{\lambda \to 0}{\sim} |\lambda|^{4(a-1) + 1}$$

and therefore develops a singularity for $a < \frac{3}{4}$.

The appearance of long range correlations can then be connected with the failure of the property of <u>complete regularity</u>, induced by a singularity in $f(\lambda)$ [4].

Complete regularity (we omit the definition because rather technical) is one of the properties which can be used to characterize the degree of dependence of a system of random variables. Other properties

are strong mixing, absolute regularity etc.[4]. As it is well known
there is a deep connection between some of these properties and limit
theorems for random processes[5].
Failure of complete regularity seems to be a necessary condition for
the existence of a non trivial limit block correlation function.

REFERENCES

(1) G. Jona-Lasinio, "The Renormalization Group : A Probabilistic View"
    to appear in Nuovo Cimento.
    G. Gallavotti, G. Jona-Lasinio; "Limit Theorems for Multidimensional
    Markov Processes", to appear in Comm. Math. Phys.

(2) B.V. Gnedenko, A.N. Kolmogorov; "Limit Distributions for Sums of
    Independent Random Variables", Cambridge, Mass. 1954, Chap.7.

(3) I.A. Ibragimov, Yu. V. Linnik; "Independent and Stationary Sequences
    of Random Variables", Wolter-Noordhoff Publishing, Groningen, 1971.

(4) I.A. Ibragimov, Y. Rozanov; "Processus Aléatoires Gaussiens",
    Editions MIR, Moscou 1974.

(5) I.A. Ibragimov, Yu. V. Linnik; "Independent and Stationary Sequences
    of Random Variables", Chap. 17, 18, 19. Wolter-Noordhoff Publishing,
    Groningen 1971.

<u>PHASE TRANSITIONS OF CONTINUOUS ORDER</u>

J. Zittartz
Institut für Theoretische Physik
Universität Köln
5 Köln 41, Germany

I. Conjecture of a New Type of Phase Transition

In discussions of phase transitions it is most useful to describe
the state of a large system by its intensive thermodynamic variables,
also called "fields". These field variables, like temperature, pressure,
magnetic field, or chemical potentials, are always identical in two co-
existing phases in contrast to the "density" variables like entropy,
particle density, or magnetization. The appropriate thermodynamic potential which is a function of the field variables is called the free energy. In the Ehrenfest scheme there is said to be a phase transition of
the n-th order at some point P in the space of field variables, if at
this point all derivatives of the free energy with respect to the variables exist continuously up to order n-1, but in the n-th order discon-
tinuities or divergencies show up. More generally a phase transition of
a thermodynamic system is associated with the singularities of the free
energy.

Usually the set of singular points forms one or several smooth hypersurfaces in the space of field variables. Singular behavior shows up if
one crosses the hypersurface at some point P by varying some locally defined variable x assumed to be orthogonal on the surface. In realistic
systems, for instance magnets or fluids, one observes a rather restricted class of phase transitions: When crossing the hypersurface, discontin-
uities in the first derivatives show up indicating a first order transition; only on the edges of the singular hypersurfaces the phase transitions may change to second order and thus give rise to critical behavior.

As an example we consider a simple ferromagnetic system. There are
two field variables: the magnetic field, $-\infty < B < \infty$, and the temperature,
$0 \leq T < \infty$. The singular hypersurface in this case is the line: $B=0$, $0 \leq T \leq T_C$.
When crossing the interior part of this line, $T < T_C$, we have the singular
behavior:

$$F(B,T)-F(0,T) = -|B| \cdot M_o(T), \quad B \to 0 \tag{1}$$

where M_o is the spontaneous magnetization ; at $T=T_c$, the edge of the singular line, one observes the well-known critical behavior.

In contrast to the usual behavior described **above**, I would like to conjecture a new type of phase transition corresponding to another simple type of singularity. This conjecture will be supported by the model calculation of the next section. Again we consider a smooth hypersurface of singular points P with the locally defined orthogonal field variable x, x=0 on the surface. Then the leading singular part of the free energy should have the form:

$$F_{sing} \sim |x|^{\kappa(P)} \quad , \quad x \to 0 \ , \tag{2}$$

where the exponent $\kappa(P)$ characterizing the power law singularity is assumed to be a continuous function on the hypersurface. For obvious reasons this new type will be called a phase transition of continuous order. If at some fixed point P: $n-1 < \kappa \leq n$, n integer, we have an n-th order phase transition at P according to the Ehrenfest classification scheme. To supplement the above definition it also seems natural to assume that the exponent κ approaches infinity, if the singular point P tends to the edge of the surface. This would correspond to a smooth "fading away" of the singularity in the free energy. A border-line case of a continuous phase transition would be the situation where the exponent κ remains "infinite" on the whole singular hypersurface. This means that the singularity of the free energy is weaker than any power.

In the example of the simple magnetic system discussed above, we would expect a singular line in the B,T-plane, $B=0$, $0 \leq T \leq T_\infty$, and the following behavior of the free energy:

$$F = F_{reg.}(B^2) + A(T)|B|^{\kappa} \ , \quad B \to 0. \tag{3}$$

Because of symmetry the regular part of the free energy is a function of B^2. The exponent $\kappa(T)$ is expected to increase monotonously from unity at $T=0$ to infinity at some finite temperature T_∞. As long as $\kappa=1$, we have a first order phase transition with nonvanishing magnetization in zero field, thus the system is ordered in the usual sense. If $\kappa > 1$, the zero field magnetization vanishes and there is no longer usual long range order; however, there is still a phase transition up to T_∞, beyond which the free energy will be an analytic function.

II. The Ferromagnetic Ising Model on a Cayley Tree

To support the above conjecture we shall now show that the ferromagnetic Ising model with nearest neighbour interaction on a Cayley tree lattice exhibits such a phase transition of continuous order[1]. A Cayley tree of connectivity K (also called branching ratio) is a continuously branching lattice without cycles. It may be constructed by connecting K+1 n-generation branches to a central site. An n-generation branch itself is defined as an initial site connected to K (n-1)-generation branches; a 1-generation branch is a single site. The thermodynamic limit corresponds to letting n go to infinity. As the linear chain (K=1) is well-known, we here consider $K \geq 2$.

The partition function of the model is given by

$$Z(t,b) = \sum_{\sigma_i = \pm 1} \exp \left\{ b \sum_i \sigma_i + t^{-1} \sum_{\langle i,j \rangle} \sigma_i \sigma_j \right\} \tag{4}$$

with a reduced temperature, $t = (\beta J)^{-1}$, where J is the nearest neighbour coupling, and a reduced field $\beta B = b$. This model is almost soluble in closed form. As there are no cycles on a Cayley tree, a path connecting two sites is unique. In performing the trace in (4) one can start with the surface sites and one may proceed step by step towards the interior of the tree. This procedure which, of course, resembles the transfer matrix method results in a recurrence relation for the partition function of an n-generation branch[2]:

$$Z_{n+1}^{\pm} = e^{\pm b} \left[Z_n^{+} \cdot e^{\pm t^{-1}} + Z_n^{-} \cdot e^{\mp t^{-1}} \right]^K, \quad Z_1^{\pm} = e^{\pm b} \tag{5}$$

where the $\pm$ sign indicates that the initial site spin is kept fixed to $\sigma = \pm 1$, respectively. Using the probability ratio $e^{2x_n} = Z_n^{+}/Z_n^{-}$, expression (5) is rewritten as

$$x_{n+1} = b + h(x_n), \quad h(x) = \frac{K}{2} \ln \frac{1 + \frac{\gamma}{K} \tanh x}{1 - \frac{\gamma}{K} \tanh x} \tag{6}$$

with $x_1 = b$ and the temperature parameter

$$\gamma(t) = K \cdot \tanh \frac{1}{t} \quad . \tag{7}$$

Using (5) the free energy per site in the thermodynamic limit can be derived as the sum:

$$f(t,b) - f(t,0) = \frac{K-1}{2} \sum_{n=1}^{\infty} K^{-n} \ln \left[1 + \frac{\sinh^2 x_n}{\cosh^2 t^{-1}} \right], \tag{8}$$

where the zero field result, $f(t,0) = \ln(e^{t^{-1}} + e^{-t^{-1}})$, is analytic in t(or T)[2]. We notice that the surface sites still contribute to the sum (the n=1 term) in the thermodynamic limit, as their ratio to the total number of sites tends to the finite value $\frac{K-1}{K}$, a consequence of the unusual topology of the pseudolattice.

Eqs. (6) and (8) cannot be solved in closed form. However, the singularities of the free energy can be discussed completely. First one notices that each x_n and the summand in (8) are analytic functions of b and t near their real axis, respectively; furthermore, both (6) and (8) are convergent. Therefore, singularities can arise only from nonuniform convergence. However, one can show[1],[3] that the convergence is uniform and thus the free energy is analytic, if the map $x_n \to x_{n+1}$ is contractive near the fix point (point of convergence), i.e.

$$|x_{n+1} - x_n| < \alpha | x_n - x_{n-1}| \;, \; \alpha < 1, \; n \geq N_o. \tag{9}$$

The quantity α may be replaced by $|h'(x_\infty)|$ where x_∞ is the fix point. Simple inspection of (6) then yields the result[1]: The free energy is analytic in t and b except for the singular line, $b=0$, $0 \leq T \leq T_\infty$, where $\gamma(T_\infty)=1$ defines the transition temperature.

The precise form of the singularities for $b \to 0$, $T \leq T_\infty$, can be studied by solving linear recurrence relations which are upper and lower bounds to (6) and which also give upper and lower bounds to (8), both displaying the same singular behavior. The final result is[1]:

$$f(t,b) = f_{reg.}(b^2) + A(T) \cdot |b|^\kappa \{1 + 0(|b|)\} \;, \; b \to 0 \tag{10}$$

where the exponent is

$$\kappa = \frac{\ln K}{\ln\gamma(t)} \tag{11}$$

which varies between unity at $T=0$ ($\gamma(0)=K$) and infinity at $T=T_\infty$ where the singularity fades away. Thus we have a continuous phase transition as explained in the first section. In particular, it turns out that for even integers, $\kappa = 21$, the singular behavior changes to $b^{21}.\ln|b|$ due to a compensation of divergencies in $A(T)$ and the corresponding coefficient in $f_{reg.}$.

As $\kappa > 1$ for $T>0$, the spontaneous magnetization vanishes. Therefore order in the usual sense is achieved only at zero temperature. One notices, however, that the susceptibility $\chi = -\frac{\partial^2 f}{\partial b^2}\Big|_{b=0}$ diverges if $\kappa \leq 2$, i.e. for $T \leq T_2$ where T_2 is defined by

$$\gamma^2 = K, \; 0 < T_2 < T_\infty \; . \tag{12}$$

For comparison, in the usual type of phase transition order sets in abruptly or continuously at the critical point T_c. In our case the line of continuous phase transitions interpolates between the high-temperature disordered state and the zero temperature ordered state; the onset of order at T_c in the usual situation is thus stretched to the line $0 \leq T \leq T_\infty$.

III. Extensions and Outlook

The above analysis has been extended to the corresponding anti-ferromagnetic situation[3]. Including a staggered field H into the discussion the free energy $F(T,B,H)$ is represented in a 3-dimensional space of field variables. The formalism is somewhat more involved, but similar as in Section II. Again the singularities can be worked out by studying recurrence relations. We obtain two pieces of singular surfaces. When crossing the first surface perpendicularly, we again find the singular behavior $|x|^\kappa$ where $\kappa(P) \geq 1$ varies continuously on the surface, going to infinity on the edge. On the second surface which in fact joins the first one, κ stays strictly infinite. This is therefore a border-line case as explained in the first section where the singularity is weaker than any power. So far we have not yet been able to determine the precise form of the weak singularity.

The continuous phase transition for the Ising model on the Cayley tree evidently is a consequence of the topology of the pseudolattice. However, its occurrence in more realistic physical situations is possible; we have at least two other situations in mind:

a) For random ferromagnets one considers a fraction p of magnetic atoms and a fraction 1-p of unmagnetic atoms distributed at random on a regular lattice. It is known that the critical temperature $T_c(p)$ is a decreasing function with decreasing p, vanishing for $p \leq p_o$, the percolation limit. However, Griffiths[4] has proven that the free energy is singular up to $T_c(1)$ for all p. The precise nature of the singularities is not yet known. Recently Domb[5] has argued that the ramified magnetic clusters are dominating the singular behavior. In view of the fact that Cayley tree graphs are special ramified clusters, it is conjectured that a random ferromagnet is another candidate for a continuous phase transition, possibly of the border-line case.

b) Two dimensional models with continuous symmetry are known not to have long range order. However, series expansions seem to indicate that the susceptibility is diverging below some temperature[7], although this fact is far from being proven. Again it is conjectured that in these systems a phase transition of continuous order may take place.

[1] E. Müller-Hartmann and J. Zittartz, Phys. Rev. Lett. 33, 893 (1974)

[2] T.P. Eggarter, Phys. Rev. B 9, 2989 (1974)

[3] E. Müller-Hartmann and J. Zittartz, to be published

[4]R.B. Griffiths, Phys. Rev. Lett. $\underline{23}$, 17 (1969)

[5]C. Domb, J. Phys. C $\underline{7}$, 2677 (1974)

[6]N.D. Mermin and H. Wagner, Phys. Rev. Lett. $\underline{17}$, 1133 (1966)

[7]H.E. Stanley and T.A. Kaplan, Phys. Rev. Lett. $\underline{17}$, 913 (1966)

ORDER OF THE PHASE TRANSITION RELATED TO THE

DEGENERACY AND INTERACTION WITH PROPOSAL FOR AN EXACT PROBLEM

Huzio Nakano,

Department of Applied Physics, Nagoya University

The system of many identical interacting units with certain
degenerate states is investigated[1]. The state of a unit system is
represented by a spin-like variable σ_i $(i=1,2,\cdots,N)$, which takes the
values $-s,-s+1,\cdots,s-1,s$ with integral or half-integral s. The
Hamiltonian of the system is expressed as a sum of pair interaction
energies, $\sum_{(i,j)} V(\sigma_i,\sigma_j)$. The state designated by σ_i is assumed
to be degenerate and its degeneracy is denoted by $D(\sigma_i)$. The
importance of symmetry properties of $D(\sigma_i)$ and $V(\sigma_i,\sigma_j)$ are stressed,
where the symmetry means that $D(\sigma_i)$ and $V(\sigma_i,\sigma_j)$ are even functions
of σ_i and σ_j : i.e. $D(\sigma_i)=D(-\sigma_i), V(\sigma_i,\sigma_j)=V(-\sigma_i,-\sigma_j)$. The
system of magnetic spins with usual exchange interactions belongs to
the symmetric category of constant degeneracy. The thermostatistical
average of any σ_i which is denoted by σ represents the order parameter
of the system. Dependences of σ upon temperature T and a certain
applied field H causing a perturbing energy $-H\sum_i \sigma_i$ are investigated.
For the present the pair interaction is confined to such a simple form
as $V(\sigma_i,\sigma_j)=-U\sigma_i\sigma_j -W(\sigma_i+\sigma_j)$ with the use of two constants U and
W. The partition function of such a system is expressed as

$$Z(T,H) = \sum_{\sigma_1} \cdots \sum_{\sigma_N} D(\sigma_1)\cdots D(\sigma_N) \exp\left(\beta V\sum_{(i,j)}\sigma_i\sigma_j + \beta H\sum_i \sigma_i \right) ,$$

where β denotes the inverse of the Boltzmann constant times temperature.
The properties of $Z(T,H)$ in relation to T and H are the object of the
present investigation.

The symmetric system with constant degeneracy is equivalent to
the magnetic spin system with exchange interactions and exhibits the
phase transition of second order. The Lee-Yang theorem[2] is valid
for such systems, where the singular line of the partition function on
the H-T plane is such as shown by the thick line in Fig.1. In the
simplest case of these, s=1/2 and the degeneracy $D(1/2)=D(-1/2)$ is
constant. We have therefore necessarily a second-order transition.
In case s=1, there exists such symmetric systems that have a non-
constant $D(\sigma_i)$. If a certain condition is satistfied by $D(\sigma_i)$, the
phase transition of first order occurs, where the singular line is given

by such a thick curve as shown in Fig.2.   Otherwise we have the second-order transition similar to the symmtric case of s=1/2.   The mean field theory leads to the condition expressed as the inequality D(1)= D(-1)$<$ D(0)/4 for the first-order transition to occur in the case s=1. There occurs also an induced transition by the applied field H between the temperatures T_c and T_c' such as shown in Fig.2.

For the asymmetric system[3][4] we get the singular lines such as shown in Figs.3 and 4 corresponding to Figs.1 and 2, respectively, where several branches represent the lines for various typical cases (not exhaustively) of the relevant parameters.   The intersection and contact points of the singular line with the temperature axis correspond to the transition temperatures of first-and second-order transitions, respectively.   The induced transition by the field can be also read from Figs.3 and 4 as well as Fig.2.

Finally the following notes are to be added.

(1)   The symmetric system on plane lattice with s=1 and D(1)=D(-1)$<$ cD(0),c being a certain constant,is proposed as a simplest model which might be exactly proved to exhibit the first-order phase transition of usual sort with such a singular line as shown in Fig.2. Accordingly the Lee-Yang theorem corresponding to Fig.1 should be replaced for such a system by the theorem corresponding to Fig.2.   The mean field approximation leads to the value 1/4 for the above critical constant c. In fact the asymmetric system with s=1/2 on plane lattice has been already proved exactly to exhibit the first order transition[4],which is, however,of a peculiar type (Fig.3).

(2)   The calculation in the present paper can be applied to see the occurence of the transition between different phases of dissipative process in the theory such as investigated recently[5].

(3)   By generalizing the state variable σ_i to a continuous one, some more specific features may be found,although the main feature will not essentially change.   Such a generalization is adequate in particular for the problem mentioned above in (2).

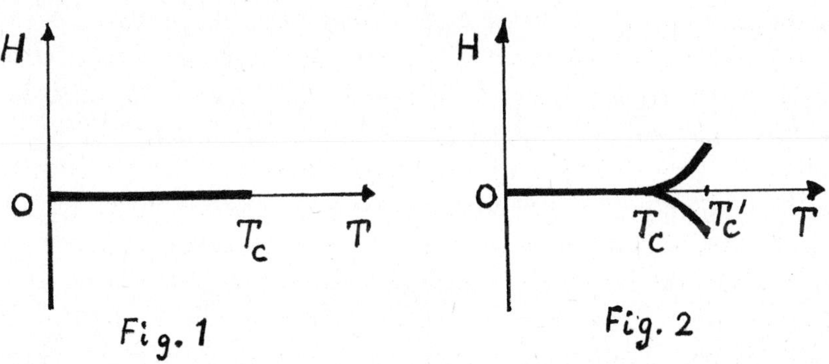

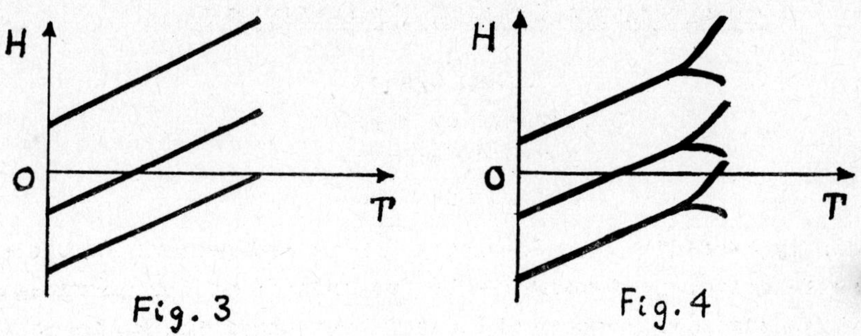

Fig. 3   Fig. 4

Peferences

1) H.Nakano, submitted to Prog. Theor. Phys..

2) C.N.Yang and T.D.Lee, Phys. Rev. 87 (1952),404,410.

   T.Asano, Prog. Theor. Phys. 40 (1968),1328.

   M.Suzuki, J. Math. Phys. 14 (1973),1088 ; see also related papers
           cited therein.

   Y.Karaki Talk in this Symposium.

3) S.Strassler and C.Kittel, Phys. Rev. 139 (1965),A758.

4) H.Nakano, Prog. Theor. Phys. 50 (1973),1510.

5) H.Nakano, Prog. Theor. Phys. 51 (1974),1279.

Discussions

Lobowitz and Kubo asked about the relation to the Lee-Yang theorem.
Further Lebowitz questioned whether the singularity shown by Fig.2 is
valid for the system with s=1/2.   Nakano denied it and clarified that
such a singularity can appear under a certain condition on the degeneracy
function D($_i$) only in the case s is equal to or more than the unity.
M.Suzuki expected his work on the Lee Yang theorem for the Ising
ferromagnets of general spin with degeneracy (J. Math. Phys. 14 (1973)
1088) might be useful.   Nakano hoped to make use of it to decide the
limit of that theorem.

THE HEISENBERG MODEL ON THE CAYLEY TREE

Tohru Ogawa

Department of Physics

Kyoto University

Kyoto, 606, Japan

Recently, Matsuda[1] and Eggarter[2] have independently pointed out that the self-consistent Bethe-Peierls treatment[3] of the ferromagnetic Ising model on the Cayley tree, which had been named the Bethe-lattice by Kurata, Kikuchi and Watari[4] and had been believed to be exact[5] does not correspond to the correct thermodynamic limit. This situation is owing to the existence of a huge number of surface atoms on a Cayley tree. The Cayley tree is one-dimensional because it contains no closed loops and at the same time rather infinite-dimensional because the ratio of the number of the surface atoms to the total does not tend to zero in the thermodynamic limit but remains finite. Thus there coexist low- and high-dimensional characters in this pseudolattice and it is interesting to see which character reflects on what physical properties.

In this paper, the ferromagnetic Heisenberg model on the finite most symmetrical Cayley tree is studied in the limit of large size. The coordination number and the number of the shells of it are assumed to be z and N respectively. The high symmetry of this model enables us to obtain the spectrum and the wavefunctions of the single-magnon states rigorously.

The Hamiltonian in the single-magnon subspace can be block-diagonalized by a proper unitary transformation into the following two types of tridiagonal matrices A and B_M: the $(N+1)\times(N+1)$-matrix A and the $M\times M$-matrix B_M are respectively defined by

$$
A = \begin{pmatrix}
z & -\sqrt{z} & 0 & & & & & \\
-\sqrt{z} & z & -\sqrt{z-1} & & & & 0 & \\
0 & -\sqrt{z-1} & z & -\sqrt{z-1} & & & & \\
& & \cdots & \cdots & \cdots & & & \\
& & & -\sqrt{z-1} & z & -\sqrt{z-1} & 0 & \\
& 0 & & & -\sqrt{z-1} & z & -\sqrt{z-1} \\
& & & & & 0 & -\sqrt{z-1} & 1
\end{pmatrix}
\tag{1}
$$

$$
B_M = \begin{pmatrix}
z & -\sqrt{z-1} & 0 & & & & & \\
-\sqrt{z-1} & z & -\sqrt{z-1} & & & & 0 & \\
0 & -\sqrt{z-1} & z & -\sqrt{z-1} & & & & \\
& & \cdots & \cdots & \cdots & & & \\
& & & -\sqrt{z-1} & z & -\sqrt{z-1} & 0 & \\
& 0 & & & -\sqrt{z-1} & z & -\sqrt{z-1} \\
& & & & & 0 & -\sqrt{z-1} & 1
\end{pmatrix}
\tag{2}
$$

where the last column and row correspond to the surface shell in the both cases. The number of the blocks with A, B_N and B_{N-t} ($1 \leq t \leq N-1$) are unity, $z-1$ and $z(z-2)(z-1)^{t-1}$ respectively. These matrices play the role of the radial wave equations for partial waves and the wavefunction of magnon itself is a product of the radial function obeying them and a factor with exponential damping toward the surface, $exp(-\kappa r)$, where $\kappa \equiv (1/2)\ln(z-1)$.

The eigenvalues of the matrix A are given by

$$\varepsilon_0 = 0 \qquad \text{and} \qquad \varepsilon_k = z - 2\sqrt{z-1}\cos k, \qquad\qquad (3),(4)$$

where k's are N roots in the range $[0, \pi]$ of the equation

$$\cot Nk = f(-1/\sqrt{z-1}, k), \qquad\qquad (5)$$
$$f(x, k) \equiv (x - \cos k)/\sin k. \qquad\qquad (6)$$

Those of the matrix B_{N-t} are in the form of the formula (4) where k's are the roots of the equation

$$\cot (N-t)k = f(\sqrt{z-1}, k), \qquad\qquad (7)$$

instead of Eq.(5), in this case. Eq.(7) has $N-t$ real roots for $N-t \leq M_c \equiv 1/(\sqrt{z-1}-1)$ and one of them becomes pure imaginary for $N-t > M_c$. The spectrum of such *surface-mode-like* states has the asymptotic form

$$\varepsilon_t \simeq (z-2)^2/(z-1)^{N-t+1}, \qquad (N-t \to \infty). \qquad\qquad (8)$$

The accumulation of them at the vicinity of zero governs the thermodynamic properties of the system at infinitesimal temperatures. When magnon is approximated as Boson, the density of magnons excited per atom is calculated as

$$\rho = \lim_{N\to\infty} (z-2)/[z(z-1)^N] \sum_m [\exp(+\beta\varepsilon_m)-1]^{-1} \simeq \lim_{N\to\infty} N/\beta, \qquad\qquad (9)$$

where use has been made of the relation

$$\Sigma[\exp(x)-1]^{-1} \simeq \Sigma x^{-1} \qquad\qquad (10)$$

which is derived from the convex property of the exponential function and is valid so long as only the order of divergence concerns. The divergence of Eq.(9) is logarithmic because the total number of atoms is of the order $\sim (z-1)^N$ and it is just like as in the case of two-dimension. Then the ferromagnetic ground state is unstable against the magnon excitation at non-zero temperatures.

Nevertheless, can things go differently at the central part of the system and some spontaneous magnetization can appear there. Because only a few number of states have non-zero wavefunctions there among the logarithmically divergent number of excited magnons. Another calculation supports this possibility. The magnon state density of the Bethe-lattice regarded as the surfaceless Cayley tree with *translational invariance* has a gap between the ground state and the first excitation.

Some devices are required to make a rigorous statement such as the Mermin-Wagner theorem[6] because the Cayley tree is an essentially inhomogeneous. However, the calculation itself on which the judgement of the existence or nonexistence of the spontaneous magnetization is directly based is intrinsically the same both in the Mermin-Wagner theory and in the magnon approximation for the homogeneous system. Accordingly, the present results are expected to be valid beyond the approximation adopted.

<u>References</u>

1) H.Matsuda, Prog.Theor.Phys. <u>51</u>(1974), 1053.
2) T.P.Eggarter, Phys.Rev. B<u>9</u>(1974), 2989.
3) H.A.Bethe, Proc.Roy.Soc., Ser.A <u>150</u>(1935), 552.
   R.Peierls, Proc.Cambr.Phil.Soc. <u>32</u>(1936), 477.
4) M.Kurata, R.Kikuchi and T.Watari, J.Chem.Phys. <u>21</u>(1953), 434.
5) C.Domb, Adv.Phys. <u>9</u>(1960) 149.
6) N.D.Mermin and H.Wagner, Phys.Rev.Letters <u>17</u>(1966), 1133.

The more detailed report of this study has been submitted to *Progress of Theoretical Physics*.

GENERALIZED GELL–MANN AND LOW GROUP TRANSFORMATIONS:

SCALING VARIABLES AND CROSS–OVER EFFECTS

C.Di Castro
Istituto di Fisica "G.Marconi"
Università degli Studi di Roma – Italy

Unità G.N.S.M. – Roma

In dealing with critical phenomena the reduction schemes commonly used in many–body theory are useless. In fact infinitely many degrees of freedom are strongly correlated among themselves, owing to the fluctuations which extend over a coherence distance ξ and a characteristic time τ_ξ, both diverging at the critical point.

However, at least for the static case where only one infinity of degrees of freedom related to the length scale is involved, the introduction of a new reduction scheme is, at least in principle, quite simple.

This is related to the existence of "classes of universality", each comprising systems with equal symmetry, whose critical behaviour is found to be same, i.e. independent of the details of each given hamiltonian. Therefore a transformation on the parameters characterizing each system, leading to the asymptotic elimination of the "irrelevant" variables, was needed.

On the contrary the identification of those parameters on which the critical indices actually depend, led on the one hand to their perturbative ($\frac{1}{n}$ or ε –expansion) evaluation and on the other to the possibility to switch from one class of universality to another through the cross–over mechanism.

After Kadanoff's[1] formulation of scaling, to achieve these goals the Gell–Mann and Low[2] renormalization group transformation was proposed,[3] while Wilson[4] constructed his own group.

By means of the ε –expansion the same results were obtained with the two groups at least in the first few orders in ε, as first indicated in[5] up to the second order and then in[6] up to the fourth order by means of the Callan and Symanzik equation.

It was then felt[7] that no special significance should be assigned to the transformation one is considering but rather to the mechanism by which the asymptotic scale invariance in the physical variables is obtained. Actually it has been quite generally stated[8] that an infinity of equivalent transformations can be generated by means of an arbitrary diffeomorphism.

The choice of the transformation depends mainly on the simplicity of the calculations for the particular problem considered. When more than one relevant fields or "coupling constants" are present simultaneously, as for instance, in the tricritical phenomena, standard field theoretic methods become very complicated, because the renormalization functions depend non-linearly on the variables. In the ordinary Gell-Mann and Low group for instance even for the isotropic critical point the parameters of the transformation depend on the mass and the renormalized coupling constant. In fact already for this simple case a new realization of the group transformation à la Gell-Mann and Low was introduced in[9]. This group is linear in the field φ and in the bare mass $\tau \sim$ T-T$_c$ and therefore the transformation parameters depend only on the renormalized coupling constant as in the Callan and Symanzik equation, and its differential equation does not have the inhomogeneous term as the Gell-Mann and Low group.

On the other hand recently groups non linear in φ have been successfully applied to lattice systems[10] in one or two dimensions, where the results of the ε -expansion cannot be extended. The properties of a particular non linear group have also been studied[11] for a continuous model. We shall introduce generalized group transformations[8], which include all those previously mentioned.

Generalized Group Transformations

According to universality we can try to discuss critical phenomena by means of an "effective hamiltonian" with short range interactions

$$H = \int d^d x \left\{ \tfrac{1}{2} (\nabla \varphi)^2 + \sum_i \tfrac{1}{i!} \mu_i(x)\, \varphi^i(x) \right\} \tag{1}$$

with $(\nabla \varphi)^2 = \sum_{\sigma=1}^{n} (\nabla \varphi_\sigma)^2$, $\mu_1 \varphi_1 = \sum_{\sigma=1}^{n} h_\sigma \varphi_\sigma$, $\varphi^2 = \sum_{\sigma=1}^{n} \varphi_\sigma^2$, $\varphi^4 = (\varphi^2)^2, \cdots$

$\mu_1(x)$ is the external field and $\mu_2 = \tau \sim \text{T-T}_c / \text{T}_c$ is the deviation from the critical temperature. The relation of this field-theory model with the standard statistical models such as the Ising, x-y or Heisemberg models has been discussed in (12). Long range forces of the type $\frac{1}{r^{d+\sigma}}$, anisotropic exchange effects, cubic anisotropy can be included by modifying the gradient term, the quadratic and quartic interaction, respectively, where as the φ^6 term will be associated to the tricritical behaviour. These features lead to cross-ever effects from one critical behaviour to another as reviewed by Fisher.[14]

The logarithm of the partition function F is a functional of the $\{\mu_i(x)\}$. By successive Legendre transformations we can eliminate these "fields" μ_i's in favour of the corresponding conjugate "density" variables ω_i's, generating a set of thermodynamic potentials which are functionals of combinations of fields and densities $\{\bar{\nu}_i\} \equiv \{\mu_\ell, \omega_j\}, \ell \neq j$. In each case successive functional derivatives generate all the physically relevant quantities.

In particular the n-th derivative with respect to one variable generates the comulant of n-th order of the conjugate variable . For instance the first Legendre transformation of F gives rise to the functional $\Gamma(\varphi, \mu_2, \dots)$, which is in turn the generator of the one-particle irreducible n-point vertex function $\Gamma^{(n)}(x_1, \dots, x_n)$

By means of a differentiable transformation labelled by a parameter λ having the dimensions of the inverse of a length we operate on the dimensionless variables $\left\{ \nu_i = \dfrac{\bar\nu_i}{\lambda^{x_i}} \right\}$, where x_i^0 is the bare dimensionality of $\bar\nu_i$ in terms of the inverse of a length ($x_\varphi^0 = 1 - \varepsilon/2$, $x_2^0 = 2$ and $x_{\mu_4}^0 = \varepsilon$ where $\varepsilon = 4 - d$).

The transformation is determined by a set of differential equations

$$\lambda' \frac{\partial \nu_i'}{\partial \lambda'} = \Psi_i\left(\{ \nu_j' \} ; \lambda' \right)$$

(2)

where the Ψ_i functions are to be specified.

We define as generalized group transformations[8] those transformations, on the variables $\{ \nu_i \}$ which leave the partition function unchanged except for an irrelevant multiplicative constant. This means that the thermodynamic potentials are the invariants of the system (2) of differential equations of the transformation and thus satisfy the linear differential equation

$$\left[\sum_i \int \Psi_i\left(\{ \nu_j' \} ; \lambda' \right) \frac{\delta}{\delta \nu_i'} + \lambda' \frac{\partial}{\partial \lambda'} \right] \Gamma = 0$$

(3)

Scaling variables

We make the further assumption that on the critical surface of the system there exists a fixed point of the trasformation i.e. a point in the parameter space where if we iterate the transformation further the $\{ \nu_i' \}$ are left unchanged. From eqs.(2) it follows that a fixed point is a solution of the equations

$$\Psi_i\left(\{ \nu_j^* \} ; \lambda^* \right) = 0$$

(4)

If we linearize eqs.(2) around the fixed point, we see immediately that the eigenfunctions of the matrix $\sigma_{ij} = - \dfrac{\delta \Psi_i}{\delta \nu_j}$, when evaluated at the fixed point, are the asymptotic scaling variables and its eigenvalues are the corresponding critical exponents. The thermodynamic potential becomes a homogeneous function of the same variables.

As originally stressed in (4), (12) and (13) all the requirements of the universality hypothesis are now satisfied. Different classes of universality correspond to different fixed points with different symmetry properties. Moreover the critical exponents are connected to the derivatives of the transformation functions evaluated at the fixed point which are non-singular. Perturbation theory can now be applied and this is at the origin of the $1/n$ and the ε-expansions.

As stressed by Jona-Lasinio[8] two transformations where the matrices σ_{ij}

are related by a similarity transformation lead to the same critical indices i.e. are equivalent. This is the case if the two transformations are differentiably conjugate.

Non-linear terms allow to discuss corrections to the asymptotic power-law behaviour and possibly the cross over from one type of critical behaviour to another i.e. from one fixed to point to another.[14]

In general for each set $\{x_i\}$ of critical exponents related to a given fixed point we could look for the scaling variables[13] $g_i'(\{v_j'\})$ which under the group transformation scale exactly as $g_i' = (\lambda'/\lambda)^{-x_i} g_i$ and therefore satisfy the equations

$$\lambda' \frac{\partial g_i'}{\partial \lambda'} = -x_i\, g_i' \quad ; \qquad \left[-\sum_i x_i \int g_i' \frac{\delta}{\delta g_i'} + \lambda' \frac{\partial}{\partial \lambda'} \right] \tilde{\Gamma}(\{g_i'\}) = 0 \tag{5}$$

The scaling variables are not unique since they depend on which group transformation is considered. More generally we can now say that equivalent group transformations lead to various sets of scaling variables with the same critical exponents.

On the contrary two sets of scaling variables $\{g_i'\}$ and $\{g_i''\}$ corresponding to two different sets of critical exponents $\{x_i\}$ and $\{x_i''\}$ cannot be equivalent i.e. the cross-over from one critical behaviour to another is controlled by a non-differentiable transformation.

In each particular case an explicit determination of the scaling variables would completely solve the problem of critical phenomena. This was done in the large n limit by Ma[15] and is not feasible in general.

A cross-over model

If the transformation is linear in φ, a linear relationship in the correlation functions is induced, which is local in the Gell-Mann and Low case[3][5][9] and non-local in the Wilson case.[12] This is the main reason why recently[16] it was possible to give a formal evidence that the two groups are asymptotically equivalent in the sense specified above, whereas the comparison with non linear groups is much more difficult.

As in (9) we now consider the simplest model Hamiltonian with only $u\varphi^4$ interaction. The original set of variables is simply $\{v_i\} \equiv \{\varphi, \tau, u\}$. The standard renormalization procedure is to express the transformation in terms of correlation functions by the choice of a normalization point. In the Gell-Mann and Low group this is possible owing to the presence of the multiplicative factor in the linear relationship for the correlation functions, by choosing

$$\left.\frac{\partial \Gamma^{(2)}}{\partial k'^2}\right|_{N.P.} = 1 \quad , \quad \left.\frac{\partial \Gamma^{(2)}}{\partial \tau'}\right|_{N.P.} = 1 \quad , \quad \left.\frac{\Gamma^{(4)}}{u'}\right|_{N.P.} = 1 , \tag{6}$$

provided the energy correlation functions also transform linearly.

Since τ' and φ' are the two obviously relevant fields, we parametrize the transformation in terms of u' only as specified in (9). Successive functional derivatives of the group equation (3) for Γ generate the corresponding equations for $\frac{\partial \Gamma^{(2)}}{\partial k'^2}, \frac{\partial \Gamma^{(2)}}{\partial \tau'}$ and $\Gamma^{(4)}$, which once specified at the N.P. determine the quantities $\psi_\varphi(u'), \psi_\tau(u')$ and $\psi_u(u')$.

Normalization conditions (6) correspond to the subtractions necessary in diagrammatic technique to perform wave function, mass and vertex renormalization.

If the transformation is non linear, differentiations with respect to φ' and τ' of eq.(3), lead to coupled equations for different correlation functions, and the choice of a normalization point may become meaningless.

Our linear transformation is now completely specified and we can study[9] the fixed-point solution at $d = 4-\varepsilon$ dimension. We obtain a trivial fixed point $u^* = 0$ with classical exponents $\{x_i^0\}$, which is unstable for $d < 4$ and a non-trivial fixed point $u^* \sim O(\varepsilon)$ with non-classical exponents $\{x_i\}$, which is stable for $d < 4$. There is an additional relevant field u in the trivial case and therefore this fixed point can be associated with tricritical behaviour.[17] u becames the cross-over field from critical to tricritical behaviour at the lowest order, with cross over exponent ϕ given by $x_u = \phi/\nu$. Since $u^* = 0$, the φ^6 term becomes essential in the description of tricritical phenomena. This model can be studied around $d = 3$ in the same way as the φ^4 model around $d = 4$. At $d = 3$ the indices are classical with logarithmic corrections as found in (17) by Wilson's method.

If we expand the group equation for τ' and u' at second order we obtain

$$\lambda' \frac{\partial \tau'}{\partial \lambda'} = \psi_\tau(u') \simeq -x_\tau^0 \, \tau' - (x_\tau - x_\tau^0)\frac{u'}{u^*} \, \tau' \tag{7}$$

$$\lambda \frac{\partial u'}{\partial \lambda'} = \psi_u(u') \simeq -x_u^0 \, u' \left(1 - \frac{u'}{u^*}\right) \tag{8}$$

The transformation becomes completely parametrized by the critical exponents $\{x_i^0\}$ and $\{x_i\}$ evaluated at the lowest order.

If we now assume that $\{x_i^0\}$ and $\{x_i\}$ are two sets of indices obtained by other sources and consider eqs.(7) and (8) as exact, this model coincides with the one hypotized by Riedel and Wegner[18] to discuss cross-over effects in various cases such as tricritical systems, the Askin and Teller model and Fisher exponent renormalization.

The two sets of scaling variables are now easily determined

$$g'_\tau = \tau' \left(\frac{u'}{u^*}\right)^x \qquad\qquad g^o_\tau = \tau' \left(1 - \frac{u'}{u^*}\right)^x$$

$$g'_u = \frac{1 - u'/u^*}{u'/u^*} \qquad\qquad g^o_u = \frac{u'/u^*}{1 - u'/u^*}$$

(9)

with $x = x_\tau - x^o_g / x^o_u$

They, asymptotically near to each/one of the fixed points, reduce to τ', $u'-u^*$ or τ' and u' respectively. They are related by the non-equivalent transformation

$$g'_\tau(\lambda') = g^o_\tau(\lambda') \left[g^o_u(\lambda') \right]^x \quad , \qquad g'_u(\lambda') = \left[g^o_u(\lambda') \right]^{-1}$$

(10)

which allows for the discussion of the whole cross-over region.[18]

As previously indicated in the discussion of the model hamiltonian, when a more involved symmetry property is taken into account a new cross-over field is present, which leads to a further renormalization condition of type (6). The related critical exponent coincides with the new cross-over index. The explicit evaluation of critical indices by means of the ε-expansion in all the cases previously indicated is described by Yamazaki in the paper presented at this conference[18] to which we also direct the reader for the original references.

In this scheme it becomes natural to consider/any transport coefficient also as an "additional field " to be renormalized. The corresponding renormalization function leads to a further anomalous dimension, which is the dynamical critical index associated with the transport coefficient under consideration. Of course, in order to introduce bare transport coefficients, one is compelled to start with an equation of motion of "the time-dependent Landau-Ginzburg" type as it was actually done in (20) and (21).

The important problem still under consideration is "if and how" is it possible to derive scaling starting from the microscopic interactions instead of performing the renormalization on the equations of motion. Quite heavy work of coarse - graining is necessary to derive these equations of motion in the very few cases where it is possible.[22] We feel that this process of introducing "hydrodynamics" into the theory is a necessary step to obtain dynamic scaling, if dynamic scaling hold at all.

References

1) L.P.Kadanoff et al., Rev.Mod.Phys. $\underline{39}$, 395 (1967)

2) M.Gell-Mann and F.Low, Phys.Rev. $\underline{95}$, 1300 (1954)

3) C.Di Castro and G.Jona-Lasinio, Phys.Lett. $\underline{29}$A , 322 (1969)

4) K.G.Wilson, Phys.Rev. B$\underline{4}$, 3174-3184 (1971)

5) C.Di Castro, Lett.Nuovo Cim. $\underline{5}$, 69 (1972)

6) E.Brezin, J.C.Le Guillou, J.Zinn-Justin and B.G.Nickel, Phys.Lett. $\underline{44}$A, 277 (1973)

   For a review of this method see J.Zinn-Justin, Cargèse Lecture Notes (1973)

7) C.Di Castro, Proceeding of the Conference on the Renormalization Group in Critical Phenomena and Quantum Field Theory, Philadelphia (1973)

8) G.Jona-Lasinio, "Proceeding of the Nobel Symposium" vol.24 Academic Press (1973)

9) C.Di Castro, G.Jona-Lasinio and L.Peliti, Ann.of Phys. $\underline{87}$, 327 (1974)

10) Th.Niemeyer and J.M.J.Van Leeuwen, Phys.Rev.Lett. $\underline{31}$, 1411 (1973); Physica $\underline{71}$, 17
    (1974); M.Nauenberg and B.Nienhuis, Phys.Rev.Lett. $\underline{33}$, 944 (1974);
    D.Nelson and M.Fisher, preprint (1974)

11) T.L.Bell and K.G.Wilson, preprint (1974)

12) K.G.Wilson and J.B.Kogut, Phys.Rep. $\underline{12}$C, 76 (1974)

13) F.Wegner, Phys.Rev. B$\underline{5}$, 4529 (1972)

14) M.E.Fisher, Rev.Mod.Phys. $\underline{46}$, Oct.1974

15) S.Ma,   Phys.Rev. $\underline{40}$A, 1818 (1974)

16) G.Benettin, C.Di Castro, G.Jona-Lasinio, L.Peliti, A.Stella, preprint (1974)

17) E.K.Riedel and F.J.Wegner, Phys.Rev.Lett. $\underline{29}$, 349 (1972)

    F.J.Wegner and E.K.Riedel, Phys.Rev. B$\underline{7}$, 248 (1973)

18) E.K.Riedel and F.J.Wegner, Phys.Rev. B$\underline{9}$, 294 (1974)

19) Y.Yamazaki paper presented at this conference and references quoted therin

20) B.I.Halperin, P.C.Hohenberg, S.Ma, Phys.Rev. B$\underline{10}$, 139 (1974)

21) B.I.Halperin, P.C.Hohenberg, E.Siggia, Phys.Rev.Lett. $\underline{32}$, 1289 (1974)

22) F.Abrahams and T.Tsuneto, Phys.Rev. $\underline{152}$, 416 (1966)

    C.Di Castro and W.Young, Nuovo Cim. $\underline{62}$B, 273 (1969)

This talk was originally given at the Congressino on "The Renormalization Group
and its applications in statistical phisics" - Trieste July 1974.

GENERALIZATIONS AND APPLICATIONS OF CALLAN-SYMANZIK EQUATIONS
TO STATISTICAL MECHANICS IN CRITICAL PHENOMENA

Yoshitake Yamazaki

Department of Applied Physics, Tohoku University
Sendai, Japan

The aim of the present talk is firstly to generalize the Callan-Symanzik(C-S) equations found in field theories, into the generalized C-S equations appropriate to study critical behaviors in nonrelativistic systems with long range interactions and/or with many coupling constants. The second aim is to apply the generalized C-S equations to various systems and to investigate the critical behaviors in the second order(SO) phase transitions, the tricritical(TC) phase transitions and the multicritical(MC) phase transitions, for the systems with the short range(SR) interaction or with the long range(LR) interactions decaying as $1/r^{d+\sigma}$ (d: dimension of space, $\sigma>0$). In case of the second order phase transitions, anisotropic cubic systems are also investigated.

CALLAN-SYMANZIK(C-S) EQUATIONS

The C-S equations in relativistic field theories were derived by Callan and Symanzik[1] The C-S equations and the renormalization-group (R-G) equations in the sense of Gell-Mann-Low[2] were instantaneously derived by means of differential vertex operations(DVO)[3] based on the Zimmermann method[4] The latter method leads to define the useful concept of normal products in a mathematically rigorous formalism and makes possible obtain the explicit solutions in momentum space of a recursion relation in the Bogoliubov-Parasiuk-Hepp(BPH) renormalization method[5] Problems in nonrelativistic field theories, mainly in statistical mechanics in critical phenomena have been recently attacked by means of renormalization theories(see the references[6]) and in particular have started to be solved regorously in terms of the C-S equations[7]

MODEL HAMILTONIAN

In order to systematically investigate critical behaviors in the various phase transitions of systems with the SR interaction or with LR ones decaying as $1/r^{d+\sigma}$, the following model Hamiltonian is adopted:

$$H = \int d^d x [\tfrac{1}{2}\hat{\partial}_\mu \phi_u \hat{\partial}_\mu \phi_u + \tfrac{1}{2}m^2_{\underset{\sim}{u}u}\phi^2_u + V[\phi_u]] \tag{1}$$

where $\phi_{u\alpha}(x)$ is an unrenormalized real field with N-components, $\hat{\partial} \equiv \partial^{\sigma/2}$, $\phi_{\sim u}^2 \equiv \sum_{\alpha=1}^{N} \phi_{u\alpha}^2(x)$, m_u an unrenormalized mass and $V[\phi_u]$ a set of interaction Hamiltonians. For example, the systems with $V[\phi_u]=V_s \equiv \frac{1}{4!}g_{us}(\phi_{\sim u}^2)^2$ stand for the selfavoiding walk model for N=0, the Ising model for N=1, the XY model for N=2, the Heisenberg model for N=3 and the spherical model for N=∞. The systems with $V[\phi_u]=V_s+V_c$ ($V_c \equiv \frac{1}{4!}g_{uc}\phi_{\sim u}^4 \equiv \frac{1}{4!}g_{uc}\sum_{\alpha=1}^{N}\phi_{u\alpha}^4(x)$) stand for anisotropic cubic systems. Critical behaviors in the TC and MC phase transitions can be investigated in the systems with $V[\phi_u]=V_s+V_3$ ($V_3 \equiv \frac{1}{6!}g_{u6}(\phi_{\sim u}^2)^3$) and $V[\phi_u]=V_s+V_3+V_4+\cdots$ ($V_4 \equiv \frac{1}{8!}g_{u8}(\phi_{\sim u}^2)^4$), respectively. The systems with the SR interaction and with the LR ones are denoted by $\sigma=2$ and $0<\sigma<2$, respectively.

GENERALIZATION OF CALLAN-SYMANZIK EQUATIONS

a TO SYSTEMS WITH LONG RANGE INTERACTIONS

For simplicity we consider the case $V[\phi_u]=V_s$. The effective Hamiltonian of (1) is given by

$$H = \int d^d x[\tfrac{1}{2}(1+b)\hat{\partial}_\mu \phi \hat{\partial}_\mu \phi + \tfrac{1}{2}(m^2-a)\phi_{\sim}^2 + \tfrac{g-c}{4!}(\phi_{\sim}^2)^2] \tag{2}$$

where the quantities $\{\phi,m,g\}$ are the renormalized quantities corresponding to the unrenormalized ones $\{\phi_u,m_u,g_u\}$. The quantities a, b and c are finite BPHZ functions of g. By taking account of the renormalized action principle, counting identities, and the Zimmermann identity *imposed on the normalization conditions* ($\Gamma^{(2)}(p,-p)|_{p^\sigma=-m^2}=0$, $\frac{d\Gamma^{(2)}(p,-p)}{dp^\sigma}|_{p^\sigma=\mu^\sigma}=-1$, $\Gamma^{(4)}(0,0,0,0)_{sp}=-g$), the generalized C-S equation and the R-G equation for the connected one-particle irreducible(1PI) n-point vertex function $\Gamma^{(n)}$ are obtained respectively as

$$[m^2\frac{\partial}{\partial m^2} + \mu^\sigma\frac{\partial}{\partial \mu^\sigma} + \beta\frac{\partial}{\partial g} - n\gamma_3]\Gamma^{(n)} = \alpha m^2 \Delta_0 \Gamma^{(n)} \quad , \tag{3}$$

$$[\mu^\sigma\frac{\partial}{\partial \mu^\sigma} + \hat{\sigma}(\frac{m^2}{\mu^\sigma},g)\frac{\partial}{\partial g} - n\hat{\tau}(\frac{m^2}{\mu^\sigma},g)]\Gamma^{(n)}(p_1,\cdots,p_n;m^2,\mu^\sigma,g) = 0 \tag{4}$$

where α, β and γ_3 are functions of the ratio m^2/μ^σ and g, and $\Delta_0=\frac{1}{2}\int d^d x \times N_{d-\sigma}[\phi^2(x)]$. In the limit m$\to$0 the functions $\hat{\sigma}$ and $\hat{\tau}$ are related as $\beta(0,g)=\hat{\sigma}(0,g)$ and $\gamma_3(0,g)=\hat{\tau}(0,g)$. For the other vertex functions $\{\Gamma_{\phi^2}^{(n)}, \Gamma_{\phi_\alpha\phi_\beta}^{(n)}, \Gamma_{\phi^2\phi^2}^{(n)}$, etc.$\}$ corresponding to the physical quantities the similar C-S and R-G equations can be derived.[8]

b TO SYSTEMS WITH MANY COUPLING CONSTANTS

In most cases of actual critical phenomena various fields and many coupling constants are concerned. Without loss of generality we can confine ourselves to a simple model whose interaction Hamiltonian $V[\phi_u]$ consists of ℓ kinds of ϕ^4-type interactions i.e.,$\{V_1,\cdots,V_\ell\}$.

The corresponding coupling constants are denoted by $g_{u1},\cdots,g_{u\ell}(\equiv\{g_u\})$. For simplicity we put $\sigma=2$. The C-S equation for the connected 1PI n-point Green's (or vertex) functions generalized for (3), is derived to be

$$[-\tfrac{1}{2}\xi\tfrac{\partial}{\partial\xi} + \sum_{j=1}^{\ell} \beta_j(g_j)\tfrac{\delta}{\delta g_j} + \tfrac{1}{2}\{d-n(\tfrac{d}{2}-1+\gamma_3(\{g\}))\}]\Gamma^{(n)}(\xi P;m^2,\{g\})$$

$$= \alpha m^2 \Delta_0 \Gamma^{(n)} \equiv \Delta\Gamma^{(n)}(\xi P;m^2,\{g\}) \tag{5}$$

where $P\equiv\{p_1,\cdots,p_n\}$ and ξ stands for a correlation length. The solution for this differential equation can be obtained. By imposing the boundary conditions the asymptotic behavior of $\Gamma^{(n)}$ for n>3 is also obtained.[9] Other functions $\{\Gamma^{(n)}_{\phi^2},\Gamma^{(n)}_{\phi_\alpha\phi_\beta},\Gamma^{(n)}_{\phi^2\phi^2},\text{etc.}\}$ can be described by similar asymptotic behaviors.

APPLICATIONS OF CALLAN-SYMANZIK EQUATIONS
a TO SECOND ORDER(SO) PHASE TRANSITIONS

Recently Brezin et al[7] investigated the critical phenomena in the isotropic systems with the SR interaction and obtained the critical exponents η and $\{\gamma,\omega\}$ in $4-\varepsilon$ dimensions for small ε, to order ε^4 and ε^3, respectively. Other exponents such as ν, δ and α are rigorously determined from η, γ and scaling laws. However, the crossover exponent ϕ must be determined independently. Its form[10] has been obtained to be

$$\phi = 1 + \varepsilon\frac{N}{2(N+8)} + \varepsilon^2\frac{N(N^2+24N+68)}{4(N+8)^3} + \varepsilon^3\frac{N}{2(N+8)^5} .$$

$$\cdot[\tfrac{1}{4}N^4+12N^3+197N^2+868N+1256-12(5N+22)(N+8)\zeta(3)] + O(\varepsilon^4) . \tag{6}$$

The most important point is that the value of ϕ for $N\to\infty$ converges to the exact value "2".

The isotropic systems[11] with LR interactions have been studied.

For the critical phenomena in anisotropic cubic systems[12] with the SR interaction the results obtained are the critical exponents

$$\eta = \varepsilon^2\frac{(N-1)(N+2)}{54N^2} + \varepsilon^3\frac{N-1}{N^4}\frac{1}{5832}[109N^3-222N^2+1728N-1696] + O(\varepsilon^4) ,$$

$$\gamma = 1 + \varepsilon\frac{N-1}{3N} + \varepsilon^2\frac{N-1}{N^3}\frac{1}{162}[7N^2+142N-212] + O(\varepsilon^3) , \text{ etc.} \tag{7}$$

and the stability of the solutions. The results to order ε^2 agree with the ones obtained by Aharony. The results of the next order in ε will be published in the near future.

Anisotropic cubic systems[13] with LR interactions have been investigated and the following exponents have been obtained:

$$\gamma = 1 + \frac{\varepsilon}{\sigma}\frac{2(N-1)}{3N} + (\frac{\varepsilon}{\sigma})^2\frac{N-1}{9N^3}[-\sigma(N^2-18N+24)\Psi+4N(N-1)] + O(\varepsilon^3),$$

$$\eta \equiv 2 - \sigma , \text{ etc. where } \Psi \equiv \psi(1) - 2\psi(\tfrac{\sigma}{2}) + \psi(\sigma) .$$

6 TO TRICRITICAL(TC) PHASE TRANSITIONS AND MULTICRITICAL(MC) PHASE TRANSITIONS

Since we have no space to describe these problems we confine ourselves to brief descriptions. For the critical behaviors of the systems with the SR interaction or with the LR ones in the TC phase transitions[14], we obtained the critical exponents to order ε^2 in the ε-expansion in 1973 but have not obtained them to ε^3 by one more step. For any order of MC phase transitions we have obtained the critical exponents in the lowest order of ε.

Finally the author would like to thank Profs. S. Katsura, M. Suzuki and S. Inawashiro for their useful discussions.

REFERENCES

1. C. G. Callan, Jr., Phys. Rev. D2, 1541 (1970), D5, 3202 (1972),
   K. Symanzik, Commun. Math. Phys. 16, 48 (1970), 18, 227 (1970).
2. M. Gell-Mann and F. Low, Phys. Rev. 95, 1300 (1954).
3. J. H. Lowenstein, Commun. Math. Phys. 24, 1 (1971),
   B. Schroer, Letter al Nuovo Cimento, 2, 867 (1971).
4. W. Zimmermann, "Lecture on elementary particles and quantum field
   theory", Vol.1, p.397, Cambridge, M.I.T. Press 1970,
   Ann. Phys. (N.Y.) 77, 536, 570 (1973).
5. see the references in 4.
6. K. G. Wilson and J. Kogut, Phys. Lett. 12C, 75 (1974),
   R. Brout, Phys. Lett. 10C, 1 (1974),
   F. J. Wegner, J. Phys. C7, 2098 (1974).
7. E. Brezin, J. C. LeGuillou and J. Zinn-Justin, Phys. Rev. D8, 434,
   2418 (1973), B8, 5330 (1973), D9, 1121 (1974),
   G. Parisi, Rome preprint (INFN#46),
   B. Schroer, Phys. Rev. B8, 4200 (1973).
8. Y. Yamazaki, in preparation.
9. Y. Yamazaki, in preparation.
   E. Brezin, J. C. LeGuillou and J. Zinn-Justin, Phys. Rev. B10, 892
   (1974).
10. Y. Yamazaki, Phys. Lett. 49A, 215 (1974).
11. M. E. Fisher, S. Ma and B. G. Nickel, Phys. Rev. Lett. 29, 917 (1972),
    M. Suzuki, Y. Yamazaki and G. Igarashi, Phys. Lett. 42A, 313 (1972),
    Y. Yamazaki and M. Suzuki, in preparation.
12. K. G. Wilson and M. E. Fisher, Phys. Rev. Lett. 28, 240 (1972),
    A. Aharony, Phys. Rev. B8, 4270 (1973), see the references therein,
    Y. Yamazaki, in preparation.
13. Y. Yamazaki, in preparation.
14. see the references in 6.
    Y. Yamazaki and M. Suzuki, Bussei Kenkyu 21, G25 (1974) and in preparation.

Comments of Dr. M. Suzuki; I've obtained C-S eq. for LR interaction by using path-integral(PI). In this case Ward-Takahashi identity can't be used because there exist some fractional derivatives in the Lagrangian. Thus the PI formulation is convenient for the SR and LR interaction.

<u>ON THE RIGOROUS DEFINITION OF SUPERFLUIDITY AND SUPERCONDUCTIVITY</u>

Takeo Izuyama
Institute of Physics, College of General Education
University of Tokyo
Komaba , Meguro-ku , Tokyo 153 , JAPAN

It is the purpose of this report to give a rigorous definition
of superfluidity and to show that neither the off-diagonal long range
order (ODLRO) nor the Bose-Einstein condensation can be a sufficient
condition for superfluidity. The sufficient condition for superfluidity
in a system possessing the ODLRO is also presented.

We start with the Hamiltonian

$$H \;=\; \sum_j \vec{p}_j^{\,2} / 2\,m \;+\; V(\,\vec{r}_1 ,\cdots, \vec{r}_N\,)$$

together with the cyclic condition

$$V(\,\vec{r}_1,\cdots, \vec{r}_{j-1}, \vec{r}_j + \vec{L} , \vec{r}_{j+1}, \cdots, \vec{r}_N\,) = V(\,\vec{r}_1, \cdots, \vec{r}_j, \cdots, \vec{r}_N\,)$$

where $\vec{L} = (\,L, 0, 0\,)$ or $(\,0, L, 0\,)$ or $(\,0, 0, L\,)$. The impurity
potential as well as the potential for interparticle forces are included
in V. Suppose a uniform external field be switched on[1]. It exerts
the force $E(t) = E \exp(\,\varepsilon\, t\,)$ $(\,\varepsilon > 0\,)$ on a particle along the
positive direction of the x-axis. The field is described by the vector
potential $\vec{F}(t) = (\,F(t) , 0, 0\,)$ with $\partial_t F(t) = E(t)$.   Then the

Hamiltonian is $\mathcal{H} = \sum_j \dfrac{1}{2\,m} (\,\vec{p}_j + \vec{F}(t)\,)^2 + V$.   In the infinite

past the system is supposed to be in the equilibrium state described by
the canonical or the grand canonical density matrix $\rho \propto \exp\{-\beta(H-\mu N)\}$.
The above situation represents the system contained in a slowly rotating
torus. We follow the logic of Kubo's linear response theory[2]. The
linear response current against the external field $F(t)$ is

$$\overline{J(t)} = \{ \frac{N}{m} - 2 \left\langle J \frac{H^\times}{(H^\times)^2 + \varepsilon^2} J \right\rangle \} F(t)$$

where $J \equiv \sum_j p_j^{\,x} / m$, $H^\times \cdots = [\,H , \cdots\,]$, $\langle \cdots \rangle = \mathrm{Tr}\,\{\cdots\}$,
and $\langle\!\langle \cdots \rangle\!\rangle$ denotes the average over impurity distributions. The presence
of impurities is crucial in a consideration on superfluidity. The re-
sponse current is expressed as $j(t) \equiv N^{-1}\overline{J(t)} \equiv j_1(t) + j_2(t)$,
where $j_1(t) = \Lambda F(t)$ and $j_2(t) = \sigma_\varepsilon E(t)$ with

$$\Lambda \equiv \frac{1}{m} - \frac{2}{N} \left\langle\!\left\langle J \frac{Q}{H}^{\times} J \right\rangle\!\right\rangle \quad \text{and} \quad \sigma_\epsilon \equiv \frac{2}{N} \left\langle\!\left\langle J \left(\frac{Q}{H}^{\times} \frac{\epsilon}{(H^\times)^2 + \epsilon^2} \right) J \right\rangle\!\right\rangle .$$

In the above the projection operator Q is defined by

$$(Q J)_{nm} = J_{nm} \quad \text{for} \quad E_n \neq E_m \quad \text{and} \quad 0 \quad \text{for} \quad E_n = E_m .$$

It is only $j_2(t)$ that corresponds to the ordinary Ohmic current. In fact, if we take the thermodynamic limit (denoted by Lim) and then put $\epsilon \to 0$, we obtain

$$\sigma = \text{Lim} \frac{2}{N} \left\langle\!\left\langle J \frac{\delta(H^\times) Q}{H^\times} J \right\rangle\!\right\rangle$$

which is a generalized version of the Greenwood formula[3] for the Ohmic conductivity. The other component $j_1(t)$ represents the persistent current, which is non-dissipative: If Λ is non-vanishing, then a finite current exists even for an infinitesimal E(t). We have the following;

$$\text{Lim } \Lambda > 0 \quad : \quad \text{Superfluidity.}$$
$$\text{Lim } \Lambda = 0 \quad : \quad \text{Normal fluidity.}$$
$$\text{Lim } \Lambda < 0 \quad : \quad \text{The assumed (ground) state is unstable.}$$

Note that the perfect conductor $\sigma = \infty$ and the superconductor $\text{Lim } \Lambda > 0$ are entirely different.

To see that the ODLRO does not necessarily lead to $\text{Lim } \Lambda > 0$, we consider here the ground state of ideal Bosons in a large one-dimensional ring with a single δ-function potential at the origin. The single particle Hamiltonian is $H_1 = - (1 / 2 m)(d/ d x)^2 + g \delta(x)$. The even parity solution of $H_1 \psi_k = E_k \psi_k$ is $\psi_k \propto \cos[k((L/2) - x)]$ for $0 < x < L$, where k is determined by $\tan(kL/2) = mg/k$ and $E_k = k^2/2m$. The odd parity solutions of $H_1 \phi_n = E_n \phi_n$ is $\phi_n \propto \sin[(2\pi n/L)((L/2) - x)]$ with $n = 1, 2, \cdots$ and $E_n = (2\pi n/L)^2 (1/2m)$. The ground state of H_1 is described by ψ_{k_0} where k_0 is the minimum value of k. For large L , $k_0 = (\pi/L) - (2\pi/L^2 mg) + \cdots$. Thus

$$\frac{2}{N} \left\langle \Psi_0 \left| J \frac{1}{H^\times} J \right| \Psi_0 \right\rangle = \frac{4}{m} \sum_n \frac{|\langle \phi_n | p^X | \psi_{k_0} \rangle|^2}{(2\pi n/L)^2 - k_0^2}$$

$$\xrightarrow{\text{Lim}} \frac{64}{\pi^2 m} \sum_{n=1}^{\infty} \frac{4 n^2}{(4n^2 - 1)^3} = \frac{1}{m} .$$

The Thomas-Kuhn sum rule $\Lambda = 0$, which holds in any non-cyclic system, has been restored by a single scatterer in the cyclic system. Our system exhibits the condensation $\left\langle\!\left\langle \Psi_0 | a_0^\dagger a_0 | \Psi_0 \right\rangle\!\right\rangle = 8N/\pi^2$ and the ODLRO $\left\langle\!\left\langle \Psi_0 | \psi^\dagger(x') \psi(x'') | \Psi_0 \right\rangle\!\right\rangle = 2 (N/L) \neq 0$, though there is no superfluidity.

Finally a system of interacting Bosons containing low concentration of impurities is considered. Suppose that the system possesses the ODLRO

$$\lim_{|\vec{r}' - \vec{r}''| \to \infty} \mathrm{Lim} \quad \langle\!\langle \psi(\vec{r}')\, \psi^\dagger(\vec{r}'') \rangle\!\rangle \equiv \langle \Phi^2 \rangle \neq 0.$$

We start with the formula[1]

$$\Lambda = -2\,N^{-1}L^2\, \langle\!\langle J(x')\,(Q/H^x)\,J(x'') \rangle\!\rangle$$

where

$$J(x') \equiv (1/2m)\,\Sigma_j\,\{\, p_j^{\,x}\,\delta(x_j - x') + \delta(x_j - x')\,p_j^{\,x}\,\}$$

and $x' \neq x''$. We consider the case $|x' - x''| \to \infty$. It is proved that

$$\langle\!\langle J(x')\,\tfrac{Q}{H^x}\,J(x'') \rangle\!\rangle = \int dy'dz' \int dy''dz''\,\lim_{\varepsilon \to +0}\, i\,\Big(\int_0^\infty - \int_{-\infty}^0\Big)dt\, e^{-\varepsilon|t|}\,\Gamma(\vec{r}'t\,|\,\vec{r}''0)$$

where

$$\Gamma(\vec{r}'\,t\,|\,\vec{r}''\,0) = (1/2m^2)\,\Big\{ \langle\!\langle (\partial'\psi^\dagger(\vec{r}'t))\,\psi(\vec{r}'t)\,\psi^\dagger(\vec{r}''0)\,(\partial''\psi(\vec{r}''0)) \rangle\!\rangle$$
$$+\ \langle\!\langle \psi^\dagger(\vec{r}'t)\,(\partial'\psi(\vec{r}'t))\,(\partial''\psi^\dagger(\vec{r}''0))\,\psi(\vec{r}''0) \rangle\!\rangle \Big\} - (1/4m^2)\,\partial'\partial''\langle\!\langle \rho(\vec{r}'t)\,\rho(\vec{r}''0) \rangle\!\rangle$$

with $\partial = \partial/\partial x$ and $\rho(x) =$ the density operator. The last term in Γ does not contribute when $|x' - x''| \to \infty$. We adopt the decoupling approximation, which is believed to be valid in this limit.

$$\Gamma(\vec{r}'\,t\,|\,\vec{r}''\,0) = (\langle \Phi^2 \rangle/2m^2)\,\partial'\partial''\{ \langle\!\langle \psi^\dagger(\vec{r}'t)\,\psi(\vec{r}''0) \rangle\!\rangle + \langle\!\langle \psi(\vec{r}'t)\,\psi^\dagger(\vec{r}''0) \rangle\!\rangle \}$$

Then

$$\langle\!\langle J(x')\,\tfrac{Q}{H^x}\,J(x'') \rangle\!\rangle = \frac{\langle \Phi^2 \rangle}{2m^2}\,\frac{1}{\Omega}\,\sum_{k \neq 0} k^2 e^{ik(x'-x'')} \int_{-\infty}^\infty \frac{\langle A(k,\omega) \rangle}{\omega}\,d\omega$$

where $k = 2\pi n/L$, $\Omega =$ volume , and

$$A(k,\omega) = \langle a_k\,\delta(H^x - \mu - \omega)\,a_k^\dagger \rangle - \langle a_k^\dagger\,\delta(H^x + \mu + \omega)\,a_k \rangle$$

with $a_k \equiv \Omega^{-1/2}\int \psi(\vec{r})\,e^{-ikx}\,d\vec{r}$.

The quantity $\int d\omega\,(\langle A(k,\omega)\rangle/\omega)$ is obtained by a general argument. The result is

$$\int_{-\infty}^\infty d\omega\, \frac{\langle A(k,\omega) \rangle}{\omega} = \frac{m^2\,\langle \Psi^2 \rangle}{k^2\,f(k)}\,,$$

where $f(k)$ may be a function of k in the impure system. In the case of pure systems considered by the Josephson-Baym theory[4,5], $f(k) = \rho_s$.

The condition for superfluidity is found to be $f(k) = \mathrm{constant} \equiv \tilde{\rho}_s$ (for small k including the smallest one, namely, $k = 2\pi/L$).

$$\therefore\ \Lambda = -\frac{m\,\langle \Phi^2 \rangle^2}{\rho\,\tilde{\rho}_s}\,\sum_{k \neq 0}\exp[\,ik(x'-x'')\,] = \frac{m\,\langle \Phi^2 \rangle^2}{\rho\,\tilde{\rho}_s} > 0$$

where $\rho = mN/\Omega$. If ρ_s is defined by the London equation, then $\tilde{\rho}_s = \langle \Phi^2 \rangle^2 / \rho_s$. The condition $f(k) \to \tilde{\rho}_s$ ($k \to 0$) is not trivial. In the system of ideal (non-interacting) Bosons containing macroscopic number of impurities, $f(k)$ diverges as k^{-2} , which is rigorously proved. In this case $\Lambda = 0$.

References 1) T. Izuyama, Progr. Theor. Phys., <u>50</u> , 841 (1973).
2) R. Kubo, Jour. Phys. Soc. Japan, <u>12</u>, 570 (1957).
3) D. A. Greenwood, Proc. Phys. Soc., <u>71</u>, 585 (1958).
4) B. D. Josephson, Phys. Letters, <u>21</u>, 608 (1966).
5) G. Baym, in Mathematical Methods in Solid State and Superfluid Theory, edited by R. C. Clark and G. H. Derrick (Oliver & Boyd , Edinburgh 1967) p.121 ∿ 156.

VARIATIONAL PROBLEMS AND FREE BOUNDARY PROBLEMS

Jacques-Louis LIONS

COLLEGE DE FRANCE

INTRODUCTION.

It is well-known that many questions of physics lead to so-called "free-boundary problems"; in these problems, the state of the system we want to study, say $w = w(x,t)$ ([1]), is given by one or several partial differential equations subject to initial and boundary conditions in one or several regions whose boundaries are unknown (the free boundaries).

Classical examples are: the Stefan's problem, flows of Bingham's fluids, problems of elasto-plasticity, water waves, porous media, unilateral constraints, etc.. .

It has been observed in [1] that questions of optimal control for distributed systems also lead to problems of the "free boundary type".

More recently, we have observed with A. Bensoussan [2] the following: if one considers systems governed by (stochastic) Ito's differential equations, where the "control variable" can be:

      (i) a measurable function, or

      (ii) a stopping time or

      (iii) impulses (or a combination of them),

and if one uses the Hamilton-Jacobi approach, then one obtains for the optimal cost function, say $u(x,t)$ ([2]), problems of the free boundary type.

In this manner, one can think of the solution w of some free boundary problems arising in physics, as the solution u of an "associated" problem of optimal control, and vice versa.

The "association" $w \leftrightarrow u$ can simply be the identity $u = w$; or w and u can be related via some integro-differential transformation.

Some of the mathematical tools for dealing with one or the other of the forms of these problems are the variational inequalities (V.I) or

---

([1])   x = geometrical variable, t = time.

([2])   x = state of the system at time t.

the quasi-variational inequalities (Q.V.I.).

We briefly present these tools on three examples, corresponding to the three possible natures of the control variables indicated before.

These remarks lead to a number of seemingly interesting questions:

(a)   what are the free surface problems which are equivalent (via "some" transformation $w \leftrightarrow u$) to "some" optimal control problem?
The answer is known in an increasing number of cases, but it is an open question for instance for free boundary problems related to water waves.

(b)   in all the known situations the equivalence, by the transformation $w \leftrightarrow u$, between the physical problem and the optimal control problem, and also with the formulation as a V.I. or a Q.V.I., has been <u>checked</u> by showing the equivalence between the various non-linear P.D.E. obtained for the characterization of w and of u; is it possible to obtain the equivalence between the free surface problem and the optimal control problem on <u>physical grounds</u>, without writing down the P.D.E.? Even a formal solution would probably give interesting insights on the whole area.

The plan is as follows:

1. Stopping times.
2. An example with a "continuous" control variable.
3. Impulse controls.

I.- <u>STOPPING TIMES</u>.

1.1. <u>Formulation of the problem</u>.

Let us consider a system whose state $y \in R^n$ at time s is given by the solution of the Ito's differential equation:

$$(1.1) \qquad dy = g(y)ds + \sigma dw(s) \ ,$$

where g is a C^1 function of $R^n \to R^n$, where σ is a positive constant (to simplify the exposition) and where w is a Wiener process in R^n; the solution of (1.1) is subject to the initial condition:

$$(1.2) \qquad y(t) = x \ ;$$

Let $\ y_{x,t}(s)\ $ be the corresponding solution.

We can also consider a similar process in a, say, bounded open set $\mathcal{O}$ of $\mathbb{R}^n$; starting of course with $x \in \mathcal{O}$, the state is given by the solution of (1.1) as long as $\ y\ $ does not reach the boundary $\Gamma = \partial\mathcal{O}$, i.e. for

$$(1.3) \qquad s \leq \theta(x,t) = \text{exit time of } \bar{\mathcal{O}} \ .$$

When one reaches $\ \Gamma\ $ the process is interrupted (other rules can be considered when one reaches $\ \Gamma\ $; these rules would lead to other boundary conditions).

The <u>cost function</u> is

$$(1.4) \qquad J_{x,t}(\tau) = E \int_t^\tau f(y_{x,t}(s),\ s)ds$$

where the control variable is $\ v = \tau\ $ (stopping time), and where $\ E\ $ denotes the expectation. In (1.4) the function $\ f(x,t)\ $ is given, say, continuous in $\ \bar{\mathcal{O}} \times \mathbb{R}_t$. The control variable is subjected to the constraints

$$(1.5) \qquad \tau \leq \min\,(\theta(x,t),\ T),$$

where $\ T\ $ is a given finite time (the horizon of the problem). We introduce now:

$$(1.6) \qquad u(x,t) = \inf_\tau\ J_{x,t}(\tau),\ \tau \text{ subject to (1.5)}.$$

<u>The problems are</u> :

(i) to characterize $\ u\ $ by analytical properties;

(ii) to derive information on the optimal stopping time (which exists) from the knowledge of $\ u$.

(This is the Hamilton-Jacobi approach to problems of calculus of variations).

1.2. <u>Partial differential inequalities</u>.

We have shown with A. BENSOUSSAN in [6] that $\ u(x,t)\ $ is characterized by the following set of inequalities and equalities:

$$(1.7) \qquad \begin{cases} u \leq 0, \\[2mm] -\dfrac{\partial u}{\partial t} + Au - f \leq 0, \\[2mm] u\left(-\dfrac{\partial u}{\partial t} + Au - f\right) = 0 \quad \text{in} \quad \mathcal{O} \times\,]t_0, T[\,, \end{cases}$$

where in (1.7) A is given by

$$(1.8) \qquad \begin{cases} Au = -\, \Sigma\, g_i(x)\, \dfrac{\partial u}{\partial x_i} - \dfrac{\sigma^2}{2}\, \Delta u\, , \quad g(x) = \{g_i(x)\}, \\[3mm] \Delta u = \Sigma\, \dfrac{\partial^2 u}{\partial x_i^2}\, , \end{cases}$$

and where in (1.7) u is subject to the <u>boundary condition</u>

$$(1.9) \qquad u = 0 \quad \text{on} \quad \partial \mathcal{O} \times\,]t_0, T[\,,$$

and to the "<u>initial</u>" condition

$$(1.10) \qquad u = 0 \quad \text{for} \quad t = T.$$

In (1.7) t_0 is any finite (or $-\infty$) number $< T$.

<u>Remark 1.1</u>:

As it is usual in this framework, the time goes "backward" in (1.7)... ...(1.9). By changing t into $T-t$ and $u(x,t)$ into $u(x,T-t)$, one obtains the (obviously equivalent !) problem:

$$(1.11) \qquad \begin{cases} u \leq 0, \quad \dfrac{\partial u}{\partial t} + Au - f \leq 0, \quad u\left(\dfrac{\partial u}{\partial t} + Au + f\right) = 0 \\[3mm] \text{in}\ \ \mathcal{O} \times\,]0, t_1[\,, \text{ where } t_1 \text{ is an arbitrary} > 0 \text{ number,} \end{cases}$$

where u is subject to (1.9) and to :

$$(1.12) \qquad u = 0 \quad \text{for} \quad t = 0.$$

1.3. <u>Connection with Stefan's free boundary problem.</u>

The problem (1.11) (1.9) (1.12) is <u>of the free boundary type</u>; indeed, by virtue of (1.11), one sees that there are two regions in the domain

$\Theta\times]o,t_1[= Q$; in one region, say Q_o one has $u = 0$, and in the other region, say Q_1, one has $u < 0$ and

$$(1.13) \qquad \frac{\partial u}{\partial t} + Au = f.$$

Let us denote by S the interface between Q_o and Q_1 and let us introduce[1]

$$(1.14) \qquad w = \frac{\partial u}{\partial t} .$$

If we assume that the interface S can be described by

$$(1.15) \qquad t = \ell(x)$$

then, if u is a "smooth" solution of (1.11) (1.9) (1.12)[2] one has

$$(1.16) \qquad u(x,\ell(x)) = 0 \quad \frac{\partial u}{\partial x_i}(x,\ell(x)) = 0, \quad \frac{\partial u}{\partial t}(x,\ell(x)) = 0.$$

Then w satisfies

$$(1.17) \qquad \frac{\partial w}{\partial t} + Aw = \frac{\partial f}{\partial t} \quad \text{in} \quad Q_1.$$

On S one has firstly

$$(1.18) \qquad w(x, \ell(x)) = 0.$$

Taking the x_i's derivative of $\frac{\partial u}{\partial x_i}(x, \ell(x)) = 0$, one obtains

$$\frac{\partial^2 u}{\partial x_i^2}(x, \ell(x)) + \frac{\partial}{\partial x_i} w(x, \ell(x)) \frac{\partial \ell}{\partial x_i}(x) = 0.$$

Therefore

$$(1.19) \qquad \Sigma \frac{\partial w}{\partial x_i} \frac{\partial \ell}{\partial x_i} = -\Delta u(x, \ell(x)) = \text{(by using (1.16) and (1.13))}$$

$$= \frac{2}{\sigma^2} f(x, \ell(x)).$$

The conditions (1.18) (1.19) are of the type of the classical conditions met in free surface problems of Stefan's type.

1.4. Variational inequalities.

---

[1] This is an example of the transformation $w \to u$ we refer to in the Introduction.

[2] We shall return to this point.

Let us now return to problem (1.11) (1.9) (1.12). We introduce

$$H^1(\mathcal{O}) = \text{Sobolev space of order 1} = \{v \,|\, v, \tfrac{\partial v}{\partial x_i} \in L^2(\mathcal{O})\} \,,$$

$$H^1_o(\mathcal{O}) = \{v \mid v \in H^1(\mathcal{O}) \,, \quad v = 0 \quad \text{on } \partial\mathcal{O}\},$$

and we define, for $u, \; v \in H^1_o(\mathcal{O})$

$$(1.20) \qquad a(u,v) = \frac{\sigma^2}{2} \sum_{i=1}^{n} \int_{\mathcal{O}} \frac{\partial u}{\partial x_i} \frac{\partial v}{\partial x_i} \, dx - \sum_{i=1}^{n} \int_{\mathcal{O}} g_i(x) \frac{\partial u}{\partial x_i} v dx,$$

and for $f, \; v \in L^2(\mathcal{O})$, we set

$$(1.21) \qquad (f,v) = \int_{\mathcal{O}} f(x)v(x)dx.$$

We denote by $u(t)$, $f(t)$ the functions $x \to u(x,t)$, $x \to f(x,t)$.
With these notations in mind, it is a simple exercise to check that the
problem (1.11) (1.9) (1.12) is - formally - equivalent to finding $u =$
$u(t)$ such that

$$(1.22) \qquad \left| \begin{array}{l} (\frac{\partial u}{\partial t}(t), \; v-u(t)) + a(u(t), \; v-u(t)) \geq (f(t), \; v-u(t)), \\[2mm] \forall v \in H^1_o(\mathcal{O}), \quad v \leq 0, \end{array} \right.$$

where

$$(1.23) \qquad u(t) \in H^1_o(\mathcal{O}) \quad u(t) \leq 0,$$

$$(1.24) \qquad u(o) = 0.$$

The problem (1.22) (1.23) (1.24) is what is called a <u>variational
inequality of evolution</u> (V.I.).

These V.I. have been introduced in [21] by STAMPACCHIA and the A. Using
general theorems (cf. [19] [20] and the bibliography therein) one obtains
<u>the existence and uniqueness of a solution of</u> (1.22) (1.23) (1.24).
If $f \in L^2(Q)$, one obtains $u \in L^2(0,t;H^1_o(\mathcal{O}))$, satisfying (1.22) in a weak
form. If moreover $\frac{\partial u}{\partial t} \in L^2(Q)$, then the solution is strong: $\frac{\partial u}{\partial t} \in L^2(0,T;H^1_o(\mathcal{O}))$.

1.5. <u>Remarks</u>:

1) The equivalence between optimal stopping times and Stefan's problem
has been observed by several authors (let us refer to [14], [17] and the
bibliography therein) without making explicit the characterization of

u in terms of inequalities equivalent of V.I. A systematic approach
of optimal stopping times via V.I. has been made in [6], [7] and
independently in [22], [23].

2) The equivalence of Stefan's problem with V.I. (without using optimal
stopping times) has been observed by G.DUVAUT [15] and extended to
multi-phase problems by several authors,in particular by DUVAUT [15] 2).

3) The _regularity_ in t and in x of the solution of the V.I. of
Section 1.4. and of more general types has been studied by H. BREZIS [13].

4) One can consider systems with a measurable control function together
with stopping times; this leads to V.I. for _non-linear_ partial differential operators: cf. [7] and Section 2 below.

5) All this extends to problems of _games_ with stopping times. Cf. [4].

6) Some problems of physics - such as, for instance, problems of flows
of Bingham's fluids - lead to V.I. of a different nature; see [16].

7) Other problems - such as, for instance, infiltration in porous media-
can be reduced to V.I. via a transformation of a much more complicated
type than (1.14). Cf. C. BAIOCCHI [1] and for the evolution problem
TORELLI [24].

8) If u is known and if one starts at time t from x such that
$u(x,t) < 0$, then the optimal stopping time is the "first" time the state
$y_{x,t}(s)$ "reaches" the free surface S (i.e. the first time $u(y_{x,t}(s),s)$
becomes zero); the set "u < 0" is called the _continuation set_.

2. <u>AN EXAMPLE WITH A "CONTINUOUS" CONTROL VARIABLE</u>.

2.1. <u>Setting of the problem</u>.

     Let us consider now a system whose state is given by (compare to
(1.1))

$$(2.1) \qquad dy = v^2(s)g(y)ds + v(s)\ \sigma dw(s),$$

where v(s) is a scalar control function subject to

$$(2.2) \qquad v(s) \in [0, 1];$$

the state y is subject to the condition

$$(2.3) \qquad\qquad y(t) = x.$$

The corresponding solution is denoted by

$$y_{x,t}(s;v).$$

If one adds the constraint for y to stay in $\overline{\mathcal{O}}$ (notation of Section 1) then s has to be $\leq \theta(x,t;v)$ = exit time which now depends on v. If we <u>do not</u> introduce stopping times, but if we set $y_{x,t}(s;v) = x$ for s $>$ t if $x \in \partial\mathcal{O}$, we can consider the <u>cost function</u>

$$(2.4) \qquad\qquad J_{x,t}(v) = E \int_{t}^{T} F(y_{x,t}(s;v), s, v(s))ds$$

where T is given finite and where $F(x,s,\lambda)$ is a given continuous function on $\overline{\mathcal{O}} \times]-\infty,T]\times[o,1]$.

We next introduce:

$$(2.5) \qquad\qquad u(x,t) = \inf J_{x,t}(v), v(s) \quad \text{subject to (2.2).}$$

2.2. <u>P.D.E. satisfied by</u> u.

Let us proceed <u>formally</u>, using BELLMAN's "optimality principle" (cf. [3]). One obtains, by taking $v(s)= \lambda (\in[o,1])$ in $(t,t+\varepsilon)$ =

$$(2.6) \qquad u(x,t) = \inf_{\lambda}[\varepsilon F(x,t,\lambda) + E\, u(x+\varepsilon\lambda^2 g(x)+\lambda\sigma w(\varepsilon), t+\varepsilon) +0(\varepsilon)]$$

and by a formal expansion

$$(2.7) \qquad 0 = \frac{\partial u}{\partial t} + \inf_{\lambda\in[o,1]} [F(x,t,\lambda) + \lambda^2(\Sigma g_i(x) \frac{\partial u}{\partial x_i} + \frac{\sigma^2}{2} \Delta u)].$$

Using the notations of Section 1, (2.7) is equivalent to

$$(2.8) \qquad \frac{\partial u}{\partial t} + \inf_{\lambda \in [o,1]} [F(x,t,\lambda) - \lambda^2 Au] = 0.$$

If we take in particular

$$(2.9) \qquad F(x,t,\lambda) = \lambda^2 f(x,t)$$

then (2.8) is equivalent to (compare to (1.7))

$$(2.10) \quad \begin{vmatrix} -\dfrac{\partial u}{\partial t} \leq 0 \ , \\[2ex] -\dfrac{\partial u}{\partial t} + Au - f \leq 0 \ , \\[2ex] (-\dfrac{\partial u}{\partial t})(-\dfrac{\partial u}{\partial t} + Au-f) = 0 \quad \text{in} \quad \mathcal{O} \times]t_o,T[\ . \end{vmatrix}$$

The boundary condition is:

$$u(x,t) = \inf \int_t^T v^2(s) \ f(x,s)ds \ , \ v(s) \in [o,1]$$

i.e.

$$(2.11) \qquad u(x,t) = \text{given function on } \Gamma$$

and the "initial" condition is

$$(2.12) \qquad u(x,T) = 0 \ .$$

As in Section 1, by changing t into $T-t$, the problem is seen to be equivalent to

$$(2.13) \quad \begin{vmatrix} \dfrac{\partial u}{\partial t} \leq 0, \quad \dfrac{\partial u}{\partial t} + Au-f \leq 0, \quad (\dfrac{\partial u}{\partial t})(\dfrac{\partial u}{\partial t} + Au - f) = 0 \\[2ex] \text{in} \quad \mathcal{O} \times]o,t_1[\end{vmatrix}$$

with (2.11) and

$$(2.14) \qquad u(x,o) = 0.$$

This is again a problem of <u>free boundary</u>, the free surface being the interface between the regions where $\dfrac{\partial u}{\partial t} = 0$ and where $\dfrac{\partial u}{\partial t} + Au-f = 0.$

2.3. <u>Formulation as a V.I.</u>

If we assume, to simplify, that $f \geq 0$, then (2.11) reduces to

$$(2.15) \qquad u = 0 \quad \text{on} \quad \Gamma \times]o,t_1[\ .$$

With the notations of Section 1, the problem (2.13)(2.14)(2.15) is seen to be equivalent to the V.I.

$$(2.16) \quad \left| \begin{array}{l} (\frac{\partial u(t)}{\partial t}, \; v-\frac{\partial u(t)}{\partial t}) + a(u(t), v-\frac{\partial u(t)}{\partial t}) \geq (f(t), v-\frac{\partial u(t)}{\partial t}) \; , \\[2mm] \forall v \in H_o^1(\mathcal{O}) \; , \; v \leq 0 \quad \text{in} \; \mathcal{O} \end{array} \right.$$

where

$$(2.17) \quad \frac{\partial u}{\partial t} \leq 0 \; , \quad \frac{\partial u(t)}{\partial t} \in H_o^1(\mathcal{O}) \; ,$$

and

$$(2.18) \quad u(o) = 0 \; .$$

V.I. of this type have been introduced in [16]. <u>One can show the exist-ence and uniqueness of a solution of</u> (2.16)(2.12)(2.18); cf. [16].

3. - IMPULSE CONTROL.

3.1. Setting of the problem.

Let us consider now a system whose state is given by (compare to (1.1) or to (2.1)):

$$(3.1) \quad dy = g(y)ds + \sigma dw(s) + \sum_{1=1}^{N(\mathcal{O})} \xi^i \, \delta(s-\theta^i) \; ,$$

$$(3.2) \quad y(t) = x \; ,$$

where

$$(3.3) \quad \left| \begin{array}{l} t < \theta^1 < \cdots < \theta^N \leq T \; , \\[2mm] N = N(\mathcal{O}) \quad \text{is } \underline{\text{not}} \text{ given} \; , \\[2mm] \xi^i \in \mathcal{U} \subset R^n \; . \end{array} \right.$$

In (3.1), $\delta(s-\theta)$ denotes the Dirac measure at time θ . The control variable is $v = \{\theta^1, \xi^1; \theta^2, \xi^2; \cdots \theta^N, \xi^N; N\}$. We denote by

$$(3.4) \quad y_{x,t}(s;v)$$

the solution of (3.1)(3.2).

The <u>cost function</u> is given by

$$(3.5) \qquad J_{x,t}(v) = E[\int_t^T f(y_{x,t}(s;v),s)ds + k\, N(\theta)]\ ,$$

where f is a given continuous function and where k is a given > 0 constant.

We define

$$(3.6) \qquad u(x,t) = \inf_v J_{x,t}(v)\ .$$

To simplify the exposition, we assume here that $x \in R^n$ and that y can describe R^n without constraints.

3.2. <u>Inequalities satisfied by</u> u.

One can show [8]...[12] that u is characterized by the set of inequalities

$$(3.7) \qquad \left|\begin{array}{l} u - M(u) \le 0, \\[2ex] -\dfrac{\partial u}{\partial t} + Au - f \le 0\ , \\[2ex] (u - M(u))(-\dfrac{\partial u}{\partial t} + Au - f) = 0 \quad \text{in}\ \ R^n \times]t_o,T[\ , \end{array}\right.$$

subject to the "initial" condition

$$(3.8) \qquad u(x,T) = 0\ ;$$

in (3.7) we have set

$$(3.9) \qquad M(u)(s,t) = k + \inf_{\xi \in \mathcal{U}} .u(x+\xi,t).$$

<u>Remark 3.1.</u>

The problem (3.7)(3.8) is again of the "free boundary type" since we have two regions where $u = M(u)$ and where $-\dfrac{\partial u}{\partial t} + Au = f$. The main new fact here is that M is a <u>non local operator</u>.

3.3. <u>Formulation as a</u> Q.V.I.

With the notations of Section 1[1], we can verify that the problem (3.7)(3.8) (we do not change here t into $T-t$) is equivalent to

$$(3.10) \quad \left| \begin{array}{l} (-\dfrac{\partial u(t)}{\partial t}, v-u(t)) + a(u(t),v-u(t)) \geq (f(t),v-u(t)) \\[2mm] \forall\, v \in H^1_o(\mathcal{O}) \quad , \quad v \leq M(u(t)) \end{array} \right.$$

and where

$$(3.11) \quad u(t) \in H^1_o(\mathcal{O}) \quad , \quad u(t) \leq M(u(t)) \ ,$$

$$(3.12) \quad u(T) = 0 \ .$$

<u>The problem</u> (3.10)(3.11)(3.12) <u>is a</u> Q.V.I. <u>of evolution.</u>

In case $\dfrac{\partial f}{\partial t} \leq 0$, $\dfrac{\partial f}{\partial t} \in L^2(\mathcal{O} \times\ ]t_o,T[\ )$, <u>one can show the existence</u> <u>and uniqueness of a solution</u> u <u>of</u> (3.10)(3.11)(3.12), satisfying

$$u \in L^\infty(0,T;H^1_o(\mathcal{O})), \ \dfrac{\partial u}{\partial t} \in L^2(\mathcal{O} \times\ ]t_o,T[\ ) \ .$$

<u>Remark 3.2.</u>

In the <u>stationary case</u>, the problem reduces to

$$(3.13) \quad \left| \begin{array}{l} u - M(u) \leq 0 \ , \\[3mm] Au - f \leq 0 \ , \\[3mm] (u - M(u))(Au - f) = 0 \end{array} \right.$$

where

$$(3.14) \quad M(u) = k + \inf_{\xi \in \mathcal{U}}\ u(x+\xi) \ .$$

Then (3.10) reduces to the Q.V.I. of elliptic type [5]

---

[1] In order not to make the exposition too technical, we artificially add at this point the constraint $x \notin \mathcal{O}$ and $u = 0$ on $\partial \mathcal{O}$. For a more natural treatment, we refer to A.BENSOUSSAN and the A.

$$(3.15) \quad \left|
\begin{array}{l}
a(u,v-u) \geq (f,v-u) \qquad v \leq M(u) \\[2ex]
u \leq M(u)
\end{array}
\right.$$

C.BAIOCCHI has shown in [2] that a problem of infiltration in porous media can be reduced (after an integro-differential transformation) to a Q.V.I. of type (3.15) with a different operator $M(u)$; essentially $M(u)$ is constructed as follows : let $x \rightarrow \mathscr{C}(x)$ be a continuous mapping from $\overline{\mathscr{O}} \rightarrow \Gamma_1$ where Γ_1 is a subset of Γ ; then

$$Mu(x) = u(\mathscr{C} x) \ .$$

(We refer to BAIOCCHI for details; the presentation given here is an over-simplification !).

For other aspects of impulse control problems, we refer to papers of A. BENSOUSSAN and the A. and to the bibliography therein.

BIBLIOGRAPHY

[1] C. BAIOCCHI, Sur un problème de frontière libre. C.R.A.S. Paris, 273 (1971), pp. 1215-1217.

[2] C. BAIOCCHI, Note C.R.A.S. Paris, t. 278 (1974).

[3] R. BELLMAN, Dynamic programming. Princeton University Press (1957).

[4] A. BENSOUSSAN and A. FRIEDMAN, Non linear variational inequalities and differential games with stopping times. J. of Functional Analysis. 16 (1974), pp. 305-352.

[5] A. BENSOUSSAN, M. GOURSAT and J.L. LIONS, Note C.R.A.S. Paris, t. 273 (1974), pp. 1279-1284.

[6] A. BENSOUSSAN and J.L. LIONS, Problèmes de temps d'arrêt optimal et inéquations variationnelles paraboliques. Applicable Analysis (1973) (3), pp. 267-294.

[7] A. BENSOUSSAN and J.L. LIONS, Inéquations variationnelles non linéaires du premier et du second ordre. Note C.R.A.S. Paris, t. 276 (1973), pp. 1411-1415.

[8] A. BENSOUSSAN and J.L. LIONS, Nouvelle formulation de problèmes de contrôle impulsionnel et applications. Notes C.R.A.S. Paris, t. 176 (1973), pp. 1189-1192; 1333-1338; t. 278 (1974) pp. 675-679; 747-751.

[9] A. BENSOUSSAN and J.L. LIONS, On some questions related to optimal control. (In Russian) Ouspechi Mat. Nauk (1974) XXIX, (176) pp. 77-85.

[10] A. BENSOUSSAN and J.L. LIONS, Nouvelles méthodes en contrôle impul-
      sionnel. Applied Mathematics and Optimization. An Inter-
      national Journal. Vol. 1 Nb.3 Springer-Verlag (1974)

[11] A. BENSOUSSAN and J.L. LIONS, Optimal impulse and continuous
      control. Method of non linear quasi-variational inequalities.
      Troudi Mat. Nauk Inst. 1975.

[12] A. BENSOUSSAN and J.L. LIONS, Differential games with impulse times.
      Conference I.E.E.E. on Decision and Control. Phoenix November
      1974.

[13] H. BREZIS, Problèmes unilatéraux. J. Math. Pures et Appl. 51
      (1972) pp. 1-168.

[14] H. CHERNOFF, Optimal stochastic control. Sankhya 30 (1968) pp.
      221-252.

[15] G. DUVAUT, 1) C.R.Acad. Sc. Paris, 1973; 2) Rio de Janeiro,
      January 1975.

[16] G. DUVAUT and J.L. LIONS, Inéquations en Mécanique et en Physique.
      Dunod 1972. English translation to appear. Springer.

[17] B.I. GRIGIELONIS and A.N. SHIRYAEV, On Stefan's problem and optimal
      stopping rules for Markov processes. Theor. Prob. Applic.
      11 (1966), pp. 541-558.

[18] J.L. LIONS, Contrôle optimal des systèmes distribués. Dunod 1968.
      (English translation: Springer 1972; Japanese translation,
      1972).

[19] J.L. LIONS, On Partial differential Inequalities (in Russian)
      Ouspechi Mat. Nauk (1971); 16, pp. 205-263.

[20] J.L. LIONS, Quelques méthodes de résolution des problèmes aux
      limites non linéaires. Dunod, 1969.

[21] J.L. LIONS and G. STAMPACCHIA, Variational Inequalities, Comm.
      Pure and Applied Math., (967), XX, pp. 493-579.

[22] T. TOBIAS, Optimal stopping rules for diffusion processes.
      Stochastic control symposium, Sept. 1974, Budapest.

[23] T. TOBIAS, (In Russian. Differential equations, IX, 4, 1973,
      pp. 702-708.

[24] A. TORELLI, Un problème à frontière libre d'évolution en
      hydraulique. C.R.A.S. PARIS, December 1974.

UNIQUENESS, ANALYTICITY AND DECAY PROPERTIES OF CORRELATIONS IN EQUILIBRIUM SYSTEMS*+

Joel L. Lebowitz
Belfer Graduate School of Science
Yeshiva University
New York, N.Y. 10033

Abstract: We give a survey of results relating to the topics mentioned in the title. We are particularly interested in the relationship between the analyticity properties of the free energy as a function of the thermodynamic parameters of the system and the clustering and uniqueness properties of the corresponding equilibrium states i.e. correlation functions. We concentrate on Ising spin systems with ferromagnetic pair interactions.

I. Introduction and Formulation

In order to keep the presentation simple I shall restrict myself, for most of this lecture, to the simplest non-trivial many-body system known to man; the <u>Ising</u> spin system with <u>finite</u> range <u>ferromagnetic</u> pair <u>interactions</u> on a ν-dimensional cubical lattice $(IFFI)_\nu$, $\nu = 1, 2, 3,\ldots$. (Interactions which fall off exponentially behave, as far as the problems discussed here are concerned, just like finite range interactions.) The Hamiltonian of this system in a finite domain $\Lambda \subset z^\nu$ with 'boundary conditions' corresponding to specifying the values of the spin variables outside Λ has the form [1,2],

$$H(\underline{\sigma}; \Lambda, b_\Lambda) = -\tfrac{1}{2} \sum_{i,j \in \Lambda} J(i-j)\sigma_i \sigma_j - \sum_{i \in \Lambda} [h + \sum_{j \notin \Lambda} J(i-j)\bar{\sigma}_j]\sigma_i, \quad (1.1)$$

where $\sigma_i = \pm 1$, $J(\underline{x}) \geq 0$, $J(\underline{x}) = 0$ for $|\underline{x}| > R$, $|\underline{x}|$ is the length of the lattice vector $\underline{x}$, h is the external magnetic field and $\{\bar{\sigma}_i\}$ is a specified set of values of σ_i for $i \notin \Lambda$, which constitute the set of boundary conditions $\{b_\Lambda\} = b$. (More generally only the 'distribution' of the σ_i for $i \notin \Lambda$ need be specified; we can also take into account 'free' boundary conditions by setting $\bar{\sigma}_i = 0$, and 'periodic' boundary conditions by modifying the definition of the $J(\underline{x})$.)

The partition function, free energy density and correlation functions of this system at temperature $T = \beta^{-1}$ are given by

$$Z(\beta,h'; \Lambda, b_\Lambda) = \sum_{\{\sigma_i = \pm 1\}} \exp[-\beta \, H(\underline{\sigma}; \Lambda, b_\Lambda)], \quad i \in \Lambda \qquad (1.2)$$

$$\Psi(\beta,h'; \Lambda, b_\Lambda) = \{\ell n \, Z(\beta,h'; \Lambda, b_\Lambda)\}/|\Lambda| \qquad (1.3)$$

*Lecture presented at International Symposium on Mathematical Problems in Theoreti- Physics in Kyoto, Japan, Jan. 1975 and Australian Summer Institute in Applied Mathematics in Sydney, Australia.
+Supported in part by AFOSR Grant #73-2430B.

$$\langle \sigma_A \rangle \, (\beta, h'; \Lambda, b_\Lambda) = \{ \textstyle\sum_{\sigma_i = \pm 1} \} \, \{ \sigma_A \exp [-\beta \, H(\underline{\sigma}; \Lambda, b_\Lambda)] \} / Z(\beta, h'; \Lambda, b_\Lambda), \qquad (1.4)$$

where $|\Lambda|$ equals the number of sites in Λ, $h' = \beta h$, and $\sigma_A = \prod_{i \in A} \sigma_i$, $A \subseteq \Lambda$. (We have absorbed a factor $-\beta$ in the free energy and shall drop the prime on h from now on.)

We are interested in the behavior of Ψ and $\langle \sigma_A \rangle$ in the thermodynamic limit, $\Lambda \to \infty$. The limit is taken in such a way that each site i eventually is (and remains) inside Λ. Let me present some questions of interest:

1. Does the limit $\Lambda \to \infty$ of $\Psi(\beta, h; \Lambda, b_\Lambda)$ exist and is the resulting function independent of the boundary conditions b and of the way in which $\Lambda \to \infty$, i.e. of $\{b_\Lambda\}$?

The answer to this question for IFFI systems is an unqualified yes; all ways of going to the thermodynamic limit lead to the same $\Psi(\beta, h)$ for all real h and $\beta \geq 0$.

(This is true generally for systems which are charge neutral [3,4]. The interesting exceptions are systems with net charge [4] and systems with dipolar interactions in an external field [4,5,6] where the thermodynamic free energy density is expected on the basis of simple magneto (electro) -static arguments, and known experimentally, to be shape dependent. The establishment of this shape dependence for dipolar systems on a rigorous mathematical basis is (to my knowledge) the last remaining interesting physical problem in "proving the existence of the thermodynamic limit of the free energy", a task begun over twenty years ago by van Hove, c.f. [1,3].)

2. Having now a unique thermodynamic free energy $\Psi(\beta, h)$ the next question is: what are its analyticity (real) properties? Non-analyticites in β, h are connected with phase transitions and are, from both a physical and mathematical point of view, the 'interesting' features of Ψ for real macroscopic systems. True singularities clearly occur only in the thermodynamic limit $\Lambda \to \infty$, their 'precursors' however are indistinguishable from them quantitatively for macroscopic systems studied experimentally, $|\Lambda| \sim 10^{23}$. The 'finite size' corrections, (surface effects) are an interesting problem in themselves which needs much further investigation, c.f. [7]. They will not be discussed here.

3. Is the thermodynamic limit of $\langle \sigma_A \rangle \, (\beta, h; \Lambda, b_\Lambda)$ unique? Since $|\langle \sigma_A \rangle| \leq 1$ we can always find subsequences b_Λ on which the limit exists for all finite sets. We denote these by $\langle \sigma_A \rangle (\beta, h; b)$. The measure 'corresponding' to any such set of correlations defines an 'equilibrium state' which satisfies the DLR (Dobrushin-Lanford-Ruelle) eqs. and vice versa, c.f. [8,9]. If the limit is not unique then there are the further questions: i) Are all the limit functions translation invariant? i.e., does $\langle \sigma_{A+\underline{x}} \rangle (\beta, h; b) = \langle \sigma_A \rangle (\beta, h; b)$? ii) How many 'extremal' translational invariant equilibrium states are there for each β, h, and a given interaction? Translation invariant equilibrium states can always be decomposed uniquely into extremal states [1,8], e.g. if there were only two such extremal states, designated by subscripts 1 and 2, then each set of translation invariant correlation functions would have the

form $\langle\sigma_A\rangle (\beta,h;b) = \alpha_b \langle\sigma_A\rangle_1 (\beta,h) + (1-\alpha_b) \langle\sigma_A\rangle_2(\beta,h)$ where $0 \le \alpha_b \le 1$ is independent of A. ('Similar' statements hold for non translation invariant equilibrium states.)

4. What are the analyticity properties of the, possible boundary dependent, $\langle\sigma_A\rangle (\beta,h;b)$? Like the free energy the finite volume correlation functions are clearly real analytic in β and h.

5. What are the cluster properties of the infinite volume correlation function, e.g. does $[\langle\sigma_A \, \sigma_{B+\underline{x}}\rangle - \langle\sigma_A\rangle \langle\sigma_{B+\underline{x}}\rangle] \to 0$ as $|\underline{x}| \to \infty$, if yes does it do so exponentially fast, etc.? This question is connected to the questions about analyticity [10,11].

2. Survey of Selected Results

We would of course like to be able to answer the above questions for any particular interaction. Even more important we want to understand, in as much generality as possible, the connection between the properties of the free energy density which is a macroscopic thermodynamic quantity, and the correlation functions which describe the microscopic structure of the system. Thus we would like to know the general class of systems for which the analyticity of $\Psi(\beta,h)$ in some region of the $R\ell\beta$ - $R\ell h$ plane implies unique equilibrium states with strong cluster properties and vice versa. While little is known about the answer to this question in general quite a lot is known for IFFI systems (there is still quite a lot unknown too). It will generally be clear which results are specific to these systems and which have analogies for other systems. (If Ψ is required to be analytic in 'all interactions' then the above question can be answered fairly generally, c.f. Ruelle [12], but our point of view here is to think of the interactions as given while β,h are 'externally controlled' variables. Finding the set of relevant variables is of course a basic part of the problem.)

Let me begin then with a theorem which relates derivatives of Ψ to spatial 'averages' of some correlation functions.

<u>Theorem 1:</u> If $\Psi(\beta,h)$ is differentiable in h at some value of β and h then the thermodynamic limit of the magnetization is equal to the derivative of Ψ,

$$m(\beta,h;b) \equiv \lim_{\Lambda \to \infty} |\Lambda|^{-1} \sum_{i \in \Lambda} \langle\sigma_i\rangle (\beta,h; \Lambda,b_\Lambda) = \partial\Psi(\beta,h)/\partial h \qquad (2.1)$$

independent of b, c.f. [1,2].

<u>Proof:</u> $\Psi(\beta,h; \Lambda,b_\Lambda)$ is a convex function of h hence so is $\Psi(\beta,h)$ and the limit of its derivative

$$m(\beta,h; \Lambda,b_\Lambda) \equiv \frac{\partial}{\partial h} \Psi(\beta,h; \Lambda,b_\Lambda) \qquad (2.2)$$

equals the derivative of its limit whenever the latter exists.

Since the limit function $\Psi(\beta,h)$ is independent of $\{b_\Lambda\}$ and because of convexity, exists for almost all h we also have

<u>Corollary 2.</u> If $m(\beta,h;b)$ depends on $\{b_\Lambda\}$ then the right and left derivatives of $\Psi(\beta,h)$ are unequal, $\dfrac{\partial\Psi(\beta,h)}{\partial h+} > \dfrac{\partial\Psi(\beta,h)}{\partial h-}$, and there exist sequences $\{h_i\}$ decreasing (increasing) to h such that

$$\lim_{i \to \infty} \frac{\partial\Psi(\beta,h)}{\partial h}\Big|_{h_i=h} = \partial\Psi(\beta,h)/\partial h_{(\overset{+}{-})} \quad .$$

Clearly similar results hold for differentiation with respect to β (or other parameters in the Hamiltonian) and also for general systems, c.f. [13]. The next theorem, however, which is a very strong 'converse' of Theorem 1 is (at the present time) essentially restricted to IFFI 'type' of systems and I don't see any way of extending it to general systems. (Indeed there are some counter examples for antiferromagnetic systems [14] unless h in Theorem 3 is replaced by the 'staggered' field).

<u>Theorem 3.</u> If $\partial\Psi(\beta,h)/\partial h$ exists (i.e. is continuous) then <u>all</u> the correlation functions are independent of boundary conditions [15], i.e.

$$\langle\sigma_A\rangle\ (\beta,h;b) = \langle\sigma_A\rangle\ (\beta,h).$$

Theorem 3 may be usefully combined with the following results about IFFI systems.

<u>Theorem 4.</u> a) $\Psi(\beta,h)$ is jointly analytic in h and (real) β for $h \neq 0$, $\beta \geq 0$, [16]. b) $\exists$ a $\beta_a > 0$ such that for $\beta \leq \beta_a$ (high temperature region) $\Psi(\beta,h)$ is real analytic in β and h [17].

Thus for $h \neq 0$ or $\beta \leq \beta_a$ there is only one equilibrium state. This state being unique is translation invariant. It is furthermore true that the correlations in this state have <u>all</u> the desired properties:

<u>Theorem 5.</u> For $h \neq 0$ or $\beta \leq \beta_a$ the $\langle\sigma_A\rangle\ (\beta,h)$ are (real) analytic in β and h and have exponential clustering properties [16,17,18]. (For even stronger cluster properties see [11]).

We thus see that for our systems everything is fine (and thus uninteresting) for $h \neq 0$ or $\beta \leq \beta_a$. We must therefore, look at h = 0 and low temperatures to have a chance of finding something interesting. Now it can be shown, [1], that in one dimension, $(IFFI)_1$, $\Psi(\beta,h)$ is (real) analytic for all β and h. We must therefore look to higher dimensional system, which is what Peierls did and found that there was indeed something interesting happening for $\nu \geq 2$ [1,2]. (Peierls' method of proof has been extremely fruitful for our understanding of phase transitions. It was invented one evening after an unnamed mathematician gave, in a lecture, an elaborate proof of the nonexistence of a phase transition in one dimension and concluded by saying that 'clearly the same is true in higher dimensions', [19]).

<u>Theorem 6.</u> If $\nu \geq 2$ and the nearest neighbor interactions do not vanish then there exists a $\beta p < \infty$ such that

$$m(\beta,0;\Lambda,b_+) \equiv -m(\beta,0;\Lambda,b_-) \geq \epsilon > 0, \quad \text{for } \beta \geq \beta p, \tag{2.3}$$

independent of Λ, where $b_+(b_-)$ corresponds to all spins outside Λ being set equal to $1(-1)$.

The exact value of β_p is not too important here. It is generally a poor lower bound on the values of β for which the inequality $m(\beta,0; \Lambda,b_+) > 0$ holds. What is important is that for $\nu \geq 2$ we have at sufficiently low temperatures, $T \leq T_p = \beta_p^{-1}$, that $m(\beta,0;b_+) \neq m(\beta,0;b_-)$, i.e. $m(\beta,0;b)$ depends on the boundary conditions. Thus according to Corollary 2, $\partial\Psi(\beta,h)/\partial h$ will be discontinuous at $h = 0$ for $\beta \geq \beta_p$. We can actually say more.

<u>Theorem 7.</u> a) $\langle\sigma_A\rangle$ $(\beta,h;\Lambda,b_\Lambda)$ is monotone nondecreasing in β and h whenever the b_Λ are such that all the external fields appearing in the Hamiltonian (1.1) are non-negative, i.e. $h_i \equiv h + \sum_{j\notin\Lambda} J(i-j) \bar{\sigma}_j \geq 0$. b) For $h \geq 0$ the correlation functions $\langle\sigma_A\rangle$ $(\beta,h;\Lambda,b_+)$ are monotone nonincreasing in Λ and their limits, $\langle\sigma_A\rangle$ $(\beta,h;b_+)$ are translation invariant <u>even</u> when there is more than one equilibrium state, i.e. at $h = 0$ and low temperatures. Furthermore

$$\lim_{h \downarrow 0} \langle\sigma_A\rangle (\beta,h) = \langle\sigma_A\rangle (\beta,0;b_+) \qquad (2.4)$$

c) $\quad m(\beta,0;b_+) = \langle\sigma_i\rangle (\beta,0;b_+) = \lim_{h \downarrow 0} \frac{\partial\Psi(\beta,h)}{\partial h} = m*(\beta) \qquad (2.5)$

and $m*(\beta)$, the 'spontaneous magnetization', is monotone nondecreasing in β.

Part a) of the theorem is a special case of the basic Griffiths inequalities [1,2]. b) and c) follow from a) and some other arguments, c.f. [15]. (Clearly $\langle\sigma_A\rangle$ $(\beta,h;\Lambda,b_+) = (-1)^{|A|} \langle\sigma_A\rangle (\beta,-h;\Lambda,b_-)$ and $\Psi(\beta,h) = \Psi(\beta,-h)$ so similar results hold for negative fields.) It follows directly from c) that we can define a unique critical temperature $T_c = \beta_c^{-1}$ for spontaneous magnetization by the relation $m*(\beta) = 0$ for $\beta < \beta_c$, $m*(\beta) > 0$ for $\beta > \beta_c$.

To summarize then: for IFFI systems we already know, i) For $h \neq 0$ or $T > T_c$ we have unique states. For $h \neq 0$ or $T > T_a$ we have in addition analyticity and exponential clustering. ii) For $h = 0$ and $T < T_c$ we have nonunique states, discontinuity of $\partial\Psi/\partial h$ and of $\langle\sigma_i\rangle$ (β,h) (and all $\langle\sigma_A\rangle$ for $|A|$ odd).

A question which naturally arises now is what happens to analyticity and cluster properties at $h = 0$ and $T_c < T \leq T_a$, e.g. is there a temperature T_b, $T_c < T_b \leq T_a$ where Ψ and/or the $\langle\sigma_A\rangle$ stop being analytic in h and or β. (The nature of the singularities occuring at T_c will not be discussed here). This is a most interesting question and the only additional information known to me about <u>general</u> IFFI systems is given in the following theorem.

<u>Theorem 8.</u> $\Psi(\beta,h)$ and $\langle\sigma_A\rangle$ (β,h) are infinitely differentiable in β and h for $\beta < \beta_F$ where $\beta_F^{-1} = T_F$ is (essentially) the 'mean field' critical temperature, defined by the relation $\sum_{\underline{x}} \tanh \beta_F J(\underline{x}) = 1$.

<u>Proof:</u> The proof of Theorem 8 is based on the following result, [10]:

Theorem 9. If the pair correlation $\langle \sigma_i \sigma_{i+\underline{x}} \rangle$ $(\beta,0)$ (independent of b for $\beta < \beta_c$) has a bound of the form

$$0 \leq \langle \sigma_i \sigma_{i+\underline{x}} \rangle \ (\beta,0) \leq K \ |\underline{x}|^{-(k\nu+\epsilon)}, \ K < \infty, \ \epsilon > 0, \ k \in \mathbb{Z}_+ \tag{2.6}$$

then $\Psi(\beta,h) \in C^{k+1}$ and $\langle \sigma_A \rangle \ (\beta,h) \in C^k$ in both variables. (The inequality on the left of (2.6) is part of Griffiths first inequality).

The existence of an exponential bound (corresponding to $k = \infty$ in (2.5)) follows from Griffiths third inequality [20]

$$\langle \sigma_i \sigma_j \rangle \ (\beta,0) \leq \sum_{k\neq i} \langle \sigma_j \sigma_k \rangle \ \tanh \ [\beta J(j-k)], \ \text{for} \ j \neq i \tag{2.7}$$

and can be improved slightly using generalizations of this inequality c.f. [21]. An alternative method which bounds the pair correlation function using self-avoiding walks is due to Fisher [22]. It can sometimes yield an exponential bound for temperatures considerably lower than the mean field critical temperature. It is not known however at the present whether even exponential decay and thus/or C^∞ persist for all $(IFFI)_\nu$ systems down to T_c.

(To obtain an explicit bound from (2.7) we note that if (2.7) is solved as an equality then, for $\beta < \beta_F$, the solution is an upper bound to $\langle \sigma_i \sigma_j \rangle$, c.f. [21]. This yields, (for simplicity use periodic boundary conditions),

$$\langle \sigma_i \sigma_j \rangle \ (\beta,0) \leq \langle y_i y_j \rangle_g(t) / \langle y_i^2 \rangle_g(t), \quad \text{for} \ \beta < \beta_F \tag{2.8}$$

where $\langle y_i y_j \rangle_g(t)$ are the correlations in a "Gaussian" model [23] with the interactions $\beta J(i-j)$ replaced by $\tanh \ [\beta J(i-j)]$,

$$\langle y_i y_j \rangle = (\underline{t}^{-1})_{ij} \ , \quad i,j \in \mathbb{Z}^\nu \tag{2.9}$$

where $\underline{t}$ is the matrix with elements

$$(\underline{t})_{ii} = 1, \quad (\underline{t})_{ij} = \tanh \ \beta J(i-j), \ i \neq j. \tag{2.10}$$

(2.9) can be readily shown to decay exponentially (use Fourier transforms).

When (2.8) is combined with Newman's inequalities [24]

$$\langle \sigma_1 ... \sigma_{2n} \rangle (\beta,0) \leq \sum_{\text{pairings}} \langle \sigma_{i_1} \sigma_{j_1} \rangle ... \langle \sigma_{i_n} \sigma_{j_n} \rangle \tag{2.11}$$

we obtain

$$\langle \sigma_A \rangle (\beta,0) \leq \langle y_A \rangle_g(t) / [\langle y_i^2 \rangle_g(t)]^{\frac{1}{2}|A|}, \quad \beta < \beta_F \tag{2.12}$$

'Similar' inequalities hold for $h \neq 0$.)

It should be noted here that while it is very natural to conjecture that analyticity persists for $(IFFI)_\nu$ systems up to $T = T_c$ any proof of it will require making

some use of the lattice structure since the statement is not true for the Bethe
lattice [25] and for random spin systems [26].

Let me come now to a discussion of what additional information is known for
$h = 0$ and $T < T_c$. I am of course considering the case where $T_c > 0$, for $\nu = 1$ we
have $T_c = T_a = 0$.

$\underline{\text{Theorem 10.}}$ $\exists$ a temperature T_e, $0 < T_e < T_c$ ($T_e \sim T_p$) such that for $h = 0$,
$0 < T \leq T_e$ the following statements are true:

a) There are only two extremal translation invariant states and these are obtained from b_+ and b_- boundary conditions.

b) The correlations in the extremal states decay exponentially.

c) $\Psi(\beta,h)$ and $\langle \sigma_A \rangle$ (β,h) are infinitely differentiable in β and h as $h \to 0$
from either side and so are $\Psi(\beta,0)$ and $\langle \sigma_A \rangle$ $(\beta,0;b_+)$.

The proof of part a) can be obtained in different ways, [27,28] and the same is
true of b), [28,29]. It follows from a) and b) that every set of translation invariant correlation functions can, for $\beta \geq \beta_e = T_e^{-1}$, be written as a linear combination
of correlation functions which have very strong cluster properties,

$$\overline{\langle \sigma_A \rangle} \ (\beta,0);b) = \alpha \ \langle \sigma_A \rangle \ (\beta,0;b_+) + (1-\alpha) \ \langle \sigma_A \rangle \ (\beta,0;b_-), \ 0 \leq \alpha \leq 1 \qquad (2.13)$$

where the bar indicates translational averaging (if necessary).

Theorem 10c) follows from b) and a variation of Theorem 9 which applies for
$h \geq 0$ when (2.6) is replaced by the condition, c.f. [30],

$$0 \leq \langle \sigma_i \sigma_{i+\underline{x}} \rangle \ (\beta,0;b_+) - [m*(\beta)]^2 \leq K \ |\underline{x}|^{-(k\nu+\epsilon)} \qquad (2.14)$$

where, by (2.5), $m*(\beta) = \langle \sigma_j \rangle \ (\beta,0;b_+)$ for all j.

(A very interesting question, related to the problems we are discussing is
whether $\Psi(\beta,h)$ and the $\langle \sigma_A \rangle$ (β,h) can be analytically continued (below T_c) across
the line $h = 0$. If such an analytic continuation were to exist one would be tempted
to identify the Ψ and $\langle \sigma_A \rangle$ so obtained with metastable states as may be done for
systems with 'very long range' potentials [31]. For this reason this question has
generated more interest in the general statistical mechanics community than most of
the other questions I have raised. The situation at present is unclear (even for-
getting about proofs). There appear to be some strong arguments for expecting an
essential singularity in Ψ and $\langle \sigma_A \rangle$ at $h = 0$, [32]. Very recently however these
arguments have been questioned [33]. What we know rigorously [8b] is that for sys-
tems like IFFI the 'equilibrium state' cannot be continued analytically across $h = 0$
for $T < T_c$. This however does not rule out <u>at all</u> the possibility that Ψ and some
$\langle \sigma_A \rangle$, say for all $|A| \leq 137$, may be continued analytically or even that Ψ and <u>all</u>
$\langle \sigma_A \rangle$ can be continued analytically but that <u>some</u> $\langle \sigma_A \rangle$ will fail to satisfy some
positivity conditions. In summary then the question is wide open.)

Again it is not known for general $(IFFI)_\nu$ systems whether Theorem 10 remains valid for all $T < T_c$ or does there exist some T_d, $T_e \leq T_d < T_c$, where matters change. More information is known about a special class of $(IFFI)_\nu$ systems, those with nearest neighbor interactions only; $J(\underline{x}) = 0$ for $|\underline{x}| > 1$, $J(\underline{1}) = J$ independent of directions. (This is actually more restrictive than necessary but we want to keep matters simple). We shall denote these systems by I_ν and their critical temperatures by $T_{c,\nu}$. As is well known I_2 was solved by Onsager [34] for $h = 0$ (the first non-trivial many body problem ever solved). Onsager obtained $\Psi(\beta,0)$ for I_2 and found that it was singular (second derivative diverges logarithmically) at β_o given by $[\sinh 2\beta_o J] = 1$. Onsager and Kaufman [34] also showed that the pair correlation (for 'periodic boundary conditions') decays exponentially above T_o and approaches (also exponentially) $m_o^2(\beta)$ below T_o with $m_o(\beta)$ the Onsager-Yang magnetization $m_o(\beta) = [1-\sinh^{-4}(2\beta J)]^{1/8}$ for $\beta \geq \beta_o$.

It follows from these (and other) facts that T_o coincides with the critical temperature for spontaneous magnetization $T_{c,2}$, [10] that $m_o(\beta) = m*(\beta)$, [35], and that Ψ and $\langle \sigma_A \rangle$ are C^∞ in β and h for $\beta < \beta_c$ and are C^∞ in β and infinitely differentiable in h as $h \to 0\pm$ for $\beta > \beta_c$ [10,30]. It is also known [36] that I_2 has <u>only</u> translation invariant states for $T \leq T'$ where $T' \sim .7T_c$. Furthermore it has been shown recently [37] that I_2 has exactly two extremal states $(+, -$ boundaries) for all $T < T_c$.

Thus of the questions raised in the beginning of my talk the only ones left unanswered for I_2 are a) whether analyticity holds for all $T > T_c$ and b) whether the nonexistence of non-translation invariant states is true for all T (The question is really just for $T' < T < T_c$). It would be most surprising indeed if the answer to both questions were not yes, [36]. (I make no conjecture about analytic continuation below T_c).

This brings me to a most interesting point; the property of I_2 of having only translation invariant states (at least at low temperatures) definitely does not hold for $\nu > 2$. This was first shown by Dobrushin [38]. The result has been improved and the proof greatly simplified by van Beijeren [39], who proved:

<u>Theorem 11.</u> For $h = 0$ and $T < T_{c,\nu-1}$, I_ν <u>has</u> non translational invariant states.

Since $T_{c,1} = 0$ this is consistent with the results for I_2. Indeed the difference between $\nu > 2$ and $\nu = 2$ with regard to the existence of non translation invariant states is very similar to the difference between $\nu > 1$ and $\nu = 1$ with regard to the existence of a $T_{c,\nu} > 0$. This becomes clearer when we consider the nature of the Dobrushin and van Beijeren proofs. They consider a sequence of ν-dimensional cubes Λ_N, $\nu \geq 2$ with $2N$ horizontal layers labeled by $k = -N, -N+1,\ldots,N$, and boundary conditions b_N corresponding to + spins, $\bar{\sigma}_i = 1$, on the 'top half' N layers and -spins, $\bar{\sigma}_i = -1$, on the bottom half layers; $\{b_N\} = b\underline{+}$.

If we let $\sigma_{k,\alpha}$ designate the spin variable at site α in the $k\underline{th}$ layer then by symmetry $\langle \sigma_{k,\alpha} \rangle(\beta,0; \Lambda_N, b_N) = -\langle \sigma_{-k,\alpha} \rangle(\beta,0; \Lambda_N,b_N)$. At $\beta^{-1} = T = 0$ we clearly have

$\langle \sigma_{k,\alpha} \rangle$ = 1 for k > 0 and all N. The question now is whether as N → ∞, and the boundaries recede, does $\langle \sigma_{k,\alpha} \rangle$ for k > 0 remain positive for some temperature T ≠ 0 and thus $\langle \sigma_i \rangle (\beta,0;b^{\pm})$ would not be translation invariant. We already know that for T > T_c, $\langle \sigma_i \rangle (\beta,0;b)$ = 0 for <u>all</u> boundary conditions. For I_2 we also know that for 0 < T ≤ T' all states are translation invariant, and thus $\langle \sigma_{k,\alpha} \rangle (\beta,0;b^{\pm})$ = $\langle \sigma_{-k,\alpha} \rangle (\beta,0;b^{\pm})$ = 0 by symmetry. For ν ≥ 3 however it is shown by van Beijeren that

$$\langle \sigma_{1,\alpha} \rangle (\beta,0;b^{\pm}) \;\geq\; m^*_{\nu-1}(\beta) > 0 \;\text{ for } T < T_{c,\nu-1} \tag{2.15}$$

(van Beijeren also shows that, as expected, $\langle \sigma_{k,\alpha} \rangle (\beta,0;b^{\pm})$ is monotone non-decreasing in k.)

Physically (2.15) means that the ν-1 dimensional surface which separates up spins from down spins and intersects the boundary of Λ_N between the k = 1 and k = -1 level (this is a connected piece of the union of 'faces' separating cells with σ_i=1 from cells with σ_i = -1) does not fluctuate too widely as N → ∞ when ν-1 ≥ 2 and T < $T_{c,\nu-1}$ (for ν - 1 = 1 it does so fluctuate).

The interesting question for I_ν, ν ≥ 3, is now whether these non-translation invariant states persist up to $T_{c,\nu}$ or is there a 'roughening temperature' $T_{r,\nu} < T_{c,\nu}$ where the dividing surface roughens and disappears, as in I_2. There appears to be some evidence, [40] based on extrapolations from low temperature expansions and numerical computations, that $T_{r,3} \sim .57 T_{c,3} \sim 1.1\, T_{c,2}$, i.e. quite close to van Beijeren's lower bound. Again however definite results are sorely lacking.

3. Concluding Remarks

We have seen that while much remains to be done all the information available for IFFI systems are consistent with our hopes for simplicity. In lattice gas language there is a line in the chemical potential μ-temperature T plane along which two phases, liquid and gas, coexist ending in a critical point. For values of μ - T not on this line we have unique equilibrium states with (very likely) exponential decay of correlations and analyticity in μ and T. On the line itself we have (very likely) only two translation invariant states. The question now is how much of this simplicity remains for even simple continuum systems, e.g. for atoms interacting with Lenard-Jones type potentials. Since the interaction potential is now long range, falling off as r^{-n}, we cannot expect and do not get exponential decay of correlations [11,18]. On the other hand we expect to have also a solid-fluid transition and a triple point. Taking these into account however does the rest of the picture remain simple or are there real surprises, e.g. two dimensional regions of the μ-T plane where the equilibrium state is not unique and/or the free energy not analytic. We don't seem to be anywhere near getting an answer to these questions. All we do know is that if we permit a large amount of arbitrariness in the potentials then we can get very strange things indeed [41], e.g. one dimensional Heisenberg spin systems with spontaneous magnetization.

References

1. Ruelle, D. *Statistical Mechanics* (Benjamin, 1969).
2. Griffiths, R.B. in *Phase Transitions and Critical Points I*, edited by C. Domb and M.S. Green, (Academic Press, 1972).
3. Fisher, M.E. Arch. Rat. Mech. Anal. 17, 377 (1964).
4. Lebowitz, J.L. and Lieb, E.H. Phys. Rev. Lett. 22, 631 (1969); Lieb, E.H. and Lebowitz, J.L. Adv. in Math. 9, 316 (1972).
5. Griffiths, R.B. Phys. Rev. 176, 655 (1968): Brown, W.F. Jr. *Magnetostatic Principles in Ferromagnetism*, (North Holland, 1962).
6. Penrose, O. and Smith, E.R. Comm. Math. Phys. 26, 53 (1972); Smith, E.R. and Perram, J.W. Phys. Lett. 50A, 294 (1974).
7. Fisher, M.E. and Lebowitz, J.L. Comm. Math. Phys. 19, 251 (1970).
8. Dobrushin, L. Funct. Anal. Appl. 2, 291 (1968); Lanford, O.E. and Ruelle, D. Comm. Math. Phys. 13, 194 (1969): Lanford, O.E. in *Statistical Mechanics and Mathematical Problems*, (Springer-Verlag, 1973).
9. Brascamp, H. Comm. Math. Phys. 18, 82 (1970); Gruber, C. and Merlini, D. Physica 67, 308 (1973); Gruber, C. and Lebowitz, J.L. Comm. Math. Phys. (to appear).
10. Lebowitz, J.L. Comm. Math. Phys. 28, 313 (1972).
11. Duneau, M., Iagolnitzer, D. and Souillard, B. Comm. Math. Phys. 35, 307 (1974) and article by Souillard in this volume.
12. Ruelle, D. Ann. of Phys. 69, 364 (1972).
13. Lebowitz, J.L. and Lieb, E.H. Phys. Lett. 39A, 98 (1972).
14. Brascamp, H. and Kunz, H. Comm. Math. Phys. 32, 93 (1973).
15. Lebowitz, J.L. and Martin-Löf, A. Comm. Math. Phys. 25, 276 (1972).
16. Lebowitz, J.L. and Penrose, O. Comm. Math. Phys. 11, 99 (1968).
17. Gallavotti, G. and Miracle-Sole, S. Comm. Math. Phys. 12, 269 (1969).
18. Lebowitz, J.L. and Penrose, O. Phys. Rev. Lett. 31, 749 (1973); Penrose, O. and Lebowitz, J.L. Comm. Math. Phys. 39, (1974).
19. Peierls, R. private communication.
20. Griffiths, R.B. Comm. Math. Phys. 6, 121 (1967).
21. Krinsky, S. and Emery, V.J. preprint.
22. Fisher, M.E. Phys. Rev. 162, 480 (1967).
23. Berlin, T. and Kac, M. Phys. Rev. 86, 821 (1952); c.f also Simon, B. The $P(\emptyset)_2$ *Euclidian Quantum Field Theory*, (Princeton University Press, 1974).
24. Newman, C.M. preprint.
25. Muller-Hartmann, E. and Zittartz, J. Phys. Rev. Lett. 33, 893 (1974) and Zittartz, J. in this volume.
26. Griffiths, R.B. Phys. Rev. Lett. 23, 17 (1969).
27. Gallavotti, G. and Miracle-Sole, S. Phys. Rev. 5B, 2555 (1972).
28. Slawny, J. Comm. Math. Phys. 34, 271 (1973).
29. Martin-Löf, A. Comm. Math. Phys. 25, 87 (1972).
30. Gallavotti, G. and Lebowitz, J.L. Physica 70, 219 (1973).
31. Penrose, O. and Lebowitz, J.L. J. Stat. Phys. 3, 211 (1971).
32. Fisher, M.E. Physics 3, 255 (1967); Langer, J. Ann. Phys. N.Y. 41, 108 (1967).
33. Domb, C. private communication.
34. c.f. McCoy, B.M. and Wu, T.T. *The Two Dimensional Ising Model* (Harvard University Press, 1973).
35. Benettin, G., Gallavotti, G., Jona-Lasinio, G. and Stella, A.L. Comm. Math. Phys. 30, 45 (1973).
36. c.f. Gallavotti, G. Comm. Math. Phys. 27, 103 (1972); Rivista Nuov. Cim. 2-2, 133 (1972): Abraham, D.B. and Reed, P. Phys. Rev. Lett. 33, 377 (1974).
37. Messager, A. and Miracle-Sole, S. Comm. Math. Phys. to appear.
38. Dobrushin, R.L. Theo. of Prob. and Appl. 17, 582 (1972).
39. van Beijeren, H. Comm. Math. Phys. to appear.
40. Weeks, J.D., Gilmer, G.H. and Leamy, H.J. Phys. Rev. Lett. 31, 549 (1973).
41. Israel, R. preprint.

ON THE PROPAGATION OF SUPPORT OF SOLUTIONS
TO GENERAL SYSTEMS OF PARTIAL DIFFERENTIAL EQUATIONS

Shigetake Matsuura

Research Institute for Mathematical Sciences
Kyoto University
Kitashirakawa, Kyoto, 606, Japan

§1. Introduction.

The purpose of the present article is to generalize a theorem of F. John [5] to general (overdetermined) systems of partial differential operators (Theorem 3.3). At the same time, we refine the results by giving rather precise estimates for the support of solutions to such systems of equations in terms of convex hulls which are new even for single equations (Theorems 3.4 and 3.5). These estimates lead us immediately to a characterization criterion for systems of equations to have solutions with supports in a certain kind of prescribed subsets of the euclidean space (Theorem 4.1). Our criterion theorem contains, as very special cases, the results of D.K.Cohoon [2] and of K.Horie [3].

The abstract physical interpretations of the results given in the symposium as well as their proofs are omitted here. The detailed treatment will be published elsewhere.

§2. Notations.

Let us fix some notations. Let $X = \mathbb{R}^{\ell}$ be the ℓ-dimensional euclidean space and $\Xi\,(\cong \mathbb{R}^{\ell})$ be its dual. When we choose a coordinate system $x = (x_1,\ldots,x_\ell)$ in X, the coordinates $\xi = (\xi_1,\ldots,\xi_\ell)$ in Ξ should be so chosen that the duality bilinear form be written as $\langle x,\xi\rangle = x_1\xi_1 +\ldots+ x_\ell\xi_\ell$. Let $\mathbb{C}[D_1,\ldots,D_\ell]$ (or simply $\mathbb{C}[D]$) be the polynomial ring of ℓ indeterminates $D = (D_1,\ldots,D_\ell)$ with complex coefficients. $D = (D_1,\ldots,D_\ell)$ will operate in the space X as the imaginary gradient, i.e. $D_j = -i\dfrac{\partial}{\partial x_j}$ $(j = 1,\ldots,\ell;\ i = \sqrt{-1})$. Thus, each polynomial $P(D) \in C[D]$ is a partial differential operator with constant coefficients.

Now, let $\wp(D)$ be an m×n matrix of partial differential operaters

$$(2.1) \qquad \wp(D) = \begin{pmatrix} P_{11}(D)...P_{1n}(D) \\ ... \\ P_{m1}(D)...P_{mn}(D) \end{pmatrix}$$

and consider the following system of equations

$$(2.2) \qquad \wp(D)U = 0$$

where $U = \begin{pmatrix} u_1 \\ \vdots \\ u_n \end{pmatrix}$ is a column vector of unknown functions (or generalized functions). For simplicity we consider mainly C^∞- functions defined in an open (or on a regular closed) subset of X.

In discussing the equation (2.2), we may always assume that

$$(2.3) \qquad m \geq n$$

by augmenting, if necessary, the number of rows in (2.1) by addition of zero entries. Then, let

$$(2.4) \qquad \{\Delta_1,...,\Delta_n\}$$

be the totality of $n \times n$ minors of the matrix (2.1). Properties of the operator (2.1) can often be reduced to those of the set (2.4) or to those of the ideal

$$(2.5) \qquad I(\wp) = (\Delta_1,...,\Delta_N)$$

of $\mathbb{C}[D]$ generated by the set of polynomials (2.4).

For two subsets A, B of $\mathbf{R}^\ell$, their vector sum (resp. their vector difference) will be denoted by $A + B$ (resp. by $A - B$). Their set difference will be denoted by $A \setminus B$. The convex hull (resp. the closed convex hull) of a set A will be denoted by ch A (resp. by $\overline{\mathrm{ch}}$ A.)

Given a vector $\vartheta \in \Xi \setminus \{0\}$, a subset of X is called a ϑ-*slab* if it can be written in the form

$$\Omega(\vartheta;I) = \{x \in X \; ; \; <x,\vartheta> \in I\}$$

for an interval I on the real line $\mathbb{R}$. We use also the simplified notations

$$\Omega(\vartheta) = \Omega(\vartheta; (0,\infty)),$$

$$\Omega(-\vartheta) = \Omega(\vartheta; (-\infty,0)).$$

Taking their closures, we get the closed half spaces

$$\overline{\Omega}(\vartheta) = \Omega(\vartheta; [0,\infty)),$$

$$\overline{\Omega}(-\vartheta) = \Omega(\vartheta; (-\infty,0]).$$

As for hyperplanes, we use the notation

$$H(\vartheta,c) = \{x \; ; \; <x,\vartheta> = c\}$$

for a given $c \in \mathbb{R}$ and the simplified notation

$$H(\vartheta) = \{x \; ; \; <x,\vartheta> = 0\}$$

for $c = 0$.

The cones in X (or in Ξ) that we consider in this article are always convex cones with their vertices at the origin 0. Thus a subset Γ of X is called a cone if, for every $x \in \Gamma$ and for every $t > 0$, $tx \in \Gamma$. A closed convex cone Γ in X is called a ϑ-*proper* cone if it is contained in $\overline{\Omega}(\vartheta)$ and if its intersection with the hyperplane $H(\vartheta)$ is reduced to the one point set $\{0\}$. (i.e. $\Gamma \cap H(\vartheta) = \{0\}$). The last condition is clearly equivalent to say that $\Gamma \cap H(\vartheta,c)$ is always compact for any $c \in \mathbb{R}$.

§3. Quasihyperbolicity, Main theorems.

For a polynomial $P \in \mathbb{C}[D]$, we denote by P^0 its principal part (i.e. its homogeneous part of the highest degree). Given a $\vartheta \in \Xi \setminus \{0\}$, we shall shortly say that a polynomial P is ϑ-*hyperbolic* if P is hyperbolic with respect to ϑ . A polynomial P is called *weakly ϑ-hyperbolic* if its principal part P^0 is ϑ-hyperbolic (c.f. [1], [4]). We define the dual cone $\Gamma(P,\vartheta)$ and the propagation cone $\Gamma^*(P,\vartheta)$ by the identifications

$$\Gamma(P,\vartheta) = \Gamma(P^0,\vartheta),$$

$$\Gamma^*(P,\vartheta) = \Gamma^*(P^0,\vartheta),$$

when P is weakly ϑ-hyperbolic (even though P might not be ϑ-hyperbolic). We note here that the propagation of a weakly ϑ-hyperbolic polynomial is always ϑ-proper closed convex cone. If P is weakly ϑ-hyperbolic and if u is a C^∞-solution to the equation

$$(3.1) \qquad\qquad P(D)u = 0$$

in the closed half space $\Omega(\vartheta; [c,\infty))$, then for the support of u we have the inclusion relation

$$(3.2) \qquad\qquad \text{supp } u \subseteq K + \Gamma^*(P,\vartheta)$$

where K denotes a closed convex set containing the support of Cauchy data of u on $H(\vartheta,c)$. (Since $P^0(\vartheta) \neq 0$, the support of Cauchy data of u coincides with $H(\vartheta,c) \cap$ supp u.)

Now we introduce the following concept.

<u>Definition 3.1</u>. A system $\wp$ of operators (2.1) is called ϑ-*quasihyperbolic* if there exists a non-constant weakly ϑ-hyperbolic polynomial P which divides all the elements of the ideal $I(\wp)$ defined by (2.5).

<u>Remark 3.2</u>. We say that a system $\wp$ is degenerate if $I(\wp) = \{0\}$, i.e. if the rank of the matrix $\wp$ is $\leq n - 1$. Degenerate systems are always ϑ-quasihyperbolic for any $\vartheta \in E \setminus \{0\}$.

When $\wp$ is non-degenerate, it is ϑ-quasihyperbolic if and only if the greatest common divisor G of the elements of the ideal is a ϑ-quasihyperbolic polynomial (i.e. G is divisible by a non-constant weakly ϑ-hyperbolic polynomial). Thus G can be factored in the form

$$G(D) = P(D)R(D)$$

where P is a non-constant weakly ϑ-hyperbolic polynomial and R is a polynomial having no irreducible weakly ϑ-hyperbolic factor in $\mathbb{C}[D]$.

We denote this factor $P(D)$ by $\chi(\mathcal{P})$. $\chi(\mathcal{P})$ and its factorization
into irreducible components

$$(3.3) \qquad \chi(\mathcal{P}) = P_1(D)^{r_1}\ldots P_s(D)^{r_s}$$

will play a central role in our present work. Each irreducible factor
P_j is of course weakly ϑ-hyperbolic.

Now we state our main results.

$\underline{\text{Theorem 3.3}}$. Given a slab domain $\Omega = \Omega(\vartheta;(c_1,c_2))$, if the equation
(2.2) has a non-trivial solution $U \in [\mathcal{D}'(\Omega)]^n$ such that the closed set
$H(\vartheta,c) \cap \text{supp } U$ has a compact connected component for some $c \in (c_1,c_2)$,
then the system $\mathcal{P}$ is ϑ-quasihyperbolic.

$\underline{\text{Theorem 3.4}}$. Suppose that a system of operators $\mathcal{P}$ is ϑ-quasi-
hyperbolic and non-degenerate. $P_1,\ldots,P_s$ be the collection of all the
distinct irreducible components in the factorization of $\chi(\mathcal{P})$. Then,
for any non-empty subset Λ of $\{1,2,\ldots,s\}$, for any $c \in \mathbb{R}$ and for any
compact convex subset K of $H(\vartheta,c)$ with non-empty interior (in $H(\vartheta,c)$),
there exists a solution $U \in [C^{\infty}(\mathbb{R}^{\ell})]^n$ to the equation (2.2) such that
the following simultaneous set equalities hold:

$$(3.4) \qquad \overline{\text{ch}}(\overline{\Omega}(\vartheta;[c,\infty))) \cap \text{supp } U = K + \sum_{\lambda \in \Lambda} \Gamma^*(P_\lambda,\vartheta),$$

$$(3.5) \qquad \overline{\text{ch}}(\overline{\Omega}(\vartheta;(-\infty,c])) \cap \text{supp } U = K - \sum_{\lambda \in \Lambda} \Gamma^*(P_\lambda,\vartheta).$$

In particular, since propagation cones are ϑ-proper, $H(\vartheta,c') \cap \text{supp } U$
is compact for all $c' \in \mathbb{R}$.

$\underline{\text{Theorem 3.5}}$. Suppose $\mathcal{P}$ be a non-degenerate system.
1) If the equation (2.2) in $\Omega(\vartheta;[a,\infty))$ admits a non-trivial
solution $U \in [C^{\infty}(\Omega(\vartheta;[a,\infty)))]^n$ such that $H(\vartheta,c) \cap \text{supp } U$ is compact for
all $c \geq a$, then the system $\mathcal{P}$ is ϑ-quasihyperbolic and the set equality
(3.4) holds for some non-empty subset Λ of $\{1,2,\ldots,s\}$ and for all
$c \geq a$ with $K = \text{ch}(H(\vartheta,c) \cap \text{supp } U)$.
2) If the equation (2.2) in $\mathbb{R}^{\ell}$ admits a non-trivial solution
$U \in [C^{\infty}(\mathbb{R}^{\ell})]^n$ such that $H(\vartheta,c) \cap \text{supp } U$ is compact for all $c \in \mathbb{R}$, then
the system $\mathcal{P}$ is ϑ-quasihyperbolic and the simultaneous set equalities
(3.4) and (3.5) hold for some non-empty Λ and for all $c \in \mathbb{R}$ with
$K = \text{ch}(H(\vartheta,c) \cap \text{supp } U)$.

As for degenerate systems, we have

<u>Theorem 3.6</u>. The equation (2.2) admits a non-trivial solution $U \in [\mathcal{D}'(\mathbb{R}^\ell)]^n$ with compact support if and only if the system $\mathcal{S}$ is degenerate. More precisely, if $\mathcal{S}$ is degenerate, then for any open set ω and for any $\varepsilon > 0$, there exists a solution $U \in [C^\infty(\mathbb{R}^\ell)]^n$ such that

$$\overline{\omega} \subseteq \text{supp } U \subseteq \omega_\varepsilon$$

where $\omega_\varepsilon = \{x \in \mathbb{R}^\ell;\ d(x,\omega) < \varepsilon\}$ (d being a metric on $\mathbb{R}^\ell$).

§4 <u>Application</u>.

We say that a pair $T = (T^{(+)}, T^{(-)})$ of closed convex subsets of $\mathbb{R}^\ell$ is $\mathcal{S}$-*proper*, if it satisfies the conditions

i) $\quad T^{(+)} \subseteq \overline{\Omega}(\mathcal{S}), \quad T^{(-)} \subseteq \overline{\Omega}(-\mathcal{S})$;

ii) $\quad T^{(+)} \cap H(\mathcal{S}), \quad T^{(-)} \cap H(\mathcal{S})$ are both compact;

iii) $\quad T^{(+)} \cap T^{(-)}$ has non-empty interior in $H(\mathcal{S})$.

For a $\mathcal{S}$- proper pair of closed convex sets $T = (T^{(+)}, T^{(-)})$, we put $|T| = T^{(+)} \cup T^{(-)}$ and define the *direction cone* $\Gamma(T)$ of T by putting

$$(4.1) \qquad \Gamma(T) = \Gamma(T^{(+)}) \cap (-\Gamma(T^{(-)})).$$

Here, in the right hand side, Γ stands for the usual direction cones. It is clear that $\Gamma(T)$ is always a $\mathcal{S}$-proper closed convex cone.

We say that a subset S of $\mathbb{R}^\ell$ is a $\mathcal{S}$-*proper set* if there exist two $\mathcal{S}$-proper pairs $T_j = (T_j^{(+)}, T_j^{(-)})$, $j = 1,2$ with common direction cone $\Gamma(T_1) = \Gamma(T_2)$ and a vector $x_0 \in \mathbb{R}^\ell$ such that the inclusion relations

$$(4.2) \qquad |T_1| \subseteq S + x_0 \subseteq |T_2|$$

hold. For such a set S, we define its direction cone $\Gamma(S)$ by putting

$$(4.3) \qquad \Gamma(S) = \Gamma(T_1) = \Gamma(T_2).$$

It is easy to see that $\Gamma(S)$ does not depend on the choice of T_j ($j = 1,2$) or of x_0.

Now we can state

<u>Theorem 4.1</u>. Given a non-degenerate system of operators $\wp$ and a ϑ-proper subset S of $\mathbb{R}^\ell$. For the system of equations (2.2) to have a non-trivial solution $U \in [C^\infty(\mathbb{R}^\ell)]^n$ with supp $U \subseteq S$, it is necessary and sufficient that $\chi(\wp)$ has an irreducible factor P_λ such that

$$(4.4) \qquad \Gamma^*(P_\lambda, \vartheta) \subseteq \Gamma(S).$$

If dim $\Gamma(S) = k$ then it is easy to see that, for a suitable coordinate system and for a number $b \geq 0$, we have $\vartheta = (1,0,\ldots,0)$ and $\Gamma(S) \subseteq \{x;\ x_1 \geq 0,\ |x_j| \leq bx_1,\ j = 2,\ldots,k,\ x_{k+1} = \ldots = x_\ell = 0\}$ since $\Gamma(S)$ is a ϑ-proper cone. Then, the principal part $P_\lambda^0(\xi)$ of the polynomial $P_\lambda(\xi)$ satisfying (4.4) depends only on $\xi_1,\ldots,\xi_k$ and all the solutions $(\xi_1,\ldots,\xi_k) \in \mathbb{R}^k$ of the algebraic equation $P_\lambda^0(\xi) = 0$ should satisfy the inequality: $|\xi_1| \leq b(\xi_2^2 + \ldots + \xi_k^2)^{1/2}$. With these remarks in mind, we get directly from the above theorem 4.1 the results in [2] by putting $m = n$, $k = 1$ and those in [3] by putting $m = n = 1$, $k = 2$.

REFERENCES

[1] Atiyah, M.F., R.Bott and L.Garding, Lacunas for hyperbolic differential operators with constant coefficients, Acta Math. 124 (1970), 109-189.

[2] Cohoon, D.K., A characterization of the linear partial differential operators $P(D)$ which admit a non-trivial C^∞ solutions with support in an open prism with bounded cross section, J. Differential Equations, 8(1970), 195-201.

[3] Horie, K., On the characterization of the linear partial differential operators of hyperbolic type, Proc. Japan Acad., 49(1973), 506-509.

[4] Hörmander, L., Linear partial differential operators, Springer 1963.

[5] John, F., Non-admissible data for differential equations with constant coefficients, Comm. Pure Appl. Math., 10(1957), 391-398.

WAVE PROPAGATION IN A NON-LINEAR LATTICE

Morikazu Toda

Institute for Optical Research

Kyoiku University, Tokyo, Japan

1. Equation of Motion

Aiming at a non-linear mechanical system which will admit exact solutions and reveal fundamental features of the propagation of anharmonic waves, I considered a one-dimensional lattice with the nearest neighbour interaction. The equation of motion is

$$(1) \qquad m \frac{d^2 y_n}{dt^2} = -\phi'(y_n - y_{n-1}) + \phi'(y_{n+1} - y_n)$$

or, in terms of the relative displacement

$$(2) \qquad r_n = y_n - y_{n-1},$$

$$(3) \qquad m \frac{d^2 r_n}{dt^2} = -2\phi'(r_n) + \phi'(r_{n+1}) + \phi'(r_{n-1}).$$

Assuming that the stress

$$(4) \qquad f_n = -\phi'(r_n)$$

can be solved for r_n,

$$(5) \qquad r_n = -X(f_n)/m,$$

we obtain

$$(6) \qquad \frac{d^2}{dt^2} X(f_n) = f_{n+1} - 2f_n + f_{n-1}.$$

(If we write $f_n = -ds_n/dt$, the s_n is the conjugate momentum to r_n.)

In the effort to find out a new system, my strategy was along the trial and error method, seeking for appropriate f_n which would yield physically significant potential $\phi(r)$, or assuming $\phi(r)$ which might yield some meaningful solutions. There were good reasons to believe that the non-linear lattice to be found would admit periodic solutions. After several months' effort, by an inspiration, I found a nice periodic wave and the interaction potential simultaneously.[1]

The periodic solution is

$$(7) \qquad e^{-br_n} - 1 = \frac{(2K\nu)^2}{ab/m} \left[dn^2 \left\{ 2(\tfrac{n}{\lambda} \pm \nu t) K \right\} - \frac{E}{K} \right]$$

with the dispersion relation

$$(8) \qquad \frac{2K\nu}{\sqrt{ab/m}} = \left\{ \frac{1}{sn^2(2K/\lambda)} - 1 + \frac{E}{K} \right\}^{-1/2}.$$

The potential is

$$(9) \qquad \phi(r) = \frac{a}{b} e^{-br} + ar + \text{const.},$$

where a and b are constants ($ab > 0$). In the limit as $b \to 0$, the system reduces to a harmonic lattice, and in the limit as $b \to \infty$ (ab=finite), it gives a system of hard spheres. Exact solutions, methods of obtaining solutions and the invariants of motion to be explained below all include these limits.[1),4),11)]

In dimensionless units, our lattice (the exponential lattice) can be expressed by the equation of motion

$$(10) \qquad \frac{d^2 Q_n}{dt^2} = e^{-(Q_n - Q_{n-1})} - e^{-(Q_{n+1} - Q_{n+1})},$$

or, in terms of the relative displacement $r_n = Q_n - Q_{n-1}$,

$$(11) \qquad \frac{d^2 r_n}{dt^2} = 2e^{-r_n} - e^{-r_{n-1}} - e^{-r_{n+1}},$$

or, in terms of the stress f_n,

$$(12) \qquad \frac{d^2}{dt^2} \ln(1 + f_n) = f_{n-1} + f_{n+1} - 2f_n.$$

2. Particular Solutions

Besides the periodic solution, I found several particular solutions. First, taking the limit as $\lambda \to \infty$, k(modulus)$\to 1$, we obtain a pulse wave[2)]

$$(13) \qquad e^{-r_n} - 1 = \beta^2 \text{sech}^2(\alpha n \pm \beta t),$$

where α is an arbitrary constant and

$$(14) \qquad \beta = \sinh \alpha.$$

This pulse is called a lattice soliton.

In general it is covenient to use the ψ_n, which is related to f_n by

$$(15) \qquad f_n = \frac{d^2}{dt^2} \ln \psi_n.$$

A 2-soliton solution is obtained by assuming [3)]

$$(16) \qquad \psi_n = \cosh(k_1 n - w_1 t) + B \cosh(k_2 n - w_2 t),$$

or

$$(17) \qquad \psi_n = 1 + A_1 e^{2(k_1 n - w_1 t)} + A_2 e^{2(k_2 n - w_2 t)} + A_3 e^{2(k_1 + k_2)n - 2(w_1 + w_2)t}.$$

The equation of motion is satisfied by putting

$$(18) \qquad w_i^2 = \sinh^2 k_i \quad (i=1,2),$$

$$(19) \qquad B = \sinh(k_1/2)/\sinh(k_2/2)$$

or

$$(20) \qquad \frac{A_1 A_2}{A_3} = \frac{(w_1 + w_2)^2 - \sinh^2(k_1 + k_2)}{\sinh^2(k_1 - k_2) - (w_1 - w_2)^2}.$$

There are two cases:

$w_1w_2 > 0$; two solitons propagating in the same direction.

$w_1w_2 < 0$; two solitons propagating in the opposite directions.

The multi-soliton solution was first studied numerically and then its analytic expression was obtained.

3. Inverse Scattering Method

As was shown by Flaschka[5], the equation of motion of the exponential lattice can be written in a matrix form as

$$(21) \qquad \frac{dL}{dt} = BL - LB$$

where L and B are matrices such that when operated on a vector v we have

$$(22) \qquad (Lv)_n = b_n v_n + a_n v_{n-1} + a_{n+1} v_{n+1},$$

$$(23) \qquad (Bv)_n = -a_n v_{n-1} + a_{n+1} v_{n+1}.$$

Eq. (21) gives

$$(24) \qquad \frac{db_n}{dt} = 2(a_{n+1}^2 - a_n^2),$$

$$(24') \qquad \frac{da_n}{dt} = a_n(b_n - b_{n-1}).$$

Our exponential lattice is given when we put

$$(25) \qquad a_n = -\frac{1}{2}e^{-(Q_n - Q_{n-1})/2},$$

$$(25') \qquad b_n = -\frac{1}{2}P_n.$$

The a_n defined above differs from that of Flaschka's in sign. The present choice seems more appropriate to stress the parallelism between the exponential lattice and the Korteweg-de Vries equation as we shall see in the continuum limit.[12]

In principle the initial value problem of the above matrix equation can be solved by applying the inverse scattering method to the equation[6]

$$(26) \qquad L\varphi = \lambda \varphi.$$

Given the initial value of the reflection coefficient $R(z,0)$ of the wave $\varphi(z;n)$ where $\lambda = -(z+z^{-1})/2$, and the normalization coefficient $c_j(0)$ of the bound state $\lambda_j = -(z_j + z_j^{-1})/2$, we set up the kernel at time t,

$$(27) \qquad F(m) = \frac{1}{2\pi i} \oint R(z,0)e^{-t(z-z^{-1})} z^{m-1}\, dz + \sum_j c_j^2(0)e^{-(z_j - z_j^{-1})t} z_j^m ,$$

and set up the equation (a discrete version of the Gel'fand-Levitan eq.)

$$(28) \qquad K(n,m) + F(n+m) + \sum_{n'=n+1}^{\infty} K(n,n')F(n'+m) = 0 \qquad (n < m).$$

When we find the solution $K(n,m)$, we make $K(n,n)$ by

$$(29) \quad \frac{1}{[K(n,n)]^2} = 1 + F(2n) + \sum_{n'=n+1}^{\infty} \mathcal{K}(n,n')F(n'+n)$$

Then the time evolution of the wave in the lattice is given by

$$(30) \quad e^{-(Q_n-Q_{n-1})} = \left\{\frac{K(n,n)}{K(n-1,n-1)}\right\}^2$$

or,

$$(31) \quad \frac{dQ_n}{dt} = s_n - s_{n+1}$$

with

$$(32) \quad s_n = \mathcal{K}(n,n-1).$$

4. Conservation Laws

The total momentum

$$(33) \quad J^{(1)} = \sum_n P_n$$

and the total energy

$$(34) \quad J^{(2)} = \sum_n (\tfrac{1}{2}P_n^2 + X_n),$$

with

$$(35) \quad P_n = \frac{dQ_n}{dt}, \qquad X_n = e^{-(Q_n-Q_{n-1})},$$

are conserved in an infinite lattice and in a cyclic lattice. In addition
to these, we have many invariants such as

$$(36) \quad J^{(3)} = \sum_n \left\{\tfrac{1}{3}P_n^3 + P_n(X_n - X_{n+1})\right\}, \qquad J^{(4)} = \ldots .$$

Arbitrary functions of these invariants are also conserved. For a cyclic
lattice of N particles, Hénon[7] showed that there are N invariants. The
trajectories in the phase space are on the hypersurfaces of these invariants, and therefore the system is non-ergodic[8],[9],[10].

5. Continuum Limit

For the waves with wavelength long compared with the spacing between
the particles in the lattice, or when the non-linearity constant b is
very small, the equation of motion (10) or (11) reduces to equations for
weakly non-linear continuum, viz. to the Boussinesq equation and to the
Korteweg-de Vries equation. The particular solutions, the inverse scattering method and the conservation laws have their counterparts in this
limit. We shall clarify some of these relations in the following.[12]

We may write L and B of (22) and (23) as

$$(37) \quad L = b_n + a_n e^{-\partial/\partial n} + e^{\partial/\partial n} a_n,$$

$$(38) \quad B = -a_n e^{-\partial/\partial n} + e^{\partial/\partial n} a_n .$$

Assuming smooth variation of a_n, b_n and φ as functions of n, thought as a continuous variable, we may expand $\exp(\pm\partial/\partial n)$ and $a_n=-\frac{1}{2}\exp(-r_n/2)$. Neglecting higher order terms, we thus get

$$(39) \qquad L=-\frac{1}{2}P_n+2\left(-1+\frac{r_n}{2}\right)+\frac{1}{4}\left(\frac{\partial}{\partial n}r_n-r_n\frac{\partial}{\partial n}\right)-\frac{1}{2}\frac{\partial^2}{\partial n^2} \ ,$$

$$(40) \qquad B=-\frac{\partial}{\partial n}+\frac{1}{4}\left(\frac{\partial}{\partial n}r_n+r_n\frac{\partial}{\partial n}\right)-\frac{1}{6}\frac{\partial^3}{\partial n^3} \ .$$

For waves propagating to the right, we use the coordinate ξ moving uniformly to the right, and the time τ defined by

$$(41) \qquad \xi=n-t,$$

$$(42) \qquad \tau=t/24.$$

Then we get

$$(43) \qquad \frac{\partial}{\partial\tau}\varphi=\tilde{B}\varphi \ ,$$

$$(44) \text{ with} \qquad \tilde{B}=-4\frac{\partial^3}{\partial\xi^3}+3\left(\frac{\partial}{\partial\xi}u+u\frac{\partial}{\partial\xi}\right) \ .$$

In this approximation, therefore, P_n can be replaced by

$$(45) \qquad P_n=\frac{\partial Q}{\partial\xi}+\frac{1}{24}\frac{\partial Q}{\partial\tau}\cong r_n=\frac{u}{2},$$

where we have introduced u by

$$(46) \qquad u=2r_n \ .$$

We have thus

$$(47) \qquad L=-\frac{1}{2}\frac{\partial^2}{\partial\xi^2}+\frac{u}{2}-1 \ .$$

The equation of motion reduces to

$$(48) \qquad L_\tau=\tilde{B}L-L\tilde{B},$$

which turns out to be the Korteweg-de Vries equation

$$(49) \qquad u_\tau-6uu_\xi+u_{\xi\xi\xi}=0.$$

In the continuum limit, we write

$$(50) \qquad K(n,m)=K'(n,m)h, \quad F(n+m)=F'(n+m)h$$

(h=lattice spacing), and eqs.(28) and (29) yield

$$(51) \qquad K'(n,m)+F'(n+m)+\int K'(n,n')F(n'+m)dn'=0,$$

$$(52) \qquad \frac{1}{[K(n,n)]^2}=1+F'(2n)+\sum_{n'}K'(n,n')F'(n'+n)dn'=0.$$

Writing

$$(53) \qquad z=e^{ik}, \quad z_j=e^{-k_j},$$

$$(54) \qquad n-t=\xi \ , \quad m-t=\eta \ ,$$

$$(55) \qquad F'(n+m)=\widetilde{F}(\xi+\eta), \quad \kappa'(n,m)=\widetilde{\kappa}(\xi,\eta),$$

and neglecting higher order terms of k and k_j , we obtain the kernel,

$$(56) \qquad \widetilde{F}(\xi+\eta)=\frac{1}{2\pi}\int R(k,0)e^{8ik^3\tau+ik(\xi+\eta)}dk$$
$$+\sum_j c_j^2(0)e^{8k_j^3\tau-k_j(\xi+\eta)} \ ,$$

for the Gel'fand-Levitan equation

$$(57) \qquad \widetilde{\kappa}(\xi,\eta)+\widetilde{F}(\xi+\eta)+\int_\xi^\infty \widetilde{\kappa}(\xi,\zeta)\widetilde{F}(\zeta+\eta)d\zeta=0.$$

Eq.(52) gives

$$(58) \qquad K(\xi,\xi)^{-2}=1+\widetilde{F}(2\xi)+\int_\xi^\infty \widetilde{\kappa}(\xi,\zeta)\widetilde{F}(\zeta+\xi)d\zeta \ .$$

Comparing this with eq.(57) we see that

$$(59) \qquad K(\xi,\xi)^{-2}=1-\widetilde{\kappa}(\xi,\xi),$$

and eq.(30) gives in this limit ($u(\xi,\tau)=2u_n(t)$, see eq.(46))

$$(60) \qquad u=-2\frac{\partial}{\partial\xi}\widetilde{\kappa}(\xi,\xi) \ .$$

Eqs.(56),(57) and (60) are well-known equations for the Korteweg-de Vries equation (49).[13]

References:

1) M. Toda: J. Phys. Soc. Japan 22 (1967) 431.

2) M. Toda: J. Phys. Soc. Japan 23 (1967) 501.

3) M. Toda: J. Phys. Soc. Japan, Suppl. 26 91969) 235.

4) M. Toda: Prog. Theor. Phys., Suppl. 45 (1970) 174.

5) H. Flaschka: Phys. Rev. B9 (1974) 1924.

6) H. Flaschka: Prog. Theor. Phys. 51 (1974) 703.

7) M. Henon: Phys. Rev. B9 (1974) 1921.

8) J. Ford, S.D. Stoddard and J.S. Turner: Prog. Theor. Phys. 50 (1973) 1547.

9) N. Ooyama and N. Saito: Prog. Theor. Phys., Suppl. 45 (1970) 201.

10) M. Toda: Phys. Letters 48 A (1974)335.

11) M. Toda: Arkiv for Det Fysiske Seminar i Trondheim No 2-1974, to be bublished in Physics Reports.

12) M. Toda: unpublished.

13) C.S. Gardner, J.M. Greene, M.D. Kruskal and R.M. Miura: Phys. Rev. Letters 19 (1967)1095.

14) P.D. Lax: Comm. Pure and Appl. Math. $\underline{21}$ (1968)467.

Discussions

R. Jackiw:

To what invariances of the system do the constants of motion correspond ?

M. Toda:

I have no idea yet. In the harmonic limit, I found that these constants of motion reduce to linear combinations of the normal mode energies.

K. Kikkawa:

Do these conserving quantities form a closed algebra with respect to Poisson bracket ?

M. Toda:

I heard that it was shown that the constants of motion of the KdV equation were in involution. I think the same applies to our lattice.

T. Niwa:

If the potential function is perturbed do the special solutions obtained conserve ?

M. Toda:

According to computer experiments and some analytical arguments, it seems quite reasonable to believe that the general features of the particular solutions will stay qualitatively unchanged by the perturbation in the potential function. The concept of soliton seems to have wide validity in non-linear wave propagation. However, if the potential has an inflexion point ($\phi''=0$), the trajectories in the phase space can have exponential instability[10].

QUANTIZATION OF NON-LINEAR WAVES

R. Jackiw

Laboratory for Nuclear Science and Department of Physics
Massachusetts Institute of Technology
Cambridge, Massachusetts 02139 U.S.A.

There exist field theories in which a symmetry is spontaneously broken by a non-vanishing vacuum expectation value of a scalar field. The signal for this phenomenon has traditionally been the existence of constant, non-vanishing solutions to the classical field equations. It is further known that in some of these models, the classical equations possess stable static solutions, with total energy a finite amount above that of the constant solution. We suggest that these solutions indicate that the particle spectrum of the theory is richer than has been heretofore assumed. In addition to the conventional particles of the theory, there appear also heavy particles, which carry a new quantum number and are stable. A systematic approximation scheme is described with which the properties of these new particles can be exposed.[1]

The framework in which we begin the analysis is that of the variational principle for the energy functional. Any quantum problem, for example one set in field theory, can be formulated variationally. Given a Hamiltonian H depending on a field operator Φ and on canonical momentum Π, we seek the state $|\psi\rangle$ for which $\langle\psi|H|\psi\rangle/\langle\psi|\psi\rangle$ is to be stationary, but the matrix elements of the quantum field and of the product of two quantum fields are held fixed.

$$\langle\psi|\Phi(t,\underset{\sim}{x})|\psi\rangle/\langle\psi|\psi\rangle \qquad = \quad \phi(\underset{\sim}{x}) \tag{1a}$$

$$\langle\psi|\Phi(t,\underset{\sim}{x})\Phi(t,\underset{\sim}{y})|\psi\rangle/\langle\psi|\psi\rangle = \quad \phi(\underset{\sim}{x})\phi(\underset{\sim}{y}) + G(\underset{\sim}{x},\underset{\sim}{y}) \tag{1b}$$

The expectation of H in this constrained state depends on ϕ and G; we call that quantity the <u>energy functional</u>, $E(\phi,G)$. The complete variational principle is now implemented by demanding that $E(\phi,G)$ be stationary against variations of ϕ and G.

$$\frac{\delta E(\phi,G)}{\delta \phi(\underset{\sim}{x})} = 0 \qquad (2a)$$

$$\frac{\delta E(\phi,G)}{\delta G(\underset{\sim}{x},\underset{\sim}{y})} = 0 \qquad (2b)$$

The advantage of this formulation is that approximations to $E(\phi,G)$ can be readily constructed, and the problem of solving the field theory reduces to solving eqs. (2). Also, once they are solved, the equilibrium forms of $\phi(\underset{\sim}{x})$ and $G(\underset{\sim}{x},\underset{\sim}{y})$ are known.

We shall consider the simplest, the semiclassical approximation to $E(\phi,G)$. For a theory described by the Lagrangian

$$\mathcal{L}(\Phi) = \tfrac{1}{2}\partial_\mu\Phi\partial^\mu\Phi - U(\Phi) \qquad (3)$$

one finds[2]

$$E_{sc}(\phi,G) = E_c(\phi) + \tfrac{1}{8}\int d\underset{\sim}{x}\,G^{-1}(\underset{\sim}{x},\underset{\sim}{x}) + \tfrac{1}{2}\int d\underset{\sim}{x}d\underset{\sim}{y}\,G(\underset{\sim}{x},\underset{\sim}{y})\frac{\delta^2 E_c(\phi)}{\delta\phi(\underset{\sim}{x})\delta\phi(\underset{\sim}{y})} \qquad (4a)$$

The first term in (4a) is the classical energy of a static field ϕ.

$$E_c(\psi) = \int d\underset{\sim}{x}\,[\tfrac{1}{2}(\underset{\sim}{\nabla}\psi)^2 + U(\psi)] \qquad (4b)$$

The remaining terms in (4a) comprise the first quantum correction. [$E_c(\phi)$ is the tree approximation to $E(\phi,G)$; the first quantum correction corresponds to one-loop graphs.]

The variational equations (2) which are implied by (4) are

$$\nabla^2\phi(\underset{\sim}{x}) = U'(\phi) + \tfrac{1}{2}G(\underset{\sim}{x},\underset{\sim}{x})U'''(\phi) \qquad (5a)$$

$$\tfrac{1}{4}G^{-2}(\underset{\sim}{x},\underset{\sim}{y}) = [-\nabla^2 + U''(\phi)]\delta(\underset{\sim}{x}-\underset{\sim}{y}) \qquad (5b)$$

Thus $\phi(\underset{\sim}{x})$ is seen to satisfy the classical, static equation of motion with a quantum correction involving $G(\underset{\sim}{x},\underset{\sim}{x})$. We approximate further by solving eqs.(5) sequentially: we set

$$\phi(x) = \phi_c(x) + \delta\phi(x) \qquad (6)$$

and replace eqs. (5) by

$$\nabla^2 \phi_c(\underset{\sim}{x}) = U'(\phi_c) \tag{6a}$$

$$\tfrac{1}{4} G^{-2}(\underset{\sim}{x},\underset{\sim}{y}) = [-\nabla^2 + U''(\phi_c)]\delta(\underset{\sim}{x}-\underset{\sim}{y}) \tag{6b}$$

$$[-\nabla^2 + U''(\phi_c)]\delta\phi(\underset{\sim}{x}) = -\tfrac{1}{2} G(\underset{\sim}{x},\underset{\sim}{x}) U'''(\phi_c) \tag{6c}$$

Once ϕ_c, a solution of the classical equation (6a), is obtained, eq.
(6b) is solved by finding a complete set of functions $\psi_n(x)$ satisfying
a Schrödinger-like equation.

$$[-\nabla^2 + U(\phi_c)]\psi_n(\underset{\sim}{x}) = \omega_n^2 \psi_n(\underset{\sim}{x}) \tag{7a}$$

$$G^{-2}(\underset{\sim}{x},\underset{\sim}{y}) = \sum_n 4\omega_n^2 \psi_n^*(\underset{\sim}{x})\psi_n(\underset{\sim}{y}) \tag{7b}$$

Substituting our solution into (4a), determines the semi-classical
energy

$$E_{sc} = E_c + \tfrac{1}{4}\int d\underset{\sim}{x}\, G^{-1}(\underset{\sim}{x},\underset{\sim}{x})$$

$$= E_c + \tfrac{1}{2}\sum_n \omega_n \tag{8}$$

[To this order $\delta\phi$ does not contribute since $E_c(\phi)$ is stationary at
$\phi = \phi_c$; hence $E_c = E_c(\phi_c)$.] The meaning of (8) is clear: the total
energy in this approximation is composed of the classical energy, plus
the zero point energy of field fluctuations.

In analyzing the vacuum sector of the theory, we seek constant,
position independent solutions for ϕ. But it is natural to inquire
whether there exist solutions to the variational equations which are
position dependent. Unlike the translationally invariant solutions,
which determine the nature of the vacuum, the translationally non-invariant solutions cannot be associated with vacuum expectation values,
since we do not expect translational symmetry to be spontaneously broken.
As we shall show below, these solutions should be interpreted as evidence
for new states in the theory.[1]

However before embarking upon a study of position dependent solutions to (6a), we must take note of a very important stability criterion.

The eigenfrequencies ω_n^2 in (7a) must be non-negative so that the energy (8) be real. But there is always a zero-frequency mode: upon differentiation of (6a) with respect to $\underset{\sim}{x}$, it is recognized that $\nabla\phi$ satisfies (7a) with $\omega_o = 0$. For stability this must be the lowest mode. However, when the dimensionality of space is greater than one, $\nabla\phi$ involves several independent functions, and the mode is degenerate. Since a degenerate state is never the lowest level of an ordinary Schrödinger equation, we conclude that stability exists only in one spatial dimension.

I do not abandon this investigation, not because one dimension holds any particular interest, but rather because it is possible to construct stable, static models in three-dimensional space, provided something more complicated than a spinless field [or a set of spinless fields] is used; for example fields with spin.[3] Moreover, the formalism and interpretation that we develop works for the realistic situation as well. Therefore I restrict the subsequent discussion to one dimension, where the equation (6a) may be integrated once to give

$$\frac{1}{2}[\phi'(x)]^2 = U(\phi) \tag{9}$$

and the classical energy becomes

$$E_c = \int dx\left[\frac{1}{2}(\phi_c')^2 + U(\phi_c)\right]$$

$$= \int dx\,(\phi_c')^2 \tag{10}$$

As an explicit example, we pick the potential $U(\Phi) = \frac{1}{2\lambda}(m^2 - \lambda\Phi^2)^2$, which gives $\phi_c(x) = \frac{m}{\lambda^{1/2}}\tanh mx$, $E_c = \frac{4}{3}\frac{m^3}{\lambda}$. One may verify that (7a) in this case has a non-negative spectrum; the lowest frequency vanishes and the corresponding normalized eigenfunction is $\frac{\phi_c(x)}{E_c^{1/2}}$.

We now make the following reinterpretation of our results. We allege that the position dependent solution of the variational equations is an artifact of the approximation; if $E(\phi,G)$ were computed exactly, then only constant solutions would be found. [As an indication of this, note that even the present approximation cannot be carried to higher order. The solution to (6c) requires $G(\underset{\sim}{x},\underset{\sim}{x}) = \sum_n \frac{1}{2\omega_n}\psi_n^*(x)\psi_n(x)$; this is infinite since $\omega_o = 0$.]

What then is the correct interpretation of the solutions we have found? We suggest the following _Ansatz_. In addition to the conventional particles of the theory, [which we call mesons] there exist heavy particles [which we call baryons]. The baryon is described by a non-normalizable _Poincaré_ convariant state $|p\rangle$; it carries momentum p and energy $\sqrt{[p^2 + M^2]}$. In weak coupling M is of order λ^{-1}, hence $E \approx M + \dfrac{p^2}{2M}$. To order λ^0, the energy previously calculated in (10) coincides with the mass of the baryon. There exist also baryon, multi-meson states $|p; \{k_n\}\rangle$, with total momentum $p + \sum_n k_n$ and energy $E(p) + \sum_n \omega(k_n)$.

To demonstrate the consistency of the _Ansatz_, we begin by studying the equation for the no-meson matrix element of Φ. The exact equation of motion is

$$\{-(p-q)^2 + (E(p) - E(q))^2 + 2m^2\}\langle p|\Phi|q\rangle = 2\lambda\langle p|\Phi^3|q\rangle \qquad (11)$$

The lowest order in λ, [this will be seen to be order $\lambda^{-1/2}$], the energy difference vanishes, since $E(p) \approx E(q) \approx M$-- we are led to a static approximation. To the same order $\langle p|\Phi|q\rangle$ depends only on the momentum difference. On the right hand side, complete sets of states are inserted between each of the Φ factors; to lowest order only the no-meson states survive.[4] Thus the equation becomes

$$\frac{d^2}{dx^2} f(x) = -2m^2 f(x) + 2\lambda f^3(x) = U'(f) \qquad (12a)$$

$$\langle p|\Phi|q\rangle = \int dx\, e^{i(p-q)x} f(x) \qquad (12b)$$

Therefore the solution to the static, classical field equations emerges not as the expectation of the quantum field, but rather as the Fourier transform of the field formfactor.

The energy may also be calculated in the same approximation.

$$H = \int dx\,[\tfrac{1}{2}\Pi^2 + \tfrac{1}{2}(\Phi')^2 + U(\Phi)], \quad \Pi = \frac{d\Phi}{dt}$$

$$\langle p|H|p'\rangle = (2\pi)\delta(p-p')\, E(p)$$

$$E(p) = \langle p|\tfrac{1}{2}\Pi^2 + \tfrac{1}{2}(\Phi')^2 + U(\Phi)|p\rangle \qquad (13)$$

To leading order, λ^{-1}, the left hand side is just M. On the right hand side, we saturate with single baryon states. The matrix element of Π^2

gives zero, since it involves energy differences which vanish in leading order. The remaining terms are easily shown to give E_c. Hence we find consistent with the <u>Ansatz</u> $M = E_c$.

To calculate in next order, the single-meson matrix element of Φ is required. As before we assume that to lowest order it depends only on the momentum differences of the baryons.

$$<p|\Phi|p';k> = \int dx e^{i(p-p'-k)x} f(k;x) \tag{14}$$

The equation for $f(k;x)$ is obtained analogously to (12a).

$$[-\frac{d^2}{dx^2} - 2m^2 + 6\lambda f^2(x)]f(k;x) = \omega^2(k)f(k;x)$$

$$[-\frac{d^2}{dx^2} + U''(f)]f(k;x) = \omega^2(k)f(k;x) \tag{15}$$

We encounter again the Schrödinger-like equation; now the wavefunctions have a physical significance. Moreover we insist that the zero-frequency solution of (15) does not correspond to a physical state and should be dropped. That the physical states are complete without the inclusion of this mode is verified by computing the baryon matrix element of the canonical commutator $i[\dot{\Phi}(t,x),\Phi(t,y)] = \delta(x-y)$, and keeping all terms of order λ^0.

In principle the calculations in the one-baryon sector of the theory can be carried out to arbitrary accuracy, by expanding all expressions in powers of λ. In such an expansion one takes the connected part of the matrix element of Φ between states containing one baryon, n mesons and one baryon, m mesons to be of order $\lambda^{1/2(m+n-1)}$. The calculations that have been performed show that our <u>Ansatz</u> is consistent and lead to a Poincaré covariant description of the baryon.

The baryon is stable. This may be understood as follows. The theory admits a conserved current $J^\mu = \epsilon^{\mu\nu}\partial_\nu\Phi$. The matrix elements of the charge, will be proportional to $<\Phi(x)>|_{x=-\infty}^{x=\infty}$. In the no-baryon sector the field tends to the same constant as $x\to\pm\infty$, hence the matrix element vanishes. In the one baryon sector, the field tends to different constants and the charge is non-zero. It is the conservation of this charge that prevents a no-baryon state from connecting to a baryon state. Thus even though the baryons are heavy, they cannot decay into mesons.

Another remarkable feature of the baryons is that they appear to be Fermions. We are led to this conclusion by noting that the form factor $\langle p|\Phi|q\rangle = \int dx\, e^{i(p-q)x}\phi_c(x)$ is antisymmetric in $p\leftrightarrow q$, since $\phi_c(x) = -\phi_c(-x)$. Presumably this is a feature of two-dimensional field theories, which do not have a spin-statistics theorem.

Evidence for our interpretation, with all its starting aspects, has been recently found in the study of the sine-Gordon equation. Here the potential is $U(\Phi) = \frac{m^2}{\lambda}(1-\cos\lambda^{1/2}\Phi)$ and one can obtain a stable static solution to the classical equations, $\phi_c(x) = \frac{4}{\lambda^{1/2}}\tan^{-1}\exp xm$. Hence the theory can be developed as above. On the other hand a careful examination of the Green's functions of the sine-Gordon theory reveals that they are identical to those of the massive Thirring model, $\mathcal{L} = i\bar\psi\gamma^\mu\partial_\mu\psi - M\bar\psi\psi - \frac{g}{2}\bar\psi\gamma^\mu\psi\bar\psi\gamma_\mu\psi$, with the identifications $4\pi/\lambda = 1 + g/\pi$, $\lambda^{1/2}\varepsilon^{\mu\nu}\partial_\nu\Phi = 2\pi\bar\psi\gamma^\mu\psi$, $(m^2/\lambda)\cos\lambda^{1/2}\Phi = -M\bar\psi\psi$.[5] Hence it is very plausible to associate the Fermion field of the Thirring model with the massive baryons of the sine-Gordon equation.

Do these new, remarkable states have any significance for practical physics? Of course the question can only be addressed to models in 4-dimensional space-time, where we have shown that fields with spin must be necessarily used. Various such candidates have been proposed but their consequences are not yet completely understood.[3] It may be, as has been long speculated, that the baryons occuring in nature will be found to coincide with the mathematical baryons which I have here discussed.

REFERENCES

1) J. Goldstone and R. Jackiw, Phys. Rev. D (in press). Similar conclusions have been obtained by R. Dashen, B. Hasslacher, and A. Neveu, Phys. Rev. D (in press). These authors use a WKB method for field theory.

2) J. Cornwall, R. Jackiw and E. Tomboulis, Phys.Rev.D $\underline{10}$, 2428 (1974).

3) H. B. Nielsen and P. Olesen, Nucl. Phys. $\underline{B61}$, 45 (1973);
G. 't Hooft, Nucl. Phys. $\underline{B79}$, 276 (1974);
T. Eguchi and H. Sugawara, Phys. Rev. D (in press);
L. D. Faddeev, Max-Planck Institute preprint;
A. M. Polyakov, Landau Institute preprint;
S. Mandelstam, Berkeley preprint.

4) The calculation of matrix elements by using equations of motion, saturating with intermediate states and performing a static approximation was suggested to us by A. Kerman and A. Klein, Phys. Rev. $\underline{132}$, 1326 (1963).

5) S. Coleman, Harvard University preprint.

ON THE NONLINEAR DIFFUSION EQUATION OF
KOLMOGOROV-PETROVSKII-PISKUNOV TYPE

Yoshinori Kametaka

Department of Mathematics

Osaka City University

Osaka, Japan

Asymptotic behavior of the solutions of the initial value problem for a nonlinear diffusion equation of KPP type is studied. There exists one parameter family of travelling waves $w_\lambda(x + 2\lambda t)$ with negative speed -2λ where $\lambda \geq \lambda_0 = \sqrt{f'(0)}$. Each travelling wave is locally stable from above and below on the phase plane. Especially the slowest one is stable from above almost in the large on the phase plane. These are the improvements of the results of KOLMOGOROV-PETROVSKII-PISKUNOV [1].

Consider the initial value problem for a nonlinear diffusion equation of KPP type

$$(1) \qquad \begin{cases} [\frac{\partial}{\partial t} - (\frac{\partial}{\partial x})^2]u = f(u), \quad 0 \leq u \leq 1, \quad (x,t) \in R^1 \times (0,\infty), \\[2ex] u(x,0) = u_0(x) \qquad x \in R^1. \end{cases}$$

We assume: $f(\xi) \in C^\infty[0,1]$, $f(0) = f(1) = 0$, $f'(0) > 0 > f'(1)$, $f(\xi) > 0$ and $f'(0)\xi - f(\xi) \geq 0$ for $\xi \in (0,1)$. Putting $u = w(x + 2\lambda t)$, (1) reduces to

$$(2) \qquad w'' - 2\lambda w' + f(w) = 0, \quad 0 \leq w \leq 1, \quad x \in R^1, \quad ' = \frac{d}{dx}, \quad \lambda \geq 0.$$

KPP(KOLMOGOROV-PETROVSKII-PISKUNOV)[1] showed the following facts:

(i) If $0 \leq \lambda < \lambda_0 = \sqrt{f'(0)}$ then (2) implies $w \equiv 0$ or 1.

(ii) If $\lambda \geq \lambda_0$ then (2) and the normalizing condition $w(0) = 1/2$ determine the unique solution $w_\lambda(x)$. It satisfies $w_\lambda'(x) > 0$ $x \in R^1$, $w_\lambda(-\infty) = 0$ and $w_\lambda(+\infty) = 1$.

(iii) If the solution $u(x,t)$ of (1) starts from the Heaviside's step function then the shape of $u(x,t)$ converges to that of $w_{\lambda_0}(x)$. When one observes $u(x,t)$ from the moving frame, at its origine the solution takes always any fixed constant value. The propagation speed of the moving frame approaches to $-2\lambda_0$ as $t \longrightarrow +\infty$.

The purpose of this report is to improve these results. First we introduce the notion of KPP transform. We define the class M of smooth functions by

$$(3) \qquad M = \left\{ u(x); \ u'(x) > 0 \ \ x \in R^1, \ \ u(-\infty) = 0, \ \ u(+\infty) = 1 \right\}.$$

For any smooth function $u(x,t)$ belonging to the class M for any fixed $t \geq 0$, we define $u^{-1}(\xi,\tau)$ by the implicit relation:

$$(4) \qquad u(u^{-1}(\xi,\tau),\tau) = \xi \qquad (\xi,\tau) \in (0,1) \times [0,\infty).$$

The new function $\hat{u}(\xi,\tau)$ given by

$$(5) \qquad \hat{u}(\xi,\tau) = u'(u^{-1}(\xi,\tau),\tau) \qquad (\xi,\tau) \in (0,1) \times [0,\infty), \qquad \left(' = \frac{\partial}{\partial x}\right)$$

is called the KPP transform of $u(x,t)$. Let us call $(\hat{u},\xi)$-plane the phase plane. If $u(x,t) = u(x)$ is independent of t, then $u^{-1}(\xi,\tau) = u^{-1}(\xi)$ and $\hat{u}(\xi,\tau) = \hat{u}(\xi)$ are also independent of τ. Hereafter we always assume that the solution $u(x,t)$ of (1) has the initial function $u_o(x)$ belonging to the class M. It is easy to see that $u(x,t)$ belongs to the class M for any fixed $t \geq 0$. The transformation

$$(6) \qquad \begin{cases} \xi = u(x,t) \\ \\ \tau = t \end{cases} \quad \text{or} \quad \begin{cases} x = u^{-1}(\xi,\tau) \\ \\ t = \tau \end{cases}$$

gives diffeomorphism: $R^1 \times [0,\infty) \ni (x,t) \longrightarrow (\xi,\tau) \in (0,1) \times [0,\infty)$. Hereafter we use the abbreviation that for any smooth functions with two independent variables ' and $\cdot$ mean the partial differentiations with respect to the first variable and the second variable respectively.

Now we state our results

__Theorem 1__ KPP transform $\hat{w}_\lambda(\xi) = w_\lambda'(w_\lambda^{-1}(\xi))$ of $w_\lambda(x)$ has the following properties:

$$(7) \qquad \hat{w}_\lambda(\xi) \in C^1[0,1] \cap C^\infty(0,1),$$

$$(8) \qquad \hat{w}_\lambda(\xi) > 0 \qquad \xi \in (0,1),$$

$$(9) \qquad \hat{w}_\lambda(0) = \hat{w}_\lambda(1) = 0, \quad \hat{w}_\lambda'(0) = \sigma_-(\lambda) = \lambda - \sqrt{\lambda^2 - \lambda_o^2} > 0,$$

$$\hat{w}_\lambda'(1) = \tau_-(\lambda) = \lambda - \sqrt{\lambda^2 - f'(1)} < 0,$$

$$(10) \qquad \lim_{\xi \to +o} \xi\sqrt{-\log \xi}\,\hat{w}_\lambda''(\xi) = 0,$$

$$(11) \qquad \hat{w}_{\lambda_2}(\xi) < \hat{w}_{\lambda_1}(\xi) \qquad \xi \in (0,1), \quad \lambda_2 > \lambda_1 \geq \lambda_o,$$

$$(12) \qquad \lim_{\lambda \to \lambda_1} \sup_{o<\xi<1} |\hat{w}_\lambda(\xi) - \hat{w}_{\lambda_1}(\xi)| = 0,$$

$$(13) \qquad \lim_{\lambda \to +\infty} \sup_{o<\xi<1} \hat{w}_\lambda(\xi) = 0.$$

<u>Theorem 2</u>(KPP) Suppose that the solution $u(x,t)$ of (1) has the initial function $u_o(x) \in M$. We assume the existence of the limit:

$$(14) \qquad \lim_{\tau \to +\infty} \hat{u}(\xi,\tau) = \hat{u}_\infty(\xi) \qquad \xi \in (0,1)$$

and the supplementary condition:

$$(15) \qquad \hat{w}_{\lambda_1}(\xi) \leq \hat{u}(\xi,\tau) \leq \hat{W}(\xi) \qquad (\xi,\tau) \in (0,1) \times [0,\infty)$$

for some $\lambda_1 \geq \lambda_o$ and $W(x) \in M$ $(W(0) = 1/2)$. Then there exists $\lambda \geq \lambda_o$ such that

$$(16) \qquad \lim_{\tau \to +\infty} \sup_{x \in R^1} |(\tfrac{\partial}{\partial x})^j(\tfrac{\partial}{\partial t})^k u(x + u^{-1}(\xi,\tau),\tau) -$$

$$- (2\lambda)^k(\tfrac{d}{dx})^{j+k} w_\lambda(x + w_\lambda^{-1}(\xi))| = 0$$

for any integers $j \geq 0$ and $k \geq 0$,

$$(17) \qquad \lim_{\tau \to +\infty} (u^{-1})^\cdot(\xi,\tau) = -2\lambda.$$

If we have the additional condition:

$$(18) \qquad \dot{\hat{u}}(\xi,\tau) \leq 0 \ (\text{or} \ \geq 0) \qquad (\xi,\tau) \in (0,1) \times [0,\infty)$$

then we have the monotonicity of the convergence in the following sense:

$$(19) \qquad (\text{sgn } x)\{u(x + u^{-1}(\xi,\tau),\tau) - w_\lambda(x + w_\lambda^{-1}(\xi))\} \searrow 0 \ (\text{or} \ \nearrow 0)$$

as $\tau \nearrow +\infty$. Here sgn $x = x/|x|$.

The crucial point is to show the conditions on the initial function $u_o(x)$ which assure (14). Now we introduce modified error function $E(x)$ by the relation:

$$(20) \qquad E(x) = \int_{-\infty}^{x} H(y,1)dy.$$

Here $H(x,t) = (4\pi t)^{-1/2}\exp(-x^2/4t)$ is the fundamental solution of the diffusion equation. It is easy to see that there exists $\delta_o > 0$ such that

$$(21) \qquad u_o(x) = E(x/\sqrt{\delta}) \qquad x \in R^1 \qquad 0 < \delta \le \delta_o$$

satisfies $\hat{u}_o(\xi) > \hat{w}_{\lambda_o}(\xi)$ and $\{\hat{u}_o'(\xi) + f(\xi)/\hat{u}_o(\xi)\}' \le 0$ for $\xi \in (0,1)$.

Theorem 3 Suppose that the solution $u(x,t)$ of (1) has the initial function $u_o(x)$ given by (21). Then we have

$$(22) \qquad \hat{u}(\xi,\tau) \ge \hat{w}_{\lambda_o}(\xi) \qquad (\xi,\tau) \in (0,1) \times [0,\infty),$$

$$(23) \qquad \dot{\hat{u}}(\xi,\tau) \le 0 \qquad (\xi,\tau) \in (0,1) \times [0,\infty),$$

$$(24) \qquad \lim_{\tau \to +\infty} \sup_{0<\xi<1} |\hat{u}(\xi,\tau) - \hat{w}_{\lambda_o}(\xi)| = 0$$

and the conclusions (16) and (17) of Theorem 2 are valid replacing λ by λ_o. Especially we have

$$(25) \qquad (\text{sgn } x)\{u(x + u^{-1}(\xi,\tau),\tau) - w_{\lambda_o}(x + w_{\lambda_o}^{-1}(\xi))\} \searrow 0 \qquad \text{as } \tau \nearrow +\infty.$$

Theorem 4 Suppose that $u(x,t)$ and $u_o(x)$ are the same as that of Theorem 3. Suppose that the solution $v(x,t)$ of (1) has the initial function $v_o(x) \in M$. If

$$(26) \qquad \hat{w}_{\lambda_o}(\xi) \le \hat{v}_o(\xi) \le \hat{u}_o(\xi) \qquad \xi \in (0,1),$$

then we have

$$(27) \qquad \hat{w}_{\lambda_o}(\xi) \le \hat{v}(\xi,\tau) \le \hat{u}(\xi,\tau) \qquad (\xi,\tau) \in (0,1) \times [0,\infty)$$

and the conclusions (16) and (17) of Theorem 2 are valid replacing $u(x,t)$ and λ by $v(x,t)$ and λ_o.

This theorem means that the slowest travelling wave $w_{\lambda_o}(x)$ is stable from above almost in the large on the phase plane.

We call that the function $u_o(x)$ belongs to the class N if and only if it belongs to the class M and $\hat{u}_o(\xi)$ satisfies

$$(28) \quad \begin{cases} \hat{u}_o(\xi) \in C^1[0,1] \cap C^2(0,1), \\[2mm] \hat{u}_o(0) = \hat{u}_o(1) = 0, \qquad \hat{u}_o{}'(0) > 0 > \hat{u}_o{}'(1), \\[2mm] \xi(1 - \xi)\hat{u}_o{}''(\xi) \text{ is bounded for } \xi \in (0,1). \end{cases}$$

Theorem 5 Let fix any $\lambda \geq \lambda_o$. Suppose that the solutions $u_k(x,t)$ $(k = 1,2)$ of (1) have the initial functions $u_{ko}(x) \in N$. We assume that

$$(29) \quad \hat{u}_{1o}{}'(0) = \hat{u}_{2o}{}'(0) = \sigma_-(\lambda) = \lambda - \sqrt{\lambda^2 - \lambda_o{}^2} \,,$$

$$(30) \quad \hat{u}_{2o}{}'(1) > \tau_-(\lambda) = \lambda - \sqrt{\lambda^2 - f'(1)} > \hat{u}_{1o}{}'(1),$$

$$(31) \quad \{\hat{u}_{2o}{}' + f(\xi)/\hat{u}_{2o}\}' \geq 0 \geq \{\hat{u}_{1o}{}' + f(\xi)/\hat{u}_{1o}\}' \qquad \xi \in (0,1).$$

Then we have

$$(32) \quad \dot{\hat{u}}_2(\xi,\tau) \geq 0 \geq \dot{\hat{u}}_1(\xi,\tau) \qquad (\xi,\tau) \in (0,1) \times [0,\infty),$$

$$(33) \quad \hat{u}_{1o}(\xi) > \hat{w}_\lambda(\xi) > \hat{u}_{2o}(\xi) \qquad \xi \in (0,1),$$

$$(34) \quad \lim_{\substack{\tau \to +\infty \\ k=1,2}} \sup_{0 < \xi < 1} |\hat{u}_k(\xi,\tau) - \hat{w}_\lambda(\xi)| = 0$$

and the conclusions (16) and (17) in Theorem 2 are valid replacing $u(x,t)$ by $u_k(x,t)$. Especially we have

$$(35) \quad (-1)^{k-1}(\operatorname{sgn} x)\{u_k(x + u_k{}^{-1}(\xi,\tau),\tau) - w_\lambda(x + w_\lambda{}^{-1}(\xi))\} \searrow 0$$

as $\tau \nearrow +\infty$.

Theorem 6 Let fix any $\lambda \geq \lambda_o$. Suppose that $u_k(x,t)$ and $u_{ko}(x)$ $(k = 1,2)$ are the same as that of Theorem 5. Suppose that the solution $v(x,t)$ of (1) has the initial function $v_o(x) \in M$. If

$$(36) \quad \hat{u}_{1o}(\xi) \geq \hat{v}_o(\xi) \geq \hat{u}_{2o}(\xi) \qquad \xi \in (0,1),$$

406

then we have

$$(37) \qquad \hat{u}_1(\xi,\tau) \geq \hat{v}(\xi,\tau) \geq \hat{u}_2(\xi,\tau) \qquad (\xi,\tau) \in (0,1) \times [0,\infty),$$

and the conclusions (16) and (17) in Theorem 2 are valid replacing $u(x,t)$ by $v(x,t)$.

The proof of these results are based on the following fundamental relations

$$(38) \quad \begin{cases} [\frac{\partial}{\partial t} - (\frac{\partial}{\partial x})^2 - c_1(x,t)](u' - \hat{w}_\lambda(u)) = 0 \\[2ex] c_1(x,t) = f'(u) + (u' + \hat{w}_\lambda(u))\hat{w}_\lambda''(u) \end{cases}$$

$$(39) \quad \begin{cases} [\frac{\partial}{\partial t} - (\frac{\partial}{\partial x})^2 - c_2(x,t)]\dot{\hat{u}}(u(x,t),t) = 0 \\[2ex] c_2(x,t) = f'(u) + 2\{u'''/u' - (u''/u')^2\} \\[2ex] \dot{\hat{u}}(u(x,t),t) = u''' + (u'')^2/u' - f(u)u''/u' + f'(u)/u' \end{cases}$$

$$(40) \quad \begin{cases} [\frac{\partial}{\partial t} - (\frac{\partial}{\partial x})^2 - c_3(x,t)](\hat{u}_k(v(x,t),t) - v'(x,t)) = 0 \\[2ex] c_3(x,t) = f'(v) + (v' + \hat{u}_k(v,t))\hat{u}_k''(v,t) \end{cases}$$

These relations and the standard comparison theorem for the diffusion equation assure the conservation of the positivity or negativity of each quantities $u' - \hat{w}_\lambda(u) = \hat{u} - \hat{w}_\lambda(\xi)$, $\dot{\hat{u}}(u(x,t),t) = \dot{\hat{u}}(\xi,\tau)$ and $\hat{u}_k(v(x,t),t) - v'(x,t) = \hat{u}_k(\xi,\tau) - \hat{v}(\xi,\tau)$. Detailed proofs of the results in this report and the other types of stability theorems may be found in [2].

Finally we show simple examples. If one applies our theorems to the case of

$$f(\xi) = \lambda_0^2 \xi(1 - \xi^n) \qquad \xi \in (0,1), \qquad \lambda_0 > 0, \qquad n = 1,2,3,\ldots$$

and

$$u_0(x) = \{1 + (2^{n/2} - 1)\exp(-n\sigma x/2)\}^{-2/n}, \qquad \sigma > 0$$

$$(\hat{u}_0(\xi) = \sigma\xi(1 - \xi^{n/2}))$$

then one may be able to have very sharp informations concerning the final states of solutions of the problem (1).

References

[1] Kolmogorov, A., I. Petrovskii and N. Piskunov, Étude de l'équa-
tion de la diffusion avec croissance de la quantité de matière et son
application à un problème biologique, Bulletin de l'Université d'État
à Moscou, Série International. vol. I Section A (1937), 1-25.
[2] Kametaka, Y., On the nonlinear diffusion equation of Kolmogorov-
Petrovskii-Piskunov type, (to appear in Osaka J. Math.)

GLOBAL SOLUTIONS TO THE BROADWELL'S MODEL OF BOLTZMANN EQUATION
FOR A SIMPLE DISCRETE VELOCITY GAS

Takaaki NISHIDA   and   Masayasu MIMURA

Department of Applied   Department of Mathematics
Mathematics and Physics   Konan University
Kyoto University   Kobe, Japan
Kyoto, Japan

ABSTRACT   The Carleman model and the Broadwell model of Boltzmann equation for the discrete velocity gas are investigated about the global solutions for the Cauchy problem.   (i) The asymptotic decay of solutions is proved on the Cauchy problem for the Carleman model.   (ii) The solutions of the Cauchy problem for the Broadwell model is shown to exist in the large in time under some conditions on the initial data.

§ 1.   Introduction

We consider two simple models of Boltzmann equation for the discrete velocity gas. Carleman model [1] is one of the simplest, which is described by the following semilinear hyperbolic equation :

$$(1) \quad \begin{aligned} u_t + cu_x &= \sigma(\, v^2 - u^2 \,) \\ v_t - cv_x &= \sigma(\, u^2 - v^2 \,) \end{aligned}$$

where $0 \le t$, $-\infty < x < +\infty$, c, $-c$ are the velocities of gas particles, u, v are the numbers of the particles with the velocity c, $-c$ respectively and σ is a positive constant.

A simple and physical model for the discrete velocity gas was given by Broadwell [2] and is described by the following semilinear hyperbolic system of equations in the one spatial dimensional case :

$$(2) \quad \begin{aligned} u_t + cu_x &= \sigma(\, w^2 - uv \,) \\ v_t - cv_x &= \sigma(\, w^2 - uv \,) \\ w_t &= \frac{\sigma}{2}(\, uv - w^2 \,) \end{aligned}$$

where $0 \le t$, $-\infty < x < +\infty$, the gas is assumed homogeneous in y, z and u, v are the numbers of particles with the velocities $(\, \pm c, 0, 0 \,)$ respectively and w is that of $(\, 0, \pm c, 0 \,)$ and $(\, 0, 0, \pm c \,)$ and the constant σ corresponds to $1/$ mean free path.

Carleman model (1) has the mass conservation law and H-theorem :

$$(3) \quad (\, u + v \,)_t + c(\, u - v \,)_x = 0$$

$$(4) \qquad (\text{ulog}u + \text{vlog}v)_t + c(\text{ulog}u - \text{vlog}v)_x$$
$$= - \sigma(u^2 - v^2)\log\frac{u}{v} \leq 0$$

Broadwell model (2) has the two conservation laws (mass and momentum) :

$$(5) \qquad (u + v + 4w)_t + c(u - v)_x = 0$$
$$c(u - v)_t + c^2(u + v)_x = 0$$

It has the H-theorem too :

$$(6) \qquad (\text{ulog}u + \text{vlog}v + 4\text{wlog}w)_t + c(\text{ulog}u - \text{vlog}v)_x$$
$$= - \sigma(w^2 - uv)\log\frac{w^2}{uv} \leq 0$$

The Cauchy problem (1) with the initial data

$$(7) \qquad u(0, x) = u_0(x), \quad v(0, x) = v_0(x), \quad - \infty < x < + \infty$$

is well considered (cf. [3] and others) and there exists uniquely the smooth solution in the large in time. Recently McKean [4] (cf. [5], [6]) has shown a central limit theorem for Carleman model (1) as $\sigma = \frac{1}{\varepsilon} \to + \infty$, which may be summarized as follows. Let the solution of (1) be $u^\varepsilon = u^\varepsilon(t,x)$, $v^\varepsilon = v^\varepsilon(t,x)$ for $\sigma = 1/\varepsilon$ and $0 < u_0(x), v_0(x) \in L'(- \infty < x < + \infty)$.

(i) Euler level
$$| u^\varepsilon - v^\varepsilon | \to 0, \quad u^\varepsilon + v^\varepsilon \to n^0$$
as $\varepsilon \to 0$ and n^0 satisfies the following equation :
$$\frac{\partial n^0}{\partial t} = 0, \quad 0 \leq t, \quad - \infty < x < + \infty$$

(ii) Navier-Stokes level

The function $U^\varepsilon = u^\varepsilon(t/\varepsilon, x)$, $V^\varepsilon = v^\varepsilon(t/\varepsilon, x)$ satisfy the following equation :

$$(8) \qquad U^\varepsilon_t + \frac{c}{\varepsilon} U^\varepsilon_x = \frac{1}{\varepsilon^2} \{ (V^\varepsilon)^2 - (U^\varepsilon)^2 \}$$
$$V^\varepsilon_t - \frac{c}{\varepsilon} V^\varepsilon_x = \frac{1}{\varepsilon^2} \{ (U^\varepsilon)^2 - (V^\varepsilon)^2 \}$$

When $\varepsilon \to 0$,
$$| U^\varepsilon - V^\varepsilon | \to 0 \quad \text{and} \quad U^\varepsilon + V^\varepsilon \to n,$$

where n satisfies the equation :

$$(9) \qquad n_t = \frac{c^2}{2} (\log n)_{xx}$$

This equation is a generalized Burgers' equation for the specific volume $n = \frac{1}{\rho}$ in the Lagrangian coordinate. In fact if we rewrite the equation (9) in the Lagrangian coordinate (t,x) to the equation in the Eulerian coordinate (τ, ξ), then it turns out to be a generalized Burgers' equation :

$$(10) \qquad \rho_\tau + (\rho u)_\xi = 0$$
$$u_\tau + u u_\xi = \frac{\mu u}{\rho}_{\xi\xi}$$

This is a compressible viscous fluid equation for the density ρ and the velocity u with the constant pressure and the constant temperature, whose global solution is considered in [7].

§ 2. Carleman model

We consider the decay of solutions for Carleman model (1) near the equilibrium state $u = v = a$ constant > 0. In this case the system (1) can be rewritten as follows :

$$(11) \quad \begin{aligned} u_t + cu_x &= \sigma \{ 2a(v - u) + v^2 - u^2 \} \\ v_t - cv_x &= \sigma \{ 2a(u - v) + u^2 - v^2 \} \end{aligned}$$

where u, v are the perturbation of solutions from the equilibrium state $u = v = a$.

Theorem 1.

Let $u_0(x)$, $v_0(x)$ be bounded continuous L^2-functions with the first derivative in $- \infty < x < + \infty$. If there exists $\delta = constant > 0$ such that

$$- a + \delta \leq u_0(x), v_0(x), \quad - \infty < x < + \infty,$$

then the solutions $u(t,x)$, $v(t,x)$ for (11) together with u_x, v_x, u_t, v_t decay to zero uniformly in x and in $L^2(R)$ as $t \to + \infty$.

This is proved by a representation of solution for (11) with the technique in [8] and by the energy estimate for its transformed single second order hyperbolic equation with the first order dissipative term :

$$(12) \quad y_{tt} - c^2 y_{xx} + 4\sigma a y_t = 16\sigma c y_x y_t \quad ,$$

where $u = y_t - cy_x$, $v = - y_t - cy_x$, $4\sigma a > 0$.

§ 3. Broadwell model

The Broadwell model is more physically interesting, but it has not well been treated mathematically. (cf. [9]) When we consider the Cauchy problem for the Broadwell model (2) with the initial data

$$(13) \quad \begin{aligned} &u(0,x) = u_0(x), \ v(0,x) = v_0(x), \ w(0,x) = w_0(x) \\ &\text{in } - \infty < x < + \infty, \end{aligned}$$

we have the following.

Theorem 2.

If the initials (13) are bounded continuous functions with the first derivative and

$$(14) \quad \begin{aligned} &0 \leq u_0(x), v_0(x), w_0(x) \leq K_0 < + \infty, \\ &\int_{-\infty}^{+\infty} u_0(x) + v_0(x) + 4w_0(x) \ dx = L_0 < + \infty, \end{aligned}$$

then there exists a constant $\sigma_0 > 0$ such that for any σ, $\sigma L_0 < \sigma_0$, we have the global smooth solution for the Cauchy problem (2) (13).

Theorem 3.

Let the initials be bounded continuous with the first derivative and

$$0 < \delta = constant \le u_0(x), \; v_0(x), \; w_0(x) \le K_0 < +\infty,$$

(15)
$$E_0 = \int_{-\infty}^{+\infty} f(\,u_0(x), \, u^0\,) + f(\,v_0(x), \, v^0\,) + 4f(\,w_0(x), \, w^0\,) \; dx < +\infty,$$

where

$$f(z,c) = z\log\frac{z}{c} - z + c,$$

$u^0, \, v^0, \, w^0$ *are the equilibrium state so that* $u^0 v^0 = (\,w^0\,)^2$.

Then there exists $\sigma_0 > 0$ *such that for any* σ, $\sigma E_0 < \sigma_0$, *we have the global smooth solution for the Cauchy problem (2) (13).*

The proof is based on the conservation of mass (5) and the H-theorem (6).

We note a decay of solution for (2) (13). When the initial data (13) are periodic in x and are near to the equilibrium state u^0, v^0, w^0 respectively ($u^0 v^0 = (w^0)^2$), the solution exists in the large in time and decays exponentially in t to that state. (cf. [10] [11])

We can get the global existence theorem for the Cauchy problem of the Broadwell model in the two spatial dimensional case analogously under the restrictions on σ and the initial data. Unfortunately these existence theorems are too poor to get the central limit theorems for the Broadwell model. The difficulty as $\sigma = 1/\varepsilon$, $\varepsilon \to 0$, is parallel to treat the shock formation in the gas dynamics. In fact if we get the limit for (2) as $\varepsilon \to 0$, by the H-theorem the limit in the compressible Euler level is supposed to satisfy a nonlinear hyperbolic system of two equations which is analogous to the system of the hydrodynamical gas motion :

(16)
$$\rho_t + (\,\rho u\,)_x = 0$$
$$(\,\rho u\,)_t + f(\,\rho, \, u\,)_x = 0 \, ,$$

where

$$f(\,\rho, \, u\,) = \frac{c^2}{3} \, \rho \left\{ 2 \sqrt{1 + 3 \frac{u^2}{c^2}} - 1 \right\}$$

$$= \frac{c^2}{3} \, \rho + \rho u^2 - \frac{3}{4} \frac{\rho u^4}{c^2} + \cdots \, .$$

The first two terms are exactly the same as that of the ideal isothermal gas motion. This system developes in general the shock waves in the course in time even for the sufficiently smooth initial data. (cf. [8] [12] [13])

References

[1]   T. Carleman, Problèmes Mathématiques dans la Théorie Cinetique des Gaz, Uppsala, 1957.

[2]   J. Broadwell, Shock structure in a simple discrete velocity gas, Phys. of Fluids, $\underline{7}$ (1964) 8.

[3]   I. Kolodner, On Carleman's model for the Boltzmann equation, in Nonlinear Problems edited by R. Langer, Madison, Univ. of Wisconsin Press, 1963.

[4]   H. McKean, Jr, Lectures on the Boltzmann equation at the Courant Institute of Math. Sci. NYU, 1972.

[5]   H. Grad, Asymptotic equivalence of the Navier Stokes and nonlinear Boltzmann equations, Proc. Sympo. in Appl. Math., $\underline{17}$ (1965), Amer. Math. Soc.

[6]   R. Ellis and M. Pinsky, Asymptotic nonuniqueness of the Navier-Stokes equation in kinetic theory, Bull. Amer. Math. Soc., $\underline{80}$ (1974) 6.

[7]   N. Itaya, On the temporally global problem of the generalized Burgers equation, J. Math. Kyoto Univ., $\underline{14}$ (1974) 1.

[8]   P. Lax, Development of singularities of solutions of nonlinear hyperbolic partial differential equations, J. Math. Phys., $\underline{5}$ (1964) 5.

[9]   S. Godunov and U. Sultangazin, On the discrete models of Boltzmann kinetic equation, Uspehi. Mat. Nauk, $\underline{26}$ (1971).

[10]   S. Ukai, On the existence of global solutions of mixed problem for nonlinear Boltzmann equation, Proc. Japan Academy, $\underline{50}$ (1974) 3.

[11]   H. Tanaka, On the Carleman model, private communication.

[12]   J. Glimm, Solutions in the large for nonlinear hyperbolic systems of equations, Comm. Pure Appl. Math., $\underline{18}$ (1965) 4.

[13]   J. Glimm and P. Lax, Decay of solutions of systems of nonlinear hyperbolic conservation laws, Memo. Amer. Math. Soc., No.101 (1970).

[14]   T. Nishida and M. Mimura, On the Broadwell's model for a simple discrete velocity gas, Proc. Japan Academy, $\underline{50}$ (1974) 10.

SCALING METHOD FOR ASYMPTOTIC EVALUATION IN
NONEQUILIBRIUM STATISTICAL MECHANICS

Hazime Mori

Department of Physics, Kyushu University, Fukuoka, Japan

A scale transformation of the nonequilibrium macroscopic system
to larger similar systems is introduced to extract the dependence of
macroscopic properties on the macroscopic characteristic lengths, and
is shown to be useful for determining possible types of fluctuations
and the Markov processes of macroscopic variables, including nonlinear
fluctuations of unstable modes in open systems far from equilibrium.

1. In this talk, I would like to discuss some basic problems which we
meet when studying unstable modes near critical points and in turbulences. The best-investigated instabilities would be the thermodynamic instabilities which occur at gas-liquid critical points and ferromagnetic critical points. Open systems far from equilibrium give us
a greater variety of instabilities; for example, the hydrodynamic instabilities give us Bénard-Rayleigh's instability on thermal convection, and homogeneous turbulences, obeying Kolmogorov's spectrum for
the turbulent energy. It is my ultimate aim to formulate dynamics of
unstable modes associated with these instabilities.

In thermodynamics and hydrodynamics, fluctuations are negligible
and the macroscopic causality holds. Fluctuations can be observed
only on the microscopic level, and they must obey a normal distribution due to the central limit theorem.

When an instability occurs, the situation changes drastically as
known in critical phenomena and turbulences. When a critical point is
approached, the damping rate of <u>unstable modes</u> λ_q tends to zero; $\lambda_q \to 0$.
This is known as the critical slowing-down. The magnitude of fluctua-
tions is described by

$$\chi_q \equiv \Omega < \delta z_q^* \delta z_q > = \int d\underline{r}' \, e^{i\underline{q} \cdot (\underline{r} - \underline{r}')} < \delta z(\underline{r}) \delta z(\underline{r}') >,$$

where $\delta z \equiv z - <z>$, Ω is the volume of the system and the angular brackets
denote the average. $z(\underline{r})$ is a local density at position $\underline{r}$, and z_q is
its Fourier component with wave vector $\underline{q}$. According to a fluctuation-
dissipation theorem, χ_q is inversely proportional to the damping rate
λ_q and tends to infinity as the critical point is approached. Thus
when an instability occurs, the fluctuations of unstable modes are

amplified and reach a macroscopic level. Then the causal description breaks down, and a <u>stochastic description</u> must be employed.

Unstable modes bring three effects: First, the fluctuations of unstable modes have a <u>long-range correlation</u> and do not obey a normal distribution. Secondly, nonlinear mode couplings of unstable modes lead to a renormalization of transport coefficients; example are the anomalous transport near thermodynamic critical points and the eddy viscosity in turbulences. Thirdly, a new pattern appears out of the fluctuations and is stabilized in the post-critical region. Once a new stable state is reached, the fluctuations regress and the causal description becomes again valid.

Thus when instabilities occur, we have two basic problems:
1) How to treat the fluctuations of unstable modes?
2) How to determine post-critical patterns?
These two problems are related to each other.

Macroscopic properties of the system depend on the long-range correlation of unstable modes. Hence the central limit theorem and its simple extension like the system-size expansion [1,2] cannot be valid. We have to find <u>a new method for asymptotic evaluation</u> for large systems which enables us to study the dependence of macroscopic properties on the correlation length. Such a method is provided by the <u>scaling method</u> [3] which I would like to outline here.

<u>2.</u> <u>Decomposition into a deterministic and a stochastic motion.</u>

The macroscopic state variables I consider in this talk are the Fourier components of mechanical local densities $X_\mu(\underline{r},t)$, $(\mu=1,2,\cdots)$ with small wave numbers;

$$X_k(t) \equiv \frac{1}{\Omega} \int d\underline{r} \, \exp(i\underline{q}\cdot\underline{r})X_\mu(\underline{r},t), \quad (q \leq q_c), \tag{1}$$

where q_c is a cutoff and k denotes the set $(\mu,\underline{q})$. Let me decompose $X_k(t)$ into a deterministic motion $y_k(t)$ and its fluctuation $Z_k(t)$,

$$X_k(t) = y_k(t) + Z_k(t), \tag{2}$$

and suppose that $y_k(t)$ is determined by the solution of a deterministic equation,

$$dy_k(t)/dt = -h_k(y), \quad (y \equiv \{y_k\}), \tag{3}$$

with macroscopic initial and boundary conditions. In normal systems, $y(t)$ represents the macroscopic motion. The fluctuation is described by a stochastic equation of motion

$$dZ_k(t)/dt = -\Delta h_k(Z,t) + R_k(t), \tag{4}$$

where

$$\Delta h_k(Z,t) \equiv h_k(y(t)+Z) - h_k(y(t)),\qquad(5)$$

and Z denotes the set $\{Z_k\}$. $R_k(t)$ is a fluctuating force which is assumed to ensure the <u>Markov condition</u>.

The decomposition (2) is essential from two reasons. One is that the deterministic motion $y(t)$ is macroscopically controllable by the choice of the macroscopic initial and boundary conditions and external forces, whereas the fluctuation $Z(t)$ is not since its origin is of microscopic nature. In nonequilibrium steady states, the macroscopic state y depends on the boundary conditions and hence on the size and shape of the system. The most striking example is the Bénard cells in thermal convection. The size and shape dependence of y differs from that of Z. As a result, y and Z have different scaling exponents. This is the second reason for the decomposition.

3. <u>Scaling method for asymptotic evaluation.</u>

Let me transform the system to a larger similar system with macroscopic distances and wave vectors given by

$$\ell_L = L\ell, \quad \underline{q}_L = \underline{q}/L, \quad \Omega_L = L^d\Omega, \quad (L \gg 1),\qquad(6)$$

where ℓ represents a macroscopic characteristic length and d is the dimensionality. The inverse of the cutoff, $1/q_c$, is the smallest distance of the spatial variation of macrovariables, and is also transformed so that the number of the macroscopic degrees-of-freedom is invariant. The microscopic characteristic lengths (e.g., the intermolecular force range r_0, the mean intermolecular distance $(\Omega/N)^{1/3}$) are, however, kept constant. Therefore one simple choice is $\ell=1/q_c$.

The following three are assumed to be scale invariants:

(1) the macroscopic flow pattern, which, in the present talk, means not only the flow pattern of fluids in coordinate and wave-number space, but also the macroscopic symmetry and shape of thermodynamic systems,

(2) the probability for fluctuations of macrovariables,

(3) fundamental material constants whose invariance is ensured by the fact that the microscopic lengths are kept constant.

Scaling exponents of the deterministic motion are defined by

$$y_k^L = L^{-\alpha_k} y_k, \quad t_L = L^\tau t.\qquad(7)$$

Then the scale invariance of the macroscopic flow pattern leads to

$$y_k(t) = \ell^{-\alpha_k} Y_k(q\ell, t/\ell^\tau).\qquad(8)$$

When applied to macroscopic fluid motions, this scaling must be consistent with the hydrodynamic similarity laws. Scaling exponents of the fluctuation are defined by

$$z_k^L = L^{-\beta_k} z_k, \quad t_L = L^\theta t. \tag{9}$$

Here $\tau \geq \theta \geq 0$ and we can always single out the θ time scale in fluctuation processes by taking large L. Hence the scale invariance of the probability leads to

$$Z_k(t) = \ell^{-\beta_k} \Psi_k(q\ell, t/\ell^\theta), \tag{10}$$

$$\chi_k(t) \equiv \Omega <\delta z_k^* \delta z_k>(t) = \ell^{\gamma_k} C_k(q\ell, t/\ell^\theta), \tag{11}$$

where $\gamma_k = d - 2\beta_k \geq 0$.

The macrovariable $X_k(t)$ is the sum of $y_k(t)$ and $Z_k(t)$. The scaling factors $L^{-\alpha_k}$ and $L^{-\beta_k}$, however, lead to

$$X_k(t) = \begin{cases} y_k(t) & \text{if } \alpha_k < \beta_k, & (12a) \\ y_k(t) + Z_k(t) & \text{if } \alpha_k = \beta_k, & (12b) \\ Z_k(t) & \text{if } \alpha_k > \beta_k, & (12c) \end{cases}$$

for large systems. Case (a) means that if $\alpha_k < \beta_k$, then the macroscopic behavior of X_k is <u>deterministic</u>. Such a situation holds for thermodynamics and hydrodynamics. In thermodynamics, y_k represent the thermodynamic state variables per unit volume and are determined by the free-energy minimum; $\partial F(y)/\partial y_k = 0$. Then y_k and fluctuation Z_k are of the order of unity and of $1/\sqrt{\Omega}$, respectively, leading to

$$\alpha_k = 0, \quad \beta_k = d/2, \quad (\gamma_k = 0) \text{ for } \ell = \Omega^{1/d} \text{ or } 1/q_c. \tag{13}$$

This will be called the thermodynamic scaling. The deviation of scaling exponents from this scaling means the dependence of y_k and $\sqrt{\Omega} z_k$ upon the <u>size</u>, <u>shape</u> and <u>intrinsic characteristic lengths</u> of the system. The Navier-Stokes equation in hydrodynamics leads to that the local fluid velocity has $\alpha_k = 1$. Hence

$$\alpha_k = 1, \quad \beta_k = 3/2, \quad (\tau = 2, \theta = 2) \tag{14}$$

in the laminar flow. Thus the condition for macroscopic causality is given by $\alpha_k < \beta_k$. Then the fluctuations can be shown to obey the normal distribution [3].

Case (b) means that if $\alpha_k = \beta_k$, then the fluctuations have the magnitude of the same order as y_k. Such a situation occurs near the thermodynamic critical point where the correlation length ξ for the fluctuations of the order parameter increases anomalously; $\xi \sim \varepsilon^{-\nu}$, $(\nu > 0)$, where $\varepsilon \equiv |T - T_c|/T_c$, T_c being the critical temperature. In the critical region where $\xi q_c > 1$, ξ is a macroscopic length and is scaled by $\xi_L = L\xi$ so that the critical temperature is approached according to $\epsilon_L = L^{-(1/\nu)}\epsilon$. This scaling reproduces the scaling laws for thermodynamic critical phenomena. In accordance with the thermodynamic and correlation scaling laws, the order parameter has, for $\ell = \xi$ or $1/q_c$,

$$\alpha_k = \beta_k = (d-2+\eta)/2, \quad (\gamma_k = 2-\eta) \tag{15}$$

in the critical region, where η is the correlation scaling exponent. The temperature dependence of macroscopic properties near the critical point is extracted by this scaling.

Next let me take fully-developed turbulences in three-dimensional Navier-Stokes fluids. Kolmogorov's spectrum $q^{-5/3}$ for the turbulent energy determines scaling exponent β_k. Then we get

$$\alpha_k = 1, \quad \beta_k = -1/3, \quad (\tau = 2, \theta = 2/3) \tag{16}$$

in the inertial subrange of wave numbers, where a cascade process of energy from small to large wave numbers is driven by the nonlinear inertial term. This is a typical example of case (c). Thus we find that $\alpha_k \geq \beta_k$ for <u>unstable modes</u> and hence their macroscopic behavior is <u>stochastic</u>.

4. <u>Normal fluctuations</u> $(\alpha_k < \beta_k)$

The ratio of the two scaling factors,

$$\epsilon_k(L) \equiv [Z_k/y_k]^L = L^{\alpha_k - \beta_k}, \quad (L \gg 1), \tag{17}$$

provides us with an <u>expansion parameter</u> for fluctuation if $\alpha_k < \beta_k$. Then (5) can be expanded in powers of ϵ and (4) leads to a linear stochastic equation of motion

$$dZ_k/dt = -\Sigma_j h_{kj}(y)Z_j + R_k(t). \tag{18}$$

The logarithm of the probability distribution function is also expandable in powers of ϵ, leading to [3]

$$P(z,t) = N(t) \exp\{-\frac{\Omega}{2} \Sigma\Sigma_{ki} \chi_{ki}^{-1}(t)[z_i^* - \bar{z}_i^*(t)][z_k - \bar{z}_k(t)]\}, \tag{19}$$

where $\bar{z}_k$ is the mean regression of fluctuations and χ_{ki} is the variance matrix. Their time evolution is given by

$$d\bar{z}_k/dt = -\Sigma_j h_{kj}(y)\bar{z}_j, \tag{20}$$

$$d\chi_{ki}/dt = -\Sigma_j [h_{kj}(y)\chi_{ji} + h_{ij}^*(y)\chi_{kj}] + \alpha_{ki*}(y), \tag{21}$$

where α_{ki} is the spectral density of the fluctuating force;

$$\Omega \langle R_i^*(t')R_k(t);y \rangle = \delta(t-t')\alpha_{ki*}(y), \tag{22}$$

the angular bracket denoting the conditional average with the value of X being fixed to be y. Thus when $\alpha_k < \beta_k$, the probability distribution is <u>normal</u>. This is a generalized central limit theorem.

In the steady state $y=y^S$, h_{kj} and α_{ki} are constants. Then, taking t infinity in (21), we obtain

$$\alpha_{ki*}(y^S) = \Omega\Sigma_j[h_{kj}(y^S)<z_i^*z_j>^S + h_{ij}^*(y^S)<z_j^*z_k>^S], \tag{23}$$

where $<\cdots>^S$ indicates the average over the stationary distribution. This gives a <u>fluctuation-dissipation theorem</u> which holds even for steady states far from equilibrium. In the representation of diagonalizing $h_{ki}(y^S)$ with eigenvalues λ_k,

$$\bar{z}_k(t) = z_{0k} \exp[-t\lambda_k], \tag{24}$$

$$\chi_{ki}(t) = \Omega<z_i^*z_k>^S\{1-\exp[-t(\lambda_k+\lambda_i^*)]\}, \tag{25}$$

where we have assumed the initial distribution $P(z,0)=\Pi_k\delta(z_k-z_{0k})$. In order to ensure $\alpha_k<\beta_k$ and hence the expansion in powers of ϵ, the real parts of λ_k's must be positive so that y^S is stable against any fluctuation with non-zero measure. This indicates that if y^S is unstable, then there must exist a subset $\{Z_j\}$ with $\alpha_j \geq \beta_j$. This defines <u>unstable modes</u> $\{Z_j\}$.

<u>5.</u> The most dominant part of the fluctuation drift term $\Delta h_k(Z,t)$ is determined by a subset $\{Z_j^0\}$ which leads to the minimum value $\Delta_k \equiv |\text{Min}_j[\beta_j-\alpha_j]|$. It turns out that

$$\left[\frac{\Delta h_k(Z,t)}{h_k(y)}\right]^L = \begin{cases} L^{-\Delta_k} & \text{if } \alpha_j < \beta_j, & (26a)\\ L^0 & \text{if } \alpha_j = \beta_j, & (26b)\\ L^{m_k\Delta_k} & \text{if } \alpha_j > \beta_j, & (26c) \end{cases}$$

where m_k is the degree of the highest-order power of $\{y_j\}$ in $h_k(y)$. We find that if $h_k(y)$ is nonlinear in a set of unstable modes, then the fluctuation drift term $\Delta h_k(Z,t)$ contains <u>nonlinear mode couplings</u> between the unstable modes. Thus unstable modes produce nonlinear mode couplings.

The asymptotic evaluation by the scaling method is also useful for determining possible types of the Markov processes of macrovariables, including nonlinear fluctuations of unstable modes, and for exploring the renormalization of transport coefficients by the nonlinear mode couplings of fluctuations [3]. It is indispensable for nonequilibrium statistical mechanics to achieve the coarse-graining in space and time in the reduction of variables. These problems were also discussed in the talk.

References
1] N.G. van Kampen, Can. J. Phys. <u>39</u> (1961) 551.

2] R. Kubo, K. Matsuo and K. Kitahara, J. Stat. Phys. $\underline{9}$ (1973) 51.

3] H. Mori, Prog. Theor. Phys. $\underline{52}$ (1974) 433; $\underline{53}$ (1975) to be published.

Discussion

<u>Bernhard Mühlschlegel</u>: 1) Did I understand you correctly that hydrodynamical similarity dictates the scaling behavior? 2) Your talk was on three dimensions. Hydrodynamics may be quite interesting in two dimensions. Is there some special scaling behavior in two dimensions?

<u>H. Mori</u>: 1) Yes, you did. Scaling exponent α_k is determined by hydrodynamical similarity. In Navier-Stokes equation, Reynolds' number $R \equiv u\ell/\nu$ must be invariant. Since ν is also invariant in my scaling, $u \sim 1/\ell$ and hence $\alpha_k = 1$, leading to α_k of (14) and (16). 2) Yes, there is. Fluctuation scaling exponent β_k depends on the dimensionality d, leading to $\beta_k = d/2$ in the laminar flow. Therefore $\beta_k = \alpha_k$ in two dimensions. Then fluctuations become nonlinear, and their nonlinear mode couplings would lead to a renormalization of transport coefficients. This must be the origin of the long-time tail in transport coefficients in two dimensions.

SELF-ENTRAINMENT OF A POPULATION OF
COUPLED NON-LINEAR OSCILLATORS

Yoshiki Kuramoto

Department of Physics, Kyushu University, Fukuoka, Japan

Temporal organization of matter is a widespread phenomenon over a macroscopic world in far from thermodynamic equilibrium. A previous study on chemical instability[1] implies that a simplest nontrivial model for a temporally organized system may be represented by a macroscopic self-sustained oscillator Q obeying the equation of motion

$$\dot{Q} = (i\omega + \alpha)Q - \beta|Q|^2 Q ,$$
$$\alpha,\beta > 0. \tag{1}$$

Consider a population of such oscillators Q_1, $Q_2,\cdots Q_N$ with various frequencies, and introduce interactions between every pair as follows.

$$\dot{Q}_s = (i\omega_s + \alpha)Q_s + \sum_{r \neq s} v_{rs}Q_r - \beta|Q_s|^2 Q_s ,$$
$$r,s = 1, 2,\cdots N . \tag{2}$$

We found that it is possible to construct from (2) a soluble model for a community exhibiting mutual synchronization or self-entrainment above a certain threshold value of the coupling strength. Such a type of phase transition has been considered by Winfree[2] without resorting to specialized models but only phenomenologically.

Our simplifying assumptions are:

(I) $v_{rs} = v/N$ independently of r and s,

(II) $\alpha,\beta \to \infty$ but α/β, ω_s, v = finite,

(III) $N \to \infty$.

Let us put $Q_s = \rho_s e^{i\varphi_s}$. Owing to the assumption (II), the amplitude ρ_s may be fixed at $\sqrt{\alpha/\beta}$. Thus we have only to consider the equation

$$\dot{\varphi}_s = \omega_s + \frac{v}{N} \sum_r \sin(\varphi_r - \varphi_s) . \tag{3}$$

As an illustration, we summarize the results obtained when the distribution of the native frequency is a Lorentzian with the peak at ω_0 and the width γ. In this case the threshold condition is

$$\eta \equiv 2|\gamma/v| = 1 . \tag{4}$$

For $\eta < 1$ the asymptotic behavior of $\varphi_s(t)$ as $t \to \infty$ has two possibilities depending on the value of its native frequency ω_s. Thus we classify the oscillators into two groups :

(A) $\left| \dfrac{\omega_s - \omega_0}{v\sqrt{1 - \eta}} \right| < 1$.

The oscillators satisfying this condition are mutually synchronized and one finds

$$\varphi_s(t) = \widetilde{\omega}_s t + \psi_s + \psi' , \tag{5}$$

$$\widetilde{\omega}_s = \omega_0 , \tag{6}$$

where ψ_s is a single-valued function of ω_s and ψ' is an arbitrary constant but should be the same for all s belonging to this group.

(B) $\left| \dfrac{\omega_s - \omega_0}{v\sqrt{1 - \eta}} \right| > 1$.

The oscillators of this group fail to synchronize, but their frequencies are shifted. We find

$$\varphi_s(t) = \widetilde{\omega}_s t + f_s(t), \tag{7}$$

where

$$\widetilde{\omega}_s = \omega_0 + (\omega_s - \omega_0)\sqrt{1 - \frac{v^2(1 - \eta)}{(\omega_s - \omega_0)^2}} \tag{8}$$

and

$$f_s(t) = f_s\left(t + \frac{2\pi}{\widetilde{\omega}_s}\right) . \tag{9}$$

If we define an order parameter σ by the number of the oscillators belonging to (A) divided by N, then we find

$$\sigma = \begin{cases} \dfrac{2}{\pi} \mathrm{Tan}^{-1} \dfrac{2\sqrt{1 - \eta}}{\eta} & (\ \eta < 1\ ) \\[2mm] 0 & (\ \eta > 1\ ) \end{cases} . \tag{10}$$

Finally the stationary distribution of the effective frequency ω may be expressed as

$$\begin{aligned} f(\widetilde{\omega}) = {}& \sigma\delta(\widetilde{\omega} - \omega_0) \\ & + \frac{\gamma}{\pi} |\widetilde{\omega} - \omega_0| / [\{(\widetilde{\omega} - \omega_0)^2 + \gamma^2 + v^2\chi\}\sqrt{(\widetilde{\omega} - \omega_0)^2 + v^2\chi}] \end{aligned} \tag{11}$$

where

$$\chi = \begin{cases} 1 - \eta & (\ \eta < 1\ ) \\[2mm] 0 & (\ \eta > 1\ ) . \end{cases}$$

An infinitely sharp peak at $\widetilde{\omega} = \omega_0$ corresponds to a macroscopic oscillation. The background represents the contribution from the oscillators which have failed to synchronize with the main oscillation. The intensity of the background near the center has been reduced drastically due to the factor $|\widetilde{\omega} - \omega_0|$. It is very interesting to notice that our spectrum resembles that of the α-rhythm of human brain wave.[3]

References
1] Y. Kuramoto and T. Tsuzuki, Prog. Theor. Phys. 52 (1974), 1399.
2] A.T. Winfree, J. Theor. Biol. 16 (1967), 15.
3] N. Wiener, "Cybernetics" MIT Press, Cambridge, Mass., 1948.

THERMODYNAMIC LIMIT OF NON-EQUILIBRIUM SYSTEMS
—— EXTENSIVE PROPERTY, FLUCTUATION
AND NONLINEAR RELAXATION ——

Masuo SUZUKI

Department of Physics, University of Tokyo

Bunkyo-ku, Tokyo, Japan

The existence of thermodynamic limits of non-equilibrium distribution function is proved under the following assumptions, namely that the initial state is described by a canonical distribution, an effective Hamiltonian of which is a sum of local Hamiltonians, and that certain averages of "local" operators are bounded. This proof yields an extension of Kubo's Ansatz on the extensive property of a macrovariable to quantal systems and to non-Markoffian macrovariables in stochastic models. A generating function (generalized thermodynamic potential) formalism is used effectively in discussing the above non-equilibrium problems. Fluctuation and Nonlinear relaxation are discussed in the above scheme. As applications of the general theory, some exact solutions are presented for generating functions of a simple (but non-trivial) stochastic linear chain and the generalized XY -- model in one dimension.

Recently Kubo[1][2] proposed the Ansatz that the distribution function $P(X, t)$ of an extensive macrovariable X at time t has the asymptotic form

$$P(X, t) = C \exp[\Omega \phi(x, t)], \tag{1}$$

for a large volume Ω with $x = X/\Omega$. This is a generalization of the concept of the extensivity in equilibrium statistical thermodynamics to non-equilibrium problems. Kubo et al.[2] proved the Ansatz (1) when X is a Markoffian macrovariable in a stochastic model, in the sense that the above type of solution of the stochastic equation of the form

$$\frac{\partial P(X, t)}{\partial t} = \Gamma_M P(X, t), \tag{2}$$

propagates in time.

The purpose of the present paper is to study <u>how and in what conditions</u> the concept of the extensive property is extended to more general cases of quantal systems and stochastic systems with non-Markoffian macrovariables. For stochastic systems, we prove the extensive property, taking the example of a kinetic model of spins which is represented by the following <u>microscopic</u> master equation

$$\frac{\partial P(\{\sigma_j\};t)}{\partial t} = \Gamma\, P(\{\sigma_j\};t), \tag{3}$$

where Γ is a temporal evolution operator and $\{\sigma_j\}$ denotes a microscopic configuration of the system ($\sigma_j = \pm 1$). The distribution function $P(X, t)$ of a macrovariable X is defined by

$$P(X,t) = \sum_{\{\sigma_j=\pm 1\}} \delta(\mathbf{X}-X)\, P(\{\sigma_j\};t), \tag{4}$$

where $\mathbf{X}$ is a stochastic variable corresponding to X. For a quantal system, we start from the Liouville equation for the density-matrix:

$$i\frac{\partial \rho(t)}{\partial t} = [\mathcal{H}, \rho(t)]\,; \quad \hbar = 1, \tag{5}$$

where $\mathcal{H}$ is the Hamiltonian of the system. The "reduced density-matrix" or the distribution function of the relevant macrovariable X is given, analogously to (4), by

$$\rho(X,t) = \mathrm{Tr}\,\delta(\mathbf{X}-X)\,\rho(t), \tag{6}$$

where $\mathbf{X}$ is the operator corresponding to the variable X. Our main problem is to find conditions to assure the extensive property of $\rho(X, t)$:

$$\rho(X, t) = C\, \exp[\Omega\,\phi(x,t)]. \tag{7}$$

For the proof of the extensive property, it is convenient to introduce a generating function $\Psi(\lambda, t)$ defined by

$$\Psi(\lambda,t) = \begin{cases} \mathrm{Tr}\, e^{\lambda \mathbf{X}} \rho(t) & \text{------(quantal)} \\ \sum_{\{\sigma_j=\pm 1\}} e^{\lambda \mathbf{X}} P(t) & \text{------(stochastic or classical)} \end{cases} \tag{8}$$

where $\rho(t)$ or $P(t)$ denotes the density-matrix of a quantal system or the distribution function of a stochastic (or classical) system. The probability distribution function $\rho(X, t)$ (or $P(X, t)$) is given by the inverse transformation

$$\rho(X,t) = \frac{1}{2\pi i}\int_{c-i\infty}^{c+i\infty} e^{-\lambda X}\, \Psi(\lambda,t)\, d\lambda. \tag{9}$$

In the following, we always assume that the initial state is described by a canonical distribution, $\rho(0) = \exp\mathcal{H}^{(i)}$ (and $\mathrm{Tr}\rho(0) = 1$) where $kT\mathcal{H}^{(i)}$ denotes an effective hermitian Hamiltonian, which may be different from the real Hamiltonian $\mathcal{H}$, and also we assume that X and $\mathcal{H}$ are hermitian.

Now, if the generating function $\Psi(\lambda, t)$ has the extensive property:

$$\Psi(\lambda, t) = C_1\, \exp[\Omega\,\psi(\lambda,t)], \tag{10}$$

then $\rho(X, t)$ is shown to have the extensive property:

425

$$\rho(x, t) = C_2 \exp[\Omega\{\psi(\lambda_0, t) - \lambda_0 x\}], \tag{11}$$

for a large Ω, λ_0 being a saddle point given by a real root of the equation

$$\frac{\partial \psi(\lambda, t)}{\partial \lambda} = x \ . \tag{12}$$

Then, we study[3] conditions under which the generating function $\Psi(\lambda, t)$ has the extensive property. We assume first that $\mathbf{X}, \mathcal{H}^{(i)}$ and $\mathcal{H}$ are sums of "local" operators $\mathbf{X}(\mathbf{r}), \mathcal{H}^{(i)}(\mathbf{r})$ and $\mathcal{H}(\mathbf{r})$;

$$X = \int X(\mathbf{r}) d\mathbf{r} \ , \ \mathcal{H}^{(i)} = \int \mathcal{H}^{(i)}(\mathbf{r}) d\mathbf{r} \ \text{and} \ \mathcal{H} = \int \mathcal{H}(\mathbf{r}) d\mathbf{r}, \tag{13}$$

where a "local" operator $Q(\mathbf{r})$ is defined by a functional of field (or spin) operators $\psi(\mathbf{r}')$; $\mathbf{r}' \in D(\mathbf{r}) \equiv$ a circle domain of finite fixed radius b at the center $\mathbf{r}$. Under these general conditions, we can prove the following theorem:

<u>Theorem I (quantal)</u> <u>If the local operators $\mathbf{X}(\mathbf{r}), \mathcal{H}^{(i)}(\mathbf{r})$ and $\mathcal{H}(\mathbf{r})$ are bounded in certain averages, then</u>

$$\lim_{\Omega \to \infty} \Omega^{-1} \log \Psi_\Omega(\lambda, t) = exist (= \psi(\lambda, t)) \text{(uniformly convergent)} \tag{14}$$

<u>for $|\lambda| \le \Lambda$ (fixed) and t finite. Therefore, $\Psi(\lambda, t)$ has the extensive property (10) and consequently so does $\rho(X, t)$.</u>

In order to prove Theorem I, we investigate systems of increasing size L_n (say, $L_n = 2^n a$, n a large integer and $\Omega_n = L_n^d$), as in static arguments.[4] Correspondingly we define $\psi_n(\lambda, t)$ by

$$\psi_n(\lambda, t) = \Omega_n^{-1} \log \Psi_{\Omega_n}(\lambda, t). \ (See \ Fig. 1) \tag{15}$$

Our main task is to prove that this series of functions $\{\psi_n(\lambda, t)\}$ satisfy Cauchy's condition on convergence. For this purpose, we divide the volume Ω_n into 2^d subdomains Ω_{n-1} as shown in Fig.2. Let us call the boundary region shaded in Fig.2 "domain Ω_2" and the rest "domain Ω_1". Correspondingly, the operators $\mathbf{X}, \mathcal{H}^{(i)}$ and $\mathcal{H}$ are separated into the form

$$X = X_1 + X_2, \ \mathcal{H}^{(i)} = \mathcal{H}_1^{(i)} + \mathcal{H}_2^{(i)} \text{ and } \mathcal{H} = \mathcal{H}_1 + \mathcal{H}_2 \ . \tag{16}$$

Now, one of our keypoints in proving the existence of the thermodynamic limit of the function $\Psi(\lambda, t)$ is to derive the following inequality

$$\left| \log \Psi_{\Omega_1 + \Omega_2}(\lambda, t) - \log \Psi_{\Omega_1} \right| \le \varepsilon_n(\lambda, t) \ . \tag{17}$$

Here, $\varepsilon_n(\lambda, t)$ is, after a lengthy calculation, obtained as

$$\varepsilon_n(\lambda, t) = (|\lambda| c_1 + c_2 + 2t c_3) \Omega_2, \tag{18}$$

where c_1, c_2 and c_3 denote upper bounds of certain averages of local operators $\mathbf{X}(r), \mathcal{H}^{(i)}(\mathbf{r})$ and $\mathcal{H}(\mathbf{r})$, respectively. Noting the ratio of Ω_2/Ω_n is given by $2^{-n}(2bd/a)$, we arrive at the inequality $|\psi_n(\lambda, t) - \psi_{n-1}(\lambda, t)| \le 2^{-n} c(\Lambda, t_0)$ for $|\lambda| \le \Lambda$ and $|t| \le t_0$, where $c(\Lambda, t_0) = (\Lambda c_1 + c_1 + 2t_0 c_3)(2bd/a)$. Repeated applications of this inequality yields

$$|\psi_{n+m}(\lambda,t)-\psi_n(\lambda,t)| \leq 2^{-n} c(\Lambda,t_0), \qquad (19)$$

for any positive integer m. This is Cauchy's condition on the <u>uniform</u> convergence of series $\{\psi_n(\lambda, t)\}$ for t finite.

For stochastic systems, we can obtain more explicit results. For brevity, we start from the following time evolution operator of single spin flips:

$$\Gamma=\sum_j \Gamma_j \; ; \; \Gamma_j P(\{\sigma_j\};t) = -W_j(\sigma_j)P(\cdots,\sigma_j,\cdots) + W_j(-\sigma_j)P(\cdots,-\sigma_j,\cdots). \qquad (20)$$

Instead of the boundedness of local operators in quantal systems, we introduce here the concept that the probability function $P(\{\sigma_j\})$ is "normal", which is defined by $P(\ldots, -\sigma_j, \ldots, t) \leq c_3 P(\ldots, \sigma_j, \ldots, t)$, for any configuration, where c_3 is a constant independent of the system size Ω. For convenience we designate this as $P \in \mathcal{N}$. This condition states that the ratio of the probability distribution does not change in order $\Omega^\alpha (\alpha>0)$ with a single flip. This is a quite natural condition which is believed to be, in general, satisfied in ordinary situations. In similar arguments, we can obtain the following theorem:

<u>Theorem II (stochastic)</u>. <u>If $P(t') \equiv e^{t'(\Gamma_1+\mu\Gamma_2)} e^{\mathcal{H}(i)} \in \mathcal{N}$ ("normal")</u> <u>for any separation of Γ into two parts Γ_1 and Γ_2 (and for $0\leq t'\leq t$ and</u> <u>$0\leq\mu\leq1$) in a stochastic system, then we have</u>

$$\lim_{\Omega\to\infty} \psi_\Omega(\lambda,t) = \psi(\lambda,t) = exist, \qquad (21)$$

<u>and the limit is uniformly convergent for $|\lambda| \leq \Lambda$ and t is finite.</u> Therefore, $\Psi(\lambda, t)$ and $P(X, t)$ have the extensive property.

Next we discuss the fluctuation and nonlinear relaxation of a macro-variable X with the use of the extensive property (7). The following theorem is easily proved:

<u>Theorem III</u>: For a large Ω, the most probable value y(t) of x is given by

$$y(t)=\left(\frac{\partial\psi}{\partial\lambda}\right)_{\lambda=0} = \langle x \rangle_t \; ; \; \langle X \rangle_t = \int X\rho(X,t)dX = Tr X \rho(t), \qquad (22)$$

and the variance $\sigma(t)$ is given by

$$\sigma(t)=\left(\frac{\partial^2\psi}{\partial\lambda^2}\right)_{\lambda=0} = \Omega^{-1}\langle(X-\langle X\rangle_t)^2\rangle_t = \Omega\langle(x-\langle x\rangle_t)^2\rangle_t. \qquad (23)$$

As was partly seen in the above discussions, the function $\Psi(x, t)$ is closely related to fluctuations of the relevant macrovariable X. Here we present some theorems on the relation between the extensive property of $\rho(X, t)$ (or the generating function $\Psi(\lambda, t)$) and cumulants[5]

$\{<x^n>_{c,t}\}$

Theorem IV: If $\Psi(\lambda, t)$ has the extensive property, then all cumulants $\{<x^n>_{c,t}\}$ are extensive (i.e., $O(\Omega)$). Conversely, if all cumulants $\{<x^n>_{c,t}\}$ are extensive and

$$\text{if } \lim_{n \to \infty} \left(<X^n>_{c,t} / n! \right)^{1/n} = \Lambda_0^{-1} < \infty, \tag{24}$$

then there exists the generating function $\Psi(\lambda, t)$ for $|\lambda| < \Lambda_0$ and it has the extensive property (and consequently so does $\rho(X, t)$).

Theorem V: The function $\phi(x, t)$ is expressed by cumulants as

$$\phi(y(t)+z) = -\sum_{n=2}^{\infty} \frac{z^n}{n!} \lim_{\Omega \to \infty} f_n \left(<X^2>_{c,t} \Omega^{-1}, \cdots, <X^n>_{c,t} \Omega^{-1} \right), \tag{25}$$

where the function f_n is defined by the identity

$$\left(\psi_2^{-1} \frac{\partial}{\partial \lambda} \right)^{n-2} \psi_2^{-1} \equiv f_n \left(\psi_2, \psi_3, \cdots, \psi_n \right) ; \quad \psi_n \equiv \frac{\partial^n \psi(\lambda)}{\partial \lambda^n} \tag{26}$$

for an arbitrary analytic function ψ, and $\psi_2^{2n-3} f_n (\psi_2, ---, \psi_n)$ is a homogeneous polynomial of order $(n-2)$ with respect to the variables $\psi_2, \psi_3, ---\psi_n$. The most probable path is designated by $x = y(t)$.

Theorem VI: When $\phi(y(t)+z, t) = \sum_{n=2}^{\infty} a_n z^n / n!$ is known, all fluctuations or cumulants $\{<x^n>c,t\}$ are obtained explicitly with the use of the following recursion formula;

$$<X^2>_{c,t} = <(X - <X>_t)^2> = \Omega \sigma(t) = -\Omega a_2^{-1},$$

$$<X^n>_{c,t} = \Omega \left\{ a_n [\sigma(t)]^n + [\sigma(t)]^{3-n} g_n \left(<X^2>_{c,t} \Omega^{-1}, ---, <X^{n-1}>_{c,t} \Omega^{-1} \right) \right\} \tag{27}$$

for $n \geq 3$ and for Ω large, where g_n is a polynomial defined by $g_n (\psi_2, --, \psi_{n-1}) = -\psi_2^{2n-3} f_n - \psi_2^{n-3} \psi_n$.

As applications of the above general theory, we study here an exactly soluble model.

The extensive property and nonlinear relaxation in the kinetic Ising chain. In general, the generating function $\Psi(\lambda, t)$ in the kinetic Ising model is expressed in the form

$$\Psi(\lambda, t) = <0| e^{\lambda X} e^{t W(\beta)} e^{\beta H + H^{(i)}} |0> Z Z_0^{-1}, \tag{28}$$

where $Z = \text{Tr} \exp(-\beta H)$, $Z_0 = \text{Tr} \exp H^{(i)}$, $W(\beta) = \rho_{eq}^{-1/2} \Gamma \rho_{eq}^{1/2}$, $\rho_{eq} = \exp(-\beta H)/Z$, and $|0>$ is the vacuum state defined by

$$|0> \equiv \rho_{eq}^{-\frac{1}{2}} |P_{eq}> \equiv \rho_{eq}^{\frac{1}{2}} \sum_{\{\sigma\}} |\sigma_1>_1 \times \cdots \times |\sigma_N>_N ; \quad \sigma_j^z |\sigma_j>_j = \sigma_j |\sigma_j>_j. \tag{29}$$

In particular, the generating function of the energy $E = H = -J \sum_{j=1}^{N} \sigma_j^z \times \sigma_{j+1}^z$ for the linear chain is easily obtained in the form $\Psi(\lambda, t) = Ce^{N\psi(\lambda, t)}$ for N large, where

$$\psi(\lambda, t) = \frac{1}{2\pi} \int_0^{\pi} \log f(q, \lambda, \mu, t) \, dq + \log \left(\frac{\cosh \beta J}{\cosh \beta_0 J} \right), \tag{30}$$

and we assumed that $H^{(i)} = -\beta_0 H$ (i.e., the initial temperature $T_0 = (\beta_0 k_B)^{-1}$ is different from T). Here, the function $f(q, \lambda, \mu, t)$ is shown to be

$$f(q,\lambda,\mu,t)=e^{-\alpha t}\{e^{t\lambda_q}(c+s\cos\psi_q)(c'+s'\cos\psi_q)+e^{-t\lambda_q}ss'\sin^2\psi_q\},\tag{31}$$

with the use of the diagonalization[6] of $W(\beta)$, where $\mu = \beta - \beta_0$,

$$\begin{cases}\lambda_q=\alpha(1-\gamma\cos q)\\ \gamma=\tanh(2J/k_BT),\end{cases}\begin{cases}c=\cosh(2J\mu)\\ s=\sinh(2J\mu),\end{cases}\begin{cases}c'=\cosh(2\lambda J)\\ s'=\sinh(2\lambda J),\end{cases}$$

$$\cos\psi_q=(\cos q-\gamma)\alpha\lambda_q^{-1}\quad\text{and}\quad\sin\psi_q=\sqrt{1-\gamma^2}(\sin q)\alpha\lambda_q^{-1}.\tag{32}$$

Thus, the extensive property is confirmed explicitly in the linear stochastic chain. The nonlinear relaxation of the energy is given by

$$y(t)=\langle\mathcal{H}\rangle_t=\left(\frac{\partial\psi}{\partial\lambda}\right)_{\lambda=0}=-J\tanh(\beta J)+Js\frac{1}{\pi}\int_0^\pi\frac{e^{-2t\lambda_q}\sin^2\psi_q}{c+s\cos\psi_q}\,dq.\tag{33}$$

The variance $\sigma_E(t)$ is expressed by the integral

$$\sigma_E(t)=2J^2\left[1-\frac{1}{\pi}\int_0^\pi\left\{\cos\psi_q+s\cdot\frac{e^{-2t\lambda_q}\sin^2\psi_q}{c+s\cos\psi_q}\right\}^2 dq\right].\tag{34}$$

It is easily found that the variance $\sigma_E(t)$ shows, in general, an enhancement of fluctuations.

For Markoffian systems, the master equation of the probability distribution $P(x, t)$ is given in the form[1]

$$\varepsilon\frac{\partial}{\partial t}P(x,t)=-\mathcal{H}\left(x,\varepsilon\frac{\partial}{\partial x},t\right)P(x,t),\tag{35}$$

$$\mathcal{H}(x,p,t)=\int(1-e^{-rp})w(x,r,t)\,dr,\tag{36}$$

where $w(x, r, t)$ denotes the transition probability. Now we define the function $\phi_\varepsilon(x, t)$ by

$$\phi_\varepsilon(x,t)=\varepsilon\log P(x,t)\quad\text{or}\quad P(x,t)=\exp\left[\frac{1}{\varepsilon}\phi_\varepsilon(x,t)\right].\tag{37}$$

Then, $\phi_\varepsilon(x, t)$ satisfies the following nonlinear differential equation

$$\frac{\partial\phi_\varepsilon(x,t)}{\partial t}+\mathcal{H}\left(x,\frac{\partial}{\partial x}\phi_\varepsilon,t\right)+\varepsilon(R_\varepsilon+S_\varepsilon)=0,\tag{38}$$

where

$$R_\varepsilon=\int r^2 w(x-\varepsilon r,r,t)e^{-r\frac{\partial}{\partial x}\phi_\varepsilon(x-\varepsilon\theta_1\cdot r,t)}\frac{\partial^2}{\partial x^2}\phi_\varepsilon(x-\varepsilon\theta_1 r,t)\,dr,\tag{39}$$

$$S_\varepsilon=\int r\,e^{-r\frac{\partial}{\partial x}\phi_\varepsilon(x,t)}\frac{\partial}{\partial x}w(x-\varepsilon\theta_2 r,t)\,dr,\tag{40}$$

with $0\le\theta_1\le1$ and $0\le\theta_2\le1$. Here we have used the mean value theorem in differential calculus. Therefore, if R_ε and S_ε are bounded, Eq.(38) approachs the equation

$$\frac{\partial}{\partial t}\phi_0(x,t)+\mathcal{H}\left(x,\frac{\partial}{\partial x}\phi_0,t\right)=0,\tag{41}$$

as ε goes to zero. This equation is shown to have a unique analytic solution when the initial function $\phi_0(x, 0)$ is analytic. The <u>solution</u> $\phi_\varepsilon(x, t)$ of (38) is also shown to approach $\phi_0(x, t)$ for an appropriate series $\varepsilon_1, \varepsilon_2, ---, \varepsilon_n, --- \to 0$, by the help of the Ascoli-Arzela theorem, and also $\phi_\varepsilon(x, t)$ is continuous at $\varepsilon = 0$ with respect to ε in the domain $D = \{(t, x); 0 \leq t \leq \ell, |x| \leq \ell - At\}$, if $|\phi(x, 0)| < \infty$, $C \equiv |R_\varepsilon + S_\varepsilon| < \infty$ and

$$\left| \frac{\partial \mathcal{H}(x,y)}{\partial y} \right| = \left| \int r e^{-ry} w(x,r) \, dr \right| \leq A$$

$$\text{or} \quad \left| \mathcal{H}(x,y) - \mathcal{H}(x,y') \right| \leq A |y - y'| . \tag{42}$$

The proof of these results will be published soon elsewhere.

The author should like to thank Professors R. Kubo and D. Fujiwara for their useful discussions and comments.

References

1) R. Kubo, in Synergetics (Proc. Symp. Synergetics, 1972, Schloss Elmau), ed. H. Haken (B. G. Teubner, Stuttgart) (1973). See also, N. G. van Kampen, Can. J. Phys. <u>39</u> (1961) 551; and in <u>Fundamental Problems in Statistical Mechanics</u> (E. G. D. Cohen, ed. Amsterdam, 1962). R. Kubo, in this volume.

2) R. Kubo, K. Matsuo and K. Kitahara, J. Stat. Phys. <u>9</u> (1973) 51.

3) M. Suzuki, Phys. Letters <u>50A</u> (1974) 47, and Prog. Theor. Phys. <u>53</u> (1975) (in press).

4) R. B. Griffiths, in <u>Phase Transitions and Critical Phenomena</u> ed. C. Domb and M. S. Green (Acad. Press, 1972) and references cited therein. D. Ruelle, <u>Statistical Mechanics: Rigorous Results</u> (W. A. Benjamin, Inc., New York, 1969). M. E. Fisher and J. L. Lebowitz, Comm. Math. Phys. <u>19</u> (1970) 251.

5) R. Kubo, J. Phys. Soc., Japan <u>17</u> (1962) 1100.

6) B. U. Felderhof, Rep. Math. Phys. <u>1</u> (1971) 215; <u>2</u> (1971) 151. B. U. Felderhof and M. Suzuki, Physica <u>56</u> (1971) 43.

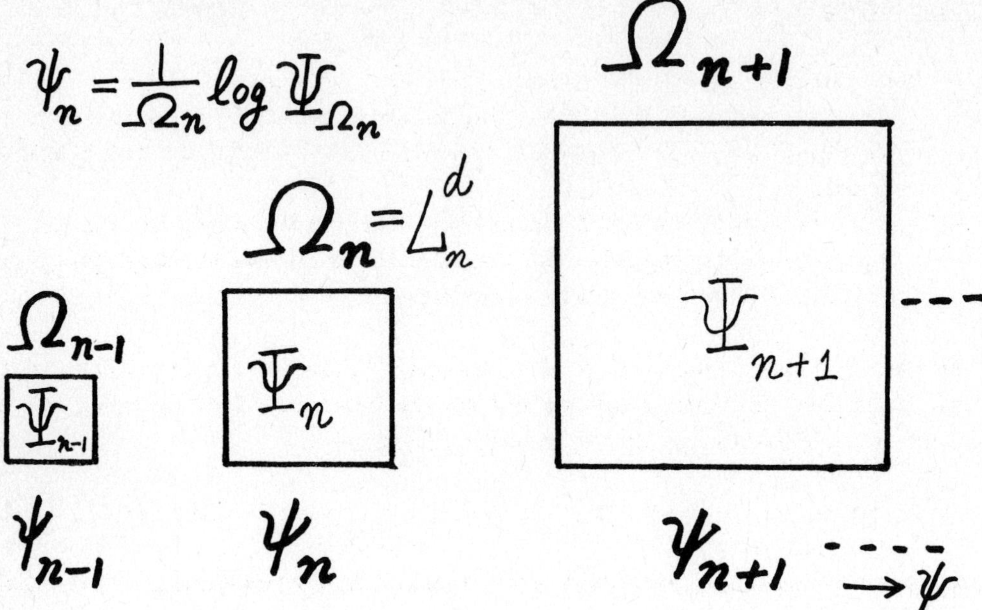

Fig. 1.   A series of systems of increasing size L_n and a functional series of ψ_n.

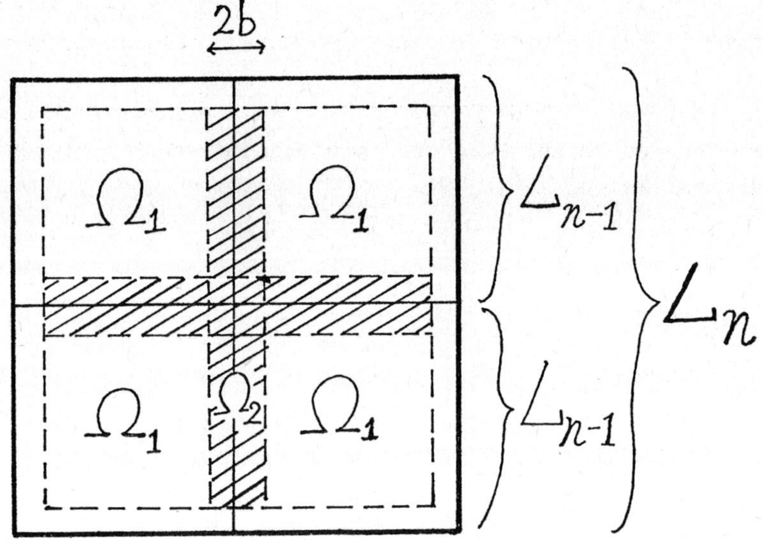

Fig. 2.   Domains Ω_1 and Ω_2 with each margin of width b inside in it; Ω_2 is the shaded region.

Discussion

B. Mühlschlegel (question): Does your $y(t)$ relaxation shown in the Figure deviate much from the corresponding relaxation obtained by the
  simple-minded Fokker-Planck treatment of the kinetic Ising-Weiss
  model?

M. Suzuki (answer): Yes, the nonlinear relaxation curve in the kinetic
  Ising chain is qualitatively different from the kinetic Ising-Weiss
  model, particularly for the variance.

B. Mühlschlegel (question): Are the conditions that $\lambda \leqslant \Lambda$ and t is
  "finite" in your stochastic theorem independent of each other?

M. Suzuki (answer): Yes, they are independent.

J. Zittarts (question): How about the limiting function $\psi(\lambda, t)$ at the
  critical point?

R. Kubo (comment): The proof of existence of thermodynamic limit does
  not guarantee the regularity of the $\psi(\lambda, t)$ (the logarithm of the
  generating function). If a phase change occurs, $\psi(\lambda)$ may be singular.

N. Saito (question): The Hamiltonian is assumed to be a sum of local
  Hamiltonians. How do you separate this in the regions of Ω_1 and
  Ω_2?

M. Suzuki (answer): As was shown in the text, we can separate these
  into the two regions, according to whether each center of local
  operators belongs to Ω_1 or Ω_2.

<u>EQUILIBRIUM STATES AND ERGODIC PROPERTIES OF</u>
<u>INFINITE SYSTEMS</u>*

Joel L. Lebowitz
Belfer Graduate School of Science
Yeshiva University
New York, N.Y. 10033

Abstract

We discuss the ergodic theoretic structure of infinite classical systems and
present results on the ergodic properties of some simple model systems, e.g. ideal
gas, Lorentz gas, Harmonic crystal. (The ergodic properties of the latter system
are shown to be related in a simple way to the spectrum of the force matrix; when
the spectrum is absolutely continuous, as in the translation invariant crystal, the
flow is Bernoulli.) We argue that ergodic properties, suitably refined by the inclusion of space translations, and other structure, are important for an understanding of non-equilibrium properties of macroscopic systems, [1-5].

Possible additional structures include requirements of stability for the stationary state. We shall present result on the classical analogue of the work by Haag,
Kastler and Trych-Pohlmeyer, Araki and others, [6-8].

The esixtence of a time evolution and equilibrium states for various anharmonic
crystal systems will also be discussed, [9].

References

1. For a general review of finite system ergodic properties see Lebowitz, J.L. and
   Penrose, O. 'Modern Ergodic Theory', Physics Today, Feb. 1973, p. 23; Lebowitz,
   J.L. 'Ergodic Theory and Statistical Mechanics' in <u>Transport Phenomena</u>, G.
   Kirczenow and J. Marro, ed. (Lecture Notes in Physics 31, Springer-Verlag, 1974);
   c.f. also article by Kubo, I. in this book.
2. For a general review of infinite system time evolution and ergodic theory see
   Goldstein, S., Lebowitz, J.L. and Aizenman, M. 'Ergodic Properties of Infinite
   Systems'; Lanford, O.E. 'Time Evolution of Large Classical Systems', to appear in
   <u>Proceedings of Battelle Rencontres on Dynamical Systems 1974</u> (Springer-Verlag,
   1975); Goldstein, S. Comm. Math. Phys. <u>39</u>, 302 (1975).
3. Goldstein, S. and Lebowitz, J.L. "Ergodic Properties of an Infinite System of
   Particles Moving Independently in a Periodic Field", Comm. Math. Phys. <u>37</u>, 1-18,
   (1974).
4. Aizenman, M., Goldstein, S. and Lebowitz, J.L. "Ergodic Properties of an Infinite
   One-Dimensional Hard Rod System", Comm. Math. Phys., <u>39</u>, 289-301 (1975).
5. Lanford, O.E. and Lebowitz, J.L. 'Time Evolution and Ergodic Properties of Harmonic Systems', to appear in <u>Proceedings of Battelle Rencontres on Dynamical Systems</u>
   <u>1974</u> (Springer-Verlag, 1975).
6. Haag, R., Kastler, D. and Trych-Pohlmeyer, E. Comm. Math. Phys. <u>38</u>, 173 (1974).
7. Araki, H. Comm. Math. Phys. <u>38</u>, 1 (1974).
8. Lebowitz, J.L., Aizenman, M. and Goldstein, S. "On the Stability of Equilibrium
   States of Finite Classical Systems", submitted to J. Math. Phys.
9. Brascamp, H., Lanford, O.E., Lebowitz, J.L., Lieb, E.H. to be published.

*Supported in part by AFOSR Grant #73-2430B

VARIATIONAL PRINCIPLES FOR MARKOV PROCESSES
AND ONSAGER PRINCIPLE

Hiroshi Hasegawa

Department of Physics, Kyoto University
Kyoto, Japan

Consider the well-known Chapman-Kolmogorov composition formula
for two conditional probabilities in continuous Markov processes in the
N-dimensional Cartesian coordinate space

$$P(x^1 t_1 | x^2 t_2) = \int P(x^1 t_1 | xt) P(xt | x^2 t_2) dx$$
$$t_2 < t < t_1,$$

(1)

and analyze the integrand on the right hand side as a function of the
space-time variables (x t). As the theory of functional integrations
indicates, the conditional probability density $P(ys|xt)$ satisfies the
usual diffusion equations (a forward and a backward Fokker-Planck equation in regard to the variables (y s) and (x t), respectively), which
we choose for simplicity as follows:

$$\frac{\partial}{\partial s} P = -\frac{\partial}{\partial y_\mu}(v_\mu(y) P) + D\frac{\partial^2}{\partial y_\mu \partial y_\mu} P \equiv AP \tag{2}$$

$$-\frac{\partial}{\partial t} P = v_\mu(x)\frac{\partial}{\partial x_\mu} P + D\frac{\partial^2}{\partial x_\mu \partial x_\mu} P \equiv A^+ P. \tag{3}$$

(We use the usual summation convention with respect to the subscript μ.)
We assume that the velocity components v_μ are arbitrary functions of
the space coordinates in so far as the linear operators A and A^+ (the
adjoint of A) generate a well behaved semigroup characteristic for the
time variables s to $+\infty$ and t to $-\infty$, which makes the process under consideration generally non-Gaussian. The corresponding Lagrangian by which
the function-integral form of the $P(ys|xt)$ can be expressed is given by

$$L(\dot{x}, xt) = \frac{1}{4D}(\dot{x}_\mu - v_\mu)(\dot{x}_\mu - v_\mu) + \frac{1}{2}\frac{\partial v_\mu}{\partial x_\mu} \tag{4}$$

which extends Onsager-Machlup Lagrangian for Gaussian processes.

The purpose of this note is to show that the integrand of Eq.(1),
a *product* of solution of the forward and backward equations, follows what
may be called "stochastic dynamics", where the pertinent evolution equa-

tion has a complete parallelism with that in quantum mechanics, and that a firm variation principle can be constructed for the evolution such that it allows to deduce the variation principle of irreversible processes, with an appropriate understanding of *indeterminateness*, predicted by Onsager[1] and elucidated by Onsager and Machlup[2]. The formal analogy between a Schrödinger equation and a Fokker-Planck equation has long been discussed[3-4] but not in the scope of variation principles, perhaps because the probability concept in quantum mechanics is so different from that in the theory of Brownian motion. We show that by considering the concept of *joint probability density* in the latter theory as the counterpart of the quantum probability density the aforementioned analogy can fully be restored, yielding a physical interpretation useful for foundation of the statistical mechanics of non-equilibrium states, so that a better understanding of the "principle of least dissipation" and "Onsager's principle[5]" may be achieved.

Let ψ be any solution of the forward Eq.(2), $\tilde{\psi}$ be any solution of the backward Eq.(3), both writtern in regard to the common space-time variables (x t), and ρ be their product:

$$\rho = \psi\tilde{\psi}. \tag{5}$$

Besides this we introduce the quantity S defined by

$$S = -\frac{1}{2}\log\frac{\psi}{\tilde{\psi}} \tag{6}$$

which corresponds to the action variable in the ordinary dynamics. One can construct a Lagrangian functional $\mathcal{L}\{\psi\tilde{\psi}\}$ or $\mathcal{L}\{\rho S\}$ for the variation in the respective functional space, starting with the "classical" Lagrangian given in Eq.(4), by the following prescription:

(a) Legendre transform :
$$L = p_\mu \dot{x}_\mu - H(p,x), \qquad p_\mu = \frac{\partial L}{\partial \dot{x}_\mu},$$

where
$$H(p,x) = D(p_\mu + \frac{v_\mu}{2D})(p_\mu + \frac{v_\mu}{2D}) - \frac{v_\mu v_\mu}{4D} - \frac{1}{2}\frac{\partial v_\mu}{\partial x_\mu},$$

(b) introduction of a conjugate momentum $\tilde{p}_\mu$ and a modification of the Lagrangian L such taht

$$L \longrightarrow L(p\tilde{p},x) = \frac{p_\mu + \tilde{p}_\mu}{2}\dot{x}_\mu - H(p\tilde{p},x),$$

where
$$H(p\tilde{p},x) = D(p_\mu + \frac{v_\mu}{2D})(\tilde{p}_\mu + \frac{v_\mu}{2D}) - \frac{v_\mu v_\mu}{4D} - \frac{1}{2}\frac{\partial v_\mu}{\partial x_\mu},$$

(c) identification of the momenta p and $\tilde{p}$ in terms of the objects ψ and $\tilde{\psi}$:
$$p_\mu = -\frac{\partial}{\partial x_\mu}\log\psi, \qquad \tilde{p}_\mu = \frac{\partial}{\partial x_\mu}\log\tilde{\psi},$$

(d)   assumption of the conservation of the probability density $\rho = \psi\tilde{\psi}$

  i.e.

$$\frac{\partial\rho}{\partial t} + \frac{\partial}{\partial x_\mu}(\dot{x}_\mu\rho) = 0.$$

One then computes $\mathcal{L}\{\psi\tilde{\psi}\}$ by an average procedure of the $L(p\tilde{p},x)$ with the probability density $\rho = \psi\tilde{\psi}$ as follows:

$$\mathcal{L}\{\psi\tilde{\psi}\} = \int L(p\tilde{p},x)\rho\,dx = \frac{d}{dt}\int(-\tfrac{1}{2}\log\tfrac{\psi}{\tilde{\psi}})\psi\tilde{\psi}\,dx$$

$$+ \int\left(\begin{array}{l}\tfrac{1}{2}\dfrac{\partial}{\partial t}(\log\psi - \log\tilde{\psi})\\[6pt] + D(\dfrac{\partial}{\partial x_\mu}\log\psi - \dfrac{v_\mu}{2D})(\dfrac{\partial}{\partial x_\mu}\log\tilde{\psi} + \dfrac{v_\mu}{2D}) + \dfrac{v_\mu v_\mu}{4D} + \dfrac{1}{2}\dfrac{\partial v_\mu}{\partial x_\mu}\end{array}\right)\psi\tilde{\psi}\,dx. \tag{7}$$

The other form of the Lagrangian functional $\mathcal{L}\{\rho S\}$ may be obtained also.

Accordingly, it can be shown that the standard set of F-P eqs.

$$\frac{\partial\psi}{\partial t} = A\psi, \qquad -\frac{\partial\tilde{\psi}}{\partial t} = A^+\tilde{\psi} \tag{8}$$

is the consequence of the variation in the functional space $(\psi\tilde{\psi})$, namely

$$\delta\int_{t_1}^{t_2}\mathcal{L}\{\psi\tilde{\psi}\}dt = 0, \qquad \left(\int\psi\tilde{\psi}dx\right)_{t_1} = \left(\int\psi\tilde{\psi}dx\right)_{t_2}, \tag{9}$$

a satisfaction of which assures the relation

$$\mathcal{L}\{\psi\tilde{\psi}\} = \frac{d}{dt}\int(-\tfrac{1}{2}\log\tfrac{\psi}{\tilde{\psi}})\psi\tilde{\psi}\,dx = \frac{d}{dt}\int S\rho\,dx. \tag{10}$$

Note that the ρ can be regarded as a joint probability density for the occurrence of two events x at t besides x^1 at t_1, namely

$$\rho(xt,\ x^1t_1) = P(x^1t_1|xt)\psi(xt), \qquad t < t_1. \tag{11}$$

We discuss in the following a physical result of the variation by means of the equivalent Lagrangian functional $\mathcal{L}\{\rho S\}$.

I.   <u>the deterministic variation principle</u>

$$\delta\int_{t_1}^{t_2}\mathcal{L}\{\rho S\}\underset{\delta\rho,\,\delta S}{dt} = 0, \qquad \int\rho\,dx = \text{const.}$$

$$\delta\int S\rho\,dx = 0 \quad \text{at } t = t_1,\ t_2, \tag{12}$$

where

$$\mathcal{L}\{\rho S\} = \int(S - S_0)(\frac{\partial\rho}{\partial t} + \frac{\partial}{\partial x_\mu}(v_\mu - D\frac{\partial}{\partial x_\mu}\log\rho_0)\rho)\,dx$$

$$-\frac{1}{4D}\int V_\mu(S)V_\mu(S)\rho\,dx + \frac{D}{4}\int X_\mu(\rho)X_\mu(\rho)\rho\,dx, \tag{13}$$

with the definitions:

$$V_\mu(S) \equiv -2D\frac{\partial}{\partial x_\mu}(S + \tfrac{1}{2}\log\rho_0) \tag{14a}$$

$$X_\mu(\rho) \equiv \frac{\partial}{\partial x_\mu}\log\frac{\rho}{\rho_0}. \tag{14b}$$

The density ρ_0 is assumed to satisfy $A\rho_0 = 0$, i.e. the steady-state solution of the forward equation (considered as a reference system for the dynamical system ρ). The resulting evolution equations for ρ and S are deterministic, the same as Bohm's equation in quantum mechanics, which are equivalent to Eqs.(8).

II. the indeterministic variation principle

$$\delta \mathcal{L}\{\rho S\}_{\delta S} = 0, \qquad \rho(x\ t) = \text{fixed}, \tag{15}$$

where $\mathcal{L}\{\rho S\}$ is the same as given by (13) and (14). This is a maximum principle, because

$$\delta^2 \mathcal{L}\{\rho S\}_{\delta S} = -D \int (\frac{\partial}{\partial x_\mu}\delta S)(\frac{\partial}{\partial x_\mu}\delta S)\,\rho dx < 0, \tag{16}$$

which just represents the "principle of least dissipation". The resulting evolution equation is that for ρ only; a Fokker-Planck type equation with fluctuating velocity defined by (14a), which may be looked as a Langevin equation (indeterministic).

As a result of the variation I one can deduce an equality similar to Eq.(10), which may be written as

$$\frac{d}{dt}\int(-\log\tilde\psi)\,\rho dx \;=\; \frac{1}{2}\frac{d}{dt}\int(-\log\frac{\rho}{\rho_0})\,\rho dx \;-\; \frac{1}{2}(\Phi\{\rho V\} + \Psi\{\rho X\}),\tag{17}$$

where

$$\Phi\{\rho V\} = \frac{1}{2D}\int V_\mu V_\mu \rho dx, \qquad \Psi\{\rho X\} = \frac{D}{2}\int X_\mu X_\mu \rho dx. \tag{18}$$

The latter two expressions correspond to the Onsager-Machlup dissipation functions expressed in terms of the flux and force (V_μ and X_μ in (14a,b)), respectively. A time integration of both hand sides of Eq.(17) then yields

$$\int(-\log P(x^1 t_1|x^2 t_2))P(x^1 t_1|x^2 t_2)\psi(x^2 t_2)dx^2 + \log\rho_0(x^1)$$
$$= \frac{1}{2}(\ S\{\rho\}_{t_2} + S\{\rho\}_{t_1})\ -\int_{t_1}^{t_2}\frac{1}{2}(\Phi\{\rho V\} + \Psi\{\rho X\})dt. \tag{19}$$

This represents Onsager's principle showing the conditional entropy related to the single-state entropy, $S\{\rho\} \equiv \int(-\log\frac{\rho}{\rho_0})\rho dx$, and the two forms of the dissipation functionals. The case for the reversed time order, $t_1 < t_2$, can be discussed also by a suitable redefinition of the entropies, from which a most general reciprocity relation can be obtained.

1) L.Onsager, Phys. Rev. <u>37</u> 405, <u>38</u> 2265 (1931).

2) L.Onsager and S.Machlup, Phys. Rev. <u>91</u> 1505 (1953).

3) R.Fürth, Zeits. für Phys. <u>81</u> 143 (1933).

4) E.Nelson, *Dynamical Theories of Brownian Motion* (1967).

5) N.Hashitsume, Prog. Theor. Phys. <u>8</u> 461 (1952).

FUNCTIONAL AVERAGES IN STATISTICAL PHYSICS

B. Mühlschlegel
Institut für Theoretische Physik
Universität zu Köln

5 Köln 41, Germany

I.

Integration in functional spaces was originally used in physics along two somewhat different roads. On the first road we find the path integrals of Feynman and of Kac-Wiener for the trajectories of particles in quantum mechanics and in Brownian motion. The second road was pursued in order to integrate the Schwinger differential equations, containing variational derivatives, of quantum field theory[1]. Problems of interacting particles in solid state physics were in the beginning seldomly treated by means of functional integrals[+ 2], rather Green functions and Feynman diagrams became widely used for studying the zero and finite temperature properties of various microscopic models. Only recently, functional integrals of a special type (partition functions for Ginzburg-Landau functionals) became a starting point for a great activity in research on critical phenomena associated with second order phase transitions; the renormalization group formulations developed here have led at the same time to interesting common features of both phase transitions in matter and quantum field theory with ϕ^4 interaction[3]. Although nothing new is added to our knowledge of critical phenomena by the work which is reported here, these recent achievements may serve as a motivation to discuss functional integrals also in connection with some other properties of many-particle systems.[4]

II.

Let us add to the Hamiltonian of free particles described by the one-body operator H_o an interaction of the form $H_1 = A \cdot B$ where A and B are also one-body operators which commute with each other, but not necessarily with H_O. The partition function of the system

---

+ With the notable exception of Feynman's work on the polaron ground state energy by means of an ingenious variational procedure for the path integral. See his book quoted in Ref.1).

$$\mathcal{Z} = \text{Trace } e^{-\beta(H_o + AB)} \tag{1}$$

$\beta = 1/k_B T$, may then easily be represented as a functional integral

$$\mathcal{Z} = \int \delta z \, e^{-\beta F(z,\beta)} \, , \tag{2}$$

$$\delta z = \delta \mathrm{Re} z \, \delta \mathrm{Im} z ,$$
$$\beta F(z,\beta) = \frac{\pi}{\beta} \int_0^\beta d\tau \, |z(\tau)|^2 -$$
$$- \log \text{Trace } e^{-\int_0^\beta d\tau \left[H_o + \sqrt{\frac{\pi}{\beta}} A \, z(\tau) - \sqrt{\frac{\pi}{\beta}} B \, z^*(\tau) \right]} \, . \tag{3}$$

Writing Eq. (2) as

$$\mathcal{Z} = \left\langle \text{Trace } e^{-\int_0^\beta d\tau \, H_\tau} \right\rangle \, , \quad H_\tau = H_o + \sqrt{\frac{\pi}{\beta}} A \, z(\tau) - \sqrt{\frac{\pi}{\beta}} B \, z^*(\tau) \tag{4}$$

where the bracket indicates a Gaussian functional average, we see at once
that by the use of this average the "easier" problem H_τ of free particles moving in a τ -dependent potential is linked to the original problem with direct interaction between particles.

The formula (3) follows by expressing the interaction as a sum of
two squares

$$AB = -\frac{1}{4}(A-B)^2 + \frac{1}{4}(A+B)^2 \, , \tag{5}$$

and by applying the identity

$$e^{\pi \ell^2} = \int_{-\infty}^\infty ds \, e^{-\pi s^2 - 2\pi \ell s} \tag{6}$$

to both squares in the exponential. This must, of course, be done with
some care since when H_o does not commute with A and B, the exponentials
in (3),(4) are ordered exponentials on the imaginary time axis; therefore the quantum mechanical change caused by adding H_1 = AB to H_o is the
reason for the appearance of the functional average.

For the usual two-body forces in extended systems the interaction
consists of a sum of terms considered so far

$$H_1 = \sum_i A_i B_i \, , \quad [A_i, B_i] = 0 \, . \tag{7}$$

Take as an example the interaction in momentum space where q stands
for i :

$$H_1 = \frac{1}{2\mathcal{V}} \sum_q \upsilon(q) \, \varrho_q^+ \, \varrho_q \, . \tag{8}$$

$\nu(q)$ and ς_q are the Fourier transforms of the two-body potential and of the density operator, resp.; Ω is the volume of the system. A similar expression holds for the Heisenberg exchange interaction between spins on a lattice. There is no difficulty to incorporate the sum in Eq. (7) into formula (3): we simply have to consider a multiple functional integral with $\delta z \longrightarrow \prod_i \delta z_i$.

   Functional averages of the form described above have been introduced to quantum statistics by Stratonovich and by Hubbard[4]. The physical idea behind Eq. (2) is to get a generalization of the Hartree or molecular field concept with particles moving independently in a fluctuating field determined by $z(\tau)$. Correlation functions can be obtained as functional averages in a similar manner.

III.

   The most obvious approximation on the functional integral of Eq. (2) is a Taylor expansion around a stationary point $\bar{z}_i$ of $F(z, \beta)$ determined by

$$\frac{\delta F}{\delta z_i} = 0 .$$

(9)

For the relevant models of statistical physics like pair forces of the type of Eq. (8), the Ising and Heisenberg model, and the BCS model of superconductivity Eq. (9) is the Hartree self consistency requirement, and the quadratic terms of the Taylor expansion add random-phase (RPA) contributions to the molecular field approximation $F(\bar{z}(\beta), \beta)$. For systems with phase transition the Taylor expansion including the quartic term gives, in a small momentum approximation, the familiar Ginzburg-Landau functional F_{GL}, when dynamical fluctuations ($\nu \neq 0$) in

$$z_q(\tau) = \frac{1}{\beta} \sum_{\nu=-\infty}^{\infty} z_{q\nu} \, e^{-2\pi i \nu \tau / \beta}$$

(10)

are neglected: $z_q(\tau) \rightarrow z_q = z_{q0}$. This then links

$$Z_{GL} = \int \delta z_q \, e^{-\beta F_{GL}}$$

(11)

used today for the study of critical phenomena with the underlying microscopic models.

   It should be borne in mind, however, that the Taylor expansion has been performed around the molecular field value $\bar{z}(\beta)$ where $\bar{z}(\beta_c^H) = 0$ leads to the transition temperature in simple Hartree approximation. It is therefore quite clear that all success in the study of critical phenomena near the phase transition gives at present no information about the transition temperature itself.

To shift β_c away from the "bare" Hartree value β_c^H is no easy task, and it may be questionable whether the Gaussian average method is superior to the more conventional procedures. Certainly, the special features of the microscopic model will be important here. Concerning Gaussian averages I may mention an approximation which is based upon a variational procedure similar to the one used by Feynman for the polaron path integral: $\bar{F}(z,\beta)$ of Eq. (3) is approximated by a trial functional F_t, in abbreviated notation

$$F_t = \alpha\,(z - a)^2 .\tag{12}$$

Rather than looking for the stationary point of F, α and a have to be determined by the stationarity of the original free energy $\sim \log Z(a,\alpha)$. This has been done for the Ising model with the result that β_c^H is shifted towards the transition temperature of the spherical model[6].

IV.

Superconductivity can be treated by means of functional integrals. The most simple version of this treatment is obtained by using a BCS interaction of the type

$$H_1 = -\frac{V}{\Omega}\,B^+ B \tag{13}$$

with

$$B^+ = \sum_k b_k^+ \quad,\quad b_k^+ = c_{k\uparrow}^+ c_{-k\downarrow}^+ \tag{14}$$

$c_{k\uparrow}^+$ creates an electron of momentum k and spin-up, b_k^+ creates a pair of electrons with opposite momenta and spins, $V>0$ is the effective interaction and Ω the volume. For large volume, $H_o + H_1$ (H_o being the Hamiltonian of free electrons) describes an extended system, and the operators B^+ and B practically commute with each other. Therefore, we can immediately use the formulas (2) and (3) for the BCS functional integral. Due to the symmetry of the interaction H_1 the minimum of $\bar{F}(z,\beta)$ is reached, not at an isolated point $\bar{z}$ but at a circle $|\bar{z}| = \bar{r}(\beta)$ (in accordance with particle number conservation). The corresponding Hartree-like thermodynamic potential

$$F_{BCS} = F(\bar{z},\beta) \tag{15}$$

gives the well known BCS-description of the superconducting phase with $\bar{z}$ being proportional to the temperature-dependent energy gap, or order parameter.

The truncated form of the interaction H_1 in Eq. (13) is responsible for the fact that the BCS thermodynamic potential becomes exact in the thermodynamic limit. We may ask in this connection what will happen to superconductivity when we don't consider bulk systems but when, instead, the volume Ω becomes small. The question is especially interesting when the linear dimensions of the system are smaller than the BCS coherence length ξ which may be of order 1000 $\overset{\circ}{A}$ or more, and which has to be considered as the natural length for all physical phenomena.

The frame work of functional averages suggests here a simple approach[7]. Clearly, a steepest-descents expansion used for the bulk with $\xi^{-3}\Omega$ being the large expansion parameter is no good for the small system. Instead we may approximate the BCS functional statically by dropping the τ-dependence in Eq. (2),(3) thus converting z into a normal two-dimensional integral z_{stat}. Of course, in the small system one has to use appropriate one-electron states labeled by α, with energy ε_α relative to the Fermi energy, and the pairing in Eq. (14) is between time reversed states $\alpha, \tilde{\alpha}$. The trace in Eq. (3) can be evaluated; moreover, putting $z = re^{i\varphi}$, the φ-integration can be performed and one obtains

$$z_{stat} = z_0 \, \frac{\beta}{g\delta} \int_0^\infty d|\psi|^2 \, e^{-\beta f\left(|\psi|^2, \beta\right)} \tag{16a}$$

with z_0 being the partition function of free electrons and

$$f(|\psi|^2, \beta) = \frac{|\psi|^2}{g\delta} - \frac{2}{\beta} \sum{}' \log \frac{\cosh \beta E_\alpha/2}{\cosh \beta \varepsilon_\alpha/2} \quad , \quad E_\alpha = \sqrt{\varepsilon_\alpha^2 + |\psi|^2} . \tag{16b}$$

Here we have introduced a dimensionless coupling constant g by putting

$$\frac{V}{\Omega} = g\delta \quad , \quad \delta = \frac{1}{N(0)\Omega} . \tag{17}$$

The important energy in (16) is δ, the reciprocal density of states of the whole system at the Fermi energy, which measures the average spacing of single-electron levels. The sum in (16b) contains a characteristical cut-off of states with energy $|\varepsilon_\alpha| > \omega_D$, and therefore the "transition temperature" T_c can be introduced as usual by

$$k_B T_c = 1.14 \, \omega_D \, e^{-1/g} . \tag{18}$$

The relevant dimensionless parameter to measure size effects is then

$$\bar{\delta} = \frac{\delta}{k_B T_c} \quad ; \tag{19}$$

performing the thermodynamic limit (going over to the bulk system)
means $\bar{\delta} \to 0$.

Various properties of small superconducting particles have been
calculated by the use of Eq. (16). In Fig. 1 and 2 the specific heat
and the spin susceptibility are shown as examples, and it is seen how
the well known bulk behavior is washed out as $\bar{\delta}$ increases. Concerning
the numerical calculations for finite $\bar{\delta}$ one has to realize that the
size effect enters Eq. (16) in a two-fold manner. First the sum is
discrete in (16b) which is taken care of by an equal-level-spacing description, and second the integration in (16a) has to be performed without steepest-descents expansion in order to include all static fluctuations.

V.

Functional integrals in connection with magnetic problems like
formation of localized magnetic moments (Anderson model) and band magnetism (Hubbard model) have been studied during the last years by several authors[8][9]. The Gaussian functional average attacks here an interaction of the type

$$H_I = U n_\uparrow n_\downarrow , \tag{20}$$

which is a Coulomb correlation between electrons in a level ε_d localized at the same site. The Anderson functional is immediately written
down by employing our basic Eqs. (1) - (4):

$$Z_{Anderson} = \left\langle Trace \, e^{-\int_0^\beta d\tau \left[E_\uparrow(\tau) n_\uparrow + E_\downarrow(\tau) n_\downarrow + W + H_0 \right]} \right\rangle , \tag{21}$$

with

$$E_\uparrow(\tau) = \varepsilon_d - \sqrt{\frac{\pi U'}{\beta}} z(\tau) \quad , \quad E_\downarrow(\tau) = \varepsilon_d + \sqrt{\frac{\pi U'}{\beta}} z^*(\tau) \quad . \tag{22}$$

H_o describes free electrons in extended conduction band states, and W
accounts for a hopping between these band states and the localized one.
Note that in (21) the localized spin-up and spin-down electrons move
independently in a time-dependent potential determined by $z(\tau) = x(\tau) + i y(\tau)$
where, according to (22) $x(\tau)$ corresponds to a magnetic field and

$i_Y(\tau)$ to an electric potential. The band states given by H_o only serve to broaden the localized levels ε_d via the hopping interaction W. In a broad-band approximation the level width becomes

$$\Gamma = \pi \, |\bar{V}|^2 N(o) \quad , \tag{23}$$

$\bar{V}$ hopping matrix element, $N(o)$:density of band states at the Fermi energy. Once this is accounted for, the extensive band part in (21) can be forgotten, and $Z_{Anderson}$ represents a truly local many-body problem.

The Hubbard model has the Coulomb correlation (20) at each lattice site, and hopping between sites is already contained in H_o in the Wannier representation. Therefore we get a similar expression as in (21) with

$$\delta z \longrightarrow \prod_i \delta z_i$$

$$Z_{Hubbard} = \left\langle Trace \, e^{-\int_o^\beta d\tau \left[\sum_{i,\sigma=\uparrow\downarrow} E_{i\sigma}(\tau) n_{i\sigma} + H_o \right]} \right\rangle \tag{24}$$

where i runs over all sites.

To both models the static approximation can be applied which neglects all dynamical fluctuations by dropping the $z_{\nu \neq 0}$ Fourier components of $z(\tau)$. For the one-site problem $Z_{Anderson}$ this leads to a smooth transition from a magnetic state (Curie law) to a non-magnetic state (temperature independent Pauli susceptibility) when the relevant parameter of the problem, $\eta = U/\pi\Gamma$, decreases. Amit and Keiter[8] have included the $z_{\nu = \pm 1}$ – fluctuations and find a similar smooth behavior. The influence of dynamical fluctuations is strongest for $\eta \approx 1$ and leads to a reduction of the effective magnetic moment.

Although the static approximation to the Anderson model becomes exact in the limiting cases $\eta = U/\pi\Gamma \rightarrow 0, \infty$ and gives the right first order corrections to these limits, it misses the Kondo effect for $\eta \gg 1$, even when the lowest dynamical fluctuations are included. It is then the easiest to rely upon the Schrieffer-Wolff transformation which transforms the Anderson model into the s-d model. We should, however, in this respect mention the further work on the functional integral by Hamann which is reviewed in Ref.[8] The interesting idea of this work is that dynamical effects are included by a hopping back and forth between the two Anderson-Hartree minima of $F(z, \beta)$.

$Z_{Hubbard}$ in static approximation is still non trivial since one faces an N-site problem. But with further approximations (CPA) a qualitative understanding of the metal-insulator transition, and a phase diagram can be obtained.[9]

VI.

The functional integrals discussed above differ from the Feynman-Kac path integrals mentioned in the beginning in the respect that these contain the whole dynamics ab initio whereas by the functional averages a direct interaction between particles is formally replaced by classical auxiliary fields. This representation is sometimes useful for its own in order to prove general relations. Concerning approximations it should be said that all results obtained so far for conventional systems have also been obtained by means of diagrams. Some approximations, however, look especially simple in the frame of functional averages, and in fact were introduced because of this simplicity.

In conclusion, I at least want to mention two areas where the physical situation is described by functional integrals. This is first the theory of disordered systems where localized states in random potentials have been considered[10], and second the study of topological problems connected with properties of polymerised materials[11]. Both problems are very interesting, and the reader may consult the original literature.

References

1) See for instance
J.M. Gelfand and A.M. Yaglom, J. Math. Phys. $\underline{1}$, 48 (1960);
R.P. Feynman and A.R. Hibbs "Quantum Mechanics and Path Integrals",
McGraw-Hill, Inc. New York 1965; "Analysis in Function Space",
W.T. Martin and J. Segal, ed., The M.I.T. Press, Cambridge, Mass.
1964

2) We should, however, mention here the early work by S.F. Edwards,
Phil. Mag $\underline{4}$, 1171 (1959)

3) K.G. Wilson and J. Kogut, Physics Reports $\underline{12C}$, 75 (1974)

4) A short review has also been given recently which may be consulted for
more references: B. Mühlschlegel, Proceedings of "International
Conference on Functional Integration and its Applications", London, April 1974, to be published.

5) J. Hubbard, Phys. Rev. Lett. $\underline{3}$, 77 (1959)

6) B. Mühlschlegel and J. Zittartz, Z. Phys. $\underline{180}$, 219 (1963)

7) B. Mühlschlegel, D.J. Scalapino and R. Denton, Phys. Rev. B$\underline{6}$, 1967
(1972)

8) Anderson model: Review by D.R. Hamann and J.R. Schrieffer, In "Magnetism" (H. Suhl ed.) Vol V, Academic Press, New York, 1973
D.J. Amit and H. Keiter, J. Low Temp. Phys. $\underline{11}$, 603 (1973)

9) Hubbard model: J.R. Schrieffer, Lecture Notes, Banff Summer School,

1969, unpublished

J.C. Kimball and J.R. Schrieffer, Int. Conference on Magnetism and
Magnetic Materials, Chicago, 1971, unpublished

M.Cyrot, J. Physique **33**, 125 (1972)

10) J. Zittartz and J.S. Langer, Phys. Rev. **148**, 741 (1966)

11) S.F. Edwards in Ref. [4]

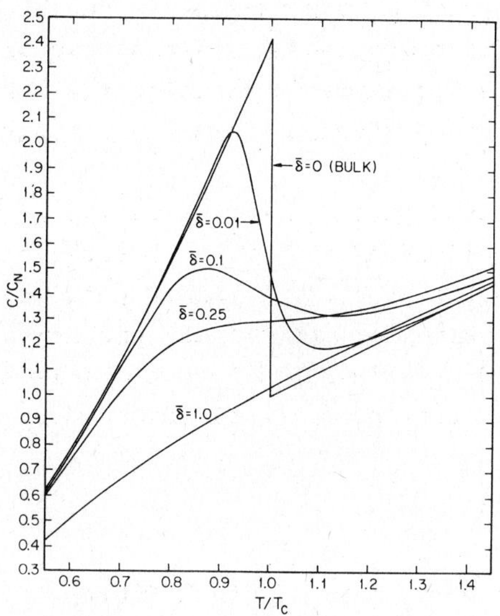

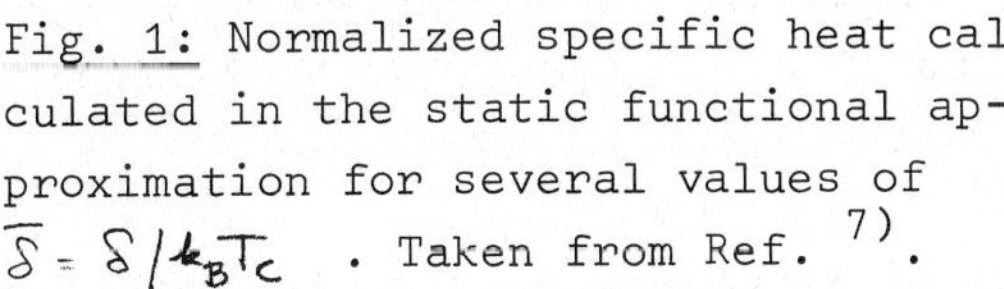

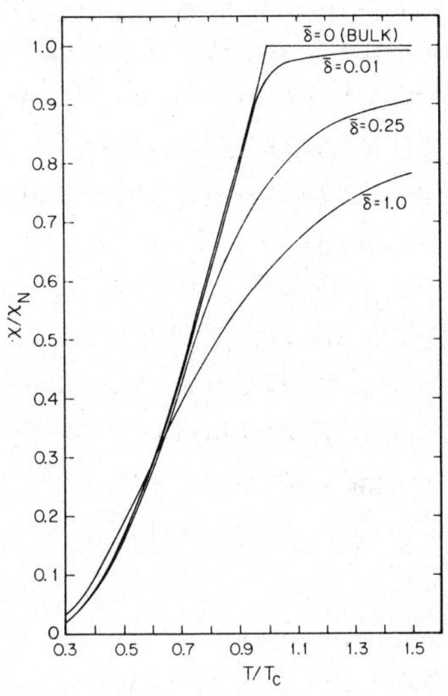

Fig. 1: Normalized specific heat calculated in the static functional approximation for several values of
$\bar{\delta} = \delta/k_B T_c$. Taken from Ref. [7].

Fig. 2: Spin susceptibility normalized to the Pauli value
$(g\mu_B)^2/2\delta$. The results
for several values of $\bar{\delta} = \delta/k_B T_c$
are shown together with the BCS
limiting value. Taken from Ref. [7].

ON THE VALIDITY OF COLLECTIVE VARIABLE
DESCRIPTION OF BOSE SYSTEMS

M. TAKAHASHI

Department of Physics, College of General Education,
Osaka University, Osaka, Japan

For many boson systems interacting with two-body potential,
simple perturbation calculation causes the difficulty of divergence
in the higher order. Bogoliubov and Zubarev (BZ) rewrote the
Schroedinger equation for this system with the density variable
and obtained a new non-Hermite Hamiltonian. Moreover they calculated
the ground state energy by this Hamiltonian through the second order.
Fortunately the perturbation terms are convergent except some very
special potentials. We point out that there is a mathematical inconsistency in the derivation of BZ Hamiltonian and show that perturbation terms for one-dimensional boson system with delta-function
potential is inconsistent with the exact solution by Lieb and Liniger.

BZ transformed the coordinate variables $r_1, \ldots, r_N$ in the Schroedinger equation

$$i\frac{\partial}{\partial t}\psi(r_1,\ldots,r_N;\ t) = (-\sum_{i=1}^{N} \Delta_i + \frac{1}{2}\sum_{i \neq j} V(r_i - r_j)\ )\ \psi\ (r_1,\ldots r_N;\ t\ )$$

$$V(r) = \frac{1}{\Omega}\sum_{k} \nu(k) e^{ikr}, \tag{1}$$

into the density variables $\rho_k = \frac{1}{\sqrt{N}}\sum_{j=1}^{N} e^{-ikr_j}$, and put $\psi = f\Psi$,

$$f \equiv \exp(\frac{1}{4}\sum_{k \neq 0} \rho_k \rho_{-k}\ )$$

The equation for Ψ is

$$i\frac{\partial \Psi}{\partial t} = (H_0 + \frac{1}{\sqrt{N}} H_1)\Psi\ ,$$

$$H_0 = E_0^{(0)} + \sum_{k \neq 0} k^2/\lambda_k\ b_k^+ b_k\ ,$$

$$H_1 = \frac{1}{4}\sum_{k,k',k+k' \neq 0}\left(\frac{\lambda_{k+k'}}{\lambda_k \lambda_{k'}}\right)^{\frac{1}{2}}(kk')\ (b_{k+k'} + b_{-k-k'}^+)((1+\lambda_k)b_{-k} + (\lambda_k - 1)b_k^+)$$

$$((1+\lambda_{k'})b_{-k'} + (\lambda_{k'} - 1)b_{k'}^+)\ . \tag{2a}$$

Here
$$E_0^{(0)} = \frac{N(N-1)}{2\Omega}\nu(0) + \frac{1}{2}\sum_{k \neq 0} (k^2/\lambda_k - k^2 - \frac{N\nu(k)}{\Omega})\ ,$$

$$\lambda_k = \sqrt{k^2/(k^2+2N\nu(k)/\Omega)}\ ,$$

$$b_k^+ = -\sqrt{\lambda_k}\frac{\partial}{\partial\rho_{-k}} + \frac{1}{2\sqrt{\lambda_k}}\rho_k\ , \qquad b_k = \sqrt{\lambda_k}\frac{\partial}{\partial\rho_k} + \frac{1}{2\sqrt{\lambda_k}}\rho_{-k}\ . \tag{2b}$$

Though the degrees of freedom of the original system is dimension x N, the number of variables in Eq.(2a) is infinity. Then there must be infinite number of constraint conditions among ρ_k's. But if we have a solution of Eq.(2b) when ρ_k's are independent, then we will have a solution of original equation (1) by substituting $\rho_k=\frac{1}{\sqrt{N}}\sum_j e^{-ikr_j}$. It is not a popular method to solve the differential equation by increasing the number of variables. However this procedure is mathematically correct. Regarding ρ_k's as independent variables we have Bose commutation relations for b_k and b_k^+ . If b_k and b_k^+ defined by Eq.(2b) are Hermite conjugate, namely,

$$(b_k\Psi,\Phi) = (\Psi,b_k^+\Phi) \tag{3}$$

is satisfied for arbitrary wave functions Ψ and Φ , perturbational expansion becomes possible. The ground state energy is expanded as

$$E_0 = E_0^{(0)} + \frac{1}{\sqrt{N}}E_0^{(1)} + \frac{1}{N}E_0^{(2)} + \frac{1}{N^{\frac{3}{2}}}E_0^{(3)} + \ \cdots \tag{4}$$

Here $E_0^{(0)}$ is given by Eq.(2b) and odd order terms are all zero. $E_0^{(2)}$ is

$$E_0^{(2)} = -\frac{1}{8}\sum_{p,q,p+q\neq0}\lambda_p\lambda_q\lambda_{p+q}(pq)(1+\lambda_p)(1+\lambda_q)\left\{(pq)(1-\lambda_p)(1-\lambda_q)\right.$$
$$- (p,p+q)(1-\lambda_p)(1-\lambda_{p+q})-(q,p+q)(1-\lambda_q)(1-\lambda_{p+q})\Big\}$$
$$\times\ (p^2/\lambda_p+q^2/\lambda_q+(p+q)^2/\lambda_{p+q})^{-1} \tag{5}$$

From the definition of inner product in $\{r_j\}$ space it should be defined in $\{\rho_k\}$ space by

$$(\Phi(\{\rho_k\}),\Psi(\{\rho_k\}))=\int\Phi^*(\{\tfrac{1}{\sqrt{N}}\textstyle\sum e^{-ikr_j}\})\Psi(\{\tfrac{1}{\sqrt{N}}\textstyle\sum e^{-ikr_j}\})f^2(\{\tfrac{1}{\sqrt{N}}\textstyle\sum e^{-ikr_j}\})dr_1\cdots$$
$$\cdots\cdots dr_N, \tag{6}$$

From this definition we see that Eq.(3) is not satisfied for some wave functions. Therefore b_k and b_k^+ are not Hermite conjugate. We can change the definition of inner product so that b_k and b_k^+ become conjugate. But in this case there is no guarantee that wave function belongs to L_2.

Let us consider the one-dimensional boson system interacting via a repulsive delta-function potential in the thermodynamic limit. We put $\nu(k)=2c$. The ground state energy per unit length e_0 can be calculated from Eqs.(2b) and (5)

$$e_0 n^{-3} = \gamma - \frac{4}{3\pi}\gamma^{\frac{3}{2}} - \left(\frac{1}{\pi^2} - \frac{1}{12}\right)\gamma^2 + O(\gamma^{\frac{5}{2}}) \ , \tag{7}$$

where n is the density of particles and $\gamma \equiv c/n$. On the other hand Lieb and Liniger showed that e_0 is given by a solution of the Fredholm type integral equation. They showed that $e_0 n^{-3}$ is given by $\gamma - \frac{4}{3\pi}\gamma^{\frac{3}{2}}$ at $\gamma \ll 1$ by the numerical calculation. To obtain the next term we calculate numerically e_0 and plot $\left\{ e_0 n^{-3} - \left(\gamma - 4/3\pi\gamma^{\frac{3}{2}}\right)\right\}\gamma^{-2}$ as a function of $\sqrt{\gamma}$. We obtain 0.0654 ± 0.0002 as the coefficient of γ^2 term, This means that

$$e_0 n^{-3} = \gamma - \frac{4}{3\pi}\gamma^{\frac{3}{2}} + 0.0654\gamma^2 + O(\gamma^{\frac{5}{2}}) \ . \tag{8}$$

This value of coefficient of γ^2 term is apparently different from the result of BZ Hamiltonian -0.01798785.

Considering that there is a mathematical inconsistency in the derivation of BZ Hamiltonian, we may conclude that their Hamiltonian is not a rigorous description of the Bose system.

There are other collective variable descriptions of Bose systems which use density, velocity or phase as variables of system. Sunakawa, Yamasaki and Kebukawa proposed a Hermite Hamiltonian. Feenberg, Lee and others gave the method of correlated bases functions. These theories can be checked with the use of exact solution by Lieb and Liniger. Details of these are found in Ref. 3.

References

1)   N.N.Bogoliubov and D.N.Zubarev, Soviet Physics 1 (1955), 83; ZETF 28 (1955),129.

2)   E.H.Lieb and W.Liniger, Phys, Rev.130 (1963),1605.

3)   M.Takahashi,(preprint), to be published in Prog. Theor. Phys. 53 No 2.

Question and answer

F. Nakano; Is it fatal for BZ theory that b_k and $b_k^{\dagger}$ are not conjugate?

M. Takahashi; It is possible that BZ theory is an approximation. However it is not a good approximation at least for one-dimensional boson system with delta-function interaction.

EQUILIBRIUM STATISTICAL MECHANICS
OF ONE-DIMENSIONAL CLASSICAL LATTICE SYSTEMS

David Ruelle

I.H.E.S., 91440 Bures-sur-Yvette, France

<u>Abstract</u>. A broad description of the equilibrium statistical mechanics of classical one-dimensional lattice systems with exponentially decreasing interactions is given. We indicate unsolved problems, and mention applications to differentiable dynamical systems.

1. Introduction.

One-dimensional systems with short range interactions are not highly considered in statistical mechanics, because they have no phase transitions. This applies particularly to classical lattice spin systems. These are nevertheless the systems which I would like to discuss here. A reason for the renewed interest in them is that they occur naturally in connection with differentiable dynamical systems, and in particular in the study of the asymptotic behaviour of solutions of some types of differential equations. The spin systems which occur in this way have exponetially decreasing interactions. The basic facts about such systems were discovered by Araki [1] incidentally in his study of one-dimensional quantum systems. In what follows I shall give a broad description of the equilibrium statistical mechanics of classical lattice systems with exponentially decreasing interactions, indicate unsolved problems, and mention applications to differentiable dynamical systems.

2. Configuration space.

At each point x of the lattice $\mathbb{Z}$ a finite number of "spin values" are allowed, forming a set Ω_o. A matrix t indexed by $\Omega_o \times \Omega_o$ and with entries 0 or 1 is also given. An allowed spin configuration on the interval $[k, \ell] \subset \mathbb{Z}$ is an element $\xi = (\xi_k, \xi_{k+1}, \ldots, \xi_\ell)$ of $(\Omega_o)^{[k, \ell]}$ such that

$$t_{\xi_k \xi_{k+1}} = \cdots = t_{\xi_{\ell-1} \xi_\ell} = 1$$

We denote by $\Omega[k, \ell]$ the set of these allowed configurations. The space of configurations on the whole lattice $\mathbb{Z}$ is

$$\Omega = \{\xi \in (\Omega_o)^{\mathbb{Z}} : t_{\xi_x \xi_{x+1}} = 1 \quad \text{for all} \quad x \in \mathbb{Z}\} .$$

Using on Ω_o the discrete topology, and on $(\Omega_o)^{\mathbb{Z}}$ the product topology, we find that Ω is compact. If we define $\tau : \Omega \to \Omega$ by

$$(\tau\xi)_x = \xi_{x+1} \tag{1}$$

then τ is a homeomorphism of Ω ; (1) also defines a homeomorphism

$\tau : \Omega_{[k,\ell]} \mapsto \Omega_{[k-1,\ell-1]}$ for any finite interval $[k,\ell]$, or similarly for a semi-infinite interval $[k,+\infty)$.

It will be convenient to assume from now on that there exists $N > 0$ such that all entries of t^n are > 0 for $n \geq N$. This amounts to requiring that the topological dynamical system (Ω,τ) is topologically mixing.

3. <u>Problem</u>.

The system (Ω,τ) is called a subshift of finite type, or a topological Markov chain. It is entirely determined by the number $|\Omega_o|$ of elements of Ω_o and by the $|\Omega_o| \times |\Omega_o|$ matrix t . Let the system (Ω',τ') be similarly constructed from Ω'_o, t' . We say that (Ω,τ) and (Ω',τ') are isomorphic if there exists a homeomorphism h of Ω' on Ω such that $h\tau' = \tau h$. When do two square matrices t,t' (in general of different orders) define isomorphic subshifts? This problem has been investigated by Williams [16], unfortunately his work is inconclusive [17], and the question remains open.

4. <u>Thermodynamic limits</u>.

An interaction Φ is a real function on the (allowed) configurations in finite intervals. We assume that it is invariant by the translation τ . Given an interaction Φ , an energy function

$$U^{\Phi}_{[a,b]} : \Omega_{[a,b]} \mapsto \mathbb{R}$$

is defined for each finite interval $[a,b]$ by

$$U^{\Phi}_{[a,b]}(\xi) = \Sigma_{k,\ell:\, a\leq k\leq \ell\leq b}\, \Phi(\xi|[k,\ell]) \ .$$

One writes then

$$Z^{\Phi}_{[a,b]} = \sum_{\xi\in\Omega_{[a,b]}} \exp[-U^{\Phi}_{[a,b]}(\xi)]$$

$$P^{\Phi}_{[a,b]} = \frac{1}{b-a} \log Z^{\Phi}_{[a,b]}$$

$$\sigma^{\Phi}_{[a,b]}\{\xi\} = (Z^{\Phi}_{[a,b]})^{-1} \exp\left[- U^{\Phi}_{[a,b]}(\xi)\right]$$

In particular $\sigma^{\Phi}_{[a,b]}$ is a measure on $\Omega_{[a,b]}$. Suppose that

$$\|\Phi\| = \sum_{\ell=0}^{\infty} (\ell+1) \sup_{\xi \in \Omega_{[0,\ell]}} |\Phi(\xi)| < +\infty \ .$$

Then the following limit exists

$$P^{\Phi} = \lim_{b-a\to\infty} P^{\Phi}_{b-a}$$

(thermodynamic limit for the pressure - this would hold also without the factor $(\ell+1)$

in the definition of $\|\Phi\|$).

There is also a unique measure σ on Ω such that for every finite interval

$[k,\ell]$,

$$\lim_{a\to-\infty, b\to+\infty} \alpha_{[k,\ell],[a,b]} \sigma^{\Phi}_{[a,b]} = \alpha_{[k,\ell],\mathbb{Z}} \sigma \tag{2}$$

where $\alpha_{[k,\ell],[a,b]} : \Omega_{[a,b]} \mapsto \Omega_{[k,\ell]}$ gives the restriction to $[k,\ell]$ of a confi-

guration in $[a,b]$ when $[k,\ell] \subset [a,b]$, and similarly for $\alpha_{[k,\ell],\mathbb{Z}}$. The probabi-

lity measure σ is the unique Gibbs state for the interaction Φ (Dobrushin [5],

Ruelle [7]) .

5. Exponentially decreasing interactions.

We say that Φ is exponentially decreasing if there exists $\theta \in (0,1)$ such

that

$$\|\Phi\|_{\theta} = \sup_{[k,\ell]} \theta^{-(\ell-k)} \sup_{\xi \in \Omega[k,\ell]} |\Phi(\xi)| < +\infty \ .$$

For fixed θ these interactions form a Banach space $\mathcal{B}^{\theta}$. Let $\mathcal{B}^{\theta}_{\mathbb{C}}$ be the correspon-

ding complex Banach space, then one can show that

$$\Phi \mapsto P^{\Phi}$$

extends to an analytic function on a neighborhood of $\mathcal{B}^{\theta}$ in $\mathcal{B}^{\theta}_{\mathbb{C}}$.

This can be proved by the transfer matrix method (Ruelle [7], Araki [1]).

We sketch the proof. For a continuous complex function A on $\Omega_{[1,+\infty)}$, let

$$\text{var}_n A = \sup\{|A(\xi)-A(\xi')|:\xi_x = \xi'_x \quad \text{for} \quad 1 \le x \le n\}$$

$$\|A\|_\theta = \sup_{n\ge 0} (\theta^{-n} \text{var}_n A) .$$

Those functions for which $\|A\|_\theta$ is finite form a Banach space $\mathcal{F}^\theta_{\mathbb{C}>}$. An operator $\mathcal{L}$ on $\mathcal{F}^\theta_{\mathbb{C}>}$ is defined by

$$(\mathcal{L}A)(\xi) = \sum_{\xi_0} A(\tau^{-1}(\xi_0,\xi)) \times \exp\Big[- \sum_{\ell\ge 0} \Phi(\xi_0,\xi_1,\ldots,\xi_\ell)\Big]$$

($\mathcal{L}$ is the "transfer matrix").

For $\Phi \in \mathcal{B}^\theta$, one can show that the number $\exp P^\Phi$ is a simple eigenvalue of $\mathcal{L}$, and the rest of the spectrum of $\mathcal{L}$ is contained in a cercle with radius $< \exp P^\Phi$ centered at the origin of $\mathbb{C}$. Since one can also show that $\Phi \to \mathcal{L}$ is an entire analytic function on $\mathcal{B}^\theta_{\mathbb{C}}$, it follows that $\Phi \to P^\Phi$ is analytic in a neighbourhood of $\mathcal{B}^\theta$, as announced.

6. <u>Problem</u>.

Surprisingly, Dobrushin [6] has obtained analyticity properties of the map $\Phi \to P^\Phi$ without assuming exponential decrease of the interaction. His proof does not use the transfer matrix method, is not very transparent, and makes assumptions which are not very natural. Can one give a natural proof of Dobrushin's results ?

7. <u>Exponential decay of correlations</u>.

For a continuous real function A on Ω , let

$$\text{var}_n A = \sup\{|A(\xi)-A(\xi')|: \xi_x = \xi'_x \quad \text{for} \quad |x| \le n\}$$

$$\|A\|_\theta = \sup_{n\ge -1} (\theta^{-2n-1}\text{var}_n A) .$$

Those functions for which $\|A\|_\theta$ is finite form a Banach space $\mathcal{F}^\theta$ for the norm $\|A\|_\theta + \|A\|$. If $A \in \mathcal{F}^\theta$ there is $\Phi \in \mathcal{B}^\theta$ such that $\Phi|\Omega_{[k,\ell]} = 0$ when $\ell-k$ is odd, and

$$A(\xi) = \sum_{\ell \geq 0} \Phi(\xi|[-\ell,+\ell]) \ .$$

One writes then

$$P(A) = P^{\Phi}$$

and it can be checked that this definition does not depend on the particular choice of Φ . Also $A \to P(A)$ extends to an analytic function in a neighbourhood of $\mathfrak{F}^{\theta}$ in $\mathfrak{F}^{\theta}_{\mathbb{C}}$ (the complex Banach space corresponding to $\mathfrak{F}^{\theta}$).

Let σ_A be the unique Gibbs state for the interaction Φ (it does not depend on the particular choice of Φ). Sinai [15] has shown that if $A,A' \in \mathfrak{F}^{\theta}$, then $\sigma_A = \sigma_{A'}$ if and only if there exist $c \in \mathbb{R}$ and $C \in \mathfrak{F}^{\theta}$ such that

$$A' - A = c + C \circ \tau - C \ .$$

Let $A \in \mathfrak{F}^{\theta}$, then there exist $a,b > 0$ such that, if $B_1,B_2 \in \mathfrak{F}^{\theta}$,

$$\left| \sigma_A(B_1 \cdot (B_2 \circ \tau^x)) - \sigma_A(B_1)\sigma_A(B_2) \right| \leq e^{a-b|x|} \|B_1\|_{\theta} \|B_2\|_{\theta}$$

(exponential decay of correlations). One can compute the derivatives of P in terms of σ_A . If $B_1,\ldots,B_{\ell} \in \mathfrak{F}^{\theta}$, let

$$D_A^{\ell}(B_1,\ldots,B_{\ell}) = \frac{d^{\ell}}{ds_1 \ldots ds_{\ell}} \, P(A + \sum_i s_i B_i) \Big|_{s_1 = \ldots = s_{\ell} = 0} \ .$$

Then

(a)
$$D_A^1(B_1) = \sigma_A(B_1)$$

(b)
$$D_A^2(B_1,B_2) = \sum_{x \in \mathbb{Z}} \left[\sigma(B_1 \cdot (B_2 \circ \tau^x)) - \sigma(B_1)\sigma(B_2) \right]$$

(c) $B_1 \mapsto D_A^2(B_1,B_1)$ is a positive semi-definite quadratic form on $\mathfrak{F}^{\theta}$. Its kernel is $\{c + C \circ \tau - C : c \in \mathbb{R}, C \in \mathfrak{F}^{\theta}\}$ and is thus independent of A . There is $R_A > 0$ such that $\left[D^2(B_1,B_1) \right]^{1/2} \leq R \|B_1\|_{\theta}$

(d) For all $p \in \mathbb{R} \pmod{2\pi}$,
$$\sum_{x \in \mathbb{Z}} e^{-ipx} \left[\sigma_A(B_1 \cdot (B_2 \circ \tau^x)) - \sigma_A(B_1)\sigma_A(B_2) \right] \geq 0 \ .$$

These results are of course not unexpected, and some of them should hold under much more general conditions. The proofs however are not as easy as one would imagine (see [10]).

8. ζ-<u>function</u>.

Write $\Omega(m) = \{\xi \in \Omega : \tau^m\xi = \xi\}$ and let $A \in \mathfrak{F}^\theta$. The power series

$$\sum_{m=1}^{\infty} \frac{z^m}{m} \sum_{\xi \in \Omega(m)} \exp \sum_{k=0}^{m-1} A(\tau^k\xi)$$

converges for $|z| < e^{-P(A)}$. One can show that there exists $R > e^{-P(A)}$ such that

$$d_A(z) = \exp[-\sum_{m=1}^{\infty} \frac{z^m}{m} \sum_{\xi \in \Omega(m)} \exp \sum_{k=0}^{m-1} A(\tau^k\xi)]$$

extends to an analytic function in $\{z: |z| < R\}$ with only one zero, this zero is simple and located at $e^{-P(A)}$.

It would be very interesting to increase the domain of analyticity (or meromorphy ?) of $d_A(z)$ because $1/d_A$ can be interpreted as a ζ-function (see Bowen [3], section 5). In the case of a lattice gas with strictly exponential pair interaction one can show that d_A is meromorphic in the entire complex plane.

9. <u>Applications to differentiable dynamical systems.</u>

In a remarkable paper, Sinai [15] has shown how to handle measure theoretical problems for a class of differentiable dynamical systems in terms of statistical mechanics of a one-dimensional lattice systems. Sinai treated Anosov diffeomorphisms and flows, using Markov partitions [13], [14]. As shown by Bowen, Markov partitions exist for the more general Axiom A diffeomorphisms [2] and flows [3]. This permits the extension of Sinai's ideas to these Axiom A diffeomorphisms (Ruelle [8] and flows (Bowen and Ruelle [4]). We cannot go here into all the necessary definitions, but mention a typical result (see [8]).

<u>Theorem</u>. Let Λ be a C^2-Axiom A attractor for a diffeomorphism f. Then for almost every x in a neighbourhood of Λ (in the sense of smooth, or "Lebesgue" mea-

sure) the following limit exists and is independent of x

$$\text{weak } \lim_{n \to \infty} \frac{1}{n} \sum_{k=0}^{n} \delta_{f^n x} = \mu \, .$$

Here δ_x is the unit mass at x . If $f|\Lambda$ is mixing and A,B are C^1 functions in a neighbourhood of Λ ,

$$\mu(A.(B \circ f^n)) \; - \; \mu(A) \, \mu(B)$$

tends exponentially fast to zero when $n \to \infty$. (This last result is obtained from the exponential decay of correlations for Gibbs states).

A similar result holds for flows [4] (i.e. solutions of differential equations), but the exponential decrease of correlations has not been proved. Does it hold in general? The question is of some interest in relation to the problem of turbulence (see Ruelle and Takens [12], Ruelle [9]).

Let us also mention that the problem of ζ-functions for flows would benefit from a better understanding of the question in Section 8.(See Bowen [3] Section 5). Finally, methods of statistical mechanics are also useful in the discussion of homology problems (see Ruelle and Sullivan [11]).

10. <u>References</u>

[1]   H. Araki. Gibbs states of a one-dimensional quantum lattice. Commun. math. Phys. <u>14</u>, 120-157(1969).

[2]   R. Bowen. Markov partitions for Axiom-A diffeomorphisms. Amer. J. Math. <u>92</u>, 725-747(1970).

[3]   R. Bowen. Symbolic dynamics for hyperbolic flows. Amer. J. Math. To appear.

[4]   R. Bowen and D. Ruelle. The ergodic theory of Axiom-A flows. To appear.

[5]   R.L. Dobrushin. The problem of uniqueness of a Gibbsian random field and the problem of phase transitions. Funkts. Analiz. Pril. <u>2</u> No. 4, 44-57(1968); English transl. Funct. Anal. Appl. 2, 302-313(1968).

[6]   R. L. Dobrushin. Analyticity of correlation functions in one-dimensional classi-
      cal systems with slowly decreasing potentials. Commun. math. Phys. $\underline{32}$, 269-289
      (1973).

[7]   D. Ruelle. Statistical mechanics of a one-dimensional lattice gas. Commun.
      math. Phys. $\underline{9}$, 267-278(1968).

[8]   D. Ruelle. A measure associated with Axiom-A attractors. To appear.

[9]   D. Ruelle. Turbulence and Axiom-A attractors. Notes from the school of mathe-
      matical Physics. Camerino (1974).

[10] D. Ruelle. Notes on classical statistical mechanics. To appear.

[11] D. Ruelle and D. Sullivan. Currents, flows, and diffeomorphisms. To appear.

[12] D. Ruelle and F. Takens. On the nature of turbulence. Commun. math. Phys. $\underline{20}$,
      167-192(1971) and $\underline{23}$, 343-344(1971).

[13] Ia. G. Sinai. Markov partitions and C-diffeomorphisms. Funkts. Analiz Pril. $\underline{2}$
      No. 1, 64-89(1968); English transl. Funct. Anal. Appl. $\underline{2}$, 61-82(1968).

[14] Ia. G. Sinai. Construction of Markov partitions. Funkts. Analiz Pril. $\underline{2}$ No. 3,
      70-80(1968); English transl. Funct. Anal. Appl. $\underline{2}$, 245-253(1968).

[15] Ia. G. Sinai. Gibbsian measures in ergodic theory. Uspekhi mat. Nauk $\underline{27}$ No. 4,
      21-64(1972); English transl. Russian math. Surveys $\underline{166}$, 21-69(1972).

[16] R.F. Williams. Classification of subshifts of finite type. Ann. of Math. $\underline{98}$,
      120-153(1973).

[17] R.F. Williams. Errata to "Classification of subshifts of finite type". Ann. of
      Math. $\underline{99}$, 380-381(1974).

A LOOK AT THE DEVELOPMENT OF SPECTRAL

AND SCATTERING THEORY IN JAPAN

Teruo Ikebe

Department of Mathematics

Kyoto University

Kyoto, Japan

This is not a complete historical survey of spectral and scattering theory in Japan. Many omissions and negligences exist. Specifically, many important contributions from outside this country will be left out.

Although it is not easy to explain in precise terms what main problems of spectral and scattering theory are, we try this taking as an example a simple scattering system. Given two self-adjoint operators H_o (unperturbed) and H (perturbed), we want to compare the spectral structure of one with that of the other. Suppose H_o is known to be absolutely continuous. (For terminology we shall mostly follow Kato[6].) What conditions on the perturbation $H - H_o$ make the spectral structures of these operators similar? More technically:

(1) Does the limiting absorption method apply to H? (Or, does the limiting value of the resolvent of H exist in a sense or other when its argument approaches a real point in the (absolutely) continuous spectrum?)

(2) Is the absolutely continuous part of H similar to H_o?

(3) What is the spectral structure of the singular part of H?

   i) Is it discrete?

  ii) Are any eigenvalues imbedded in the absolutely continuous spectrum?

(4) Does an eigenfunction expansion theorem hold for H provided it does for H_o?

(5) Do the wave operators $W_\pm = \text{strong lim}_{t \to \pm\infty} e^{itH} e^{-itH_o}$ exist?

(6) Are the wave operators complete? Is the scattering operator $S = W_+^* W_-$ unitary?

(7) Does the invariance principle of the wave operators hold?

And many other more particular problems might be enumerated.

Toward the above-mentioned problems there are, roughly speaking, two approaches: one is abstract, and the other concrete. The abstract approach deals

with operators in an (or more than one) abstract Hilbert space.   Concerning this
Professor Kuroda will give a more detailed account.   The second approach deals mostly with concrete (partial) differential operators appearing in classical as well as
modern physics, such as Maxwell's equations, Schrödinger equations, Dirac equations,
etc..   Of course, there are some which may be located in between, like Friedrichs'
model.

    <u>1</u>.   The opening of mathematical theory of scattering in this country was,
perhaps, called in 1957 by Kato's work[2, 3] on finite-dimensional (or degenerate)
and trace-class perturbations.   Then Kuroda's work[1, 2, 3] follows.   These
pieces of work are in an abstract setting, and indicate fundamental ideas and methods
of how to solve scattering problems.   Stimulated by them and Povzner's 1953 paper
(Mat. Sb. 32(74) (1953), 109-156) Ikebe[1] studied in 1960 the Schrödinger operator
$H = -\Delta + V(x)$ in the three-dimensional Euclidean space assuming that $V(x) =$
$= O(|x|^{-\alpha})$ with $\alpha > 2$.   The before-mentioned problems except (7) (while Problem
(5) had been treated by Kuroda[1]) were attacked and solved.   It should be noted
that the basic problem of self-adjointness of the Schrödinger operator had been answered affirmatively almost 10 years before by Kato[1] and there had been an important contribution also by Kato[4] to Problem (3-ii).

    After the appearance of Ikebe's paper[1] many an effort has been devoted,
in and outside of Japan, to decrease the exponent α which comes out in the poten-
tial $V(x) = O(|x|^{-\alpha})$.   An important step is made by S. Agmon in 1970 (Actes
Congrès intern. Math. t. 2 (1971), 679-683), and Saitō[2, 3], Mochizuki[7] and Kuroda
[9, 10] succeeded with $\alpha > 1$.   Soon afterward an attempt came to be made at pene-
trating the Coulomb barrier ($\alpha = 1$).   Although the exact solution of the eigenvalue problem for the purely Coulomb potential had been long known, it was not known until the appearance of a paper of Dollard (J. Math. Phys. 5 (1964), 729-738) on modified wave operators whether or not the usual wave operators existed for the Coulomb
potential.   Problems (1), (2) and (3) concerning long-range potentials ($\alpha > 0$),
under an additional assumption on the asymptotic behavior of $V(x)$, were solved by
Ikebe-Saitō[1] and also by Lavine (J. Functional Anal. 12 (1973), 30-54).   Very
recently, a solution to Problem (4) has been obtained by Ikebe[8, 9] and Saitō[6].
But Problem (6) still remains open, while Problem (5) (and also Problem (7)) has already been settled by a time-dependent method.

    <u>2</u>.   Problem (3-ii) is a hard problem, having a close bearing upon the
unique continuation theorem for elliptic differential equations.   The first important contribution was, as mentioned above, by Kato[4] who grasped it as a problem in
a neighborhood of the point at infinity, and obtained asymptotic growth estimates for
solutions to the eigenvalue problem $-\Delta u + V(x)u = \lambda u$ ($\lambda > 0$).   There have been
many contributions from abroad to this problem.   Rather recently, Ikebe-Uchiyama

[1], Masuda[3] and Uchiyama[5] obtained some significant results.   This problem displays a greater difficulty if it is not considered in a *full* neighborhood of infinity.   Konno[2] and Tayoshi[1, 2] contributed in this direction.

3.   The Schrödinger operator might be considered too simple.   But methods and techniques developed for Schrödinger operators are applicable to second- and higher-order elliptic differential operators.   Moreover, it can be noticed that no essential difficulty arises in the treatment of exterior problems (with compact obstacles) for elliptic operators.   For these problems one can count a rather big bunch of researches in this country.   E.g., Shizuta[1], Ikebe[3, 4], Oeda[1], Konno[1], Mochizuki[7], Uesaka[1], Ikebe-Tayoshi[1], Ushijima[3], Kuroda[10, 11] and Kako[2].   The work of Oeda, Konno and Kako is interesting in that an exterior problem is viewed as the limiting case of a sequence of whole-space problems.

4.   While the limiting absorption method directly asks for the boundary values of the resolvent, there is a method, called the limiting amplitude method, in which one is involved with the asymptotic behavior of solutions to the time-dependent equation of Schrödinger's type: i du/dt = Hu.   To this sort of problem contributed Mizohata-Mochizuki[1], Kiyama[1], Iwasaki[1], Kubota-Shirota[1] and others.   In this connection one cannot forget an important contribution of Masuda[1, 2] on the relation between exponential decay and a certain spectral property of H.

5.   The Dirac operator $-i \sum_j \alpha_j \partial/\partial x_j + \beta + V(x)$ describing a relativistic system can also be treated in almost the same way as the Schrödinger operator. Mochizuki[1] and Yamada[1] studied the spectral property of the Dirac operator with a short-range potential $(V(x) = O(|x|^{-\alpha}), \alpha > 1)$.   Very recently, Yamada has succeeded in applying the limiting absorption method to the Dirac operator with a long-range potential.

6.   The operator $-iE(x) \sum_j A_j \partial/\partial x_j$, with E(x) and A_j matrices, appears in classical wave propagation problems.   While a number of papers were published by Wilcox and Schulenberger, the whole-space, exterior and half-space problems were dealt with by Mochizuki[5, 6], Matsumura[1], Wakabayashi[1, 2, 3], Suzuki[1], Yajima[1, 2] and Ikebe[6, 10].   In the whole-space problem the main assumption on E(x) (which describes the medium of propagation) is that E(x) - I be short-range. No long-range theory seems to exist.

7.   Scattering theory of Lax-Phillips' type (Scattering theory, Academic Press, 1967) has not been studied very strenuously in this country.   But Iwasaki [2] made a notable contribution to this theory.   By his result it has been made possible to treat wave equations in even-dimensional spaces along the line laid by

Lax and Phillips, while their original theory was limited to odd-dimensional spaces.

    8.   It is usual that operators (differential or of other types) describing physical processes are given *formally*. Thus it is important to set up suitable Hilbert spaces and show the (essential) self-adjointness of the operators under consideration. It was in 1951 that Kato[1] published a basic result on this problem for Schrödinger operators. An important progress since then was made by Ikebe-Kato[1] about a decade later. Today we have more sophisticated results by Kato, Simon, Walter, etc.. Arai[2] discussed this problem for Dirac operators.

    9.   Sometimes it is hard to distinguish the absolutely continuous spectrum from the singular spectrum. In such a case we have to content ourselves with a rougher classification of the spectrum —— into essential and discrete spectra. Essentially following the technique of Žislin, Uchiyama[1] proved that atomic systems (atoms, ions, molecules) with magnetic external fields have a definite threshold dividing the essential and discrete spectra. On the other hand, Arai[1] obtained a proof that the Stark effect on atomic systems through an electric field whose potential is a linear function of the coordinates entails the essential spectrum extending over the whole real line.

    10.   There are several pieces of work concerning the discrete eigenvalues below the continuous spectrum. Uchiyama[1, 2, 3, 4] gave criteria for the finiteness of eigenvalues of the many-body problem below the threshold. Konno-Kuroda [1] proposed an abstract criterion for the finiteness of perturbed eigenvalues which is applicable to the one-body Schrödinger operator. Setô[1] extended Bargman's inequality to the n-dimensional space in the spherically symmetric potential case. Tamura[1, 2] considered the asymptotic distribution of the negative eigenvalues of Schrödinger (and elliptic) operators in the neighborhood of 0. Among other results related with somewhat general spectral properties we mention here Ichinose's work[1] on the essential spectrum of the tensor product of linear operators.

    11.   The Friedrichs model is semi-abstract. To this one can find contributions of Mochizuki[2] and Ushijima[2]. The techniques developed for self-adjoint Schrödinger operators can also be applied without any essential alterations to some physical systems governed by non-self-adjoint operators. Mochizuki[3, 4], Ikebe[7] and Saitō[4, 5] are examples.

References

The following list contains, besides the papers quoted above, those which are to be quoted in S. T. Kuroda's article given in these Proceedings.

The writer is much indebted to Professor S. T. Kuroda for his painstaking work of preparing these references.

Arai, M.
[1] On the essential spectrum of the many-particle Schrödinger operator with combined Zeeman and Stark effect. Publ. RIMS, Kyoto Univ. Ser. A 3 (1968), 271-287.
[2] On essential self-adjointness of Dirac operators. Preprint (1974).

Asano, K.
[1] Notes on Hilbert transforms of vector valued functions in the complex plane and their boundary values. Proc. Japan Acad. 43 (1967), 572-577.

Ichinose, T.
[1] Tensor product of linear operators and the method of separation of variables. Hokkaido Math. J. 3 (1974), 161-189.

Ikebe, T.
[1] Eigenfunction expansions associated with the Schrödinger operators and their applications to scattering theory. Arch. Rational Mech. Anal. 5 (1960), 1-34.
[2] On the phaseshift formula for the scattering operator. Pacific J. Math. 15 (1965), 511-523.
[3] On the eigenfunction expansion connected with the exterior problem for the Schrödinger equation. Japan. J. Math. 36 (1967), 33-55.
[4] Scattering for Schrödinger operators in an exterior domain. J. Math. Kyoto Univ. 7 (1967), 93-112.
[5] Wave operators for $-\Delta$ in a domain with non-finite boundary. Publ. RIMS Kyoto Univ. Ser. A 4 (1968), 413-418.
[6] Scattering for uniformly propagative systems. Proc. Intern. Conf. Functional Anal. Related Topics, Tokyo, 1969, Tokyo Univ. Press, 1970, 225-230.
[7] A spectral theory for the reduced wave equation with a complex refractive index. Publ. RIMS Kyoto Univ. 8 (1972/73), 579-606.
[8] Spectral representation for Schrödinger operators with long-range potentials. Preprint (1974).
[9] Spectral representation for Schrödinger operators with long-range potentials, II — perturbation by short-range potentials. Preprint (1974).
[10] Remarks on non-elliptic stationary wave propagation problems. Preprint (1974).

Ikebe, T. and T. Kato
[1] Uniqueness of the self-adjoint extension of singular elliptic differential operators. Arch. Rational Mech. Anal. 9 (1962), 77-92.

Ikebe, T. and Y. Saitō
[1] Limiting absorption method and absolute continuity for the Schrödinger operator. J. Math. Kyoto Univ. 12 (1972), 513-542.

Ikebe, T. and T. Tayoshi
[1] Wave and scattering operators for second-order elliptic operators in R^3. Publ. RIMS Kyoto Univ. Ser. A 4 (1968), 483-496.

Ikebe, T. and J. Uchiyama
[1] On the asymptotic behavior of eigenfunctions of second-order elliptic opera-

tors. J. Math. Kyoto Univ. 11 (1971), 425-448.

Inoue, A.
  [1] An example of temporally inhomogeneous scattering. Proc. Japan Acad. 49 (1973), 407-410.
  [2] Wave and scattering operators for an evolving system $d/dt - iA(t)$. J. Math. Soc. Japan 26 (1974), 608-624.

Iwasaki, N.
  [1] On the principle of limiting amplitude. Publ. RIMS Kyoto Univ. Ser. A 3 (1968), 373-392.
  [2] Local decay of solutions for symmetric hyperbolic systems with dissipative and coercive boundary conditions in exterior domains. Publ. RIMS Kyoto Univ. Ser. A 5 (1969), 193-218.

Kako, T.
  [1] Scattering theory for abstract differential equations of the second order. J. Fac. Sci. Univ. Tokyo Sect. IA 19 (1972), 377-392.
  [2] Approximation of exterior Dirichlet problems; convergence of wave and scattering operators. Publ. RIMS Kyoto Univ. 10 (1974/75) (to appear).
  [3] Spectral and scattering theory for the J-selfadjoint operators associated with the perturbed Klein-Gordon type equations. Preprint (1974).

Kato, T.
  [1] Fundamental properties of Hamiltonian operators of Schrödinger type. Trans. Amer. Math. Soc. 70 (1951), 195-211.
  [2] On finite-dimensional perturbation of self-adjoint operators. J. Math. Soc. Japan 9 (1957), 239-249.
  [3] Perturbation of continuous spectra by trace class operators. Proc. Japan Acad. 33 (1957), 260-264.
  [4] Growth properties of solutions of the reduced wave equation with a variable coefficient. Comm. Pure Appl. Math. 12 (1959), 403-425.
  [5] Wave operators and similarity for non-selfadjoint operators. Math. Ann. 162 (1966), 258-279.
  [6] Perturbation theory for linear operators. Springer Verlag, 1966.
  [7] Some results on potential scattering. Proc. Intern. Conf. Functional Anal. Related Topics, Tokyo, 1969, Tokyo Univ. Press, 1970, 206-215.

Kato, T. and S. T. Kuroda
  [1] Theory of simple scattering and eigenfunction expansions. Functional Analysis and Related Fields, F. E. Browder ed., Springer Verlag, 1970, 99-131.

Kiyama, S.
  [1] On the exponential decay of solutions of the wave equation with the potential function. Osaka J. Math. 4 (1967), 15-35.

Konno, R.
  [1] Approximation of obstacles by high potentials; convergence of scattering waves. Proc. Japan Acad. 46 (1970), 668-671.
  [2] Non-existence of positive eigenvalues of Schrödinger operators in infinite domains. J. Fac. Sci. Univ. Tokyo Sect. IA 19 (1972), 393-402.

Konno, R. and S. T. Kuroda
  [1] On the finiteness of perturbed eigenvalues . J. Fac. Sci. Univ. Tokyo Sect. I 13 (1966), 55-63.

Kubota, K. and T. Shirota
  [1] The principle of limiting amplitude. J. Fac. Sci. Hokkaido Univ. Ser. I 20 (1967), 31-52.

Kuroda, S. T.
  [1] On the existence and the unitary property of the scattering operator. Nuovo

Cimento Ser. X 12 (1959), 431-454.

[2] Perturbation of continuous spectra by unbounded operators, I. J. Math.
     Soc. Japan 11 (1959), 247-262.
[3] Perturbation of continuous spectra by unbounded operators, II. J. Math.
     Soc. Japan 12 (1960), 243-257.
[4] On a stationary approach to scattering problems. Bull. Amer. Math. Soc. 70
     (1964), 556-560.
[5] Perturbation of eigenfunction expansions. Proc. National Acad. Sci. USA 57
     (1967), 1213-1217.
[6] An abstract stationary approach to perturbation of continuous spectra and
     scattering theory. J. d'Anal. math. 20 (1967), 57-117.
[7] A stationary method of scattering and some applications. Proc. Intern. Conf.
     Functional Anal. Related Topics, 1969, Tokyo, Tokyo Univ. Press, 1970,
     231-239.
[8] Some remarks on scattering for Schrödinger operators. J. Fac. Sci. Univ.
     Tokyo Sect. IA 17 (1970), 315-329.
[9] Scattering theory for differential operators, I, operator theory. J. Math.
     Soc. Japan 25 (1973), 75-104.
[10] Scattering theory for differential operators, II, self-adjoint elliptic
     operators. J. Math. Soc. Japan 25 (1973), 222-234.
[11] Scattering theory for differential operators, III, exterior problems. Spec-
     tral Theory and Differential Equations, W. N. Everitt ed., Springer Verlag,
     1974, 229-243.

Kuroda, S. T. and H. Morita
    [1] in preparation.

Masuda, K.
    [1] On the exponential decay of solutions for some partial differential equa-
         tions. J. Math. Soc. Japan 19 (1967), 82-90.
    [2] Asymptotic behavior in time of solutions for evolution equations. J. Func-
         tional Anal. 1 (1967), 84-92.
    [3] Positive eigenvalues of Schrödinger operators and the unique continuation
         theorem. Math. Seminar Notes RIMS No. 160 (1972), 61-67.

Matsumura, M.
    [1] Comportement des solutions de quelque problèmes mixtes pour certains sys-
         tèmes hyperboliques symétriques à coefficients constants. Publ. RIMS Kyoto
         Univ. Ser. A 4 (1968), 309-359.

Mizohata, S. and K. Mochizuki
    [1] On the principle of limiting amplitude for dissipative wave equations. J.
         Math, Kyoto Univ. 6 (1966), 109-127.
    [2]

Mochizuki, K.
    [1] On the perturbation of the continuous spectrum of the Dirac operator. Proc.
         Japan Acad. 40 (1964), 707-712.
    [2] On the large perturbation by a class of non-self-adjoint operators. J. Math.
         Soc. Japan 19 (1967), 123-158.
    [3] Eigenfuntion expansions associated with the Schrödinger operator with a
         complex potential and the scattering inverse problem. Proc. Japan Acad. 43
         (1967), 638-643.
    [4] Eigenfunction expansions associted with the Schrödinger operator with a
         complex potential and the scattering theory. Publ. RIMS Kyoto Univ. Ser. A
         4 (1968), 419-466.
    [5] Spectral and scattering theory for symmetric hyperbolic systems in an ex-
         terior domain. Publ. RIMS Kyoto Univ. Ser. A 5 (1969), 219-258.
    [6] The principle of limiting amplitude for symmetric hyperbolic systems in an
         exterior domain. Publ. RIMS Kyoto Univ. Ser. A 5 (1969), 259-265.
    [7] Spectral and scattering theory for second order elliptic differential oper-
         ators in an exterior domain. Lecture Notes, Univ. Utah, 1972.

Oeda, K.
[1] Approximation of obstacles by high potentials; convergence of eigenvalues.
     Proc. Japan Acad. 46 (1970), 663-667.

Saitō, Y.
[1] Eigenfunction expansions associated with second-order differential equations
     for Hilbert space-valued functions. Publ. RIMS Kyoto Univ. 7 (1971/72),
     1-55.
[2] The principle of limiting absorption for second-order differential equations
     with operator-valued coefficients. Publ. RIMS Kyoto Univ. 7 (1971/72), 581-
     619.
[3] Spectral and scattering theory for second-order differential operators with
     operator-valued coefficients. Osaka J. Math. 9 (1972), 463-498.
[4] The principle of limiting absorption for the non-selfadjoint Schrödinger
     operator in R^N (N ≠ 2). Publ. RIMS Kyoto Univ. 9 (1974), 397-428.
[5] The principle of limiting absorption for the non-selfadjoint Schrödinger
     operator in R^2. Osaka J. Math. 11 (1974), 295-306.
[6] Spectral theory for second-order differential operators with operator-valued
     coefficients and their applications to the Schrödinger operators with long-
     range potentials, I (limiting absorption principle); II. Preprints (1974).

Setô, N.
[1] Bargmann's inequalities in spaces of arbitrary dimensions. Publ. RIMS Kyoto
     Univ. 9 (1974), 429-461.

Shizuta, Y.
[1] Eigenfunction expansion associated with the operator -Δ in the exterior
     domain. Proc. Japan Acad. 39 (1963), 656-660.

Suzuki, T.
[1] The limiting absorption principle and spectral theory for a certain non-
     selfadjoint operator and its application. J. Fac. Sci. Univ. Tokyo Sect.
     IA 20 (1973), 401-412.

Tamura, H.
[1] The asymptotic eigenvalue distribution for non-smooth elliptic operators.
     Proc. Japan Acad. 50 (1974), 19-22.
[2] The asymptotic distribution of the lower part eigenvalues for elliptic
     operators. Proc. Japan Acad. 50 (1974), 185-187.

Tanaka, S.
[1] Modified Korteweg-de Vries equation and scattering theory. Proc. Japan
     Acad. 48 (1972), 466-469.
[2] Analogue of Fourier's method for Korteweg-de Vries equation. Proc. Japan
     Acad. 48 (1972), 647-650.
[3] Korteweg-de Vries equation: Construction of solutions in terms of scatter-
     ing data. Osaka J. Math. 11 (1974), 49-59.

Tayoshi, T.
[1] On the spectrum of the Laplace-Beltrami operator on a non-compact surface.
     Proc. Japan Acad. 47 (1971), 187-189.
[2] The asymptotic behavior of the solutions of (Δ + λ)u = 0 in a domain with
     the unbounded boundary. Publ. RIMS Kyoto Univ. 8 (1972), 375-391.
[3] Spectra of elliptic operators with unbounded boundary. Preprint (1974).

Uchiyama, J.
[1] On the discrete eigenvalues of the many-particle system. Publ. RIMS Kyoto
     Univ. Ser. A 2 (1966), 117-132.
[2] Finiteness of the number of discrete eigenvalues of the Schrödinger opera-
     tor for a three particle system. Publ. RIMS Kyoto Univ. Ser. A 5 (1969),
     51-63.

[3] Finiteness of the number of discrete eigenvalues of the Schrödinger operator for a three particle system, II. Publ. RIMS Kyoto Univ. 6 (1970), 193-200.

[4] On the existence of the discrete eigenvalues of the Schrödinger operator for the negative hydrogen ion. Publ. RIMS Kyoto Univ. 6 (1970), 201-204.

[5] Lower bounds of growth order of solutions of Schrödinger equations with homogeneous potentials. Preprint (1974).

Uematsu, H.

[1] Spectral theory of Laplacian in an asymptotically cylindrical domain. J. Fac. Sci. Univ. Tokyo Sect. IA 18 (1972), 525-536.

Uesaka, H.

[1] The Green function for $\Sigma \, (i\partial/\partial x_n - b_n)^2 u = \lambda u$ in the exterior domain and its application to the eigenfunction expansion. Publ. RIMS Kyoto Univ. 8 (1972/73), 393-418.

Ushijima, T.

[1] Spectral theory of the perturbed homogeneous elliptic operator with real constant coefficients. Sci. Papers Coll. Gen. Educ. Univ. Tokyo 16 (1966), 27-42.

[2] On the spectrum of some Hamiltonian operators. Sci. Papers Coll. Gen. Educ. Univ. Tokyo 16 (1966), 127-133.

[3] Note on the spectrum of some Schrödinger operators. Publ. RIMS Kyoto Univ. Ser. A 4 (1968/69), 497-509.

Wadati, M.

[1] The modified Korteweg-de Vries equation. J. Phys. Soc. Japan 34 (1973), 1289-1296.

Wakabayashi, S.

[1] Eigenfunction expansions for symmetric systems of first order in the half-space R^n_+. Proc. Japan Acad. 50 (1974), 6-10.

[2] Eigenfunction expansions for symmetric systems of first order in the half-space R^n_+. Preprint (1974).

[3] The principle of limit amplitude for symmetric hyperbolic systems of first order in the half-space R^n_+. Preprint (1974).

Yajima, K.

[1] The limiting absorption principle for uniformly propagative systems. J. Fac. Sci. Univ. Tokyo Sect. IA 21 (1974), 119-131.

[2] Eigenfunction expansions associated with uniformly propagative systems and their applications to scattering theory. Preprint (1974).

Yamada, O.

[1] On the principle of limiting absorption for the Dirac operator. Publ. RIMS Kyoto Univ. 8 (1972/73), 557-577.

[2] Eigenfunction expansions for Dirac operators. Preprint (1974).

ON THE BORN-OPPENHEIMER APPROXIMATION [*]

J.M. COMBES

Département de Mathématiques
Centre Universitaire de Toulon

83130 LA GARDE - FRANCE

Abstract

We discuss the validity of the Born-Oppenheimer approximation for molecular levels. It is shown that asymptotic expansions for these levels exist in power of the B.O. parameter $\kappa = (\frac{m}{M})^{1/4}$ and prescriptions are given for their calculation. In particular to fourth order some adiabatic corrections to Born-Oppenheimer formula are found.

Introduction : One of the most famous "golden rules" of physics is the one proposed by Born and Oppenheimer [1] for the interpretation of molecular spectra. From them we learned that :

a) levels of diatomic molecules can be indexed by vibrational (n) and rotational (1) quantum numbers.

b) Perturbation expansions of these levels should be made in powers of $\kappa = (\frac{m}{M})^{1/4}$ (m : electron mass, M : reduced mass of the nuclei).

c) To fourth order

$$(1) \qquad E_{n,1} (\kappa) \simeq V_o + \kappa^2 (2n + 1) \nu + \kappa^4 \frac{1(1 + 1)}{R_o^2}$$

Here V_o is the minimal electronic energy, ν some proper frequency and R_o the internuclear distance for the equilibrium configuration.

Despite the experimental success of this semi-classical formula some problems naturally arise.

First one knows that even to fourth order this formula is not exact ; corrections have been proposed until very recently (see e.g [2]) second since κ is not a small parameter higher order terms should be relevant ; prescriptions are thus needed for their calculation together with the control of the asymptotic nature of expansions so obtained. Here of course, one is faced with the problem of finding a mathematical justification to Born-Oppenheimer ansatz on the use of the parameter κ.

In the first section of this paper we will show where the complexity of these problems comes from. In sect. 2 we will use a projection operator technique to reduce them to a semi-classical like perturbation problem for a one-body Schrödinger hamiltonian with an energy dependent interaction. In sect. 3 we will describe the results.

I - Spectral properties of molecular hamiltonians.

We consider a diatomic molecule with N electrons and nuclei A, B. The corresponding quantum mechanical hamiltonian is given by :

$$H(M) = -\frac{\Delta_A}{2M_A} - \frac{\Delta_B}{2M_B} + V_{AB} + H_{el}$$

where V_{AB} is the internuclear interaction and H_{el} the Hamiltonian of the electronic system. We will restrict it to the sector of zero total momentum. Formaly as the reduced mass M tends to infinity H(M) tends to

$$(2) \quad H(\infty) = H_{el} + V_{AB} = \int^{\oplus} H(X_A, X_B)\, dX_A dX_B$$

where

$$H(X_A, X_B) = \sum_{i=1}^{N} \frac{-\Delta_i}{2m} + \sum_{i<j\,=1}^{N} \frac{1}{|\xi_i - \xi_j|} - \left| \sum_{i=1}^{N} \frac{Z_A}{|\xi_i - X_A|} + \frac{Z_B}{|\xi_i - X_B|} \right| + \frac{Z_A Z_B}{|X_A - X_B|}$$

Global invariance properties imply that the spectrum $\sigma(R)$ of $H(X_A, X_B)$ only depend on $R = |X|$ where $X = X_A - X_B$. Namely one has the following covariance relations

$$P(X_A + a, X_B + a) = T^{-1}(a) P(X_A, X_B) T(a)$$
$$(3)$$
$$P(\Lambda X_A, \Lambda X_B) = U^{-1}(\Lambda)\, P(X_A, X_B)\, U(\Lambda)$$

where $P(X_A, X_B)$ is an eigenprojector for $H(X_A, X_B)$ and T, U are translation and rotation operators on the electronic wave-function space. Let us decompose $\sigma(R)$ into it's discrete and essential parts

$$\sigma(R) = \left\{ V_o(R), V_1(R), \ldots \right\}\ U\sigma_e(R)$$

A typical structure for the electronic terms $V_i(R)$ is depicted in fig. 1, as a result of the Coulomb singularity at R = 0, qualitative analyticity properties [3] and the fact that as $R \to \infty$, $V_j(R) \to \Lambda_{A,j}(\infty) + \Lambda_{B,j}(\infty)$ where $-\Lambda_{A,j}$, $-\Lambda_{B,j}$ are binding energies for atoms with nuclei A, B in the infinite mas limit. As a consequence of (2) one has $\sigma(H(\infty)) = \underset{R}{U}\ \sigma(R) = [V_o, \infty)$ and the spectrum of $H(\infty)$ is purely continuous. On the other hand, the following is known [4] about the spectrum $\sigma(H(M))$ of $H(M)$:

$$\sigma_e(H(M)) = [\Lambda(M), \infty)$$

where $\Lambda(M) = \Lambda_A(M) + \Lambda_B(M)$ is the lowest two-body threshold of the system, i.e. the sum

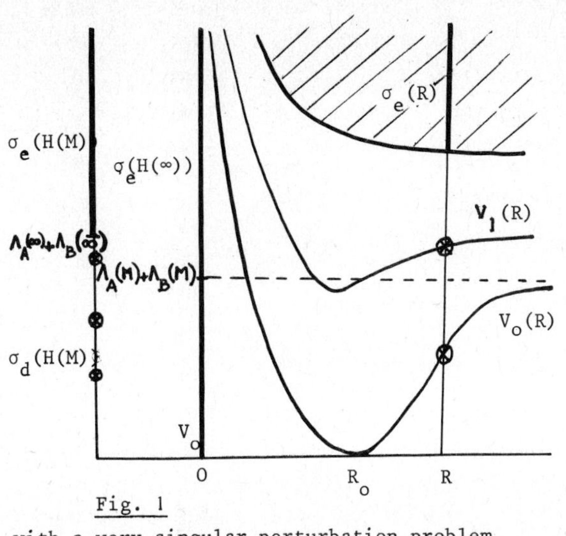

<u>Fig. 1</u>

of the two ground-state energies for separate atoms with nuclei A, B and finite masses M_A, M_B. As $M \to \infty$ $\Lambda_A(M)$ and $\Lambda_B(M)$ decrease to $\Lambda_A(\infty)$, $\Lambda_B(\infty)$ (fig.1) which shows that $\sigma_e(H(M))$ does not tend towards $\sigma(H(\infty))$. The lacking part of $\sigma_e(H(\infty))$ however is obtained from an accumulation of eigenvalues in the interval $\left[V_0 \, , \, \Lambda_A(\infty) + \Lambda_B(\infty)\right]$. The fact,which can be understood on the basis of Born-Oppenheimer formula, that in the limit $M = \infty$ a discrete spectrum turns to a continuous one shows that we are faced with a very singular perturbation problem.

<u>2 - Adiabatic wave-functions and reduction to a one-body problem.</u>

Let $P_n = \int^{\oplus} P_n(X_A, X_B) \, dX_A dX_B$ where $P_n(X_A,X_B)$ in the eigenprojector for $H(X_A, X_B)$ associated to the level $V_n(R)$. In the zero total momentum sector P_n can be identified with the projector operator on states of the form

$$\phi_n(X_A, X_B \, ; \, \xi_{el}) = C(X) \, e_n(X_A, X_B \, ; \, \xi_{el}), \quad C \in L^2(R^3), \text{ where}$$

$$P_n(X_A, X_B) \, e_n(X_A, X_B) - e_n(X_A, X_B)$$

or on linear combinations of such states if the level is degenerate.

States in the range of P_n are adiabatic states ; the electronic system is in the $n^{\underline{th}}$ excited state and adapts "instantaneously" to the nuclear configuration. This is of course only an approximation of physical states of the molecule, based on the fact that the nuclear motion is slow as compared to the electronic one. We will see below that it is good to the order κ^4. Notice the relations :

$$P_n P_m = \delta_{nm} \, P_n$$

$$H(\infty)P_n = A_n P_n \qquad (A_n = \text{multiplication operator by the function } V_n(\cdot))$$

<u>THEOREM</u>

Let $E(M) \in \left[V_0, \, V_1\right]$, where $V_1 = \min V_1(R)$, and $H(M)\psi = E(M)\psi$. Then

a) $(P_0 H(M)P_0 - U(E(M)))P_0 \, \psi = E(M)\psi$ where

$U(E(M)) = P_0 H(M)Q_0 (Q_0 H(M)Q_0 - E(M))^{-1} Q_0 H(M)P_0$ and $Q_0 = I - P_0$

b) $Q_0 \psi = (Q_0 H(M)Q_0 - E(M))^{-1} Q_0 H(M)P_0 \psi$.

This theorem is obtained immediately by projecting the Schrödinger equation and noticing that $\sigma(Q_0 H(M) Q_0) \subset \sigma(Q_0 H(M) Q_0) = [V_1, \infty)$.

REMARKS :

1.) For higher energies E(M) one needs to consider simultaneously many electronic levels and to replace P_0 by $\sum_{j=0}^{n} P_j$ with n large enough.

2) From the above theorem b), one sees that the adiabatic part of the wave-function determines it completely and that the non-adiabatic part is of order κ^4 as $Q_0 H(M) P_0$.

3) Using the covariance properties (3) one can rewrite (a) as an equation for C :

$$(4) \quad \left[- \frac{\Delta_x}{M} + V_0(X) + \frac{\alpha(X)}{M} - U(E(M)) \right] C = E(M) C$$

where $\alpha(X) = \dfrac{M_A M_B}{(M_A + M_B)^2} < e(X_A, X_B), - P_{el}^2 \, e \, (X_A, X_B) > + <e(X_A, X_B), (-\Delta_x e)(X_A, X_B)>.$

(Here P_{el} is the total momentum operator of the electronic system). No explicit form is available for U(E(M)) but one can show $0 \leqslant U(E(M)) \leqslant M^{-2} D$ where D is a second order differential operator with regular coefficients. If many electronic levels had to be considered simultaneously $(E(M) > V_1)$ one would get instead of (4) a set of coupled Schrödinger equations.

3 – Perturbation theory for Eq.(4).

Modulo the energy-dependent term in (4) one has to deal with the classical limit problem $(\frac{h^2}{M} \to o)$ for the Schrödinger equation. Eq.(4) leads to the consideration of a family :

$$h(E,M) = - \frac{\Delta_x}{M} + V_0(X) + \frac{\alpha(X)}{M} - U(E)$$

Denote by $\omega_n(E,M)$ the $n^{\underline{th}}$ eigenvalue of H(E, M) ; then solutions E(M) of eq.(4) are given by :

$$\omega_n(E,M) = E.$$

The following can be shown about $\omega_n(E,M)$ ([5])

a) They have asymptotic expansions in κ to any order. Prescriptions can be given for computation as Rayleigh-Schrödinger coefficients of related operators. For example, up to order 4, they coincide with R.S. coefficients for

$$V_0 + \kappa^2 \left[- \frac{d^2}{dR^2} + \kappa^2 \left(\frac{1(1+1)}{R^2} + \alpha(R_0 + \kappa R) \right) + \omega(\kappa,R) \right]$$

where $\omega(\kappa, X) = \kappa^{-2} |V(R_0 + \kappa R) - V_0|$.

This gives various corrections to Born – Oppenheimer formula (1).

b) $\omega_n(\kappa, E)$ are analytic in E in a neighbourhood of V_0.

From a) and b) one easily deduces existence of asymptotic expansions to any order for the molecular levels E(M).

[1] M. BORN and R. OPPENHEIMER : Zur Quantentheorie der Molekeln - Ann. der Physik - 84, 457 (1927) -

[2] R. PACK and J. HIRSHFIELDER : Energy corrections to the Born-Oppenheimer approximation. Jour. Chem. Phys. 52, 521 (1970).

[3] AVENTINI and R. SEILER : On the electronic spectrum of the diatomic molecular ion. Preprint Frei Univ. Berlin (1974).

[4] W. HUNZIKER : On the spectra of Schrödinger Multi-particle Hamiltonians. Helv. Phys. Acta 39, 451 (1966).

[5] J.M. COMBES : Asymptotic expansions for quantum mechanical bound-states energies near the classical limit.
Preprint. Univ. de Toulon (1975) (To appear in"Proceedings of the Symposium on Spectral and Seattering theory and related topics" RIMS, Kyoto (75)).

* This paper is a summary of a joint work with P. AVENTINI, P. DUCLOS, A. GROSSMAN and R. SEILER.

QUESTION   :

H. ESAWA : You have restricted your problem to a center - of - mass system. Is that the center - of - mass system of the whole system or that of the two nuclei only ?

J.M. COMBES : It is the center of mass of the whole system. The adiabatic wave-functions used belong to the sector of the Hilbert space with zero total momentum.

ABSTRACT APPROACHES TO SPECTRAL AND SCATTERING THEORY,
CONTRIBUTIONS FROM JAPAN

S. T. Kuroda
Department of Pure and Applied Sciences
University of Tokyo
Komaba, Tokyo, 153 Japan

As is mentioned by Professor Ikebe in his lecture in this Symposium, there are roughly speaking, two approaches to mathematical theory of scattering: concrete and abstract. A general view including both approaches having been given by Professor Ikebe, here I shall be concerned more specifically with abstract approaches and try to explain some developments of the abstract approach made in this country.

In general, in an abstract approach one tries to attack a problem by paying main attention to that nature of the problem which can be formulated in general terminology of spectral theory of linear operators, while in concrete approaches one tries to attack more directly a given problem of, say, partial differential operators. Of course, results in spectral theory are utilized in a concrete approach, and conversely an abstract approach is often guided by concrete problems at hand. Two approaches may not be so sharply distinguished from each other.

$\underline{1}$. Abstract methods may be further classified into two types: time-dependent method and time-independent method. In time-dependent methods one tries to prove the existence and the completeness of wave operators $W_{\pm} = \operatorname*{s-lim}_{t \to \pm\infty} e^{itH} e^{-itH_o} P_o$, where H_o and H are unperturbed and perturbed Hamiltonians and P_o is the projection on the absolutely continuous subspace of H_o .   (This is Problems (5) and (6) of Ikebe's paper in these proceedings. Hereafter, we shall simply write Problem (n) to quote Problem (n) of Ikebe's paper.)

In time-independent methods, or in stationary methods as they are sometimes called, one seeks for a "mathematical legalization" of a formula like

$$\Omega_{\pm} = \int \frac{dE(\lambda)}{d\lambda}\{1+V(H_o-(\lambda\pm i0))^{-1}\}d\lambda P_o = \int \{1-(H-(\lambda\mp i0))^{-1}V\}\frac{dE_o(\lambda)}{d\lambda}\, d\lambda P_o$$

Here $H = H_o + V$ and E and E_o are spectral resolutions of H and H_o , respectively.

It is to be noted that both methods are contained in early works of

Kato[2,3]. (To avoid duplication the list of references of this paper is combined with that of Ikebe's paper in these proceedings and given at the end of Ikebe's paper.) Because of the involvement of $(H_o-(\lambda\pm i0))^{-1}$ or $(H-(\lambda\mp i0))^{-1}$ the stationary methods have later become more and more related to the limiting absorption method.

Kato[3] solved Problems (5) and (6) for $H = H_o + V$, with V belonging to the trace class. Kuroda[2,3] generalized his results to unbounded V and applied them to Schrödinger operators $-\Delta + V(x)$ with $V \in L^2 \cap L^1$, or roughly $V(x) = O(|x|^{-\alpha})$, $\alpha > 3$. The above mentioned results of Kato can be applied to examples in which the singular continuous spectrum of H does actually exist. In other words, time-dependent method of this type may not be effective to solve Problem (3).

<u>2</u>. In the second phase of the development much effort has been turned to abstract stationary methods. Among earlier contributions from mathematicians of this country one may cite works of Ushijima[1,2] concerning applications of Friedrichs model and of Kuroda[4] on an attempt to make a stationary treatment of trace class perturbation. In connection with the latter Asano[1] gave a criterion for the existence of the boundary values of the Cauchy integral of vector valued functions.

The Friedrichs model has evolved into the notion of gentle perturbation and, in somewhat different context, the notion of smooth perturbation was introduced by Kato in 1966[5]. Since then much efforts have been made to construct such stationary methods which enable us to handle most of Problems (1) - (7). Kuroda[6] gave an intermediate result and Kato-Kuroda[1] gave a general theory which is applicable to various situations including the trace class perturbations as well as smooth type perturbations. However, the knowledge about the singular spectrum (Problem (3)) was incomplete even in the case of smooth perturbation. For Schrödinger operators Kato[7] solved Problem (6) for $V(x)$ $= O(|x|)^{-\alpha})$, $\alpha > 1$, and Problem (3) for $\alpha > 5/4$. After having been stimulated by Agmon's work in 1970, in which Problem (3) was solved for $\alpha > 1$, Kuroda[9] proved the discreteness of singular spectrum of H in an abstract setting. Eigenfunction expansions can be handled in these frameworks to some extent, as in e.g. Kuroda[5], Kato-Kuroda[1].

Another way of approach to Schrödinger operators, which may be considered as semi-abstract, is to regard it as an "ordinary" differential operator with operator valued coefficients. Starting from works of Jäger (Math. Z. 113(1970), 68-98) Saito[1, 2, 3, 6] pushed this way and obtained limiting absorption principle, eigenfunction expansions etc. up to long range potentials.

<u>3</u>. In each stage of development the abstract results were applied to some problems of partial differential operators, e. g., by Kuroda[8, 10, 11] to the Schrödinger operator and higher order selfadjoint elliptic operators including exterior problems, by Suzuki[1] to uniformly propagative systems of Wilcox, by Uematsu[1] to Laplacian in an asymptotically cyrindrical domain, and by Kuroda[7] rather incompletely to a problem with periodic potentials. Works of Yamada[1], Yajima[1, 2] and Tayoshi [3], which are moré closely related to Agmon's method and are of more concrete nature, may be cited here, too.

<u>4</u>. The scattering theory for wave equations (or for uniformly propagative systems) gives rise to scattering theory involving two Hilbert spaces. The above cited works of Suzuki and Yajima has some connection with this. Two Hilbert space character is more heavily involved in works of Kako[1, 3] concerning wave or Klein-Gordon equations with an indefinite energy norm.

<u>5</u>. Friedrichs model for non-selfadjoint perturbation was treated by Mochizuki[2]. The above cited work of Suzuki and Kako may be regarded as "weakly non-selfadjoint" problem. There, the existence of symmetry with respect to another metric helps to solve Problem (3).

<u>6</u>. So far we have not mentioned much on time-dependent methods for solving Problem (5) only. This is called Jauch-Cook method. In an early stage Problem (5) for Schrödinger operators was solved by this method under the condition that $\alpha > 1$. The wave operators in a domain with infinite boundary was discussed by Ikebe[5]. Recently, there has appeared, here and abroad, some interests on temporarily inhomogeneous Schrödinger equations and associated wave operators. Works of Inoue [1, 2] and Kuroda-Morita[1] may be cited. Some results on the completeness of wave operators have been obtained.

<u>7</u>. Expressions for scattering matrix like the phase shift formula or a formula such as

$$\mathcal{S}_{kk'}(\lambda) = \delta(k-k') - 2\pi i \langle \psi_+(\lambda;k)|V|\psi_0(\lambda;k')\rangle$$

can be derived rigorously by either concrete or abstract methods. See, e.g., Ikebe[2], Kuroda[8, 9].

<u>8</u>. Inverse problems themselves have not been studied much in this

country. However, there are a number of contributions by Tanaka[1, 2, 3], Wadati[1], and others to the investigation of the KdV equation by means of the inverse problem.

<u>REFERENCES</u> are included in the list at the end of the paper of T. Ikebe in these proceedings.

<u>EIGENFUNCTION EXPANSIONS FOR DIFFERENTIAL OPERATORS</u>
<u>WITH OPERATOR-VALUED COEFFICIENTS AND THEIR APPLICATIONS</u>
<u>TO THE SCHRÖDINGER OPERATORS WITH LONG-RANGE POTENTIALS</u>

Yoshimi Saitō

Department of Mathematics
Osaka City University
Sugimoto-cho, Sumiyoshi-ku, Osaka, Japan

§1. Introduction. This report is concerned with the differential operator

$$L = - \frac{d^2}{dr^2} + B(r) + C(r) \qquad (r \in (0, \infty) = I).$$

Here $B(r)$, $r \in I$, is a non-negative, self-adjoint operator in a Hilbert space X with its domain $\mathscr{D}(B(r)) = D$ which does not depend on $r \in I$. And for each $r \in I$ $C(r)$ is a bounded, self-adjoint operator on X. It should be noted that there are some partial differential operators which are converted into the operators of the form L. Let us consider, for example, the Schrödinger operator $- \Delta + Q(y)$ in R^n. Set $X = L_2(S^{n-1})$, S^{n-1} being the $(n-1)$-sphere, and define a unitary operator U from $L_2(R^n, dy)$ onto $\mathscr{H} = L_2(I, X, dr)$ by

$$U\colon L_2(R^n, dy) \ni f(y) \longmapsto r^{(n-1)/2} f(r\omega) \in L_2(I, X, dr) = \mathscr{H}$$

$$(r = |y|, \ \omega = y/r \in S^{n-1}).$$

Here and in the sequel for a real number ρ we denote by $L_2(I, X, (1+r)^\rho dr)$ the Hilbert space of all X-valued functions $u(r)$ on I such that

$$\int_I |u(r)|^2 (1+r)^\rho \ dr < \infty,$$

where $|\ |$ means the norm of X. The inner product $(\ ,\ )_\rho$ and norm $\|\ \|_\rho$ of $L_2(I, X, (1+r)^\rho dr)$ are defined by

$$(u, w)_\rho = \int_I (u(r), w(r))(1+r)^\rho \ dr$$

and

$$\|u\|_\rho = \left[(u, u)_\rho\right]^{1/2},$$

respectively, where $(\ ,\ )$ means the inner product of X. Then we have

$$\begin{cases} U(-\Delta + Q(y))U^* = -\dfrac{d^2}{dr^2} + B(r) + C(r) & (r \in I), \\[2mm] B(r) = r^{-2}(-\Lambda_n + (n-1)(n-3)/4), \qquad C(r) = Q(r\omega)X, \end{cases}$$

Λ_n being the Laplace-Beltrami operator on S^{n-1} and U^* denoting the adjoint of U, and hence the Schrödinger operator $-\Delta + Q(y)$ is unitarily equivalent to the operator of the form L.

Let us impose some asymptotic conditions on the coefficients $B(r)$ and $C(r)$. $C(r)$ can be decomposed as $C(r) = C_0(r) + C_1(r)$ with bounded, self-adjoint operators $C_0(r)$, $C_1(r)$ $(r \in I)$ on X. Here $C_0(r)$ and $C_1(r)$ are assumed to satisfy the following two conditions (C_0) (C_1):

$$(C_0) \qquad \begin{cases} \| C_0(r) \| \leqq c_0(1+r)^{-\varepsilon} & (r \in I), \\[2mm] \| \frac{d}{dr}C_0(r) \| \leqq c_0(1+r)^{-1-\varepsilon} & (r \in I), \end{cases}$$

$$(C_1) \qquad \| C_1(r) \| \leqq c_0(1+r)^{-1-\varepsilon} \qquad (r \in I),$$

where the norm $\| \ \|$ means the operator norm and c_0 and ε are positive constants. As for $B(r)$ let us assume

$$(B) \qquad -\frac{d}{dr}(B(r)x, \ x) \geqq \frac{\beta}{r}(B(r)x, \ x) \qquad (r \geqq R_0, \ x \in D)$$

with constants $R_0 > 0$ and $\beta > 1$. In addition to these conditions we have to assume some conditions such as some local smoothness conditions on $B(r)$ and $C(r)$ etc. Under these assumptions we can develope an eigenfunction expansion theory (or, to be more exact, an eigenoperator expansion theory) for L. This is an extension of Jäger [2] and Saitō [3], [4], where the case that the long-range term $C_0(r)$ is identically zero has been treated.

§2. The limiting absorption principle. Now let us state the limiting absorption principle for L which is our main tool.

Theorem 1 (limiting absorption principle). Let $k \in \mathbb{C}_+ = \{ k \in \mathbb{C} \ / \ \mathrm{Im}\ k \geqq 0, \ \mathrm{Re}\ k \neq 0 \}$ and let $f(r)$ be an X-valued function on I with $f \in L_2(I, \ X, \ (1+r)^{2\delta}dr)$, where δ is a constant such that $1/2 < \delta < (2+\varepsilon)/4$. Then there exists a unique solution $v = v(k, f)$ of the equation

$$(*) \quad \begin{cases} (L - k^2)v = f, \\[4pt] v \in L_2(I,\ X,\ (1+r)^{-2\delta}dr), \\[4pt] v' - ikv \in L_2(I,\ X,\ (1+r)^{2\delta-2}dr)\ \ (\text{"radiation condition"}), \\[4pt] v(0) = 0. \end{cases}$$

The mapping $(k,\ f) \longmapsto v(k,\ f)$ is continuous as the mapping from $\mathbb{C}_+ \times L_2(I,\ X,\ (1+r)^{2\delta}dr)$ into $L_2(I,\ X,\ (1+r)^{-2\delta}dr)$.

We shall make some remarks on this theorem. Let Φ be all (X-valued) test functions on I, that is, the element $\emptyset$ of Φ is an X-valued, sufficiently smooth function on I with compact support in I. Since $L_{/\Phi}$, the restriction of L onto Φ, is a symmetric and lower semi-bounded operator in $\mathfrak{h} = L_2(I,\ X,\ dr)$, the Friedrichs extension T of $L_{/\Phi}$ is well-defined. T can be considered as a self-adjoint realization of L in $\mathfrak{h}$. Let $z = k^2 \pm i\alpha$, where $\alpha > 0$ and k is a real number with $k \neq 0$. Then we have

$$v(\sqrt{k^2 \pm i\alpha},\ f) = R(z)f, \qquad\qquad R(z) = (T - z)^{-1}.$$

The limiting absorption principle guarantees that $R(z)f$ can be continuously extended to the solution $v(\pm k,\ f)$ of the equation $(*)$ as α tends to 0, i.e., we have

$$R(z)f \longrightarrow v(\pm k,\ f)\ \ (\alpha \downarrow 0) \quad \text{in } L_2(I,\ X,\ (1+r)^{-2\delta}dr).$$

Let us make one more remark. Let $\mathcal{B}$ be a Hilbert space obtained by completion of Φ by the norm

$$\| u \|_{\mathcal{B}} = \left[\int_I \left\{ |u'(r)|^2 + |B^{1/2}(r)u(r)|^2 + |u(r)|^2 \right\} dr \right]^{1/2},$$

the norm $|\ |$ being the norm of X, and let ℓ belong to the conjugate space $\mathcal{B}^*$ of $\mathcal{B}$, i.e., ℓ is a continuous linear functional on $\mathcal{B}$,

$$\ell : \mathcal{B} \ni u \longmapsto \langle \ell,\ u \rangle \in \mathbb{C}.$$

Then we can replace the inhomogeneous term $f(r)$ in the equation $(*)$ by $\ell \in \mathcal{B}^*$. In this case the first relation of $(*)$ should be rewritten in a weak form

$$(v,\ (L - k^2)\emptyset)_0 = \langle \ell,\ \emptyset \rangle \qquad (\emptyset \in \Phi),$$

where $(\ ,\ )_0$ means the inner product of $\mathfrak{h} = L_2(I,\ X,\ dr)$. It can be shown that this new equation has a unique solution $v = v(k,\ \ell)$, too, if

$$|\ell|_\delta = \sup\left\{|\langle \ell, (1+r)^\delta \phi \rangle| \,/\, \|\phi\|_{\mathcal{B}} = 1\right\} < \infty.$$

Thus we have obtained an extension of Theorem 1.

Here and in the sequel let us assume that $\varepsilon > 1/2$ in the conditions (C_0) and (C_1). Then, using Theorem 1, we can show

Theorem 2 (the asymptotic behavior of $v(k, f)$). Let k be a real number with $k \neq 0$, $f(r) \in L_2(I, X, (1+r)^2 dr)$ and let $v = v(k, f)$ be a unique solution of the equation $(*)$. Then the strong limit

$$\text{s-}\lim_{n\to\infty} e^{-i\mu(r_n, k)} v(r_n)$$

exists in X, where

$$\mu(r, k) = kr - (2k)^{-1} \int_0^r C_0(t)\, dt,$$

and $\{r_n\}$ is a sequence such that $r_n \uparrow \infty$ and $v'(r_n) - ikv(r_n) \longrightarrow 0$ in X as $n \longrightarrow \infty$. The limit is independent of the choice of the sequence $\{r_n\}$.

§3. Eigenfunction expansion. Let us assume that $\varepsilon > 1/2$ in (C_0) and (C_1). Set $\bar{I} = [0, \infty)$ and let $\mathbb{C}_+$ be as in Theorem 1. First we shall define the Green kernel $G(r, s, k)$ $(r, s \in \bar{I}, k \in \mathbb{C}_+)$ for the operator L. $G(r, s, k)$ is a bounded linear operator on X such that $v = G(\cdot, s, k)x$ is the solution of the equation $(*)$ with $f(r)$ replaced by $\ell[s, x] \in \mathcal{B}^*$ for all $x \in X$, where $\ell[s, x]$ is a continuous linear functional over $\mathcal{B}$ defined by

$$\langle \ell[s, x], \phi \rangle = (x, \phi(s)) \qquad\qquad (\phi \in \mathcal{B}),$$

$(,)$ being the inner product of X. The existence of $G(r, s, k)$ follows from the second remark after Theorem 1. The Green kernel $G(r, s, k)$ can be easily seen to be the resolvent kernel of the self-adjoint realization T of L, which was defined in §2, i.e., we have

$$R(z)f(r) = \int_I G(r, s, \sqrt{z})f(s)\, ds \qquad (z \in \mathbb{C} - \mathbb{R},\ R(z) = (T - z)^{-1}).$$

Therefore, denoting by $E(\cdot)$ the spectral measure associated with T, we obtain

$$E((a,\ b))f(r) = \int_a^b d\lambda \int_I \left\{ G(r,\ s,\ \sqrt{\lambda}) - G(r,\ s,\ -\sqrt{\lambda}) \right\} f(s)\ ds$$

$$(0 < a < b < \infty).$$

On the other hand by the use of Theorem 2 it can be shown that there exists the strong limit

$$\eta(r,\ k) = \text{s-lim}_{s \to \infty} e^{-i\mu(s,\ k)} G(r,\ s,\ k) \qquad (k \in \mathbb{R} - \{0\}).$$

Here $\mu(s,\ k)$ is as in Theorem 2. For each pair $(r,\ k) \in \bar{I} \times (R - \{0\})$ $\eta(r,\ k)$ is a bounded linear operator on X and we have

$$\eta^*(0,\ k) = 0, \qquad (L - k^2) \eta^*(\cdot,\ k) x = 0 \qquad (x \in X),$$

where $\eta^*(r,\ k)$ is the adjoint of $\eta(r,\ k)$. $\eta(r,\ k)$ is called the eigen-operator for L. It follows from the Green formula that

$$(\{G(r,\ s,\ k) - G(r,\ s,\ -k)\} x,\ y) = 2ik(\eta(s,\ k)x,\ \eta(r,\ k)y)$$

$$(x, y \in X,\ k \in \mathbb{R} - \{0\},\ r, s \in \bar{I}).$$

These relations are combined to give

Theorem 3 (expansion theorem). The generalized Fourier transforms $\mathcal{F}_\pm$ from $\mathfrak{h} = L_2(I,\ X,\ dr)$ into $\hat{\mathfrak{h}} = L_2((0,\ \infty),\ X,\ dk)$ are well-defined by

$$(\mathcal{F}_\pm f)(k) = \underset{N \to \infty}{\text{l.i.m.}} \int_0^N \eta_\pm(r,\ k) f(r)\ dr \qquad \text{in } \hat{\mathfrak{h}},$$

where

$$\eta_\pm(r,\ k) = \pm\sqrt{\tfrac{2}{\pi}} ik \eta(r,\ \pm k) \qquad (k > 0).$$

For any Borel set $B \subset (0,\ \infty)$ we have

$$\mathcal{F}_\pm^* \chi_{\sqrt{B}} \mathcal{F}_\pm = E(B),$$

$\mathcal{F}_\pm^*$ being the adjoints of $\mathcal{F}_\pm$, respectively, and $\chi_{\sqrt{B}}$ being the charac-teristic function of $\sqrt{B} = \{k > 0 \ / \ k^2 \in B\}$.

Further, we can show

Theorem 4. The generalized transforms $\mathcal{F}_\pm$ are orthogonal, or equivalently, $\mathcal{F}_\pm$ transform $\mathfrak{h}$ onto $\hat{\mathfrak{h}}$.

§4. The Schrödinger operator in R^n. As has been remarked in §1, the Schrödinger operator $-\Delta + Q(y)$ in R^n is converted into the form L by the unitary operator $U = r^{(n-1)/2}$. Let us assume that the potential $Q(y)$ the following condition (Q):

(Q)　　　　$Q(y)$ is a real-valued function on R^n $(n \neq 2)$. $Q(y)$ can be decomposed as $Q(y) = Q_0(y) + Q_1(y)$ with real-valued functions $Q_0(y)$ and $Q_1(y)$ on R^n such that

$$
\begin{cases}
Q_0(y) = O(|y|^{-(1/2)-\alpha}), \\[2ex]
\dfrac{\partial Q_0}{\partial y_j} = O(|y|^{-(3/2)-\alpha}), \\[2ex]
\dfrac{\partial^2 Q_0}{\partial y_j \partial y_m} = O(|y|^{-2-\alpha}), \\[2ex]
Q_1(y) = O(|y|^{-(3/2)-\alpha})
\end{cases}
$$

$$(|y| \longrightarrow \infty , \quad j,m = 1,2, \cdots, n)$$

with $\alpha > 0$.

Then all the results obtained in §2~§3 can be applied to the operator $-\Delta + Q(y)$. As is well-known the restriction of $-\Delta + Q(y)$ onto C_0^∞ is essentially self-adjoint in $L_2(R^n, dy)$ with a unique self-adjoint extension H. The spectral measure associated with H will be denoted by $\tilde{E}(\cdot)$.

Theorem 5. There exist the eigenoperators $\tilde{\eta}_\pm(|y|, |\xi|)$, $y, \xi \in R^n$, which are bounded linear operators on $L_2(S^{n-1})$, and the generalized Fourier transforms $\tilde{\mathcal{H}}_\pm$ from $L_2(R^n, dy)$ onto $L_2(R^n, d\xi)$ are well-defined by

$$(\tilde{\mathcal{H}}_\pm F)(\xi) = \underset{N \to \infty}{\text{l.i.m.}} \int_{r<N} (\tilde{\eta}_\pm(r, |\xi|)F(r\cdot))(\omega') \, r^{n-1} \, dr$$

$$(F \in L_2(R^n, dy), \quad \omega' = \xi/|\xi|)$$

in $L_2(R^n, d\xi)$. For any Borel set $B \subset (0, \infty)$ we have

$$\tilde{E}(B) = \tilde{\mathcal{H}}_\pm^* \chi_{\sqrt{B}} \tilde{\mathcal{H}}_\pm .$$

Let us consider the case of $Q(y) = 0$, i.e., the case of the Laplacian. In this case our generalized Fourier transforms become equal to the ordinary Fourier transforms in $L_2(R^n)$. Therefore the ordinary Fourier transforms are a special case of our generalized Fourier transforms.

§5. Concluding remarks. Ikebe [1] has shown that the Schrödinger operator $-\Delta + Q(y)$ can be treated directly by the use of essentially the same ideas as ours. In his treatment the condition $n \neq 2$ is unnecessary.

As for the short-range term $C_1(r)$ the condition that $\varepsilon > 1/2$ can be replaced by a weaker condition that $\varepsilon > 0$. But, since Theorem 2 does not seem to be valid in this case, we need some new devices such as the ones used in Saitō [4]. Moreover $C_1(r)$ may admit some local singularities in the interval I.

For the proof of the theorems given in this report see Saitō [5].

References

[1]   T. Ikebe: Spectral representations for the Schrödinger operators with long-range potentials, I, II, Preprint, 1974.

[2]   W. Jäger: Ein gewöhnlicher Differentialoperator zweiter Ordnung fur Funktionen mit Werten in einem Hilbertraum, Math. Z. 113 (1970), 68-98.

[3]   Y. Saitō: The principle of limiting absorption for second-order differential equations with operator-valued coefficients, Publ. RIMS 7 (1972), 581-619.

[4]   Y. Saitō: Spectral and scattering theory for second-order differential operators with operator-valued coefficients, Osaka J. Math. 9 (1972), 463-498.

[5]   Y. Saitō: Spectral theory for second-order differential operatore with operator-valued coefficients and their applications to the Schrödinger operators with long-range potentials I, II, Preprint, 1974.

<u>ASYMPTOTIC COMPLETENESS IN THREE-PARTICLE</u>

QUANTUM MECHANICAL SCATTERING

Lawrence E. Thomas

Mathematics Department
University of Virginia
Charlottesville, Virginia 22903

The purpose of this note is to report on some recent results in three-particle non-relativistic spinless quantum mechanical potential scattering [1,2,3]. The principal result of these investigations can be formulated in the following theorem: Let

$$H = H_o + \Sigma V = \sum_{i=1}^{3} \frac{p_i^2}{2m_i} + \sum_{i<j} V_{ij}(\underset{\sim}{x}_{ij})$$

acting in $L^2(\mathbb{R}^6)$ be the Hamiltonian for the three-particle system with center of mass motion removed, where (i) $V_{ij}(\underset{\sim}{x}) \epsilon L^{3/2+\delta}(\mathbb{R}^3) \cap L^{3/2-\delta}(\mathbb{R}^3)$ for some $\delta > 0$; (ii) the two body Hamiltonians

$$h_{ij} = \frac{k_{ij}^2}{2m_{ij}} + V_{ij}(\underset{\sim}{x}_{ij}),$$

acting in $L^2(\mathbb{R}^3)$, have only strictly negative energy eigenstates $\{\phi_{ij}^n\}_n$; and (iii) the operator-valued functions $W_{ij}(x\pm i0)$ $= |V_{ij}|^{1/2} \hat{r}_{ij}(x\pm i0)|V_{ij}|^{1/2}, x\epsilon \mathbb{R}$, acting in $L^2(\mathbb{R}^3)$ are bounded. Here,

$$\hat{r}_{ij}(z) = r_{ij}(z) - \sum_{n} \frac{|\phi_{ij}^n \rangle \langle \phi_{ij}^n|}{\lambda_{ij}^n - z},$$

$r_{ij}(z)$ is the resolvent for h_{ij}, and λ_{ij}^n is the eigenvalue corresponding to ϕ_{ij}^n. Then the wave operators exist, and are complete in the sense that the direct sum of the ranges of the wave operators is the absolutely continuous subspace of the Hilbert space with respect to H. The scattering operator is therefore unitary. (Condition (iii) is assured if in addition to conditions (i) and (ii) being satisfied, V_{ij} is of the form $V_{ij}(\underset{\sim}{x}) = (1+|\underset{\sim}{x}|)^{-1-\epsilon}(v_{ij}^p(\underset{\sim}{x})+v_{ij}^\infty(\underset{\sim}{x}))$ with $\epsilon > 0$, $v_{ij}^p \epsilon L^p(\mathbb{R}^3)$, p>3/2, and $v_{ij}^\infty \epsilon L^\infty(\mathbb{R}^3)$, and the operators $W_{ij}(0)$ exist [4,5].)

These results generalize the work of Faddeev on the three body problem because more singular local behavior and less restrictive long range behavior of the potentials are both accommodated. A similar theorem holds for space dimensions n>3 although no analogous theorem is known for one or two space dimensions. An additional result of Ginibre and Moulin establishes that the negative spectrum of H contains no singular continuous part [1]. The positive singular spectrum is contained in a closed set of measure zero, but the existence of positive singular continuous spectrum with these interactions remains an open question.

The proof of the theorem is carried out primarily in configuration space using time independent techniques. Weighted Hilbert spaces [1] or L^p spaces [2] replace the Banach spaces of Hölder continuous functions of momenta employed by Faddeev [6]. Well established methods (e.g. criteria for compactness of operators in L^p, Kato's theory of smooth operators [7] and Agmon's theory using weighted Hilbert spaces [4]) become available with a consequent simplification of the analysis.

In outline, the proof begins by considering Faddeev's equations, which are modified appropriately to exploit the fact that the operator valued functions $|V_{ij}|^{1/2}(H_o-z)^{-1}|V_{ik}|^{1/2}: L^2(\mathbb{R}^6) \to L^2(\mathbb{R}^6), (ij)\neq(ik)$ are compact and norm Holder continuous in z for $z = x\pm i0$, $x\varepsilon \mathbb{R}$. This follows easily from the estimate [8],

$$|| |V_{ij}|^{1/2} e^{-itH_o} |V_{ik}|^{1/2} || < c(q) t^{-3/q} || |V_{ij}|^{1/2} ||_{L^q(\mathbb{R}^3)} || |V_{ik}|^{1/2} ||_{L^q(\mathbb{R}^3)}.$$

The components of the solution to these equations, which are operator valued analytic functions of z for Im $z \neq 0$, may be continued to the real axis where they exist as bounded operators in appropriate weighted Hilbert spaces or L^p spaces for $z = x\pm i0$, $x\varepsilon \mathbb{R}$ and x away from a closed set of measure zero. The components are then used to construct the full resolvent R(z) and give a meaning to it for $z = x\pm i0$ away from the singular set.

In reference [2], the wave operators are defined in terms of spectral integrals, i.e. Riemann Stieltjes integrals, with operator-valued integrands and projection valued measures. For example, the wave operators relating to the free channel are given by

$$\Omega_\pm^o = 1 - \underset{\varepsilon\to 0}{\text{s-lim}} \int R(\mu\pm i\varepsilon) \sum_{i<j} V_{ij} dE_o(\mu)$$

with $E_o(\mu)$ the spectral family corresponding to H_o, cf. [9]. The control over R(z) for z near the real axis obtained through the modified Faddeev equations is sufficient to imply the existence of these expressions along with the completeness of the wave operators.

REFERENCES

[1] J. Ginibre and M. Moulin, "Hilbert space approach to the quantum mechanical three-body problem," preprint, Laboratoire de Physique Theorique et Hautes Energies, Orsay, France.

[2] L.E. Thomas, "Asympotic completeness in two- and three-particle quantum mechanical scattering," to appear in Ann. Phys., (1975).

[3] J. Howland, "Abstract stationary theory of multichannel scattering," to appear in J. Functional Analysis.

[4] S. Agmon, Lectures at the conference "Mathematical theory of scattering," Oberwolfach, (1971).

[5] R. Lavine, "Absolute continuity of positive spectrum for Schrödinger operators with long-range potentials," J. Functional Analysis 12 (1973) 30-54.

[6] L.D. Faddeev, Mathematical Aspects of the Three-Body Problem in the Quantum Scattering Theory, Israel Program for Scientific Translations, Jerusalem, Israel, (1965).

[7] T. Kato, "Wave operators and similarity for some non-self-adjoint operators", Math. Ann. 162 (1966), 258-279.

[8] R.J. Iorio, Jr. and M.O'Carroll, Asympotic completeness for multi-particle Schrodinger operators with weak potentials," Comm. Math. Phys. 27 (1972), 137-145.

[9] T. Kato and S.T. Kuroda, "The abstract Theory of scattering," Rocky Mountain J. of Math., 1, (1971) 127-171.

DECAY AND ASYMPTOTICS FOR WAVE EQUATIONS WITH DISSIPATIVE TERM
==

Kiyoshi MOCHIZUKI

Yoshida College, Kyoto University
Kyoto, Japan

The purpose of this report is to consider the asymptotic behavior as $t \to \infty$ of solutions $w = w(x, t)$ for the wave equation

$$(1) \qquad w_{tt} - \Delta w + c(x, t)|w_t|^{\rho-1}w_t = 0 \qquad (t \geq 0)$$

in R^n $(n \geq 2)$, where $w_t = \partial w/\partial t$, $w_{tt} = \partial^2 w/\partial t^2$, Δ is the Laplacian in $x \in R^n$, $\rho \geq 1$ and $c(x, t)$ satisfies the following conditions:

(C.1) $c(x, t) \geq 0$ and is bounded measurable in x and t.

(C.2) $|c_t(x, t)| + |\nabla c(x, t)| \leq \text{const } c(x, t)$ (∇ is the gradient).

In what follows, we use the following notations: L^p $(p \geq 1)$ is the space of measurable functions in R^n such that

$$\|f\|_{L^p} = (\int_{R^n} |f(x)|^p dx)^{1/p} < \infty ;$$

H^k $(k=1, 2, \cdots)$ is the Sobolev space, that is, for functions $f \in H^k$

$$\|f\|_{H^k} = (\sum_{|\alpha| \leq k} \int_{R^n} |\nabla^\alpha f(x)|^2 dx)^{1/2} < \infty,$$

where α are the multi-indices; E is the space of pairs $f = \{f_1, f_2\}$ of functions such that

$$\|f\|_E = (\frac{1}{2} \int_{R^n} \{|\nabla f_1(x)|^2 + |f_2(x)|^2\} dx)^{1/2} < \infty;$$

If X is a Banach space and I is a real closed interval, then by $L^p(I; X)$ we mean the space of strongly measurable functions f on I into X such that

$$\int_I \|f(t)\|_X^p dt < \infty \qquad \text{if} \quad 1 \leq p < \infty$$

and

$$\text{ess sup } \{\|u(t)\|_X, \; t \in I\} < \infty \qquad \text{if} \quad p=\infty .$$

Following the definition given by Lions-Strauss [3], we shall say that $w(t)$ is a solution on $[0, \infty)$ of equation (1) if

(i) $w(t) \in L^{\infty}([0, \infty); E)$, $w_t(t) \in L^{\rho+1}([0, \infty); L^{\rho+1})$

(ii) $w(t)$ <u>satisfies the equation</u>

$$(2) \quad \int_0^{\infty} \{ -(w_t(t), v_t(t))_{L^2} + (\nabla w(t), \nabla v(t))_{L^2} + b(t, w_t(t), v(t)) \} dt = 0$$

$$b(t, w_t(t), v(t)) = \int_{R^n} c(x, t) |w_t(x, t)|^{\rho-1} w_t(x, t) \overline{v(x, t)} \, dx$$

<u>for any test function</u> $v(t) \in L^1([0, \infty); E) \cap L^{\rho+1}([0, \infty); L^{\rho+1})$ <u>with compact support in</u> $(0, \infty)$.

The existence and uniqueness theorem for the initial value problem of equation (1) is studied in [3] (cf., also Amerio-Prouse [1]) and can be summarized as follows:

<u>Theorem</u> (Lions-Strauss). <u>Suppose</u> <u>that</u> $c(x, t)$ <u>satisfies</u> (C.1) <u>and</u> (C.2). <u>Then, given arbitrary initial conditions</u>

$$w(0) = w_1 \in H^2, \quad w_t(0) = w_2 \in H^1 \cap L^{2\rho},$$

(1) <u>has a unique solution</u> $w(t)$ <u>which also satisfies</u>

(iii) $w(t) \in L^{\infty}([0, T]; L^2)$ <u>for any</u> $T > 0$,

(iv) $w_{tt}(t)$, $\Delta w(t)$, $\nabla w_t(t) \in L^{\infty}([0, \infty); L^2)$.

Further, by a theorem of Strauss [4], we observe

(v) $w(t)$ <u>is</u> E-<u>valued continuous function of</u> t <u>and satisfies the energy equation</u>

$$(3) \quad \|w(t)\|_E^2 + \int_0^t b(\tau, w_t(\tau), w_t(\tau)) d\tau = \|w(0)\|_E^2.$$

We define the energy in the region $\Omega \subset R^n$ at time t for the solution as

$$\|w(t)\|_{E,\Omega}^2 = \frac{1}{2} \int_{\Omega} (|\nabla w(t)|^2 + |w_t(t)|^2) \, dx.$$

Then we can prove the following theorem which asserts the local energy decay of the solution $w(t)$.

<u>Theorem 1.</u> <u>Suppose</u> <u>that</u> $c(x, t)$ <u>satisfies</u> (C.1), (C.2) <u>and</u>

(C.3) $\quad c(x, t) \leq \mathrm{const}(1+|x|)^{-\gamma}, \quad \gamma + \frac{1}{2}(n-1)(\rho-1) > 1 \quad (\rho \geq 0),$

where "const" is independent of t. Then
(a) for any finite region Ω in R^n

$$\|w(t)\|_{E,\Omega} \to 0 \quad \text{as} \quad t \to \infty ,$$

(b) there exists a (non-empty) class of initial data $\{w_1, w_2\} \in H^2 \times (H^1 \cap L^{2\rho})$ for which the total energies $\|w(t)\|_E^2$ of solutions do not decay.

Let $w_0(t)$ be the solution of the equation

(4) $\quad w_{0tt} - \Delta w_0 = 0 \quad \text{in} \quad R^n$

with initial data $\{w_0(0), w_{0t}(0)\} \in C_0^\infty(R^n) \times C_0^\infty(R^n)$. Then it follows that

(5) $\quad w_{0t}(t) \in L^1([0, T]; E) \cap L^{\rho+1}([0, T]; L^{\rho+1}),$

$\quad\quad w_{0tt}(t), \Delta w_0(t) \in L^1([0, T]; L^2) \quad$ for any $T > 0.$

Thus we have from (2) and (iv)

(6) $\quad (w(t), w_0(t))_E + \frac{1}{2} \int_0^t b(\tau, w_t(\tau), w_{0t}(\tau)) d\tau = (w(0), w_0(0))_E$

for any $t > 0$.

The above theorem can be proved based on this equation and the following two lemmas.

Lemma 1. Let $w_0(t)$ be as given above. Then we have

(7) $\quad \|w_0(t)\|_E = \|w_0(0)\|_E \quad$ for any $t \in R^1,$

(8) $\quad (w_0(t), f)_E \to 0 \quad \text{as} \quad t \to \infty \quad$ for any $f \in E,$

(9) $\quad \int_0^\infty b(t, w_{0t}(t), w_{0t}(t)) dt < \infty .$

Lemma 2. The set $\{w(t); t \geq 0\}$, where $w(t)$ is the solution of (1), is pre-compact with respect to each local energy norm $\|\cdot\|_{E,\Omega}$ (Ω being bounded in R^n).

Next, we restrict ourselves to equation (1) in R^3 with ρ such that $1 \leq \rho \leq 3$. In this case, the solution $w(x, t)$ satisfies the equation

$$(10) \qquad w(x, t) = w_0(x, t) -$$

$$- \frac{1}{4\pi} \int_0^t \int_{|x-y|=t-\tau} \frac{1}{t-\tau} c(y, \tau) |w_t(y, \tau)|^{\rho-1} w_t(y, \tau) dS_y d\tau.$$

Here $w_0(x, t)$ is the solution of (4) with initial data $w_0(0) = w(0)$ and $w_{0t}(0) = w_t(0)$.

Using this equation, we can prove the

Theorem 2. Suppose that $c(x, t)$ satisfies (C.1), (C.2) and

(C.4) $\quad c(x, t) \le \text{const}(1+|x|)^{-\gamma}, \qquad \gamma > 4-\rho.$

Then every solution $w(t)$ of (1) in R^3 satisfies $\|w(t)\|_{L^2} = O(1)$ as $t \to \infty$ and there exists a solution $w_0^+(t)$ of equation (4) such that

$$\|w(t) - w_0^+(t)\|_{L^2} \to 0 \quad \text{as} \quad t \to \infty.$$

Remark. If $w_t(t)$ and $w_{tt}(t)$ belongs to $L^{\rho+1}([0, T]; L^{\rho+1})$ for any $T > 0$, then we see that $w_0^+(t)$ is E-valued continuous function of t and

$$\|w(t) - w_0^+(t)\|_E \to 0 \quad \text{as} \quad t \to \infty.$$

Corollary. Suppose that $\rho = 1$ (that is, (1) is a linear equation) and $c(x, t)$ satisfies (C.1), (C.2) and (C.4). Then for every solution $w(t)$ of (1), there exists a solution $w_0^+(t)$ of (4) such that $w_0^+(t)$ is E-valued continuous function of t and

$$\|w(t) - w_0^+(t)\|_E \to 0 \quad \text{as} \quad t \to \infty.$$

The operator which maps $w(t)$ to $w_0^+(t)$ is called the wave operator. The above corollary guarantees the existence of the wave operator which is contraction in the energy space E.

Finally we remark that the existence and properties of the wave operator have been developed by Lax-Phillips [2] for wave equation: $w_{tt} = \Delta w$ in an exterior domain with lossy boundary conditions: $w_n + \alpha(x)w_t = 0$, $\alpha(x) \ge 0$.

References

[1] Amerio, L., and Prouse, G., Almost-periodic functions and function-al equations, Van Nostrand Reinhold Company, 1971.

[2]   Lax, P. D., and Phillips, R. S., Scattering theory for dissipative hyperbolic systems, J. Functional Analysis $\underline{14}$, 172-235 (1973).

[3]   Lions, J. L., and Strauss, W. A., Some non-linear evolution equations, Bull. Soc. Math. France $\underline{93}$, 43-96 (1965).

[4]   Strauss, W. A., On continuity of functions with values in various Banach spaces, Pacific J. Math. $\underline{19}$, 534-551 (1966).

TREATMENT OF THREE BODY PROBLEMS IN COORDINATE SPACE

Tatuya Sasakawa
Department of Physics, Tohoku University
980-Sendai, Japan

In an attempt of solving the three-body problem in coordinate
space, we face with three difficulties ; (1) to impose the elastic
and the break-up boundary conditions at the same time, (2) to find
the Fredholm solution of the system and (3) to find a practical way
of solving the problem. In this paper, we will report how to
overcome (1) and (2). Also the analytic structure of the solution
at low energies is discussed. By analytic continuation of the low
energy formula to the complex momentum plane, the presence of
virtual state in ^{2}S and a resonance in ^{4}P are predicted. The Efimov
effect is interpreted from the analytical view point.

§1. Correct asymptotic behavior

In solving the three body problem in coordinate space, we must
impose the correct asymptotic behavior as the boundary condition.
We designate by α the pair of particles 1 and 2, r_α their relative
distance, ρ_α the distance of the particle 3 relative to the center
of mass of the pair α. The asymptotic behavior of the three body
system reads

$$\psi \sim \sum_\alpha \varphi(r_\alpha)\,[e^{i\vec{k}\cdot\vec{\rho}_\alpha} + \frac{e^{ik\rho_\alpha}}{\rho_\alpha}\,f(\theta_\alpha)] + \frac{e^{i\sqrt{E}\,R}}{R^{5/2}}\,F(r/\rho) \ , \quad (1)$$

where $\varphi(r_\alpha)$ denotes the wave function of deuteron, k the wave
number $\sqrt{(2\mu/\hbar^2)(E+|E_d|)}$ ($|E_d|$; The binding energy of deuteron),
$f(\theta_\alpha)$ the scattering amplitude of neutron-deuteron channel, R the
hyper-radius, r/ρ the hyper-angle and $F(r/\rho)$ the break-up amplitude.
The boundary condition (1) must be imposed not only for E>0 but also
for E<0 , because the Schrödinger equation says nothing about the
breaking of continuity of analyticity of the solution.

§2. How one can impose the correct boundary conditions

We demonstrate for the potential scattering how one can impose
the correct boundary condition[1]. Let us introduce the functions
$|u\rangle$ and $|w\rangle$ by

$$|u\rangle = \frac{1}{\sqrt{k}} \sin kr \ , \quad |w\rangle = \frac{1}{\sqrt{k}} e^{ikr} \quad . \tag{2}$$

The Lippmann-Schwinger equation reads

$$|\phi\rangle = |u\rangle + G_0 V |\phi\rangle \ , \tag{3}$$

where G_0 is the Green's function

$$G_0 = \frac{1}{E - H_0 + i\xi} = - |w\rangle\langle u| + g. \tag{4}$$

In Eq.(4), g is the real Green's function which vanishes at large distances from the origin

$$g \ ; \ \text{real}, \quad g \xrightarrow[r\to\infty]{} 0 \quad . \tag{5}$$

Let T stand for the scattering amplitude defined by

$$T = \langle u|V|\phi\rangle \quad . \tag{6}$$

Eq.(3) reads then

$$|\phi\rangle = |u\rangle \ - \ |w\rangle T + gV|\phi\rangle = \omega \left[|u\rangle - |w\rangle T \right] \quad , \tag{7}$$

where

$$\omega = (1 - gV)^{-1} \quad . \tag{8}$$

Since

$$\omega \xrightarrow[r\to\infty]{} 1 \quad , \tag{9}$$

the asymptotic form of $|\phi\rangle$ is

$$|\phi\rangle \sim |u\rangle - |w\rangle T \quad . \tag{10}$$

To obtain the scattering amplitude in a solved form, we put Eq.(7) in Eq.(6)

$$T = \langle u|V|\phi\rangle = \langle u|V\omega|u\rangle - \langle u|V\omega|w\rangle T \quad .$$

Therfore,

$$T = \frac{\langle u|V\omega|u\rangle}{1 + \langle u|V\omega|w\rangle} = \frac{\operatorname{Im} J(k,0)}{J(k,0)} \quad . \tag{11}$$

$J(k,0)$ is known as the Jost function. $J(k,0)$ implies the properties:[2]

(1) $J(k,0) = \det \left[1 - G_0 V \right]$. Therefore, Eq.(7) is the Fredholm solution of Eq.(3).

(2) $J(k,0)$ has the correct analytic property that it is the entire function on the upper half of the complex k-plane.

(3) The S-matrix defined by

$$S = 1 - 2iT \tag{12}$$

satisfies the unitarity.

(4) The iteration of Eq.(8) ;

$$= 1 + gV + gVgV + \cdots\cdots \qquad (13)$$

converges without regards to the magnitude of the potential, if it is local.

(5) Since $\omega|u\rangle$ or $\omega|w\rangle$ satisfies the Volterra equation, e.g.

$$\omega|u\rangle = |u\rangle - \frac{1}{k} \int_r^\infty \sin k(r-r')V(r')\omega|u(r')\rangle\, dr' \quad , \quad (14)$$

it is very easily solved numerically.

§3. Faddeev equation and Alt-Grassberger-Sandhas equation

Faddeev[3] has shown that the amplitude of the process (three free particles) to (three free particles) satisfies the equation

$$T = T_\alpha + T_\beta + T_\gamma \quad ,$$

$$T_\alpha = t_\alpha + t_\alpha G_0(T_\beta + T_\gamma) \quad . \qquad \text{(The Faddeev equation)}$$

Here t_α is the two-body scattering matrix in the three-body space. Also Faddeev has shown that the amplitude of the process (n-d) to (n,d) + (break up) is the residue of the Faddeev equation. The equation for (n,d) to (n,d) is

$$U = G_0^{-1}\, \overline{I} + \overline{I}tG_0 U \quad , \qquad (15)$$

where

$$\overline{I} = \begin{pmatrix} 0 & 1 & 1 \\ 1 & 0 & 1 \\ 1 & 1 & 0 \end{pmatrix} \quad , \qquad t = \begin{pmatrix} t_\alpha & 0 & 0 \\ 0 & t_\beta & 0 \\ 0 & 0 & t_\gamma \end{pmatrix}$$

The amplitude for (n,d) to (break up) is then

$$T = tG_0 U \qquad (16)$$

Eqs.(15) and (16) are called A-G-S equation[4].

§4. Fredholm solution of three-body problem with correct boundary conditions

We adopt the same method as in §2 to the three-body problem.

(I) **Preliminary**   Let $|\phi\rangle$ and V stand for matrices

$$|\phi\rangle = \begin{pmatrix} \phi_\alpha \\ \phi_\beta \\ \phi_\gamma \end{pmatrix} \quad , \qquad V = \begin{pmatrix} V_\alpha & 0 & 0 \\ 0 & V_\beta & 0 \\ 0 & 0 & V_\gamma \end{pmatrix} \quad . \qquad (17)$$

Then the three-body Schrödinger equation reads

$$(E - H_0 - V)|\phi\rangle = V\bar{1}|\phi\rangle . \tag{18}$$

Let $|u\rangle$ ($|w\rangle$) be the product of the deuteron wave function and the plane wave (outgoing wave) of the incoming (outgoing) particle. Let G be

$$G = (E - H_0 - V + i\varepsilon)^{-1} . \tag{19}$$

We define $\tilde{t}$ by

$$t = -V|w\rangle\langle u|V + \tilde{t} . \tag{20}$$

We can show that $\tilde{t}$ satisfies

$$\tilde{t} = V + VG_0\tilde{t} , \tag{21}$$

$$G_0 t = -|w\rangle\langle u|V + G_0\tilde{t} . \tag{22}$$

Let $|f\rangle$ ($|h\rangle$) be the regular solution of $(E-H_0)|f\rangle = 0$ (outgoing wave, that is irregular) in the hyper-spherical coordinate. Then we can write G_0 as

$$G_0 = -|h\rangle\langle f| + G^O , \tag{23}$$

Here G^O plays the role of g in §2. G^O is real, and vanishes at large distances from the origin in the hyper-spherical coordinate. We define a real matrix R by

$$R = V + VG^O R . \tag{24}$$

Then we can show that

$$\tilde{t} = R(1 - |h\rangle\langle f|\tilde{t}) . \tag{25}$$

(II) <u>Fredholm solution</u>   We use Eqs. (22), (23) and (25) in the Lippmann-Schwinger type equation of (18) to obtain

$$|\phi\rangle = |u\rangle + G_0 t\bar{1}|\phi\rangle = \Omega\left[|u\rangle - |w\rangle T^{(e)} - (1+G^O R)|h\rangle T^{(B)}\right] \tag{26}$$

where

$$T^{(e)} = \langle u|V\bar{1}|\phi\rangle , \quad \text{the elastic amplitude} , \tag{27}$$

$$T^{(B)} = \langle f|\tilde{t}\bar{1}|\phi\rangle = \langle f|t\bar{1}|\phi\rangle , \tag{28}$$

and

$$\Omega = \frac{1}{1 - G^O R\bar{1}} , \quad \Omega \xrightarrow[R\to\infty]{} 1 . \tag{29}$$

By virtue of (26) and (29), we see that $|\phi\rangle$ satisfies the correct asymptotic behavior

$$|\phi\rangle \sim |u\rangle - |w\rangle T^{(e)} - |h\rangle T^{(B)} . \tag{30}$$

The sum of $T^{(B)}_\alpha$, $T^{(B)}_\beta$ and $T^{(B)}_\gamma$ makes up the break-up amplitude

$$\bar{T}^{(B)} = T^{(B)}_\alpha + T^{(B)}_\beta + T^{(B)}_\gamma . \tag{31}$$

If we use Eq.(26) in Eqs.(27) and (28), we obtain

$$T^{(e)} = \langle u|J\bar{I}|u\rangle - \langle u|J\bar{I}|w\rangle\ T^{(e)} - \langle u|J|h\rangle\ \bar{T}^{(B)} \qquad (32)$$

and

$$\bar{T}^{(B)} = \langle f|(1 + \bar{I}^{-1})J\bar{I}|u\rangle - \langle f|(1 + \bar{I}^{-1})J\bar{I}|w\rangle\ T^{(e)}$$
$$- \langle f|(1 + \bar{I}^{-1})J|h\rangle\ \bar{T}^{(B)}, \qquad (33)$$

where J satisfies the A-G-S type equation

$$J = G_0^{-1}\ \frac{1}{1-G^0\bar{I}R} = G_0^{-1} + \bar{I}RG^0 J \quad . \qquad (34)$$

We can show term by term that the denominator of $T^{(e)}$ and $\bar{T}^{(B)}$ yields the Fredholm determinant.

§5. Exchange Singularity

If we neglect the break-up channel, the n-d elastic amplitude is obtained from the solution of Eq.(32) as

$$T_{n,d} = T_{\alpha,\alpha} + T_{\beta,\alpha} + T_{\gamma,\alpha} \qquad (35)$$

For the partial wave ℓ it reads

$$T_{n,d}^{(\ell)} = (\text{Im }\bar{J}_\ell)/\ \bar{J}_\ell = e^{i\,\delta_\ell}\ \sin\delta_\ell \quad , \qquad (36)$$

From which we obtain

$$\cot\delta_\ell = -\ \text{Re}\bar{J}_\ell\ /\text{Im}\bar{J}_\ell \quad . \qquad (37)$$

Here

$$\bar{J}_\ell = 1 + \langle u_\alpha|J|w_\alpha\rangle_\ell + 2\langle u_\alpha|J|w_\beta\rangle_\ell \quad . \qquad (38)$$

We parametrize $\bar{J}_\ell$ as

$$\bar{J}_\ell = 1 + \bar{H}_\ell(g_\ell') + \lambda_\ell + a_\ell' z^{2\ell+1}(b_\ell' + c_\ell' z^2) \qquad (39)$$

where

$$\alpha = 0.2315 \times 10^{-13} \text{cm}^{-1} \quad , \quad \alpha = \sqrt{(m/\hbar^2)|Ed|} \quad (m : \text{nucleon mass})$$
$$z = k/\alpha \quad .$$

For S-wave $\bar{H}_0(g_0')$ reads

$$\bar{H}_0(g_0') = -(g_0'/z)[\tan^{-1}(3z/2) - \tan^{-1}(z/2)$$
$$+ 0.5\ i\ell n(1 + 9z^2/4)/(1 + z^2/4))] \quad . \qquad (40)$$

For p-wave $H_1(g_1')$ is

$$\bar{H}_1(g_1') = -(g_1'/z)[\{(1/z^2)(1+5z^2/4)(\tan^{-1}(3z/2)-\tan^{-1}(z/2))-1/z\}$$
$$+ i\{(1/2z^2)(1+5z^2/4)\ell n((1+9z^2/4)/(1+z^2/4))-1\}]\ . \qquad (41)$$

These terms are due to the exchange singlarity in $\langle u_\alpha|J|w_\beta\rangle$.

After renormalization of parameters

$$g_\ell = g_\ell'/(1 + \lambda_\ell), \quad a_\ell = a_\ell'/(1 + \lambda_\ell) \quad \text{etc.} \tag{42}$$

we obtain the low energy formula

$$k^{2\ell+1}\cot \delta_\ell = -\alpha^{2\ell+1}\frac{1 + \text{Re}\,\overline{H}_\ell\,(g_\ell) + a_\ell z^2}{\dfrac{1}{z^{2\ell+1}}\,\text{Im}\,\overline{H}_\ell(g_\ell) + b_\ell + C_\ell z^2} \tag{43}$$

On the real axis, we determine parameters form phase shifts as

	g	a	b	c
^{2}S	0.5374	-0.1015	0.6070	0.0451
^{4}P	4.0447	3.5275	-16.806	0.1229

The function $\overline{J}_\ell$ has the branch points of the logarithmic type at $-\frac{2}{3}\alpha i$ and $-2\alpha i$ and nowhere else. Solving the equation $\overline{J}_\ell = 0$, we have found the poles at $k = -0.1256i(10^{13}\text{cm}^{-1})$ for ^{2}S and at $k = 0.0326 - 0.0909\,i(10^{13}\text{cm}^{-1})$ for ^{4}P. Since the cut is due to the logarithmic singularity of $\overline{J}_\ell$, we have infinite number of poles, each one lying on each Riemann sheet. These poles accumulate at the origin when α tends to 0. This is the analytic explanation of the Efimov effect[5].

1) T.Sasakawa, Progress, Theor. Phys. Supplement <u>27</u>(1963),1.

2) R.G.Newton, Scattering Theory of Waves and Particles (McGraw-Hill, New York, 1966)

3) L.D.Faddeev, Mathematical Aspects of the Three-Body Problem (Israel Program for Scientific Translations, Jerusalem, 1965)

4) E.O.Alt, P.Glassberger and W.Sandhas, Nucle. Phys. <u>B2</u>(1967), 167.

5) V.Efimov, Soviet Journ. Nucle. Phys. <u>12</u>(1971), 589; Phys. Lett. <u>33B</u>(1970),563.

<u>LINKS BETWEEN DECAY PROPERTIES OF CORRELATIONS AND</u>

<u>ANALYTICITY OF THE PRESSURE AND CORRELATION FUNCTIONS</u>

Bernard SOUILLARD

Centre de Physique Théorique de l'Ecole Polytechnique
91120 Palaiseau - France

ABSTRACT : I will present here some results obtained in collaboration with
Michel Duneau and Daniel Iagolnitzer.

Rigorous results on "strong clustering" of the correlations are indicated
here and precise links between decay properties of correlations and analyticity properties of the pressure and correlation functions are exhibited.

I. <u>DECREASE OF CORRELATIONS - DEFINITION OF THE STRONG CLUSTER PROPERTIES</u>.

For simplicity, I will restrict myself in this talk to lattice systems with finite-range or exponentially decreasing potentials, but the results also apply to a large extend to other potentials and to continuous
systems, with some adaptations.

Let us consider the correlation functions $\rho_\Lambda(X)$ (for a finite box Λ)
and $\rho(X)$ (for the infinite volume limit) which are the probabilities of finding N particles at points $(x_1, \ldots, x_N) = X$. The truncated (or connected)
correlation functions are defined by substracting from the correlations, in
all possible ways, all the correlations involving only subgroups of points :

$$\begin{cases} \rho^T(x) = \rho(x) \\ \rho^T(X) = \rho(X) - \sum_{\pi_1 \ldots \pi_k} \prod_{i=1}^{k} \rho^T(X_{\pi_j}) \end{cases}$$

where the sum $\sum_{\pi_1 \ldots \pi_k}$ runs over all non trivial partitions of the set X into
subsets of points and X_{π_j} denotes the points of the set π_j.

Under appropriate conditions, the <u>non-connected</u> correlations are
physically expected to satisfy <u>factorization properties</u> when subsets X_1, X_2 of

points are taken apart from each other :

$$\rho(X) - \rho(X_1)\rho(X_2) \to 0 \qquad \text{like} \qquad e^{-d(X_1,X_2)}$$

The factorisation property is equivalent to the following "<u>weak</u>" <u>decrease of</u> <u>the truncated correlations</u> :

$$\rho^T(X) \to 0 \qquad \text{like} \qquad e^{-\sup_{X_i,X_j} d(X_i,X_j)}$$

where the supremum is taken over all partitions of X into two subsets X_i, X_j.

     This type of decrease has been proved since a long time in various situations (Onsager [1] when there is a gap in the spectrum of the transfer matrix, Lebowitz-Penrose[2] at low activity using series expansions, Gallavotti Miracle-Sole [3] at high temperature and arbitrary activity,...)

     However, the fall-off of the $\rho^T(X)$ is in actual fact expected to be "stronger" [4], namely to depend on the separation of all points $x_1,\ldots,x_N$ with respect to each other and not only on the separation of two clusters, and this will turn out to be of importance for the study of the links between decay of correlations and analyticity properties.

     For example, the 3-point function is expected to decrease not only with respect to $d(x_2 x_3) = \text{Sup } d(X_i X_j)$ nor even with respect to the diameter of the set (ie. $d(x_1 x_3)$), but with respect to $d(x_1 x_2) + d(x_2 x_3)$, or possibly like the following distance :

$$\text{Inf}_y [d(x_1,y) + d(x_2,y) + d(x_3,y)]$$

where y is an arbitrary supplementary vertex.

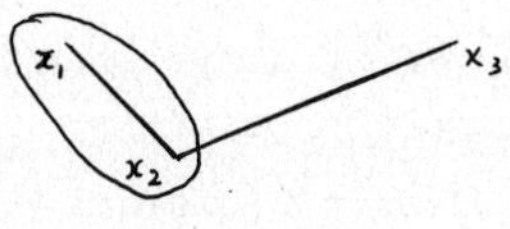

Fig. 1

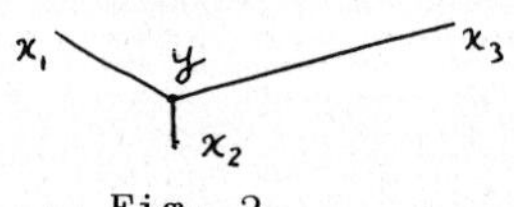

Fig. 2

     Then we say the truncated correlations satisfy a <u>strong cluster</u> <u>property</u> (S.C.P.) if there exists a function u (exponential in the case of exponentially decreasing potentials) such that :

$$\left|\rho^T_\Lambda(X)\right| < \sum_{\mathcal{T}} \prod_{\ell \in \mathcal{T}} u(\ell) = c^N \sum_{\mathcal{T}} e^{-\chi L_{\mathcal{T}}(X)}$$

where the sum $\sum$ runs over all tree $\mathcal{T}$ (i.e. graph without closed loops) over X; the product $\prod$ is taken over all lines of the tree $\mathcal{T}$, and the function u or χ and C are independent of the size of the box Λ, of the configuration X and of the number N of points of X, but depend on Φ, z, β...

Other forms of S.C.P. which express the same physical ideas are also useful :

$$|\rho_\Lambda^T(X)| < C^N \, N_1! \dots N_p! \; e^{-\chi' L(X)}$$

where the factor $N_1! \dots N_p!$ appears when the points X occupy only p different positions, $p < N$, occuring $N_1 \dots N_p$ times, and where L(X) is now the minimal length of all trees constructed on X and possibly additional points y (see Fig. 2 for an example).

A third expression of interest is :

$$|\rho_\Lambda^T(X)| < \sum_C \prod_{\ell \in C} u'(\ell)$$

where the sum Σ runs now over all <u>chains</u> constructed on the points X.

These three bounds are <u>equivalent</u> in the sense that they imply each other, with possibly a change in the rate of fall-off or in the constants.

II. <u>SITUATIONS WHEN S.C.P. HAVE BEEN PROVED</u> [5]

1) FERROMAGNETS – The S.C.P. are proved :
- outside and inside the Lee and Yang circle i.e. for $|z| < 1$ and $|z| > 1$.
- at $z = 1$ ($h = 0$) at high temperature and also as soon as the partition function $Z_\Lambda(z) \neq 0$ in some complex neighbourhood of $z = 1$ independant of Λ

2) ARBITRARY POTENTIAL – The S.C.P. are proved :
- in the Kirkwood-Salzburg region $|z| < r(\beta)$ (where the correlations are known to be analytic).
- at arbitrary z, for high temperature.
- in all the region containing $z = 0$ where $Z_\Lambda(z_1, \dots, z_\Lambda) \neq 0$ independently of $\Lambda(z_1, \dots, z_\Lambda$ being specified activities at the various points of the box Λ).
- more generally in the whole region containing $z = 0$, where the correlations are analytic with respect to z (or z and β) if some appropriate bound is known to hold. Cf. paths 1 and 2 .

Such results are proved from theorems of the following type :

<u>Theorem</u> : Let $\rho_\Lambda^T(X;\beta,z)$ be analytic in a connected domain $\mathfrak{D}$ in complex z-space containing the point $z = 0$ (or alternatively some point where S.C.P. is known to hold), and let the following bound be satisfied :

$$|z^{-N} \, \rho_\Lambda^T(X \, ; \, \beta,z)| < C^N N_1! \dots N_p!$$

with C independent of X, N, Λ and of z in $\mathfrak{D}$. Then S.C.P. hold for all z in $\mathfrak{D}$.

Another version of interest of this theorem, in which the assumption is essentially analyticity with respect to all perturbations of the field, is :

__Theorem__ : Let $\rho_\Lambda^T(X;\beta,z_1,...,z_\Lambda)$ be analytic w.r. to $z_1,...,z_\Lambda$ for all z_i in a connected domain $\mathfrak{D}$ in complex z-space containing $z = 0$, and suppose the one-point function satisfy there the bound $|\rho_\Lambda^T(x)| < C$, where C is independent of X, N, Λ and z in $\mathfrak{D}$.
Then the S.C.P. hold for all $z_i \in \mathfrak{D}$.

These theorems can be proved by using series expansions of the $\rho^T(X)$ at $z = 0$ or alternatively by the method of subharmonic functions due to Lebowitz-Penrose [7]. (However this last method does not provide detailed expression on the rate of fall-off).

Moreover some of the proofs for ferromagnets make use of the Lieb-Ruelle theorem [8].

III. FROM DECAY PROPERTIES OF CORRELATIONS TO ANALYTICITY PROPERTIES OF THE THERMODYNAMIC AND CORRELATION FUNCTIONS. AN EQUIVALENCE THEOREM.

The fact that decay properties of correlations imply regularity properties of the state in the physical region (real points) has first been emphasized by Lebowitz [9] which has proved infinite differentiability from "weak" decay properties.

The first result stated below shows that the S.C.P. (at real points) moreover imply __analyticity__. The second one shows that there is as a matter of fact an equivalence.

__1st result__ : If S.C.P. hold at some (real) point (β_0, z_0) around which the thermodynamic limit exists,
then the pressure and all correlation functions are analytic at point (β_0, z_0) with respect to the activity z and more strongly to all z_i.

__Problem__ : Let us consider the two-dimensional ferromagnetic nearest neighbour Ising model at h = 0. It is known [1] that there is a gap in the spectrum of the transfer-matrix for T larger than the critical temperature T_c. But analyticity is only known for $T > T_a > T_c$. Using the gap and various inequalities on correlations, Lebowitz [9] has proved that all the functions are infinitely differentiable for $T > T_c$. Only C^∞ is obtained because the cluster property obtained from transfer-matrix are "weak". Then is it possible to prove S.C.P. using only the gap in the spectrum of the transfer-matrix ?
This problem is also related to the question of analytic continuation through the phase transition, since it is also known [10] that there exists a gap in each pure phase, below T_c.

<u>2nd result</u> : S.C.P. at (real) points $z \in]0, z_o]$ is <u>equivalent</u> to analyticity with respect to z in some complex neighbourhood of $]0, z_o]$, plus the bounds

$$\left| \rho_\Lambda^T(X ; \beta, z) \right| < C^N \, N_1! \ldots N_p!$$

The equivalence can also be formulated with the analyticity with respect to all perturbations of the field :

$$\text{S.C.P. at real points } z \in]0, z_o] \Leftrightarrow \begin{vmatrix} \text{Analyticity w.r. to all } z_i \text{ in some} \\ \text{complex neighbourhood of }]0, z_o] \text{ plus} \\ \text{the bound } \left| \rho_\Lambda(x ; \beta, z) \right| < C. \end{vmatrix}$$

(The additional condition $\left| \rho_\Lambda(x ; \beta, z) \right| < C$ plays the role of uniformity of the analyticity of the various functions).

IV. <u>GENERALISATIONS</u>

We have presented here only results for lattice systems with exponentially decreasing interactions, and we have only considered analyticity w.r. to z or w.r. to the z_i.

♦ Part of the results apply also to systems with potentials decreasing like a power i.e. like r^{-s} ; in that case the power of the decrease of the correlations can be shown to be constant in the low activity region. One may think that it should also remain constant in all the analyticity region, but it has not been possible so far to prove this result.

♦ Results apply also to continuous systems : all results described hold also for the truncated correlation functions smeared with some appropriate test function. Most of the results can also be achieved for the non-smeared correlations, using technical improvements.

♦ The last extension of interest [11] concerns the analyticity w.r. to other variables : one is interested not only in analyticity w.r. to z or z_i, but w.r. to β or the interactions J_{ij} or other coupling constants.

To handle such analyticity properties in this framework one is led to consider S.C.P. for correlations, which are connected only with respect to subgroups of points ; for example the 4-point function connected only w.r. to the subgroups $(x_1 x_2), (x_3 x_4)$ is

$$\rho^{T\left[(x_1 x_2), (x_3 x_4) \right]}(x_1, x_2, x_3, x_4) = \rho(x_1, x_2, x_3, x_4) - \rho(x_1, x_2)\rho(x_3, x_4)$$

In that case, S.C.P. involve sums on trees between the various clusters of points.

Then one can prove also for example that :

$$\text{S.C.P.} \Leftrightarrow \text{analyticity w.r. to all } J_{ij}.$$

More generally, one would expect an assertion of the type :

$$\text{S.C.P.} \quad \Leftrightarrow \quad \text{analy. w.r. to all perturbations of}$$

of all   (partially)   the general potential Φ (including
connected correlations   the activity terms).

REFERENCES

[1] Onsager, L., Phys. Rev. **65**, 117 (1944).

[2] Lebowitz, J.L. and Penrose, O., Commun. Math. Phys. **11**, 99 (1968).

[3] Gallavotti, G. and Miracle-Sole, S. Commun. Math. Phys. **12**, 269 (1969).

[4] Duneau, M., Iagolnitzer, D., and Souillard, B., Commun. Math. Phys. **31**, 191 (1973).

[5] .Duneau, M., Iagolnitzer, D., and Souillard, B., Commun. Math. Phys. **35**, 307 (1974).

.Duneau, M., Iagolnitzer, D., and Souillard, B., Decay of correlations for infinite-range interactions, to be published in J. Math. Phys.(1975).

[6] This method has first been used in ref. [2], [3] and has been adapted in ref. [5] to the proof of S.C.P.

[7] .Lebowitz, J.L. and Penrose, O., : Phys. Rev. Letters **31**, 749 (1973).

.Lebowitz, J.L. and Penrose, O., : Commun. Math. Phys. **39**, 165 (1974).

[8] .Lieb, E.H. and Ruelle, D., J. Math. Phys. **13**, 781 (1972).

.Ruelle, D., Commun. Math. Phys. **31**, 265 (1973).

[9] Lebowitz, J.L., Commun. Math. Phys. **28**, 313 (1972).

[10] Abraham, D. and Martin-Löf, A., Comm. Math. Phys. **32**, 245 (1973).

[11] Ref. [4] and Duneau, M.,   Souillard, B., to be published.

DISCUSSION

Lebowitz : In the low fugacity region where Groeneveld proved decay of $\rho_2^T(r)$ as r^{-s} what happens to the higher order truncated functions ?

Souillard : For potentials decreasing like r^{-s}, we can prove strong decay properties ; however the decrease obtained so far for the correlations is of the type $r^{-s/2}$ or $r^{-(s-\nu)}$, ν being the dimension of the space, according to the method used, to the form of the decay properties considered, and to the system studied.

Ikeda   : You said that, in proving your first theorem, you used series
              expansion. Would you please explain how you considered the
              problem of convergence of the series ?

Souillard   : When the correlations are analytic in some domain $\mathfrak{D}$, one may
              for instance consider, following ref. 3, the conformal mapping
              $z \rightarrow t(z)$ from $\mathfrak{D}$ to the unit circle $|t| < 1$. The series expansion
              obtained with respect to t is convergent in this circle.

Ruelle   : Do you expect that when several phases coexist , each one may
              exhibit the strong cluster property ?

Souillard   : We have the feeling that the same physical reasons as before
              should indeed imply S.C.P. in each pure phase ; however this
              would imply analytic continuations through the phase transition,
              which some physicists think unreasonable.

<u>LINEAR RESPONSE , STABILITY , CLUSTER PROPERTIES</u>

A. Verbeure; Universiteit Leuven, Belgium

In conventional statistical mechanics an equilibrium state of a finite system is given by a Gibbs state. States of infinite systems (in the thermodynamic limit) are no longer of this type. It is now widely accepted that an equilibrium state of an infinite system should be described by a state satisfying the KMS-condition[1]. It was immediately realized that a better understanding of this condition should lead to better support of this idea. For classical and quantal lattice systems it is proved that the KMS-condition is equivalent with a variational principle.[2] Recently the KMS-condition has been linked to a notion of stability.[3] Meanwhile, this notion of stability has been studied also for classical infinite systems,and for finite classical and finite quantum systems.[4] All these notions of stability, have technically this in common, that they contain a condition on the existence of a new perturbed equilibrium state (existence of a time limit) and a condition on the two-point function, hence, a condition on the linear response functions.

This was our motivation to review linear response theory from the point of view of KMS-states. At the same time we generalize to infinite equilibrium states rigorously some aspects of linear response theory, as introduced by KUBO and MORI[5].

The following is based on joint work together with J. Naudts and R. Weder[6], where all details can be found.

Let $\mathcal{A}$ be the C*-algebra of observables and $t \to \alpha_t^o$ a strongly continuous group of * - automorphisms. Let ω_o be a state on $\mathcal{A}$, satisfying the KMS-condition [1] at a temperature $\beta = 1/kT = 1$. With respect to the evolution α_t^o .

Consider the GNS-triplet $(\pi, \mathcal{H}, \Omega)$ induced by the state ω_o, i.e. π is a cyclic representation of $\mathcal{A}$ on the bounded operators $\mathcal{B}(\mathcal{H})$ of the Hilbert space $\mathcal{H}$

with cyclic vector Ω , such that

$$\omega_o(x) = (\Omega, \pi(x)\,\Omega) \qquad\qquad x \ \varepsilon \ \mathcal{A}$$

$$\tau_t^o \ \pi(x) = e^{itH_o} \ \pi(x) \ e^{-itH_o} = \pi(\alpha_t^o(x))$$

H_o is a self-adjoint operator on $\mathcal{H}$.

In a standard way we extend the KMS-condition to the von Neumann algebra

$\mathcal{M} = \pi(\mathcal{A})''$ and we get the following result.

The vector state $\omega_o(x) = (\Omega, x\,\Omega)$, $x \ \varepsilon \ \mathcal{M}$ satisfies the KMS-condition at inver-

se temperature $\beta = 1$ i.e. for any pair (x,y) of $\mathcal{M}$, there exists a complex

function $F_{x,y}(z)$, defined, bounded and continuous on the strip $-1 \leqslant \text{Im} z \leqslant 0$

and analytic inside, with boundary values: $F_{xy}(t) = \tilde{\omega}_o(x_t y)$,

$F_{xy}(t-i) = \tilde{\omega}_o(y\,x_t)$ where $x_t = \tau_t^o x$

It follows also that the cyclic vector Ω is separating for the von Neumann al-

gebra $\mathcal{M}$ and that $\tilde{\omega}_o$ is stationary i.e. $\tilde{\omega}_o(x) = \tilde{\omega}_o(\tau_t^o(x))$ for all $x \ \varepsilon \mathcal{M}$ and

$t \ \varepsilon \ R$. Furthermore there exists an operator Δ , called the modular operator,

such that, if $H_o = \int_{-\infty}^{\infty} \lambda \, d\,E(\lambda)$ is the spectral resolution of the Hamiltonian H_o,

then $\Delta = \int_{-\infty}^{\infty} e^{-\lambda} dE(\lambda)$.

Let us now consider a perturbation of the evolution . Take $V = V^* = \pi(y)$ for some

$y \ \varepsilon \mathcal{A}$, then a new evolution can be defined $\tau_t^V(x) = e^{itH} x \, e^{-itH}$ for $x \ \varepsilon \ \pi(\mathcal{A})$

and where $H = H_o + \lambda V$, $\lambda \ \varepsilon \ R$.

It can easily be shown that the following sequence converges uniformly

$$\tau_t^V(x) = \alpha_t^o(x) + \sum_{n \geqslant 1} (\pm i)^n \lambda^n \int \cdots \int_{0 \leqslant s_n \leqslant \cdots \leqslant s_1 \leqslant t} ds_1 \ldots ds_n$$

$$[\tau_{s_1}^o(v) , [\cdots [\tau_{s_n}^o(v) , \tau_t^o(x)] \cdots]$$

where $\pm$ depending on $t \lessgtr 0$.

We are interested in the limit $\lim_{t \to \pm\infty} \tilde{\omega}_o(\tau_t^V(x))$ for all $x \ \varepsilon \ \pi(\mathcal{A})$

$V = V^* \varepsilon \ \pi(\mathcal{A})$ up to the first order in λ , i.e. in linear approximation or in

linear response. Hence we are interested in

$$\lim_{t\to\pm\infty} \tilde{\omega}_o(\tau_t^{V,1}(x)) = \lim_{t\to\pm\infty} \{ \tilde{\omega}_o(x)$$

$$\pm\ i\lambda \int_0^t ds\ \tilde{\omega}_o([V_s, x_t])\ \}$$

From this formula it is seen that the study is reduced to the study of the following functions: for $x,y\ \varepsilon\ \mathcal{M}$

- response function: $\phi_{xy}(t) = \tilde{\omega}_o([x_t,y])$

- correlation function: $\Psi_{xy}(t) = \tilde{\omega}_o(\{x_t,y\}_+)$

- relaxation function: $\Phi_{xy}(t) = \lim_{\varepsilon\to o^+} i \int_t^\infty \phi_{xy}(t')e^{-\varepsilon t'}\ dt'$

       if the limit exists

- admittance: $\chi_{xy}(z) = i \int_0^\infty dt\ e^{\mp izt}\ \phi_{xy}(t)$; $Imz \lessgtr 0$.

    All this functions are bilinear functionals on $\mathcal{M}$ and if we want to prove some properties on the dynamical system $(\mathcal{R}, \alpha_t^o, \omega_o)$ we must prove the property for all x,y of $\mathcal{R}$ or all x,y of $\mathcal{M}$. Therefore we look for an operator representation of these functions, which is particularly simple on a Hilbert space which we construct now.

Define the unbounded, positive, self-adjoint operator T on $\mathcal{H}$ by

$$T = \int_{-\infty}^\infty (\frac{1 - e^{-\lambda}}{\lambda})^{1/2}\ dE(\lambda)$$

As $\mathcal{M}\Omega \subset \mathcal{D}(T)$ domain of T, we define the following sesquilinear form on $\mathcal{M}$

$$(x,y)_\sim = (Tx\Omega, Ty\Omega)$$

As zero is not an eigenvalue of T and the cyclic vector Ω is separating, this form is non-degenerated, hence it is a scalar product.

Denote by $\tilde{\mathcal{H}}$ the closure of $\mathcal{M}$ with respect to this scalar product.

The space $\mathcal{H}$ may look rather ad hoc, however it is a straightforward generalization to KMS-states of the one introduced by KUBO, MORI and others in linear response theory. We list some of its properties:

(i)   for $x,y\ \varepsilon\ \mathcal{M}$

$$i \frac{d}{dt} (y_t,x)_\sim = \tilde{\omega}_o([y_t,x])$$

(ii)　　$(\dot{y},x)_{\sim} = \displaystyle\int_{-1}^{0} F_{y^{*},x}(it)\, dt$

where　$F_{xy}(z) = \begin{cases} \tilde{\omega}_{o}\,(x\ \ \overset{o}{\tau}_{-z}\ (y)\,)\ \text{for}\ \ 0 \geqslant \mathrm{Im}z \geqslant -\,1/2 \\[4pt] \tilde{\omega}_{o}\,(y\ ,\overset{o}{\tau}_{1+z}\ (x)\,)\ \text{for}\ -\,1/2 \geqslant \mathrm{Im}z \geqslant -\,1 \end{cases}$

(iii)　The Bogoliubov inequality for KMS-states can be seen as the Schwartz-
inequality with respect to the scalar product of $\widetilde{\mathcal{H}}$.

A first result is given by

Theorem 1.-

The operator U from $\widetilde{\mathcal{H}}$ into $\mathcal{H}$, defined by $Ux = Tx\Omega$ for all $x\ \varepsilon\ \mathcal{M}$,
extends to a unitary operator from $\widetilde{\mathcal{H}}$ to $\mathcal{H}$.

Hence $\mathcal{H}$ and $\widetilde{\mathcal{H}}$ are unitarily equivalent. Let $\hat{H}_{o} = U^{-1} H_{o}U$ then for $x\ \varepsilon\ \widetilde{\mathcal{H}}$:
$x_{t} = e^{it\hat{H}_{o}}x$; $\hat{H}_{o}$ is the image of the Hamiltonian H_{o} under the unitary map and
spectrum H_{o} = spectrum $\hat{H}_{o}$.

In view of this result the following operator representation is interesting:

$\phi_{x^{*}y}(t) = (x,\, e^{-it\,\hat{H}_{o}}\,\hat{H}_{o}y)_{\sim}$　　　　　$x\ \varepsilon\ \mathcal{M},\ y\ \varepsilon\ \mathcal{D}\ (\hat{H}_{o})$

$\psi_{y^{*}x}(t) = (y,\, e^{-it\,\hat{H}_{o}}\,C(\hat{H}_{o})x)_{\sim}$　　　　$y\ \varepsilon\ \mathcal{M},\ x\ \varepsilon\ \mathcal{D}\,(C(\hat{H}_{o})\,)$

where　$C\,(x) = x\,\coth\dfrac{x}{2}$

$\chi_{y^{*}x}(z) = (y\ ,\, \hat{H}_{o}\,\hat{R}(-z)\,x)_{\sim}$　　　　$x,y\ \varepsilon\ \mathcal{M}\ ,\ \mathrm{Im}z\ \neq\ 0$
　　$\hat{R}(-z) = (\hat{H}_{o} + z)^{-1}$

$\Phi_{x^{*}y}(t) = \lim_{\varepsilon \to 0^{+}}\ (x,\, i\, e^{-t(\varepsilon+i\,\hat{H}_{o})}\ \dfrac{\hat{H}_{o}}{\varepsilon + i\,\hat{H}_{o}}\, y)_{\sim}\ ;\ x,y\ \varepsilon\ \mathcal{M}.$

Furthermore we have also the existence of the relaxation function for all observables . The following theorem tells that the limit $\varepsilon \to 0$ always exists and it
gives also the explicit form of the function.

Theorem 2.-

Let $\hat{E}_{o}$ be the orthogonal projection on the null space of $\hat{H}_{o}$, then for all $x, y \ \varepsilon \ \mathfrak{M}$:

 (i) $\phi_{x \ y}(t) = (x, e^{-it\hat{H}_o} (1 - \tilde{\hat{E}}_o) \ y \ )_{\sim}$

 (ii) $\phi_{xy}(0) = \chi_{xy}(0)$

  where $\chi_{xy}(o) = \lim_{Z \to 0^+} \chi_{xy}(z)$

For gibbs states or thermal averages it is well known that there exists relations among these functions. They are usually expressed as fluctuation-dissipation theorem. A number of these relations are expressed in the following proposition for general KMS-states.

<u>Theorem 3.</u>

 (i) For $x, y \ \varepsilon \mathfrak{M}$, $t \ \varepsilon \ R$, then

$$i \frac{d}{dt} \ \overline{\Psi}_{x,y}(t) = \phi_{xy}(t)$$

 (ii) For $y \ \varepsilon \mathfrak{M}$, $x \ \varepsilon \mathfrak{D}(\hat{H}_o)$

$$(y, \ C \ (\hat{H}_o) \ \text{th} \ \frac{\hat{H}_o}{2} \ x)_{\sim} = (y, \ \hat{H}_o \ x)_{\sim}$$

 (iii) For all $x, y \ \varepsilon \ \mathfrak{M}$

$$\phi_{xy}(0) = - i \ P \int_{-\infty}^{\infty} dt \ \Psi_{xy}(t) \ \frac{2}{e^{\pi t} - R^{-\pi t}} \ ; \ (P \text{ is the principal value})$$

Finally we call a dynamical system $(\mathfrak{A}, \ \omega_o, \ \overset{\circ}{\alpha}_t)$ stable in linear response if the limit $\omega_{\pm}^{V,1}(x) = \lim_{t \to \pm\infty} \ \omega_o(\tau_t^{V,1} \ (x) \ )$ exists for all x and $V = V^*$ of $\mathfrak{A}$. If the system is stable in linear response, then $\lim_{\lambda \to 0} \ \omega_{\pm}^{V,1}(x) = \omega_o(x)$ for all $x, V = V^* \ \varepsilon \mathfrak{A}$, which is in agreement with the global definition of stability. Furthermore we have the following properties.

<u>Theorem 3.</u>

 (i) If the system $(\mathfrak{A}, \ \omega_o, \overset{\circ}{\alpha}_t)$ is stable, then $\omega_{\pm}^{V,1}(x) = \omega_o(x) \mp \lambda(x^*, (1 - \tilde{\hat{E}}_o)V)_{\sim}$

 (ii) If zero is the only eigenvalue of $\hat{H}_o$ and the singular continuous spectrum of $\hat{H}_o$ is empty, then the system is stable.

(iii)   A dynamical system ($\mathcal{O}$, ω_o, α_t^o) is strongly clustering if and only if it is weakly clustering and stable in linear response.

Where the system is called weakly clustering if for any mean $\mathcal{M}_t$ over t $\mathcal{M}_t$ $\omega_o(x\ \alpha_t^o(y)\ ) = {}^\omega{}_o(x)\ {}^\omega{}_o(y)$; $x,y\ \varepsilon\ \mathcal{O}$ and strongly clustering if $\lim_{|t|\to\infty}\ \omega_o(x\ \alpha_t^o\ (x)\ ) = \omega_o(x)\ \omega_o(y)$ for $x,y\ \varepsilon\ \mathcal{O}$.

References

[1]   R. Haag, N.M. Hugenholtz, M. Winnink ,Comm. Math. Phys.$\underline{5}$,215(1967)

[2]   e.g. D. Ruelle; Comm. Math. Phys. $\underline{5}$, 324 (1967)
        D.W. Robinson; Comm. Math. Phys. $\underline{7}$, 337 (1968) .
        H. Araki; Comm. Math. Phys. $\underline{38}$, 1 (1974).

[3]   R. Haag, D. Kastler, E. Trych-Polmeyer ,
        Comm. Math. Phys. $\underline{38}$, 173 (1974).

[4]   J. Lebowitz, private communication.

[5]   R. Kubo, J. Phys. Soc. Japan, $\underline{12}$, 570 (1957)
        H. Mori, Progr. Theor. Phys. $\underline{33}$, 423 (1965).

[6]   J. Naudts , A. Verbeure, R. Weder; Linear response theory and the
        KMS-condition; preprint KUL-Leuven , Belgium, november 1974.

        A. Verbeure, R. Weder; Stability in Linear response theory and
        cluster properties; preprint KUL-Leuven, Belgium, november 1974.

Acknowledgement

Thanks are due to H. Borchers and M. Winnink for pointing out that the presentation should benefit from taking the algebra of observables a C^*-algebra.

EQUILIBRIUM AND METASTABLE STATES OF CLASSICAL SYSTEMS

G.L.   Sewell
Department of Physics
Queen Mary College

Mile End Road
London E.1.

The object of this note is to formulate a characterisation of both equilibrium and metastable states of classical systems in terms of certain global and local stability conditions. It will be found that, according to this characterisation, a class of systems with appropriately weakly tempered or long range forces can support metastable states. This result may be regarded as complementary to that of Lanford and Ruelle $[1]$, concerning the absence of metastable states in systems with suitably strongly tempered forces. A preliminary version of the contents of the present note is to be found in Ref. $[2]$.

We shall restrict our formulation here to classical lattice systems: hard-core continuous systems may be similarly formulated within the scheme of Ref. $[3]$.

Let Σ be an assembly of identical, mutually interacting particles on a lattice $T = Z^d$, each site of which is occupiable by at most one particle. The states and forces in Σ may be specified within the framework of Ref. $[4]$. Accordingly, we represent a particle configuration for Σ by a subset x of T, consisting of the occupied sites. The family X of all subsets of T thus corresponds to a phase space for Σ. For $t \in T$, we define the cylinder set $\sigma_t = \left\{ x \in X \mid t \in x \right\}$, and equip X with the topology generated by $\left\{ \sigma_t, \ X \backslash \sigma_t \mid t \in T \right\}$, thereby rendering X compact. We define Y to be the subspace of X whose elements are finite point subsets of T; and we define the state space, Ω, for Σ to be the set of all Radon probability measures on X. Space translations may be represented in X, Y, Ω in obvious fashion. We denote by Ω_T the set of translationally invariant states of Σ.

Let A be the set of all finite point subsets of T. For each $\alpha \in A$, we denote $T \backslash \alpha$ by α_c. We define an equivalence relation $\mathcal{R}(\alpha)$ in X by specifying that $x \mathcal{R}(\alpha) x'$ means that $x \cap \alpha = x' \cap \alpha$; and define $X_\alpha =$

$X/\mathcal{R}(\alpha)$. Thus, each element x_α of X_α is a cylindrical set in X which corresponds to a unique particle configuration in α. We define $\mathcal{B}(\alpha_c)$ to be the σ-algebra generated by the family of cylindrical sets $\{x_{\alpha'} \in X_{\alpha'} \mid A \ni \alpha' \subset \alpha_c\}$. We denote the restriction of $\omega(\in \Omega)$ to $\mathcal{B}(\alpha_c)$ by ω_{α_c}. If g is a semi-bounded Borel function from X into the extended real line, we denote by $E_\omega(g/\alpha_c)$ the conditional expectation of g with respect to $\mathcal{B}(\alpha_c)$ for the state ω. If χ_B is the characteristic function for a Borel set B in X, we denote $E_\omega(\chi_B/\alpha_c)$ by $\omega(B/\alpha_c)$.

Let $\mathcal{F}$ be the set of continuous, translationally invariant functions ϕ on Y, such that $\phi(\varnothing) = 0$; and let $\|\cdot\|$ and $\|\cdot\|_1$ be the norms on $\mathcal{F}$ defined by the equations

$$\|\phi\| \equiv \sum_{o \in y} \frac{|\phi(y)|}{N(y)} \quad ; \quad \|\phi\|_1 \equiv \sum_{o \in y} |\phi(y)| \quad , \tag{1}$$

where $N(y)$ is the number of sites in y. We define Φ (resp. Φ_1) to be the Banach space $\{\phi \in \mathcal{F} \mid \|\phi\| \ (\text{resp. } \|\phi\|_1) < \infty\}$. Thus $\Phi_1 \subset \Phi$. An interaction potential for Σ is taken to be an element ϕ of Φ, with the interpretation that the potential energy of a finite system of particles occupying the point set y is $\sum_{y' \subset y} \phi(y')$. It follows from

the definition of Φ that such interactions are both stable and tempered. It may be generally assumed that the chemical potential $\mu \ (\in R)$ is absorbed into ϕ, i.e.

$$\phi(y) = \phi'(y) - \mu \, \delta_{1, N(y)} \quad , \tag{2}$$

where $\phi' \ (\in \Phi)$ is independent of μ.

For each $\phi \in \Phi$, we define a real valued function H_α^ϕ on X, representing the energy of interaction between particles in α, by the formula

$$H_\alpha^\phi(x) = \sum_{y \subset x} \phi(y). \tag{3}$$

We define Φ_2 to be the set of elements ϕ of Φ such that, for each $\alpha \in A$, there exists a lower-bounded Borel function $\tilde{H}_\alpha^\phi : X \to R \cup \{\infty\}$ representing the energy of interaction of the particles in α both with one another and with those in α_c, as specified by the following conditions.

(a) $\tilde{H}_\alpha^\phi(x \cap a) - \tilde{H}_\alpha^\phi(x' \cap a) = \left[\sum_{y \subset x \cap a; \ y \cap \alpha \neq \varnothing} - \sum_{y \subset x' \cap a; \ y \cap \alpha \neq \varnothing} \right] \phi(y)$

$$\forall A \ni a \supset \alpha, \ x \cap \alpha_c = x' \cap \alpha_c \tag{4}$$

(b) $\quad \tilde{H}_\alpha^\phi (x) = \lim_{a \to \infty} \tilde{H}_\alpha^\phi (x \cap a) \quad \forall\ x \in X,$ \hfill (5)

the limit being taken over any increasing absorbing sequence of finite
point subsets of T.

(c) For each $\quad x_c \subset \alpha_c$, $\quad \inf_{x \cap \alpha_c = x_c} \tilde{H}_\alpha^\phi (x) < \infty.$

The conditions (a) – (c) specify $\tilde{H}_\alpha^\phi$ to within a class of functions
which differ from one another by bounded $\mathcal{B}(\alpha_c)$ – measurable functions
on X. It may easily be seen that the theory which follows is independent
of the choice of $\tilde{H}_\alpha^\phi$ from this class. It may easily be verified that

$$\Phi \supset \Phi_2 \supsetneq \Phi_1.$$

We define the (global) free energy density functional $f_\theta^\phi : \Omega_T \to R$,
corresponding to the interaction potential ϕ $(\in \Phi)$ and temperature
θ $(\in R_+)$ by the formula (cf $[4]$)

$$f_\theta^\phi (\omega) = \lim_{\alpha \to \infty} \frac{1}{N(\alpha)} \left[\int H_\alpha^\phi \, d\omega + k\theta \sum_{x_\alpha \in X_\alpha} \omega(x_\alpha) \ln \omega(x_\alpha) \right],$$ \hfill (6)

where k is Boltzmann's constant and the limit is taken in Fisher's
sense. For $\phi \in \Phi_2$, $\theta \in R_+$ and $\alpha \in A$, we define the conditional

free energy functional $F_{\theta, \cdot}^\phi (\alpha/\alpha_c)$ on Ω by the equation

$$F_{\theta, \omega}^\phi (\alpha/\alpha_c) = \int d\omega_{\alpha_c} \left[E_\omega (\tilde{H}_\alpha^\phi /\alpha_c) + k\theta \sum_{x_\alpha \in X_\alpha} \omega(x_\alpha/\alpha_c) \ln \omega(x_\alpha/\alpha_c) \right].$$ \hfill (7)

Thus, $F_{\theta, \omega}^\phi (\alpha/\alpha_c)$ corresponds to the free energy of the 'open'
system of particles in α , interacting via the potential ϕ both with
one another and with the particles in α_c.

<u>Definition 1.</u> We define $K_L (\theta, \phi)$, the set of locally stable states
of Σ , corresponding to the interaction ϕ $(\in \Phi_2)$ and temperature θ,
to consist of those states ω such that, if $\alpha \in A$ and $\omega' \in \Omega$, with

$$\omega'_{\alpha_c} = \omega_{\alpha_c} \quad , \qquad \text{then } F_{\theta, \omega}^\phi (\alpha/\alpha_c) \leq F_{\theta, \omega'}^\phi (\alpha/\alpha_c).$$

<u>Note.</u> Our local stability conditions are equivalent to the Dobrushin-
Lanford - Ruelle (D L R) conditions $[1,5]$, as generalised from Φ_1 to
Φ_2 - class interactions. However, for our purposes it is essential
that $K_L (\theta, \phi)$ be regarded as the set of locally stable states, whereas
in Refs. $[1, 5]$ the D L R conditions were taken to define equilibrium
states.

<u>Definition 2</u>. Let X^O be a closed subspace of X, which is stable under space translations, and let Ω^O (resp. Ω^O_T) $= \{ \omega \in \Omega$ (resp. Ω_T)| supp $\omega \subset X^O \}$.

(i) Then we define K_G (θ, ϕ) (resp K^O_G (θ, ϕ)), the set of globally stable (resp. globally X^O - stable) translationally invariant states of Σ, corresponding to the interaction ϕ and temperature θ, to consist of those elements ω of Ω_T (resp. Ω^O_T) which absolutely minimise f^ϕ_θ (resp. $f^\phi_{\theta}|\Omega^O_T$).

(ii) We define the thermodynamic functions P, P^O, from $R_+ \times R$ into R, by the formulae

$$P(\theta, \mu) = -(k\theta)^{-1} \inf_{\omega \in \Omega_T} f^\phi_\theta (\omega) \; ; \; P^O(\theta, \mu) = -(k\theta)^{-1} \inf_{\omega \in \Omega^O_T} f^\phi_\theta (\omega), \tag{8}$$

where ϕ is assumed to be expressed in terms of the chemical potential μ according to equation (2). Thus, P represents the pressure of Σ and P^O represents its pressure when constrained to the reduced phase space X^O.

<u>Note</u>. Since, as noted above, K_L (θ, ϕ) consists of the states which satisfy the D L R conditions, it follows from Ref $[1: \quad$ Theorem 3.2$]$ and Defs. 1, 2 (i) that K_G (θ, ϕ) $=$ K_L $(\theta, \phi) \cap \Omega_T$ if $\phi \in \Phi_1$.

However this is not generally true for $\phi \in \Phi \setminus \Phi_1$, as the following example shows.

<u>Example</u>. Let ϕ' be a non-positive-valued element of $\Phi \setminus \Phi_1$. Then it follows from equations (1), (2), (4), (5), together with our definitions of Φ, Φ_1 and Φ_2, that $\phi \in \Phi_2$, with

$$\tilde{H}^\phi_\alpha (x) = - \sum_{y \cap \alpha_c \subset x \cap \alpha_c \, ; \, y \not\subset x} \phi (y). \tag{9}$$

Let σ be the pure state corresponding to the configuration x^σ in which all sites are occupied; and let σ' be an arbitrary state such that $\sigma'_{\alpha_c} = \sigma_{\alpha_c}$. Then it follows from equations (1) and (9) that $E_{\sigma'} (\tilde{H}^\phi_\alpha / \alpha_c) = 0$ if $\sigma' = \sigma$ and $= \infty$ otherwise. Hence, by equation (7) and Def. 1, σ is locally stable for all values of θ and μ. On the other hand, it follows from equations (2), (6) and (7) that, for sufficiently large values of $-\mu$, the global free energy density of σ exceeds that of the state in which all sites are unoccupied. Hence, for such values of μ, ω is locally, but not globally, stable. Further (and trivially!), it is always globally $\{x^\sigma\}$ - stable.

This example reveals the possibility that systems with suitable weakly-tempered (i.e. $\Phi \setminus \Phi_1$ - class) forces can support states which are locally, but not globally, stable. Accordingly, we propose the following definitions of equilibrium and metastable states. The motivation for these definitions, as well as their implications, will be discussed below.

<u>Definition 3.</u>   (i) We define the set K_E (θ, ϕ) of translationally invariant equilibrium states of Σ, corresponding to (θ, ϕ), to be K_G $(\theta, \phi) \cap K_L$ (θ, ϕ).

   (ii) We define the set K_M^O (θ, ϕ) of translationally invariant X^O - metastable states, corresponding to (θ, ϕ), to be K_L $(\theta, \phi) \cap K_G^O$ $(\theta, \phi) \setminus K_G$ (θ, ϕ). Thus, in this definition, meta-stability is associated with some specific reduced phase space X^O.

<u>Comments.</u> We envisage that Def. 3 may lead to a theory of metastable states possessing very long lifetimes and 'good' thermodynamical be-haviour, as observed experimentally, for the following reasons.

(1) We conjecture, for want of an adequate kinetic theory, that locally stable states are dynamically stable against perturbations arising from the coupling of Σ to a thermal reservoir. In this case, the K_M^O (θ, ϕ) -class states have infinite lifetimes.

(2) By equation (8), P^O corresponds to the pressure of a system Σ^O, say, with phase space X^O and interaction potential ϕ. Thus, P^O serves to generate the thermodynamical laws for Σ^O, just as P does for Σ. Hence the K_M^O (θ, ϕ) - class states exhibit 'good' thermodynamical be-haviour, as generated by P^O.

(3) Suppose that, at temperature θ, the system Σ possesses the following properties, which are realisable in certain models (cf. Conclusion (III) below).

(a) The globally X^O - stable states are metastable or true equilibrium states, according to whether $\mu <$ or $>$ some value $\mu_o(\theta)$; and (b) P^O is analytic in μ in some open neighbourhood of $\mu_o(\theta)$. Then it follows from Defs. 2, 3 that, under these conditions, the pressure function P^O for the X^O - metastable phase is an analytic continuation in μ of the pressure P for the equilibrium phase.

<u>Conclusions</u>   (I) By Def. 3(ii) and the note following Def. 1, Σ has no metastable states if $\phi \in \Phi_1$.

(II) On the other hand, the example following that note shows that Σ may support metastable states if $\phi \in \Phi \setminus \Phi_1$.

(III) We have obtained similar results for the physically more inter-esting case of hard-core continuous systems. In particular, we have shown that the Fisher-Felderhof cluster model $[6]$ has a metastable phase corresponding to a superheated liquid, whose pressure function

is a smooth continuation of that of the equilibrium liquid phase. These
properties stem from the fact that the model satisfies the conditions
specified in the above Comment (3), with X^o the space of single cluster
configurations.
(IV) By extending the present formalism to mean field theories, we have
shown that the Van der Waals fluid model, with infinite range Kac potential, has a metastable phase corresponding to a supercooled gas, whose
pressure is an analytic continuation of that for the stable gaseous
phase (cf. also [7,8]).

References

1. Lanford, O.E. , Ruelle, D. : Commun. Math. Phys. 13, 194 (1969).
2. Sewell, G.L. : Lett. Nuov. Cim. 10, 430 (1974).
3. Gallavotti, G., Miracle-Sole, S. : Ann. Inst. H. Poincaré 8,
       287 (1968).
4. Ruelle, D. : Commun. Math. Phys. 5, 324 (1967).
5. Dobrushin, R.L. : Theory of Prob. Appl., 13, 197 (1968).
6. Fisher, M.E., Felderhof, B.U. : Ann. of Phys. 58, 176 (1970).
7. Emch, G.G. : J. Math. Phys. 8, 19 (1967).
8. Penrose, O., Lebowitz, J.L. : J. Stat. Phys. 3, 211 (1971).

DYNAMICAL THEORIES OF BROWNIAN MOTION

J. T. Lewis[*]

DIAS, Dublin 4, Republic of Ireland

and

J. V. Pulè

Royal University, Malta.

1. Introduction

Nelson (1) has shown how to derive the Einstein-Smoluchowski
theory of Brownian motion from the Ornstein-Uhlenbeck theory which
starts from the Langevin equations on phase space.   We aim to derive
the Ornstein-Uhlenbeck theory from the Maxwell-Boltzmann theory by
making a mechanical model of the stationary Ornstein-Uhlenbeck process.
The essential ideas of this treatment are contained in the paper of Ford,
Kac and Mazur (2), but our treatment lays emphasis on the underlying
mathematical structure.   The case in which the restoring force is linear
in the displacement has been considered in (3) where the connection
with (2) is spelt out.

2. Dynamics of a Conservative Linear System (4)

We take the phase space of the system to be a real vector space Γ_1
and suppose that the total energy $E(\gamma)$ of the system corresponding to
the state $\gamma \varepsilon \Gamma_1$ is given by $E(\gamma) = mc^2 H(\gamma)$ where $\gamma \rightarrow H(\gamma)$ is a strictly positive quadratic form on Γ_1, and mc^2 has the dimensions of energy.
Choose a norm $| \cdot |$ on Γ_1 by putting $| \gamma |^2 = 2H(\gamma)$.   Let Γ be the
completion of Γ_1 with respect to the metric got from $| \cdot |$, so that
Γ is a Hilbert space with inner product $< \cdot, \cdot >$ given by $2<\gamma_1,\gamma_2> =$
$|\gamma_1 + \gamma_2|^2 - |\gamma_1|^2 - |\gamma_2|^2$.   Let $t \rightarrow \gamma_t = T_t\gamma$ be a flow on Γ leaving
H invariant.   Then there is a linear operator D on a dense domain in
which is skew-symmetric with respect to the inner product and is such
that $T_t = \exp(tD)$.   Assume that $- D^2$ is strictly positive (it is certainly non-negative since $- D^2 = D*D$).   Then D can be written uniquely
as JA where J is a complex structure and A is strictly positive.   Then
for γ in the domain of D we have

$$\frac{d}{dt}\gamma_t = JA\gamma_t \; . \tag{2.1}$$

There is a pair E, F of closed subspaces of Γ invariant under $\exp(-tA)$

------

*Speaker

(t > 0) such that F = JE and Γ = E + F.   When dim Γ = 2n < ∞ there is
an orthonormal basis $\{e_i: i = 1, .., n\}$ for E such that $Ae_i = \alpha_i e_i$;
then $\{f_i = Je_i: i = 1, .., n\}$ is an orthonormal basis for F and
$Af_i = \alpha_i f_i$.   Putting $Q_t^i(\gamma) = <e_i, T_t^*\gamma>$ and $P_t^i(\gamma) = <f_i, T_t^*\gamma>$ we find
that (2.1) yields

$$\frac{d}{dt} Q_t^i(\gamma) = \alpha_i P_t^i(\gamma) \quad \frac{d}{dt} P_t^i(\gamma) = -\alpha_i Q_t^i(\gamma) , \tag{2.2}$$

3. Dynamics of Dissipative Linear Systems

The dissipative linear systems we consider are got from conservative
linear systems by adding a damping term to the generator: let Γ, H, T_t
be as in 2, let B be an operator commuting with J such that B is zero
on E and strictly positive on F, put G = JA - B and S_t = exp(tG), t $\geqslant$ 0.

LEMMA. $\{S_t: t > 0\}$ *is a strongly continuous semi-group of contractions on* Γ
such that $S_t x \to 0$ *as* t $\to \infty$ *for each* x *in* Γ .
Suppose that dim Γ = 2n < ∞ and, with the notation of 2, put
$\beta_{ij} = <f_i, Bf_j>$ and $Q_t^i(\gamma) = <S_t e_i, \gamma>$, $P_t^i(\gamma) = < S_t f_i, \gamma>$; then

$$\frac{d}{dt} Q_t^i = \alpha_i P_t^i , \frac{d}{dt} P_t^i = -\alpha_i Q_t^i - \sum_j \beta_{ij} P_t^i , t > 0 . \tag{3.1}$$

We see that our assumptions about S_t amount to the inclusion in the
equations of motion of phenomenological damping terms linear in the
velocities.   Such terms arise in Stoke's theory of viscous damping
and in Rayleigh's theory of Doppler drag.

4. Embedding Dissipative Linear Systems

Can a dissipative linear system Γ^0, H^0, S_t = exp(tG) be embedded
isometrically in a conservative linear system Γ, H, T_t in such a way
that the flow S_t is the restriction of the flow T_t to Γ^0 ? This is
answered in the affirmative by the following version of the Sz.-Nagy
Dilation Theorem:

THEOREM 1 (see (4), (5) and (6))   *Let* $\Gamma^0 = E^0 + F^0$ *be a phase space*
with dissipative flow S_t = exp(tG), G = $J^0 A^0$ - B. *Then there exists an*
isometric embedding j: $\Gamma^0 \to \Gamma L^2(R; F^0)$, *a conservative flow* T_t = exp(tJA)
on Γ *and a projection* P *on* Γ *such that* j o S_t = PT_t o j.

Sketch of proof: let $(T_t f)(s) = f(s - t)$,

$$(Pf)(s) = \left\{ \begin{array}{ll} f(s) & , \ s < 0, \\ 0 & , \ s > 0, \end{array} \right. \quad \text{and} \quad (jx)(s) = \left\{ \begin{array}{ll} (2B)^{\frac{1}{2}} S_{-s} x & , \ s < 0, \\ 0 & \ s > 0. \end{array} \right.$$

Warning: $j \ \Gamma^0$ is not contained in $D(JA)$ and subsequent computations depend crucially on the following theorem:

THEOREM 2 (see (6)) *For every* x *in* $D(G) \subset \Gamma^0$ *we have*

$$(T_t \ o \ jx)(\cdot) = (jx)(\cdot) + \int_0^t (T_s \ o \ jGx)(\cdot) ds + \chi_{(o,t)} (\cdot) \ (2B)^{\frac{1}{2}} \ x.$$

Using the notation of 3 in the case where dim $\Gamma^0 = 2$ but this time putting $Q_t(\gamma) = \ <je, \ T_t^* \ \gamma>$, $P_t(\gamma) = \ <jf, \ T_t^* \ \gamma>$ we have

$$\frac{dQ}{dt} = \alpha P_t \ , \ \frac{dP}{dt} t = - \alpha Q_t - \beta P_t + (2\beta)^{\frac{1}{2}} \gamma(t)$$

Here the damping term arises from the flow of energy into the reservoir whose phase space is $(j\Gamma^0)^{\perp}$.   In addition there is a driving term which leads us to identify the function $t \rightarrow \gamma(t)$, $t > 0$, with the future history of the driving of the system by the reservoir.

5.   Embedding a Dissipative Non-Linear System

Consider a non-linear system with equations of motion

$$\frac{d}{dt}\overline{Q}_t = \alpha \overline{P}_t, \quad \frac{d\overline{P}}{dt} t = -\alpha \overline{Q}_t - \beta \ \overline{P}_t - \alpha \ V'(\overline{Q}_t). \tag{5.1}$$

We embed it in a conservative system by perturbing the linear system used in 4.   Put $\overline{Q}_t(\gamma) = \ <je, \ \overline{T}_t \gamma>$,$\overline{P}_t(\gamma) = \ <f, \overline{T}_t \gamma>$ where $t \rightarrow \overline{T}_t \gamma$ is the non-linear flow on Γ given by

$$\overline{T}_t \ \gamma = \ T_t^* \ \gamma - \int_0^t \ T_{t-s}^* \ o \ jGe \ V'(<je, \ \overline{T}_s \gamma >) \ ds \ . \tag{5.2}$$

(Standard methods give the existence and uniqueness of this flow under appropriate conditions on V'.)   Then we have

$$\frac{d\overline{Q}}{dt} t = \alpha \overline{P}_t \ , \ \frac{d\overline{P}}{dt} t = - \alpha \overline{Q}_t - \beta \overline{P}_t - \alpha V'(\overline{Q}_t) + (2\beta)^{\frac{1}{2}} \gamma(t) \tag{5.3}$$

and the flow $t \rightarrow \overline{T}_t$ preserves the total energy which is now given by

$$mc^2 \ \{\tfrac{1}{2} \ |\gamma|^2 + V(<je,\gamma>)\}.$$

6. Statistical Mechanics of Linear Systems

We return to the conservative systems considered in 2. When $\dim \Gamma = 2n < \infty$ the Maxwell-Boltzmann prescription puts a measure μ_T on Γ having density $(2\pi\, kT/mc^2)^{-n} \exp(-mc^2 H(\gamma)/kT)$ with respect to Lebesque measure. Then for each ξ in Γ we have $E \exp(i<\cdot,\xi>) = \int_\Gamma \exp(i<\gamma,\xi>) d\mu(\gamma) = \exp(-\tfrac{1}{2}\sigma^2 |\xi|^2)$ where $\sigma^2 = kT/mc^2$, so that $<\cdot,\xi>$ is a Gaussian random variable with $E<\cdot\ \xi> = 0$, $E <\cdot,\xi>^2 = \sigma^2 |\xi|^2$. Thus for linear systems the Maxwell-Boltzmann prescription is entirely equivalent to the following: let Ω, P be a probability space and let $\phi : \Gamma \to L^2(\Omega, P)$ be a linear mapping such that for each ξ in Γ the random variable $\phi(\xi)$ is Gaussian with $E \phi(\xi) = 0$, $E\phi(\xi)^2 = \sigma^2 |\xi|^2$. In this case, of course, we may take $\Omega = \Gamma$ and $P = \mu_T$. This prescription makes sense even when $\dim\Gamma = \infty$ and in this case we adopt it as the Maxwell-Boltzmann prescription, even though it is no longer possible to take $\Omega = \Gamma$. However in the situation described in 4 we have $\Gamma = L^2(R; F^0)$ and there is a unique probability measure μ_T on $\Gamma^* = \mathcal{S}'(R;F^0)$ such that, for each ξ in $_*\Gamma = \mathcal{S}(R; F^0)$, $E \exp(i<\cdot, \xi>) = \exp(-\tfrac{1}{2}\sigma^2|\xi|^2)$ where now $<x,\xi>$ is the pairing between Γ^* and $_*\Gamma$. We can take ϕ to be the unique continuous extension of the map $\xi \to <\cdot\ \xi>$ to a map from Γ to $L^2(\Gamma^*,\mu_T)$. T_t leaves $_*\Gamma$ invariant so its restriction $_*T_t$ is well-defined and induces an adjoint action T_t^* on Γ^*. Note that now $\phi(\cdot) = \sigma\, W(\cdot)$ where $W(\cdot)$ is the Wiener Stochastic Integral. (For a fuller account of all this see Hida (7).)

7. The Ornstein-Uhlenbeck Process

Returning to 4 we put $Q^t(x) = < T_t^* x, je>$, $P_t(x) = < T_t^* x, jf>$ and then we have

THEOREM 3 $Q_t(\cdot)$, $P_t(\cdot)$ *is a Gaussian stochastic process satisfying the Langevin equation*

$$dQ_t(\cdot) = \alpha P_t(\cdot), \quad dP_t(\cdot) = -\alpha Q_t(\cdot) - \beta P_t(\cdot) + (2\beta kT/mc^2)^{\frac{1}{2}} W_t(\cdot)$$

where $W_t(\cdot)$ *is a Wiener process such that* $EW_t = 0$, $EW_s W_t = \min(s, t)$.

References

(1) Nelson, E., Dynamical Theories of Brownian Motion, Princeton (1967)

(2) Ford, G. W., M. Kac and P. Mazur, J. M.P. 6,505-515 (1965)

(3) Lewis, J. T. and L. C. Thomas, Functional Integration, Oxford (1975)

(4) Lax, P. D. and R. S. Phillips, Scattering Theory, New York (1967)

(5) Sz.-Nagy, B. and C. Foias, Harmonic Analysis..., Amsterdam (1970)

(6) Lewis, J. T. and L.C. Thomas, Z. Wahrschein. 30, 45-55 (1974)

(7) Hida, T., Stationary Stochastic Processes, Princeton (1970)

DISTRIBUTION OF ZEROS AND THE EQUATION OF STATE

KAZUYOSI IKEDA

Department of Applied Physics, Osaka University
Yamada-Kami, Suita, Osaka, Japan

§1. As was pointed out by Yang and Lee,[1] the distribution of zeros z_1, $z_2, \ldots, z_M$ of the grand partition function

$$\Xi_\Omega(z) = 1 + \sum_{N=1}^{M} Z_N(\Omega, T) z^N = \Pi_{j=1}^{M}(1 - z/z_j) \quad [Z_N: \textit{partition function}] \quad (1)$$

for a system of interacting particles in the complex z (activity) plane in the limit of Ω (volume) $\to \infty$ is connected with the equation of state $[p = p(\rho)$, p: pressure, ρ: density] and condensation; $p/kT = \lim[\Omega \to \infty](1/\Omega) \cdot \cdot \ln\Xi_\Omega(z)$, $\rho = (1/kT)z \, dp/dz$. It is one of the interesting problems of statistical mechanics to obtain the distribution of zeros for a given systems. However, for continuous gases such calculations are difficult because of the complicated character of their partition functions. Hemmer and Hauge[2] et al. have attempted to obtain the distribution of zeros for some continuous gases, starting from the equation of state.

§2. In this lecture, we first derive equations of state from some examples of distribution of zeros. We assume that in the limit of $\Omega \to \infty$ the zeros are distributed on the circle of radius a with centre at the origin; the distribution function for zeros is denoted by $g(\theta)$ [θ being the argument of a point on the circle] and we have $g(-\theta) = g(\theta)$ and $2\int_0^\pi g(\theta) d\theta = c$ $\equiv \lim[\Omega \to \infty](M/\Omega)$. Example (i): $g(\theta) = (c/4\pi)(2 - \cos\theta)$. Example (ii): $g(\theta) = (c/8\pi)\{3 + (\pi - \theta)\sin\theta - 2\cos\theta\}$ $(0 \leq \theta \leq \pi)$. Example (iii): $g(\theta) = c\lambda/\alpha$ $(0 \leq \theta \leq \alpha)$, $= 0$ $(\alpha < \theta < \beta)$, $= c(\pi - \beta)^{-1}(1/2 - \lambda)$ $(\beta \leq \theta \leq \pi)$, where $0 < \alpha < \beta < \pi$ and $0 < 2\lambda \leq [1 + \{(\pi - \beta)/\alpha\}\tan(\beta/2)/\tan(\alpha/2)]^{-1}$. In these examples, the density ρ as a function of z is given by (3.10), (3.12), (3.15) of reference 3, respectively, and $\rho(z)$ (and its analytical continuation) is shown in Fig. 1 (i), (ii), (iii), respectively. The equation of state is expressed by OP-P'L. In (ii), the condensation point P is an "analytical" singularity ($\bullet$);[4,5]

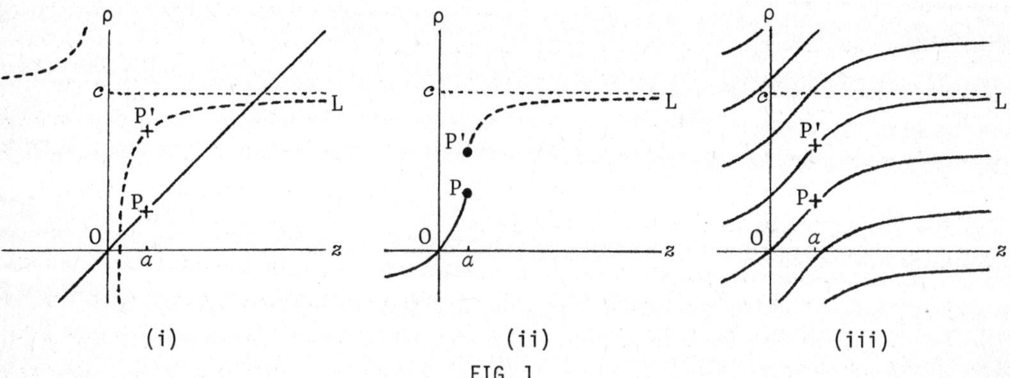

FIG. 1

in (i) and (iii), P is a "non-analytical" singularity (+).[4,5] In (i) and
(ii), the function representing gas and the function representing liquid
are different analytic functions (—— and ----); in (iii), they are
different branches of one and the same analytic function. Thus, by giving
some examples of distribution of zeros, we can see various types[5] of analytical behaviour of the functions describing condensation; (i), (ii),
(iii) belong, respectively, to types (d), (c), (b) defined in reference 5.

§3. Next we show the non-uniqueness of the derivation of the distribution
of zeros from a given equation of state. For example, if all zeros are at
one point $-d$ on the negative real axis [case (i)], the equation of state
(on the positive real axis) is given by $W(z)\,[\equiv p/kT] = c\ln\{(z + d)/d\}$ [from
(1)]. The same equation of state is obtained, (ii) by distributing the
zeros uniformly on a circle C with centre $-d$ and radius
r_0, (iii) by distributing the zeros uniformly inside the
circle C, or (iv) by distributing the zeros on the imaginary axis with distribution function $g(y) = cd/\pi(y^2 + d^2)$.
[In case (ii), W inside C is given by another analytic
function $\tilde{W} = \text{const} = c\ln(r_0/d)$; in case (iii), W' inside C
is not an analytic function; in case (iv), W when Re $z < 0$
is given by another analytic function $\tilde{W} = c\ln\{(d - z)/d\}$.]
There are infinitely many possibilities of distribution

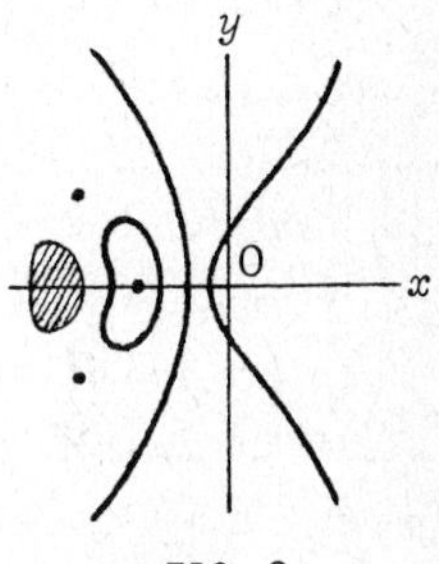

FIG. 2

of zeros leading to the same equation of state [e.g. Fig. 2]. To obtain
a unique distribution of zeros, we make the following assumptions:[6]

  (i) *The zeros are distributed on lines at most (not over domains).*

  (ii) *The zeros are so distributed that one can make as far-reaching ana-*
    lytical continuation of W(z) as possible from the positive real axis to
    the upper half and to the lower half of the complex z plane.

According to these assumptions, case (i) will be realized for the given
equation of state $W(z) = c\ln\{(z + d)/d\}$, since in this case the analytical
continuation of $W(z)$ is the most far-reaching. The validity of these assumptions will be discussed in the future. For the present we can only say
that they are based on a philosophical principle: "Nature likes economy."

§4. On the above assumptions, we derive the distribution of zeros for
gases obeying Tonks' equation of state $[p = kT/(\rho^{-1} - b)]$ and van der Waals'
equation of state $[(p + a\rho^2)(\rho^{-1} - b) = kT]$, i.e. for one-dimensional systems
of hard rods with no attraction and with infinitesimal attraction of infinite range, which are the only examples of continuous gases for which
the equation of state is exactly obtained.[7] We construct the Riemann surface of the function $W(z)\,[\equiv p/kT]$, and derive the "field plane" (i.e. the
part of the Riemann surface covered by our analytical continuation), and
from it we determine uniquely the line of zeros as "jumping line", across

which we jump from one Riemann sheet to another and the real part of $W(z)$
is continuous. The distribution function

$$g(s) = (1/2\pi)(dV_R/ds - dV_L/ds) \tag{2}$$

for zeros on the line is calculated, V_R and V_L denoting the values approached by the imaginary part of $W(z)$ from the right and left sides of
the line, respectively (s is the length of an arc on the line). Note that
a jumping line is different from a branch cut, across which the function
is analytically (smoothly) continued from one sheet to another. The following figures show the Riemann surfaces [where E (i.e. $z = -1/e$), P_1, P_2,
P_3 and O (i.e. $z = 0$) are branch points] and the field planes [where bold
lines represent jumping lines (i.e. zero lines) and usual lines represent
branch cuts]. (In Figs. 4a, 5a, 6a the infinitely many sheets concerning
O are omitted.) In Fig. 4b, K is the point of intersection of the zero
line and the positive real axis; thus K is the condensation point.

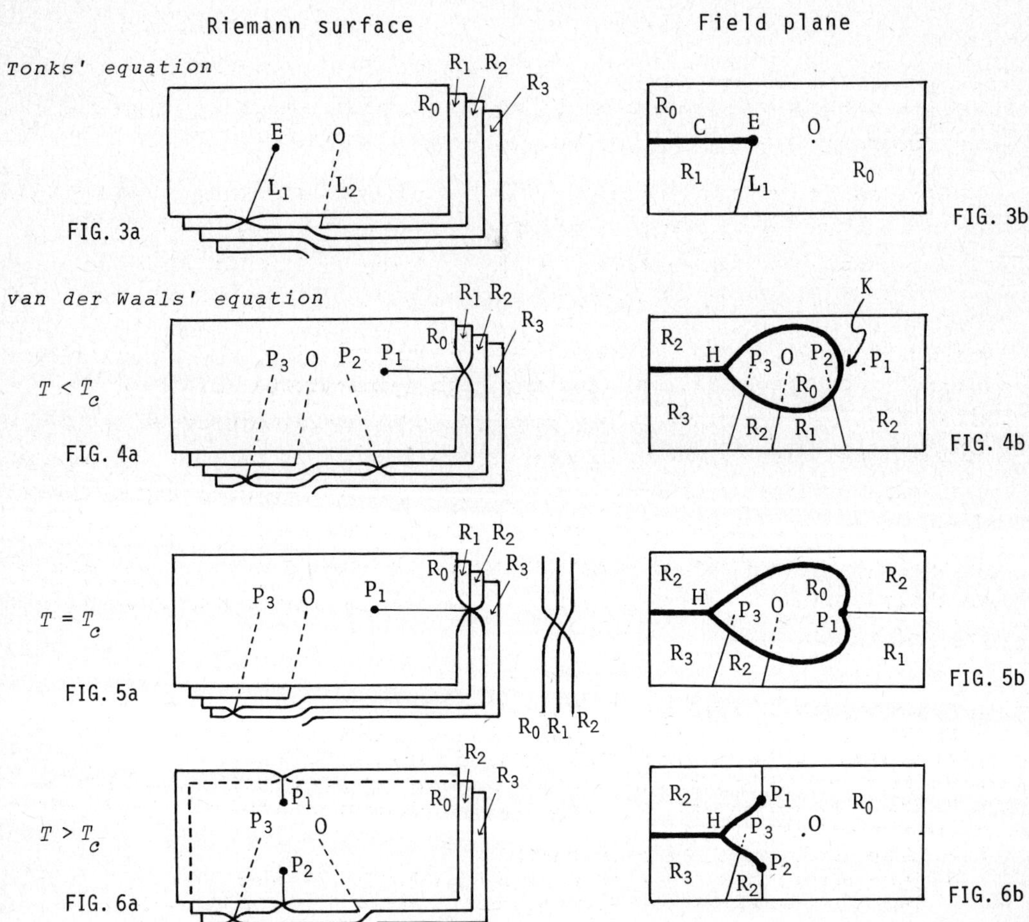

§5. For Tonks' equation (from which $z = We^W$, with $b = 1$) we confirm Hauge-Hemmer's results[2] by our method of argument (Figs. 3a, 3b). For van der Waals' equation, from which we have (on putting $\alpha \equiv a/b^2kT$)

$$W = \rho/(1-\rho) - \alpha\rho^2, \quad z = \{\rho/(1-\rho)\}\exp\{\rho/(1-\rho) - 2\alpha\rho\} \quad \text{(with } b = 1\text{)}, \quad (3)$$

we obtain[6] the equation for the zero line enclosing the origin at low temperatures (cf. Figs. 4a, 4b)

$$\begin{aligned}
r = \alpha e^{-\alpha}[&1 - \alpha^{-1} + (-1/2 + \theta^2/2)\alpha^{-2} + (-5/6 + 3\theta^2/2)\alpha^{-3} \\
&+ (-43/24 + 21\theta^2/4 - \theta^4/8)\alpha^{-4} + (-529/120 + 229\theta^2/12 - 61\theta^4/24)\alpha^{-5} \\
&+ (-8501/720 + 1131\theta^2/16 - 363\theta^4/16 + \theta^6/16)\alpha^{-6} + O(\alpha^{-7}) \\
&+ \{\alpha - 1 + \theta^2\alpha^{-1} + (1/3 + 2\theta^2)\alpha^{-2} + (5/3 + 5\theta^2)\alpha^{-3}\}e^{-\alpha}\cos\theta + O(\alpha^{-4}e^{-\alpha})],
\end{aligned} \quad (4)$$

and the distribution function for zeros on this line

$$\begin{aligned}
g(\theta) = (2\pi)^{-1}\alpha^{-1}e^{\alpha}[&1 + (-2 - \theta^2/2)\alpha^{-2} + (-37/6 + \theta^2/2)\alpha^{-3} \\
&+ (-103/8 + 47\theta^2/4)\alpha^{-4} + (-2681/120 + 395\theta^2/6 - 241\theta^4/48)\alpha^{-5} \\
&+ (-29807/720 + 6973\theta^2/24 - 2027\theta^4/48 - 121\theta^6/32)\alpha^{-6} + O(\alpha^{-7}) \\
&+ \{-2\alpha + 7\alpha^{-1} + (79/2 - 2\theta^2)\alpha^{-2} + (501/8 - 47\theta^2 + 3\theta^4/2)\alpha^{-3}\}\cdot \\
&\cdot e^{-\alpha}\cos\theta + \{-\alpha^{-1} - 4\alpha^{-2} + (-33/2 + \theta^2)\alpha^{-3}\}\theta e^{-\alpha}\sin\theta + O(\alpha^{-4}e^{-\alpha})],
\end{aligned} \quad (5)$$

and the distribution function for zeros on the part of the negative real axis between $-\infty$ and H at low temperatures

$$\begin{aligned}
g(x) = -x^{-1}q^{-2}[&1 + 2q^{-1}\ln q - 3q^{-1} + 6\alpha q^{-2} + 3q^{-2}(\ln q)^2 + (-11q^{-2} + 24\alpha q^{-3})\ln q \\
&+ (6 - \pi^2)q^{-2} - 38\alpha q^{-3} + 40\alpha^2 q^{-4} + 4q^{-3}(\ln q)^3 + (-25q^{-3} + 60\alpha q^{-4})(\ln q)^2 \\
&+ \{(35 - 4\pi^2)q^{-3} - 214\alpha q^{-4} + 240\alpha^2 q^{-5}\}\ln q + (-10 + 25\pi^2/3)q^{-3} \\
&+ (138 - 20\pi^2)\alpha q^{-4} - 388\alpha^2 q^{-5} + 224\alpha^3 q^{-6} + O\{\alpha^4(\ln\alpha)^4\}],
\end{aligned} \quad (6)$$

with $q \equiv \ln(-x) + 2\alpha$. Calculations are also made[6] for the critical temperature T_c (cf. Figs. 5a, 5b) and for high temperatures (cf. Figs. 6a, 6b). From these calculations we may conclude that the existence of repulsions between particles leads to distribution of zeros on part of the negative real axis, whereas the existence of attractions causes distribution of zeros on a curve (open or closed) crossing the negative real axis, and brings about condensation when the curve is closed to cross the positive real axis; these features might hold for the general systems of interacting molecules.

References

1. C. N. Yang and T. D. Lee, Phys. Rev. 87 (1952), 404.
2. E. H. Hauge and P. C. Hemmer, Physica 29 (1963), 1338.
   P. C. Hemmer and E. H. Hauge, Phys. Rev. 133 (1964), A1010.
3. K. Ikeda, *Modern Developments in Thermodynamics*, B. Gal-Or, ed. (Wiley, 1974), p. 311.
4. K. Ikeda, *Proceedings of the International Conference of Theoretical Physics, Kyoto and Tokyo* (1953), p. 544.
5. K. Ikeda, Prog. Theor. Phys. 16 (1956), 341.
6. K. Ikeda, Prog. Theor. Phys. 52 (1974), 54, 415, 840; 53 (1975), 66.
7. M. Kac, G. E. Uhlenbeck, and P. C. Hemmer, J. Math. Phys. 4 (1963), 216.

NEW THEOREMS ON ALGEBRAIC EQUATION AND ITS APPLICATION
TO STATISTICAL PHYSICS

Yukihiko Karaki
the Computer Centre, University of Tokyo
Tokyo, Japan.

Three themes are proposed. The first is new rigorous theorems on
Ising ferromagnets. The second is new theorems on algebraic equation.
The last is its application to antiferromagnetic Ising models.

I. New Rigorous Theorems on Ising Ferromagnets.

A new representation of spin-spin correlation functions of Ising model
was discovered. Using that, Lee-Yang theorem[1-4] was extended to correlation functions of Ising ferromagnets.

<u>Definitions</u> 1. Hamiltonian of Ising model:

$$\mathcal{H}_N = -(\vec{\mu}\,\hat{J}\,\vec{\mu}) - \sum_{k=1}^{N} m_k h_k \mu_k \quad ,$$

where $\hat{J}$ is the interaction matrix ($\hat{J}_{ij}=J_{ij}/2$, J_{ij} being the coupling
parameter between the i-th and the j-th spins, and $J_{ij}=J_{ji}$), $\vec{\mu}$ the
spin variable vecter ($\vec{\mu}=(\mu_1,\mu_2,\ldots,\mu_N)$, μ_k being the spin vaiable at
the k-site, and $\mu_k=\pm1$), ($\cdots$) the scalar product, m_k the magnetic
constant, h_k the external field, and N the total number of spins.
2. Generalized Partition Function:

$$F_N(z_1,z_2,\ldots,z_N,\beta,\hat{J}) = \mathrm{Tr}\, e^{-\beta\mathcal{H}_N} = \sum_{\mu_1}\sum_{\mu_2}\cdots\sum_{\mu_N} e^{\beta(\vec{\mu}\hat{J}\vec{\mu})} \prod_{k=1}^{N} z_k^{\mu_k} \quad ,$$

where z_k^2 is the 'fugacity' of the k-th spin, $z_k=e^{\beta m_k h_k}$.
3. Spin Function of the n-th order:

$$f_n(z_1,z_2,\ldots,z_N,\beta,\hat{J};i_1,i_2,\ldots,i_n) = \mathrm{Tr}\left(\prod_{k'=1}^{n}\mu_{i_{k'}}\right) e^{\beta(\vec{\mu}\hat{J}\vec{\mu})} \prod_{k=1}^{N} z_k^{\mu_k}.$$

<u>Lemma</u> Any spin functions of Ising model are generated from the general-
ized partition function (directly).

Proof: We introduce the fundamental identities such as,

$$\mu_k = -i \times i^{\mu_k} , \quad i = \sqrt{-1} \quad .$$

These can easily be found to hold. Using these, the spin function of the
n-th order can be expressed as follows,

$$f_n(z_1,z_2,\ldots z_N,\beta,\hat{J};i_1,i_2,\ldots,i_n) = (-i)^n F_N(\tilde{z}_1,\tilde{z}_2,\ldots,\tilde{z}_N,\beta,\hat{J}) \quad ,$$

where $\tilde{z}_k=z_k$ for $k\notin\{i_1,i_2,\ldots,i_n\}$, and $\tilde{z}_{k'}=i\,z_{k'}$ for $k'\in\{i_1,i_2,\ldots i_n\}$.
(Q.E.D.)

Theorem 1 All 'fugacity' zeros of spin functions of Ising ferromagnets ($s=1/2$) lie on the unit circle of the complex plane. (Generalized Lee-Yang theorem)

 Proof: By the well known Lee-Yang lemma, we know that if
$F_N(z_1,z_2,\ldots,z_N,\beta,\hat{J})=0$ and $|z_1|\geq 1, |z_2|\geq 1,\ldots, |z_N|\geq 1$ hold, then
$|z_1| = |z_2| = \cdots = |z_N| =1$ for $\hat{J}\geq 0$, that is, $\forall J_{ij}\geq 0$. As easily seen, $|i\,z| = |z|$.
Using these, we immediately find that if $f_n(z,z,\ldots,z,\beta,\hat{J};i_1,i_2,\ldots,i_n)=0$
($z=e^{\beta mh}$) and $|z|\geq 1$, then $|z|=1$ for $\hat{J}\geq 0$.(See the previous Lemma) (Q.E.D.)

Theorem 2 The spin functions, $f_n(z,\ldots,z,\beta,\hat{J};i_1,i_2,\ldots,i_n)$ and
$f_{n+1}(z,\ldots,z,\beta,\hat{J};i_1,i_2,\ldots,i_n,i_{n+1})$, have no common 'fugacity' zeros
for $\hat{J}\geq 0$.   (Proof: omitted)

Theorem 3 If the interaction matrix, $\hat{J}$, is reducible, then the generalized partition function is also reducible as a function of N variables,
$z_1,z_2,\ldots,$and z_N.   (Proof: omitted)

Corollary 1 The 'fugacity' zeros of the spin function of the n-th order of Ising ferromagnets distribute for $\beta =0$ (infinite temperature) as follows; the z^2-zeros degenerate at $z^2=1$ (n-fold) or at $z^2=-1$ ((N-n)-fold).

Corollary 2 The 'fugacity' zeros of the spin function of the n-th order of Ising ferromagnets (irreducible) distribute for $\beta = \infty$ (zero temperature) as follows; if n is even, the z^2-zeros are as $z_s^2 = e^{i\pi(2s+1)/N}$
(s=1,2,..,N) and that if n is odd, we obtain $z_s^2 = e^{i\pi 2s/N}$ (s=1,2,..,N).

Corollary 3 The ordinary spin-spin correlation functions of Ising model under the uniform field are written by using the generalized partition function as,

$$\langle \mu_{i_1}\mu_{i_2}\cdots\mu_{i_n}\rangle = (-i)^n \frac{F_N(z,\ldots,z,\overset{i_1}{\downarrow}iz,z,\ldots,\overset{i_2}{\downarrow}iz,z,\ldots,\overset{i_n}{\downarrow}iz,z,\ldots,z,\beta,\hat{J})}{F_N(z,z,\ldots,z,\beta,\hat{J})} .$$

II. New Theorems on Algebraic Equation.

 The algebraic equation was discovered in which the distribution of roots depends only on the 'shape' of coefficients, but not on the degree of equation (Topological invariance of roots about degree). According as the sign of the parameter of coefficients, the topology of roots belong to the Lee-Yang type (on Ising ferromagnets) or not.

Theorem 1 If the polynomial is written as,

$$f^{(2N)}(z,\delta) = (g^{(N)}(z))^2 + \delta z (h^{(N-1)}(z))^2 ,$$

where $g^{(N)}(z)$ and $h^{(N-1)}(z)$ are the real polynomials of the N-th and the N-1-th degree respectively, all zeros of whose polynomials lie on the unit circle of the complex plane and separated each by each, and

$g^{(N)}(1) \cdot h^{(N-1)}(1) \neq 0$, then the topology of zeros of $f^{(2N)}(z,\delta)$ is as follows; if $0 \gtrless \delta \gtrless -(g^{(N)}(1)/h^{(N-1)}(1))^2$, all z-zeros lie on the unit circle of the complex plane, that if $\delta > 0$, none of zeros lie on the unit circle, and that if $\delta \leq -(g^{(N)}(1)/h^{(N-1)}(1))^2$, two zeros lie on the positive real axis, and others on the unit circle. (Proof:omitted)

__Theorem 2__ If the polynomial is written as,

$$f^{(2N)}(z,\delta) = (g^{(N)}(z))^2 + \delta (g_d^{(N)}(z))^2 ,$$

where $g_d^{(N)}(z) = (z\frac{d}{dz} - N/2) g^{(N)}(z)$, $g^{(N)}(1) \neq 0$, and all zeros of the real polynomial, $g^{(N)}(z)$, lie on the unit circle, then the topology of zeros of $f^{(2N)}(z,\delta)$ is as follows; if $\delta \gtrless 0$, all zeros lie on the unit circle, and if $\delta < 0$, there always exist the zeros, not lying on the unit circle. (Proof: omitted)

__Example__ The following equation satisfies the condition of the theorem 1.

$$f^{(2N)}(z,\{a_k\}) = \sum_{k=0}^{2N} a_k z^k = 0, \quad a_{2N-k} = a_k, \quad \text{and} \quad a_k = (1+\delta)k+1 \ (0 \leq k \leq N).$$

III. Application to Antiferromagnetic Ising Models.

Various Ising models are classified through the type of the interaction matrix of the system (after appropriate permutations of lattice sites). We now consider the two cases.

__Case 1__ The system composed of equivalent ferromagnetic sublattices.

The interaction matrix is, for example, written as,

$$\hat{J} = \begin{pmatrix} \hat{J}_A & \hat{J}_{AB} \\ \hat{J}_{BA} & \hat{J}_B \end{pmatrix}$$

, where $\hat{J}_A$ (or $\hat{J}_B$) is the interaction matrix on the sublattice A (or B) (the sublattice-Hamiltonian as $\mathcal{H}_A$ (or $\mathcal{H}_B$)) and $\hat{J}_{AB}$ (or $\hat{J}_{BA}$) is the one between the sublattices A and B.

If each sublattice has translational symmetry, the partition function of the total system is perturbation-theoretically written as,

$$F_N(z,\beta,\hat{J}) = F_{N_A}(z,\beta,\hat{J}_A) F_{N_B}(z,\beta,\hat{J}_B)$$

$$+ \frac{\beta}{N_A N_B} ((\vec{1}_A \hat{J}_{AB} \vec{1}_B) + (\vec{1}_B \hat{J}_{BA} \vec{1}_A)) (z\frac{d}{dz} F_{N_A}(z,\beta,\hat{J}_A)) (z\frac{d}{dz} F_{N_B}(z,\beta,\hat{J}_B)) ,$$

where $F_{N_A}(z,\beta,\hat{J}_A) = \text{Tr} \, e^{-\beta \mathcal{H}_A}$, N_A the number of spins on the sublattice A, etc.. $(^t\vec{1}_A = \underbrace{(1,1,\ldots,1)}_{N_A},$ etc.)

Using the theorem 2 of II, the topology of the z-zeros is characterized as follows; it belongs to the Lee-Yang type if $(\vec{1}_A \hat{J}_{AB} \vec{1}_B) \gtrless 0$, and otherwise to the non Lee-Yang one. If the system is composed of three (or more than three) equivalent sublattices, the 'fugacity' zeros of the total partition function may lie on the three (or more than three) curves.

<u>Case 2</u> The system composed of ferromagnetic sublattice and weak impurity spins.

The interaction matrix is written as,

$$\hat{J} = \begin{pmatrix} \hat{J}_A & \hat{J}_{AI} \\ \hat{J}_{IA} & \hat{J}_I \end{pmatrix} \quad , \text{ where A is the ferromagnetic sublattice and I is the lattice of impurities.}$$

Now we only consider the case of $\hat{J}_I=0$. The partition function is perturbation-theoretically written as,

$$F_N(z,\beta,\hat{J}) = F_{N_A}(z,\beta,\hat{J}_A)\,Z_I + \beta\left((\vec{1}_A\hat{J}_{AI}\vec{1}_I) + (\vec{1}_I\hat{J}_{IA}\vec{1}_A)\right)\frac{z^2}{N_A N_I}\left(\frac{dF_{N_A}}{dz}\right)\left(\frac{dZ_I}{dz}\right)$$

where $Z_I=(z+1/z)^{N_I}$ (N_I being the total number of impurities).

From the above form of the partition function, we can easily find that the topology of 'fugacity' zeros belongs to the Lee-Yang type whether $(\vec{1}_A\hat{J}_{AI}\vec{1}_I) \gtreqless 0$ or not (within a certain approximation).

The detail will be published later.

The author thanks Professor S. Ono for helpful advices.

References

1) T.D.Lee and C.N.Yang, Phys.Rev. <u>87</u> (1952) 410.
2) T.Asano, J.Phys.Soc.Japan <u>29</u> (1970) 350.
3) P.B.Griffiths, J.math.Phys. <u>10</u> (1969) 1559.
4) M.Suzuki, Progr.theor.Phys. <u>41</u> (1969) 1438.

OPERATOR-VALUED-MEASURE APPROACH TO SPECTRA OF
TWO-DIMENSIONAL CLASSICAL HARMONIC LATTICES

Masahisa Fukushima

Science Laboratory, Soka University
1-236 Tangi-cho, Hachioji, Tokyo 192, Japan

The frequency distribution function (f.d.f., in short) of finite-ly-spreaded crystal lattice system is well defined by the distribution function of eigenvalues of operator associated with the system. As to infinitely-spreaded lattice system, the f.d.f. is not necessarily obtained by taking the limit of that in a corresponding finite system, because there does not exist, in general, such a limit in L_2-space. However, the support of f.d.f. in finite system tends to that of spectral function associated with the corresponding operator in infinite system. Furthermore, it is known that the support of spectral function is independent of the choice of real-coordinate system adopted to obtain a representation of resolution of identity, and hence the spectra (support of spectral function) is considered to be the quantities essential to the system.

On the other hand, when we consider an ensemble of random lattices, the situation may be slightly simplified: the set of random variables concerned with the system can be assumed to be mutually independent from the physical point of view, so that one can define the f.d.f. of the ensemble itself, just as in regular system. In such a case, there exists almost certainly a unique limit of the above-stated distribution function of eigenvalues in finite system in the stochastic sense. As for a single lattice system, however, there can not be found any reasonable f.d.f. in case of infinite system, without assuming the spacially translational invariance property of the system. In the present work, we confine ourselves to the spectra of infinite lattice system; it is left open to further investigation to find a f.d.f. for the system, which is reasonable.

As to dimensionality of lattice systems, ASAHI [1] has handled a two-dimensional classical lattice with nearest neighbor interactions. He reduces a partial difference equation obtained from the equation of motion of the system to ordinary one having operator matrices, where he imposed a restriction on the system to be bounded in one direction, in

order to avoid complexities caused by the L_2-space property in extending
the system infinitely in the other direction.   On the other hand, HORI
& FUKUSHIMA [2] have also investigated impurity problems of the same
lattice system by using certain matrices of order as high as the finite
number of atoms arraying in y-direction.

Henceforth the present author intends to make their results complete by extending the system to infinitely-spreaded system in both directions in case of bounded operators.   The present discussion is concerned only with eigenfunction expansion associated with two-dimensional
formally self-adjoint partial difference operators of the second order
in x-direction and of any even order in y-direction.

The present method is based on the spectral theory of singular
formally self-adjoint ordinary differential operators, initiated by WEYL
and STONE for the system of the second order, and completed by KODAIRA
[3,4], and independently by TITCHMARSH [5].   Though the operator-coef-
ficient difference equation is treated over a (pseudo-) Hilbert space
originated by BEREZANSKII [6], the present discussion is performed along
essentially with KODAIRA's method, partly in GOVINDARAJU's style [7].
A brief sketch of the present work is given as follows (For details,
see [8]).

Let us consider a square lattice with the nearest neighbor central force and non-central forces arriving at p site away from an atom.
Then the basic equation describing this system is given by

$$-M(x,y)\omega^2 u(x,y) = K_1(x,y)\{u(x+1,y)-2u(x,y)+u(x-1,y)\}$$

(1)
$$+ \sum_{j=1}^{p} K_{2j}(x,y)\{u(x,y+j)-2u(x,y)+u(x,y-j)\},$$

where $M(x,y)$ stands for the atomic mass at the integral point (x,y);
$-\infty \leqq a_1 \leqq x \leqq b_1 \leqq +\infty$, $-\infty \leqq a_2 \leqq y \leqq b_2 \leqq +\infty$.   Introducing a vector $\tilde{u}(x)=(\cdots,u(x,y),$
$u(x,y+1),\cdots)^t$, and infinite-dimensional matrices $\tilde{M}(x)=\mathrm{diag}(\cdots,M(x,y),$
$M(x,y+1),\cdots)$, $\tilde{K}_1(x)=\mathrm{diag}(\cdots,K_1(x,y),K_1(x,y+1),\cdots)$ and $\tilde{\Gamma}(x)$ whose y-
the row is given by $(\cdots,0,K_{2p}(x,y),\cdots,K_{21}(x,y),K_0(x,y),K_{21}(x,y),\cdots,$
$K_{2p}(x,y),0,\cdots)$, with $K_0(x,y)=-2\sum_{j=1}^{p}K_{2j}(x,y)$, we can rewrite Eq.(1) in
the matrix form

(2) $\qquad -\omega^2\tilde{u}(x)=[\tilde{M}(x)]^{-1}[\tilde{K}_1(x)\{\tilde{u}(x+1)-2\tilde{u}(x)+\tilde{u}(x-1)\}+\tilde{\Gamma}(x)\tilde{u}(x)].$

A self-adjoint matrix $\tilde{\Gamma}_x(x)$ is obtained from the matrix $\tilde{\Gamma}(x)$ by specifying the defining domain of $\tilde{\Gamma}(x)$ to be a subspace H_x of a Hilbert space,
which was discussed in the previous paper [9].   In many physical situations, we can assume that the space H_x is independent of x, in which

case we write simply $\overset{\nu}{\Gamma}(x)$ and H instead of $\overset{\nu}{\Gamma}_x(x)$ and H_x respectively.

In more general setting, we deal with a formally self-adjoint operator with operator-valued coefficients given by

$$(3) \qquad (LU)(x)=[Q(x)]^{-1}[P_1(x)U(x+1)+P_0(x)U(x)+P_1(x-1)U(x-1)],$$

where we have assumed that $Q(x)$, $P_1(x)$ and $P_0(x)$ are self-adjoint operators on a Hilbert space H for every x, and that $Q(x)$ and $P_1(x)$ have inverses. In order to handle Eq.(3) by a method developed for ordinary difference equation [9], we have to introduce another Hilbert space, in which space operators in H may behave as scalars. For this purpose a pseudo-Hilbert space introduced by BEREZANSKII [6] would be suitable.

Denote by L(H,H) the set of all bounded linear operators in H, and put $H_0 = \sum_{x=a_1}^{b_1} L(H,H)$. If we define, following BEREZANSKII [6], a certain inner product and a strong operator topology on H_0, we can see that the space H_0 becomes a Hilbert space, in which the elements of L(H,H) behave as scalars. Thus the operator L in Eq.(3) is regarded as a linear operator on the Hilbert space H_0 and U as an element of H_0.

Now we consider the equation $(LU)(x,\ell)=U(x,\ell)\hat{\ell}$, where $\hat{\ell}=\ell E_H$, E_H being the identity operator on H and $\ell \in \underline{C}$. Then we can construct a set of solutions $S_j(x,\ell)$, $j=1,2$, of this equation under certain conditions, which we call a <u>canonical system of fundamental solutions</u>.

If we impose real and self-adjoint boundary conditions at both boundary points, which are defined in the sense of strong operator topology in case where one or both of the boundary points are at infinity, then we can discuss the WEYL's classification of operators: we have four classes, limit point (or circle) case at a_1 and limit point (or circle) case at b_1. In many physical systems of crystal lattices, the limit point case at infinity may occur. In the study of the effect of surfaces of a crystal lattice, for instance, the operator L should be of the limit circle type at this boundary (surface).

We can construct a characteristic matrix $M(\ell)$ in terms of the canonical system of fundamental solutions, and of the boundary conditions if necessary, to obtain the spectral function

$$(4) \qquad [\rho^{jk}(\lambda)]=-\lim_{\varepsilon \to +0} \frac{1}{2\pi i} \int_0^\lambda [M(\mu+i\varepsilon)-M(\mu-i\varepsilon)]d\mu.$$

When the limit point case at point a_1 (or b_1) occurs, calculations of some factors in the elements of $M(\ell)$ become easier, because these factors are independent of boundary conditions. After constructing from L a self-adjoint operator $\mathcal{H}$ by specifying the defining domain $\mathcal{D}_0$, we finally arrive at the following eigenfunction expansion formula by using abstract theory of spectral decomposition of $\mathcal{H}$ and by solving a integro-difference equation:

$$(5) \qquad U(x) = \sum_{y=a_1+1}^{b_1} \int_{-\infty}^{\infty} \sum_{j,k=1}^{2} S_j^*(x,\ell)\, d\rho^{jk}(\lambda)\, S_k(y,\lambda)\, U(y), \qquad U \in \mathcal{D}_0.$$

Finally it is remarked that, in order to obtain the spectral function, it is necessary to construct a canonical system or at least to know the behaviors of the solutions at boundary points.   The application of this theory to physical system will be discussed in the future.   Finally we comment that, although the present work is concerned only with the partial difference equation of order 2 in x-direction, we can easily extend the theory to the case of any even order.

References

[1] ASAHI, T.. Application of the method of transfer matrix to two- and three-dimensional lattices with defects, Prog. Theor. Phys., Suppl. No.23(1963)59-78.

[2] HORI, J. & M. FUKUSHIMA, Application of the phase theory to the problem of impurity modes in higher-dimensional lattices, Jour. Atomic Ener. Res. Inst. Japan 1113(1966)55-64.

[3] KODAIRA, K., The eigenvalue problem for ordinary differential equations of the second order and Heisenberg theory of S-matrices, Am. J. Math. 71(1949)921-945.

[4] KODAIRA, K., On ordinary differential equation of any even order and the corresponding eigenfunction expansions, Am. J. Math. 72(1950) 502-544.

[5] TITCHMARSH, E. C., *Eigenfunction Expansion Associated with Second-Order Differential Equation, Part I,* (Oxford U. P., London, 1946)

[6] BEREZANSKII, Ju. M., *Expansions in Eigenfunctions of Selfadjoint Operators,* (A. M. S. Trans. Math. Mono. Vol.17, 1968)

[7] GOVINDARAJU, R. V., Boundary conditions in the infinite interval and some related results, Trans. A. M. S. 154(1971)113-128.

[8] FUKUSHIMA, M., Operator-valued-measure approach to a bounded two-dimensional partial difference equation of the second order, Soka Univ. Sci. Lab. Mimeo-Series No.14(1974)

[9] FUKUSHIMA, M., On the eigenvalue problem of the difference operators of any even order and the corresponding eigenfunction expansions, I. General theory, Soka Univ. Sci. Lab. Mimeo-Series No.2(1971).

MODELS FOR RELATIVISTIC STATISTICAL MECHANICS

Meinhard E. Mayer

Departments of Mathematics and Physics

University of California, Irvine, CA 92664, USA

Since these are the last fifteen minutes (or the last three pages) of our $M \cap \Phi$ and presumably everybody is already tired of exact definitions, theorems lemmas and proofs, I shall not attempt to bore you with these. Neither will I go into all the intricacies of physical interpretation of relativistic statistical mechanics. I will rather concentrate on some more or less realistic models I have been considering with my students. These may turn out to be useful in future more realistic attempts at formulating relativistic statistical mechanics.

As a general reference, containing a complete bibliography which I found extremely helpful, I would like to recommend the recent report by Jürgen EHLERS: "Progress in Relativistic Statistical Mechanics, Thermodynamics and Continuum Mechanics", Talk at GR 7, Tel Aviv, June 1974.

I will be mostly interested in general-relativistic statistical mechanics (GRSM), will briefly mention the problems of special-relativistic statistical mechanics (SRSM), and would only like to note briefly that the best developed area of this field is relativistic kinetic theory (RKT), which is treated in detail in Ehlers' reviews.

1. <u>Why Relativistic Statistical Mechanics?</u> We might as well ask why not? In addition to its academic interest (even the question of how to define equilibrium becomes difficult if time is not uniquely determined) and its potential heuristic value in learning how to quantize gravitation, astrophysics poses some "practical" problems requiring concepts from GRSM (albeit on a cosmic scale one can hardly talk of an equilibrium SM). There are also problems in relativistic plasma theory which might require the use of curvilinear coordinates. Last, and not least, some of the techniques used (at least in our models) bear a strong resemblance to "gauge theory" methods in quantum field theory, and might lead to the development of useful mathematical tools for the latter.

2. <u>GRSM Without SRSM?</u> Since locally general relativity reduces to Minkowski space, one might ask whether it makes any sense to attempt a GRSM before SRSM is on a sound footing. Any straightforward attempt to formulate classical statistical mechanics in a Poincaré invariant manner runs into the difficulty with the famous "no-interaction" theorems encountered in relativistic Hamiltonian mechanics. Non-Hamiltonian approaches (e. g., Hakim's formulation) use measures on world lines and

do not lend themselves too easily to generalizations to curved spaces.

We will therefore attempt to discuss models in which GRSM is in a
sense "locally nonrelativistic", but evolves on a curved Riemannian
manifold, thus simulating a gas in an external prescribed gravitational field. No attempt has been made yet to attack the self-consistent
problem of including the gravitational field among the observables (in
particular, the connection coefficients, which seem to be more suitable
for this than the components of the metric tensor). Another model under study, which is locally "relativistic", is modeled on relativistic
field theory. Throughout, when I talk of observables and states I do
not necessarily exclude classical statistical mechanics. The models
are designed to be vague enough to accomodate classical observables
(abelian algebras, states are measures) and quantum systems (operator algebras and their locally normal states, i. e., local density matrices). Detailed results will be published elsewhere.

3. Models. A. "Locally Nonrelativistic" GRSM. Due to the need to
define equilibrium in terms of a unique "time" variable we shall consider an Einstein space with a synchronous comoving coordinate system,
such that the metric has the form

$$ds^2 = dt^2 - g_{ik}dx^i dx^k \quad (i,k = 1, 2, 3). \qquad (1)$$

This choice of a single "time" may introduce some fictitious "potentials"
which affect the definition of equilibrium; a more refined discussion
of this point is necessary, as well as the notion of time evolution and
equilibrium). The system under consideration "lives" on the 3-dimensio-
nal manifold t = const., with the metric determined by the second term
of the r.h.s. of Eq. (1). In the tangent space to each point of this
manifold we consider an algebra of observables (either a local phase
space with its functions, or an algebra of observables generated by
creation-annihilation operators). It is important to note that in this
model the isotony requirement for the quasilocal algebra is imposed in
the tangent space, not the curved manifold, and thus the problem of algebras associated to "large" sets is circumvented. Instead there arises the problem of comparing algebras and their observables and states
in the tangent spaces to different points of the manifold. We propose
to use the connection associated to the metric for transporting observables and states. In particular, transport around a closed loop in
the manifold induces an automorphism in the local algebra (which for
obvious reasons is called a holonomy automorphism), and a transposed
action on the states, such that expectation values are left invariant.
In the simplest examples of homogeneous spaces the holonomy automorphism
may turn out to be trivial, yielding no new restrictions on the model.

$\underline{B.}$ $\underline{\text{Locally Relativistic GRSM.}}$ This case is considerably more difficult, since it assumes that the local (tangent-space) theory is in-
variant under the Poincaré group. A model of this type could be obtained by considering quasilocal algebras based on "causally convex charts"
of a four-dimensional pseudoriemannian manifold (roughly, a causally
convex chart can be made to look like an ordinary double cone by means
of a special choice of coordinates). The main problem here is **how** to
characterize equilibrium states, i. e., whether one can impose reasonable conditions leading to something akin to a "local KMS property", and
what the meaning of the appropriate β parameter is. It is worth recal-
ling that in a relativistic Gibbs enesemble the reciprocal temperature
becomes a four-vector by multiplication with the four-velocity, the Hamiltonian being replaced by the appropriate integral over the energy-
momentum tensor, etc. In this model too, observables tangent to different points should be compared by means of the connection, one must investigate the action of holonomy automorphisms (at least for spacelike
loops) and expectation values should be invariant. So far no definite
results have been obtained, but the reward for constructing a meaningful model of this kind are great: it might be a prelude to an "algebraic" approach to quantum gravitation. Relations to the approaches of
Arnowitt-Deser-Misner, De Witt and Faddeev-Popov will be explored.

$\underline{\text{4. RKT.}}$ As already mentioned, this area is well covered in the
literature. There remain problems (e. g., deriving a BBGKY hierarchy
and solving various kinetic equations) which are under investigation.

$\underline{\text{5. Conclusion.}}$ In this brief review I have managed only to point to
a few of the difficult problems encountered in this subject. The models
discussed are not very physical and the presentation given here, at
least, would not qualify as highly mathematical. Therefore this talk
could be classified in the category $\overline{M} \cup \overline{\Phi}$ (the bar means $\underline{\text{not}}$) the dual
of the proposition $M \cap \Phi$. But since the first statement is not the
whole Universe, by duality we have another proof that $M \cap \Phi \neq \emptyset$!

In conclusion I would like to thank our Japanese hosts on behalf
of all of us for a wonderful conference: ARIGATOH GOZAIMAS!

This work was partially supported by NSF Grant GP 43969.

Q: (R. Arens): Does the statement about non-interaction in relativistic mechanics remain meaningful in quantum theory?

A: In my opinion, no.

<u>PATH INTEGRALS IN RIEMANNIAN MANIFOLDS</u>

Cecile DeWitt-Morette
Department of Astronomy and Center for Relativity
The University of Texas at Austin, 78712

I. Introduction

The configuration space $\mathbb{M}$ of a physical system is rarely $\mathbb{R}^n$. Indeed the
configuration spaces of such common systems as a particle with spin, the harmonic
oscillator, several hard spheres, a system of indistinguishable particles, etc. ...
are Riemannian multiply connected manifolds [1]. The global properties of $\mathbb{M}$ contain
much information on the generic properties of the physical system; nevertheless,
they have received little attention. Why? Possibly because global problems are
difficult, but also because physical laws have, since Newton, been largely stated
as differential equations and investigated locally.

The Feynman formalism is the only global formalism of physics, but its global
aspect is somewhat blurred in the original definition of path integrals. In this
definition a path q mapping an interval $\mathbb{T} = [t_a, t_b]$ into $\mathbb{M}$ is replaced by p of its
values $q^i = q(t_i)$ for a time subdivision of the interval $\mathbb{T}$

$$t_a = t_0 < t_1 < \ldots \, t_p < t_{p+1} = t_b \text{ with } q(t_a) = a, \; q(t_b) = b$$

The path integral over the space $\mathbb{F}(a,b)$ of all possible paths of the system between
its states $A = (a, t_a)$ and $B = (b, t_b)$ is replaced by an integral over the space $\mathbb{R}^{pn}$
of the pn tuples $\{q^{i\alpha}; \; i = 1 \ldots p, \; \alpha = 1 \ldots n\}$ where n is the dimension of $\mathbb{M}$,
i.e., the number of degrees of freedom of the system. If it exists, the limit of the
integral over $\mathbb{R}^{pn}$ when $p \to \infty$ is called the Feynman path integral. No unique prescription has been given for the $\mathbb{R}^{pn}$ integral beyond the requirement that the results
agree with those obtained with the Schrödinger differential equation; the mathematical
difficulties involved in defining the $\mathbb{R}^{pn}$ integral and its limit, and the computational
difficulties in finding their values, whether unique or not! ..., has plagued both
mathematicians and physicists. Thus most of the effort has been spent in showing
that, locally, the Feynman formalism is equivalent to the Schrödinger formalism and
in presenting the Feynman path integral as the solution of the Schrödinger equation
satisfying some boundary conditions and other extraneous conditions as required by
the given problem, such as the symmetry or antisymmetry property of the wave function.
In this approach, the global properties of the space of all possible paths on $\mathbb{M}$
hardly enter the picture and much of the original beauty and power of the Feynman
formalism is lost.

Recently [2] a new definition of the Feynman path integral which does not rest on the above limiting procedure has been proposed. It focuses on the space $\mathbb{F}(a,b)$ of all possible paths of the system from A to B. Besides a different emphasis, this definition leads to new techniques for computing path integrals. These techniques give readily results obtained laboriously by other methods; they are particularly well suited to the study of systems whose configuration space is not flat where other techniques are often ambiguous and encounter a great deal of difficulties.

In this paper we shall consider configuration spaces $\mathbb{M}$ which are Riemannian multiply connected manifolds with metric g and fundamental group π. We shall illustrate the propositions with the example of a free particle whose state at time t is $q(t) \in \mathbb{M}$. The action of this system is:

$$S(q) = \frac{m}{2} \int_{\mathbb{T}} \| \dot{q}(t) \|^2 \, dt = \frac{m}{2} \int_{\mathbb{T}} \left(\frac{Dq(t)}{dt} , \frac{Dq(t)}{dt} \right)_g \, dt$$

II. Pseudomeasures

The theory of promeasures (cylindrical measures) provides the framework for integration on function spaces; more precisely it is the basis for integration on Hausdorff topological vector spaces $\mathbb{X}$, locally convex. A promeasure is a family of bounded measures defined on a family of finite dimensional spaces suitably related to $\mathbb{X}$, satisfying some coherence conditions. The restriction to bounded measures makes it impossible to use the theory of promeasures for Feynman integration. However there is a one to one correspondence between the set of promeasures on $\mathbb{X}$ and their Fourier transforms on its dual $\mathbb{X}'$. One can thus define a promeasure by its Fourier transform and states the coherence conditions as conditions satisfied by the Fourier transforms. At this point it is possible to remove the restriction to bounded measures and to generalize the concept of promeasure: Indeed, the Fourier transforms of measures, considered as distributions of order zero, are defined for all measures, bounded or not. This new concept, given for convenience a name "pseudomeasure" and a symbol "w", enters our work only by its Fourier transform $\mathcal{F}w$. It is not known whether or not the mapping $w \to \mathcal{F}w$ of the set of pseudomeasures on $\mathbb{X}$ into the set of functions on $\mathbb{X}'$ is injective; i.e., whether or not a pseudomeasure w is uniquely defined by its Fourier transform $\mathcal{F}w$.

In the present study we consider only complex gaussian pseudomeasures; a gaussian pseudomeasure is a pseudomeasure whose Fourier transform is:

$$\mathcal{F}w = \exp(-iW/2)$$

where W is a quadratic form on X'. When $\mathbb{X}$ is the space of continuous paths x on $\mathbb{T}$, $\mathbb{X}'$ is the space of measures μ on $\mathbb{T}$ and

$$W(\mu) = \underline{W}(\mu,\mu) = \int_{\mathbb{T}} d\mu_\alpha(r) \int_{\mathbb{T}} d\nu_\beta(s) \, G^{\alpha\beta}(r,s)$$

W is the variance of the gaussian, G its covariance. A normalized gaussian pseudomeasure is uniquely defined by its covariance.

<u>Proposition 1.</u> (Transformation of a Gaussian pseudomeasure under a linear mapping)[2,3]
Let X and Y be two Hausdorff, topological vector spaces, locally convex, let X'
and Y' be their topological duals; let P be a linear continuous mapping from X into
Y, let $\tilde{P}$ be the transposed mapping from X' to Y' defined by

$$< \tilde{P}y', x > = < y', Px >$$

Let w be a Gaussian pseudomeasure on X of variance W. The image of w under P is a
Gaussian pseudomeasure w_P on Y whose Fourier transform is

$$\mathcal{F}w_P = \exp(-iW_P/2) \qquad \text{with} \qquad W_P = W \cdot \tilde{P}$$

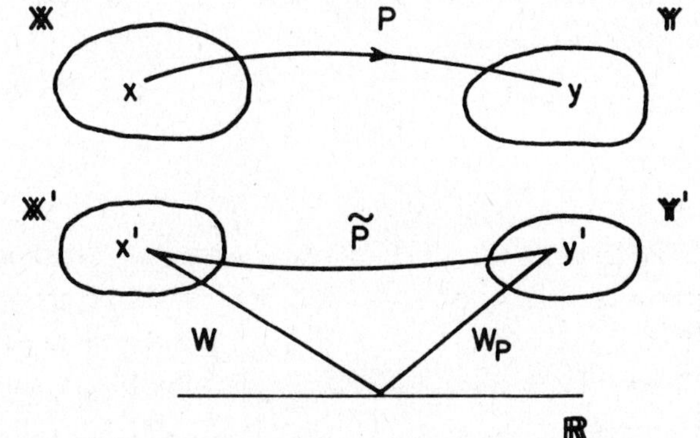

This proposition, together with the equation

$$\int_Y F(y)\,dw_P(y) = \int_X F \bullet P(x)\,dw(x)$$

makes it possible to compute many Feynman integrals.

Example: Let P : $X \to \mathbb{R}^{pn}$ by $x \mapsto y$ where y is the pn tuple $\{y^{i\alpha} = <\mu_{i\alpha}, x>\}$, then:

$$W_P(y') = y'_{i\alpha}\,W^{i\alpha j\beta}\,y'_{j\beta}$$

where $\qquad W^{i\alpha j\beta} = \underline{W}(\mu_{i\alpha}, \mu_{j\beta})$

and $\qquad dw_P(y) = (2\pi i)^{-pn/2}\,(\det W^{-1})^{1/2}\,\exp(\tfrac{i}{2}\,y^{i\alpha}\,(W^{-1})_{i\alpha j\beta}\,y^{j\beta})\,dy$

The integrand of a Feynman path integral is often a function F of a set $\{<\mu_{i\alpha}, x>\}$,
i.e., a function F $\bullet$ P of x. The path integral over X is then equal to the integral
of F(y) over $\mathbb{R}^{pn}$ with respect to the measure w_P computed in this example.
When $\mu_{i\alpha}$ is equal to the vector valued Dirac measure at t_i having only an α component
$W^{i\alpha j\beta} = G^{\alpha\beta}(t_i, t_j)$ and one can use the mapping P : $X \to Y$ by $x \mapsto <\delta^\alpha_{t_i}, x> = x^\alpha(t_i)$
to compare the two definitions of Feynman integral.
For mappings P other than $X \to \mathbb{R}^{pn}$ see [3].

III. The Feynman Green Function

Classical Physics is dominated by the Euler-Lagrange equation and Quantum
Physics by the small disturbance equation, i.e., respectively by the first and the
second variation of the action S. In this section we shall state the properties of
the second variation, old and new, necessary to compute the propagation kernel.

The expansion of S around the classical path $\bar{q}$ can be written

$$S(q) = S(\bar{q}) + \frac{1}{2} S''(\bar{q})xx + \Sigma(x)$$

In the example considered $\bar{q}$ is the geodesic from a to b; the second variation $S''(\bar{q})$,
also called the hessian of S at $\bar{q}$, is:

$$S''(\bar{q})xy = - \underset{t}{\Sigma}(y(t), \Delta\dot{x}(t))_g - \int_T (y(t), \ddot{x}(t) + R(\dot{\bar{q}}(t),x(t))\dot{\bar{q}}(t))_g dt$$

where Δ means "difference at a discontinuity": $\Delta\,\dot{x}(t) = \dot{x}(t^+) - \dot{x}(t^-)$ and where R is
the Riemann tensor; x is a vector field along $\bar{q}$, it is an element of the tangent
space at $\bar{q}$ of the space $\mathbb{F}$ of paths q : $\mathbb{T} \to \mathbb{M}$; i.e., $x \in \mathbb{TF}_{\bar{q}}$ and $x(t) \in \mathbb{TM}_{\bar{q}(t)}$
x is called a Jacobi field if and only if

$$S''(\bar{q})xy = 0 \qquad \text{for every} \quad y \in \mathbb{TF}_{\bar{q}}$$

A Jacobi field is a C^∞ - differentiable solution of the small disturbance equation,
it will be denoted $\bar{x}$. The small disturbance equation, also called the Jacobi differential equation has 2n linearly independent solutions.
The Jacobi fields can be obtained by an m-parameter variation through geodesics [5].
An m-parameter variation $\bar{\alpha}$ of a path q is a mapping

$$\bar{\alpha} : \mathbb{U} \subset \mathbb{R}^m \to \mathbb{F} \quad \text{such that} \quad \bar{\alpha}(0) = \bar{q}$$

It is convenient to introduce the mapping α, also called m-parameter variation of
q:

$$\alpha : \mathbb{U} \times \mathbb{T} \to \mathbb{M} \quad \text{by} \quad \alpha(u,t) = \bar{\alpha}(u)(t)$$

An m-parameter variation defines m "variation vector fields" $\{x_i\}$ by

$$x_i(t) = \frac{\partial\alpha}{\partial u^i}(0,t)$$

conversely

$$\alpha(u,t) = \exp_{\bar{q}(t)}(\Sigma u^i x_i(t))$$

When the family $\{\bar{\alpha}(u)\}$ is a family of geodesics, an m-parameter variation $\bar{\alpha}$ is called
a variation through geodesics.

Two points $\bar{q}(t_a) = a$ and $\bar{q}(t_b) = b$ are said to be conjugate along $\bar{q}$ if there exists a non zero Jacobi field $\bar{x}$ along $\bar{q}$ vanishing at t_a and t_b. The dimension of the vector space of all such Jacobi fields is called the multiplicity of the conjugate points. Lemma: Two points a and b are conjugate along $\bar{q}$ if and only if the mapping $\exp_a$ is critical at $(t_b - t_a)\, \dot{\bar{q}}(t_a)$; i.e., if its derivative mapping at the critical point is not one-one. Indeed, let $\dot{x}(t_a)$ be the covariant derivative of the non-zero Jacobi field vanishing at t_a and t_b.

$$\exp'_a((t_b - t_a)\dot{\bar{q}}(t_a))\,(t_b - t_a)\,\dot{x}_a = 0$$

A Jacobi field is determined by its values at any two non-conjugate points. Thus the Jacobi field equal to x_a at t_a and x_b at t_b is

$$\bar{x}(t) = -J(t,t_b)\, M(t_b,t_a) x_a - J(t,t_a)\, M(t_a,t_b) x_b$$

where $M(t_b,t_a)$ is the inverse of $J(t_b,t_a)$ and where $J(t,t_b)$ is the antisymmetric Jacobi two point vector along $\bar{q}$ such that

$$\begin{cases} J(t_b,t_a) = 0 \\ \dfrac{DJ}{dt} = (t = t_b,t_b) = g^{-1}(\bar{q}(t_b)) \text{ which will be abbreviated to } g^{-1}(t_b) \end{cases}$$

J is also known as the commutator function.
It has been shown [6] that M is the Van Vleck matrix:

$$M(t_b,t_a) = \partial^2 \bar{S}(a,b)/\partial b \partial a \quad \text{where} \quad \bar{S}(a,b) = S(\bar{q}) \quad .$$

When the end points are conjugate, we determine the Jacobi field by its Cauchy data

$$\bar{x}(t) = J(t,t_a)\, g(t_a)\dot{x}(t_a) + K(t,t_a)g(t_a)x_a$$

where $K(\cdot,t_a)$ is a Jacobi field along $\bar{q}$ defined by

$$K(t_a,t_a) = g^{-1}(t_a)$$

$$\frac{DK}{dt}(t = t_a,t_a) = 0 \quad .$$

Proposition 2. The end points a and b are conjugate along $\bar{q}$ if and only if $\det J(t_b,t_a) = 0$. The conjugate points are degenerate if and only if $\det K(t_b,t_a) = 0$. This proposition provides a convenient criterion for the onset of catastrophes in path integrals, namely:
For a fixed point $a \in M$ the catastrophe set of points $b \in M$ satisfies:

$$\begin{cases} \det J(t_b,t_a) = 0 \\ \det K(t_b,t_a) = 0 \end{cases}$$

The proof of proposition 2 and the study of catastrophes is the subject of another paper.

The Feynman Green function is the Green function of the small disturbance operator that vanishes on the boundary.

Proposition 3: The Feynman Green function is equal to

$$G(r,s) = \frac{\hbar}{m} Y(r-s)J(r,t_b)M(t_b,t_a)J(t_a,s) - \frac{\hbar}{m} Y(s-r)J(r,t_a)M(t_a,t_b)J(t_b,s)$$

where Y is the Heaviside step function equal to unity for positive argument and zero otherwise.

Proof: see [6]

When a and b are conjugate along $\bar{q}$, the Feynman Green function is not defined, and we shall need another Green function G_- of the small disturbance operator to define the propagation kernel:

$$G_-(r,s) = \frac{\hbar}{m} Y(r-s)K(r,t_b)N(t_b,t_a)J(t_a,s) - Y(s-r)J(r,t_a)N(t_a,t_b)K(t_b,s)$$

where $N(t_b,t_a)$ is the inverse of $K(t_b,t_a)$ defined in the previous paragraph.

IV. The Propagation Kernel $K(B;A) = K(b,t_b;a,t_a)$

The propagation kernel of a system from a state A to a state B is the probability amplitude for the transition $A \to B$. It gives the wave function Ψ_{t_b} at t_a in terms of the wave function Ψ_{t_a} at t_a:

$$\Psi_{t_b}(b) = \int_{\mathbb{M}} K(b,t_b;a,t_a) \, \Psi_{t_a}(a) (\det g(a))^{1/2} da$$

According to Feynman's original definition

$$K(B;A) = \lim_{p \to \infty} \int_{\mathbb{R}^{pn}} K(B;p)K(p;p-1) \ldots K(1;A) dq^1 \ldots dq^p \quad \text{where [4]}$$

$$K(k+1;k) = (\frac{1}{2\pi i})^{n/2} (\det \frac{-1}{\hbar} \frac{\partial^2 \bar{S}(k+1;k)}{\partial q^{k+1} \partial q^k})^{1/2} \exp(\frac{i}{\hbar} \bar{S}(k+1;k))$$

$\bar{S}(k+1;k)$ is a function of q^{k+1}, t_{k+1}, q^k, t_k such that

$$\lim_{\substack{p \to \infty}} \sum_{k=1}^{p} \bar{S}(k+1;k) = S(q) \overset{\text{def}}{=} \int_{\mathbb{T}} L(q(t), \dot{q}(t),t) \, dt$$

where L is the lagrangian of the system and S its action. We can write $\bar{S}(k+1;k) = \int_{t_k}^{t_{k+1}} L(q(t), \dot{q}(t),t) dt$ provided we give some prescription for the path $q: [t_k,t_{k+1}] \to M$.

A natural prescription for $q \mid [t_k,t_{k+1}]$ is the classical path from q^k to q^{k+1}.

We shall give the new definition of the propagation kernel in three steps:

a) The paths $q \in F(a,b)$ map $\mathbb{T}$ in a geodesically convex neighborhood $\mathbb{N}$ of $\mathbb{M}$.

b) The end points a and b may be conjugate points along $\bar{q}$, but there is no other conjugate point along $\bar{q}$; the conjugate points are non-degenerate.

c) There are several conjugate points along $\bar{q}$; they are non-degenerate; the end points are not conjugate.

<u>Proposition 4a.</u> When $q : \mathbb{T} \to \mathbb{N}$, the propagation kernel is

$$K(B;A) = \exp \frac{i}{\hbar} S(\bar{q}) \int_{\mathbb{X}} \exp \frac{i}{\hbar} \Sigma(x) \, dw(x)$$

where w is the gaussian pseudomeasure on $\mathbb{X}$ defined by its Fourier transform

$$\mathcal{F}w = w(\mathbb{X}) \exp(-iW/2)$$

$$\underline{W}(\mu,\nu) = \int_{\mathbb{T}} d\mu_\alpha(r) \int_{\mathbb{T}} d\nu_\beta(s) \, G^{\alpha\beta}(r,s)$$

G is the Feynman Green function; $w(\mathbb{X}) = (\det M(t_b,t_a))^{1/2}/(2\pi i)^{n/2}$

Proof and applications: see [6] and [3]

This definition is equivalent to the original one when the original one is unambiguous, it does not require an ad hoc prescription for the $\mathbb{R}^{pn}$ integral, it is defined for a large class of physical systems, and leads to simple and powerful computational techniques.

<u>Proposition 4b.</u> Let $\mathbb{Y}$ be the space of vector fields along $\bar{q}$ which vanishes at t_a but take arbitrary values at t_b. Let w_- be the gaussian pseudomeasure of covariance G_- normalized to

$$w_-(\mathbb{Y}) = (\det g^{-1}(t_b) \, N(t_b,t_a))^{1/2}$$

Then w_- on $\mathbb{Y}$ induces w on $\mathbb{X} \subset \mathbb{Y}$:

$$\int_{\mathbb{Y}} x(y) <\delta_r,y> <\delta_s,y> \, dw_-(y) \Big/ \int_{\mathbb{Y}} x(y) \, dw_-(y) = iG(r,s)$$

$$\int_{\mathbb{Y}} x(y) \, dw_-(y) = \begin{cases} \int \int_{\mathbb{X}} dw(x) & \text{for non conjugate end points} \\[2ex] w_-(\mathbb{Y}) \, \delta & \text{for conjugate non-degenerate end points} \end{cases}$$

where δ is the Dirac distribution at the origin of $\mathbb{R}^n$.

The propagation kernel is equal to

$$K(B;A) = \exp \frac{i}{\hbar} S(\bar{q}) \int_{\mathbb{Y}} \mathcal{X}(y) \, \exp \frac{i}{\hbar} \Sigma(y) \, dw_-(y)$$

where $\mathcal{X}(y)$ is the characteristic of function of $\mathbb{X} \subset \mathbb{Y}$.

The proof is given in [6]. This proposition serves two purposes:

It provides the formalism for integrating over paths with one fixed end point and one arbitrary point.

It makes it possible to approach a point b conjugate to a along $\bar{q}$ from a point b_u not conjugate to a along the geodesic $\bar{q}_u$ from a to b_u; the parameter $u \in \mathbb{R}^m$, where m is the multiplicity of the conjugate points, defines an m-parameter variation of $\bar{q}$ through geodesics such that $b_u = \exp_b u y_b$. One obtains in particular:

$$\int_{\mathbb{Y}} \mathcal{X}(y;y(t_b) = y_b) \, dw_-(y)$$

$$= \frac{1}{(2\pi i)^{n/2}} \det{}^{1/2}(M(t_b,t_a)) \exp(-\frac{i}{2} y_b g(t_b) K(t_b,t_a) M(t_a,t_b) y_b)$$

giving the value of $\int_{\mathbb{Y}} x(y) \, dw_-(y)$ stated in the proposition

Schulman [7] has examined propagation kernels between conjugate points in terms of the eigenvalues of the small disturbance operator.

Proposition 4c. There may be several non degenerate conjugate points along $\bar{q}$.

Let the end points a,b be two points in $\mathbb{M}$ which are not conjugate along any geodesic. Let π be the fundamental group of a countable CW-complex which contains one cell of dimension λ for each geodesic from a to b of Morse index λ.

Let $K_\alpha(B;A)$ be the partial propagation kernel for all the paths from a to b in the same homotopy class α, computed according to proposition 1. Then the absolute value of the propagation kernel is

$$\mid K(B;A) \mid = \mid \sum_{\alpha \varepsilon \pi} \chi(\alpha) \, K_\alpha(B;A) \mid$$

where $\{\chi(\alpha); \alpha \varepsilon \pi\}$ is the set of characters of the fundamental group.

Proof: Because there is no unique way to label the homotopy classes by the elements α of the fundamental group, K(B;A) is determined only modulo an overall unobservable phase factor [1]. The proof rests on the fundamental theorem of Morse theory [5] which states the homotopy type of $\mathbb{F}(a,b)$ and on the theorem giving the propagation kernel on a multiply connected space [1]. The kernel K has been computed by Gutzwiller [8]. Proposition 4c and its proof give a rigorous derivation for his expression.

I am grateful to Bryce S. DeWitt for discussions of his unpublished work on the small disturbance equation. Approaching the problem from a different perspective they have led to several fights beneficial to the author.

[1] Laidlaw, M.G.G., DeWitt-Morette, C.: Phys. Rev. D3, 1375-1378 (1971). Laidlaw, M.G.G.: Ph.D. Thesis, University of North Carolina at Chapel Hill, 1971.

[2] DeWitt-Morette, C.: Commun. math. Phys. 28, 47-67 (1972).

[3] DeWitt-Morette, C.: Commun. math. Phys. 37, 63-81 (1974).

[4] Morette, C.: Phys. Rev. 81, 848-852 (1951).

[5] Milnor, J.: Morse Theory. Princeton, N. J.: Princeton University Press 1963.

[6] DeWitt-Morette, C.: "Path Integrals on Riemannian Manifolds" Preprint 1975.

[7] Schulman, L.S.: "Caustics and Multivaluedness: Two results of Adding Path Amplitudes" to appear in the Proceedings of the International Conference on Functional Integration and its Applications, London 1974.

[8] Gutzwiller, M.C.: J. Math. Phys. 8, 1979-2000 (1967).

BOSE QUANTUM FIELD THEORY AS AN ISING
FERROMAGNET: RECENT DEVELOPMENTS

Barry Simon*
Physics and Mathematics Departments
Princeton University
Princeton, New Jersey
08540

One of the main new tools made available in constructive field
theory by the Euclidean revolution of 1971-73 is the approximation of
scalar field theory by Ising models and the resulting correlation inequalities. There has been quite active development of this line in the
past year and our goal here is to review some of these developments.
While we will have something to say about the basic ideas of the Ising
approximations, we will suppose that the reader has some previous ex-
posure to them, either from the original papers [Guerra, Rosen, Simon
(1973) and Simon, Griffiths (1973)] or from the pedagogic reviews [
Simon (1974 b,c) or the contributions of Guerra, Rosen, and Simon to
Velo-Wightman (1973)].

§1. The Basic Framework

The basis of the Ising methodology is two approximations: the
lattice approximation [Guerra, Rosen, Simon (1973)] and the classical
Ising approximation [Simon-Griffiths (1973)]. Rather than state precise
technical results, we paraphrase the original results in two metatheorems
below. By a "generalized Ising model" we mean a finite collection of
random variables ("spins") whose joint probability distribution has the
form

$$\exp\left(+ \sum_{i<j} a_{ij} x_i x_j\right) d\nu_1(x_1) \ldots d\nu_n(x_1)$$

* A. Sloan fellow, research partially supported by U.S.N.S.F. under grant
GP-39048
Supported in part by AFOSR under Contract F44620-71-C-0108.

for measures ν_i on R. By a "classical Ising model", we mean a model where each "spin", x_i can only take the values ± 1. If each $a_{ij} \geq 0$, the model is called "ferromagnetic".

<u>Metatheorem 1</u> (the "lattice approximation") Any $P(\phi)_2$ Euclidean field theory is a limit of generalized ferromagnetic Ising models and if P is even, each ν_i is even.

<u>Metatheorem 2</u> (the "classical Ising approximation") Any $P(\phi)_2$ Euclidean field theory with $P(X) = aX^4 + bX^2 - \mu X$ is a limit of ferromagnetic classical Ising models with $d\nu_i(x_i) = D_i \exp(+c_i \mu x_i)[\delta(x_i+1) + \delta(x_i-1)]$ for suitable positive constants D_i, c_i.

<u>Remarks</u> 1. As noted in the original papers, these theorems extend <u>formally</u> to three and four space-time dimensions; (for three dimensions, see below).

2. The classical Ising approximation proceeds in two steps. One first passes to the lattice approximation and then approximates the spins of the lattice approximation by classical Ising spins. The latter part is O.K. in three or four dimensions so the removal of "formal" in Remark 1 is only a question of convergence of the lattice approximation.

3. We have not been too explicit about what we mean by "limit". What is critical is that the Euclidean fields are limits of positive linear combinations of spins in a way that allows one to immediately extend any multilinear inequality ("correlation inequality") from the spin systems to the field theories.

There have been several extensions of the general formalism in the past year:

<u>Theorem 1.1</u> [Park (1974)] The lattice approximation converges for the finite volume, free boundary condition $(\phi^4)_3$ field theory.

<u>Theorem 1.2</u> [Dunlop-Newman (1975)] The $(\phi_1^2 + \ldots + \phi_n^2)^2$ field theory in two dimensions is a limit of ferromagnetic magnetic "spin" systems where the spins take values on the n-1 dimensional sphere (i.e. plane rotor if $n=2$; classical Heisenberg if $n=3$).

<u>Theorem 1.3</u> [Guerra-Rosen-Simon (1975)] The $P(\phi)_2$ lattice approximation in a box with periodic or Neumann boundary conditions is convergent.

<u>Remarks</u> 1. Thus far, Park's result has found no applications. This situation should change with the control of the infinite volume $(\phi^4)_3$ theory (see Osterwalder's contribution to these proceedings). It would be useful to have Park's result extended to Dirichlet boundary conditions.

2. Dunlop-Newman have also proven Lee-Yang theorems for $n=2,3$ rotors [the $n \geq 0$ case is under active investigation (H. Mahler (private communication))] and thus one has analyticity of the pressure in certain multicomponent field theories.

3. The box lattice approximation with periodic B.C. of course has spins coupled at opposite boundary points. Neumann B.C. involves dropping a term $\frac{1}{2}(q-q')^2$ from the action for any pair of nearest neighbors with q in the region and q' out (for comparison, with Dirichlet B.C., one only drops the $-qq'$ term).

4. Baker (1974a) has also discussed the Neumann lattice theory (which he unfortunately calls "free boundary conditions").

5. Correlation inequalities between Dirichlet and Periodic states have been exploited by Guerra, Rosen, and Simon (1974) [see §5, below].

§2. Some New Correlation Inequalities

Many of the recent applications of correlation inequalities in field theory have made use of inequalities proven in the past year and a half and, in turn, many of these inequalities have been proven by authors interested in field theory applications. The three new classes of inequalities are:

<u>Theorem 2.1</u> [Lebowitz (1974)] In a classical ferromagnetic Ising model with positive external field (i.e. $d\nu_i = D_i e^{+\mu_i x_i}[\delta(x_i+1) + \delta(x_i-1)]$ for any finite sets A and B:

$$\sum_{C \subset A; D \subset B} (-1)^{|C|+|D|} \, T(C,L;A \backslash C, B \backslash D) \geq 0$$

$$\sum_{C \subset A; D \subset B} (-1)^{|C|} \, T(C,D;A \backslash C, B \backslash D) \leq 0$$

where

$$T(C,D;E,F) = \langle S^{C \Delta D} \rangle \, \langle S^{E \Delta F} \rangle - \langle S^C \rangle \, \langle S^D \rangle \, \langle S^E \rangle \, \langle S^F \rangle$$

with $S^C = \prod_{i \in C} S_i$.

<u>Theorem 2.2</u> [Newman (1975a)] Let Λ be any set. Let $\mathscr{P}$ be a family of partitions of Λ into two disjoint subsets Λ_1, Λ_2 so that any partition of Λ into pairs (or if $\#(\Lambda)$ is odd into pairs plus one one-element set) is a refinement of some $\{\Lambda_1, \Lambda_2\} \in \mathscr{P}$. Then, in any classical ferromagnetic Ising model with positive external field:

$$\langle \sigma^\Lambda \rangle \leq \sum_{\{\Lambda_1, \Lambda_2\} \in \mathscr{P}} \langle \sigma^{\Lambda_1} \rangle \langle \sigma^{\Lambda_2} \rangle$$

<u>Theorem 2.3</u> [Cartier (1974), Percus (1974), Sylvester (1974)] Define u_n by:

$$u_n(i_1, \ldots, i_n) = \frac{\partial^n}{\partial \mu_1 \ldots \partial \mu_n} \langle \exp(\mu_1 \sigma_{i_1} + \ldots + \mu_n \sigma_{i_n}) \rangle \Big|_{\mu_i = 0}$$

Then, in any <u>zero</u> field ferromagnetic classical Ising model:

$$u_6(i_1, \ldots, i_6) \geq 0$$

<u>Remarks</u> 1. Lebowitz inequalities are usually written in terms of "Percus variables" and look simpler in that form.

    2. The rather complex condition on $\mathscr{P}$ in Newman's inequality is not only sufficient for the inequalities in question to hold but also necessary in that should $\mathscr{P}$ not obey that condition, there are Ising ferromagnets where the inequality fails.

    3. By the "dummy spin trick", one need only prove Newman's inequality in zero field with $\#(\Lambda)$ even. Newman does this by a graphical expansion.

    4. If $\#(\Lambda)$ is even, Newman's inequalities imply by induction [Newman (1975a)]:

$$\langle \sigma^\Lambda \rangle \leq \sum_{\text{pairings}} \langle \sigma^{i_1 j_1} \rangle \ldots \langle \sigma^{i_n j_n} \rangle$$

where the sum runs over all ways of writing Λ as n disjoint pairs (and a similar inequality by the dummy spin method if $\#(\Lambda)$ is odd). Earlier Glimm-Jaffe (1974b) have proven an inequality of this form from Lebowitz inequalities but with a constant of order $n!$ in front of the sum.

    5. The GKS inequalities assert that in positive field $u_1 \geq 0$,

$u_2 \geq 0$ and GHS imply that $u_3 \leq 0$; Lebowitz' inequalities imply $u_4 \leq 0$ at zero field. It is conjectured that $u_{2n}(-1)^{n+1} \geq 0$ at zero field.

6. As of yet, no application of $u_6 \geq 0$ is known. Feldman (1974) has remarked that the general inequality $u_{2n}(-1)^n \geq 0$ at zero field would yield a proof of a mass gap result that has now been proven by Spencer using alternate means (see §5).

7. These inequalities all extend to ϕ^4 Euclidean field theories (and the first two to $\phi^4 - \mu\phi$ theories) if σ_i is replaced by $\phi(x_i)$.

§3. Construction of States

Most of the recent applications of the Ising approximation in field theory have extended trends set by the very earliest applications (Guerra, Rosen, Simon (1973), Simon, Griffiths (1973), Nelson's contribution to Velo-Wightman (1973), and Simon (1973, 1974a)) which in turn followed trends set by the applications of correlation inequalities in statistical mechanics.

One of Griffiths' original applications of his inequalities was to prove the existence of the infinite volume limit for the correlation functions of a spin system. Nelson extended this idea to construct an infinite volume $P(\phi)_2$ theory for $P =$ even poly $- \mu X$ (the half-Dirichlet state). Glimm and Jaffe (1974 b) have recently combined correlation inequalities with the cluster expansion (see the contribution of Glimm, Jaffe, and Spencer to Velo-Wightman (1973)) to construct an infinite theory for these P ("weak coupling boundary conditions") which they then show is Nelson's state. One advantage of this construction is that it provides a proof of $:\phi^j: (j \leq \deg P)$ bounds for Nelson's state. These bounds have been used by Glimm-Jaffe (1973) and by Fröhlich (1974b).

§4. Domination by the Two-Point Function

One of the earliest results obtained using Ising methods in field theory [Simon (1973)] asserts that in any $P(\phi)_2$ theory the mass gap is determined by the falloff of the truncated two-point Schwinger function. Newman (1975a) has found a new proof of this for even ϕ^4 theories.

For as a special case of his inequalities (Theorem 2.2), he considers $\Lambda = \{1,\ldots,n; 1',\ldots,m'\}$ with $n+m$ even, his inequalities include:

$$(1\ldots n\ 1'\ldots m') \leq (1\ldots n)\ (1'\ldots m')$$
$$+ \sum_{i,j} (ij')\ (1\ldots\hat{i}\ldots n\ 1'\ldots\hat{j}\ldots m') \qquad (*)$$

which implies the stated falloff.

Glimm and Jaffe (1974b) have shown that the two-point function dominates more than just rate of falloff. Using Lebowitz' inequality inductively they prove bounds on $S_n \equiv \langle\phi(x_1)\ldots\phi(x_n)\rangle$ which include:

Theorem 4.1 [Glimm-Jaffe (1974b)] In any $\phi^4 - \mu\phi$ field theory, for any positive $f_1,\ldots,f_n$

$$0 \leq S_n(f_1,\ldots,f_n) \leq 2^{n-1}(n-1)!\ \|f_1\| \cdots \|f_n\|$$

where $\|f\| = S_2(f,f)^{1/2}$.

Remarks 1. Letting $\|\|f\|\| = S_2(|f|,|f|)^{1/2}$ one has that

$$|S_n(f_1,\ldots,f_n)| < 2^{n-1}(n-1)!\ \|\|f_1\|\| \cdots \|\|f_n\|\|$$

2. The $2^n(n-1)!$ in these bounds can be replaced by $\left[\frac{n+1}{2}\right]!$ (see theorem 4.3 below).

3. The bound in Remark 1 is precisely of the form needed to be able to recover the Minkowski region according to the (revised) Osterwalder-Schrader axioms [Osterwalder-Schrader (1975)].

4. Formally, these bounds hold in four dimensions. To be of use there, S_2 must be a distribution at coincident (i.e. an L^1 function). It is order by order in perturbation theory which is suggestive that it will be also in the actual theory.

Newman (1975 a,b) has extended Theorem 4.1 in two different ways:

Theorem 4.2 (Newman (1975b)) In any $\phi^4 + a\phi^2 - \mu\phi$ $(\mu\geq0)$, field theory, for any positive test function, f:

$$\langle e^{\phi(f)}\rangle \leq \exp[S_1(f) + \tfrac{1}{2} S_2^T(f,f)]$$

where

$$S_2^T(x,y) = S_2(x,y) - S_1(x)S_1(y)$$

<u>Theorem 4.3</u> (Newman (1975a)) In any even ϕ^4 theory:

$$0 \leq S_{2n}(x_1,\ldots,x_{2n}) \leq \sum_{\text{pairs}} S_2(x_{i_1},x_{j_1})\ldots S_2(x_{i_n},x_{j_n})$$

<u>Remarks</u> 1. The proof of Theorem 4.2 is surprisingly short and simple considering its detailed information. Let $F(\mu) = \ell n<e^{\mu\phi(f)}>$. Then GHS directly implies that $d^3F/d\mu^3 \leq 0$ for all $\mu > 0$! Since $F(0) = 0$, $F'(0) = S_1(f)$ and $F''(0) = S_2^T(f,f)$ we immediately find that

$$F(\mu) \leq \mu S_1(f) + \frac{1}{2}\mu^2 \, S_2^T(f,f)$$

2. Theorem 4.3 is just the translation of one of Newman's inequalities to field theory (see Remark 4 in §2). There are also inequalities for $\phi^4-\mu\phi$ theories.

3. Theorem 4.3. has a rather dramatic sounding restatement: the Schwinger functions of any ϕ^4 theory are dominated by those of the generalized free field with the same two-point function.

4. Theorem 4.3 is quite directly an extension of Theorem 4.1 improving the constant in front of the product of norms. Theorem 4.2 implies bounds on $S_n(f)$ via two remarks. First, by GKS and $|<A>| \leq <|A|>, G(\mu) \equiv <e^{\mu\phi(f)}>$ obeys $|G(\mu)| \leq G(|\mu|)$. Next, Cauchy estimates on the entire function $G(\mu)$ implies bounds on $G^{(m)}(0)$.

5. (This remark is due independently to the author and J. Fröhlich) In one sense, Theorem 4.2 is a very important improvement of Theorem 4.1. For, if S_2^T has very weak falloff properties (and such falloff is to be expected if $\mu > 0$; see §5 below; also note that ϕ bounds for some $\mu > 0$ implies ϕ bounds for $\mu = 0$ by GKS) then Theorem 4.2 implies that

$$<\exp(\phi(h \otimes \chi_{(0,T)}))> \leq \exp(cT)$$

at least under some weak regularity assumption on S_2^T at the coincidence singularity. Modulo technical details, it is a basic result of Fröhlich (1974a) (see Simon (1974b) for additional discussion) that such bounds imply ϕ-bounds in the sense of Glimm-Jaffe (1972).

§5. Mass Gap and Spectral Results

Cluster expansions and Bethe-Salpeter equations have proved a power-ful tool for studying the mass spectrum in weakly coupled (and large fugacity) theories   (see Glimm's contribution to these Proceedings). Thus far, correlation inequalities have provided the only tool for studying these questions in the strong coupling regime where they provide much less information than is available in the weak coupling regime. Some new information on these questions has been obtained in the past year:

Theorem 5.1 [Guerra, Rosen, and Simon (1974)] There is a mass gap (i.e. 0 is an isolated simple eigenvalue of the Hamiltonian) in any $(a\phi^4 + b\phi^2 - \mu\phi)_2$ field theory with $\mu \neq 0$.

Theorem 5.2 [Spencer (1974)] In any even $(\phi^4)_2$ theory, the first excited even state has an energy at least twice as large as the first excited state.

Remarks 1. These extend earlier results which used correlation inequali-ties, namely those of Simon (1974a) and of Glimm, Jaffe, and Spencer (their contribution to Velo-Wightman (1973)) respectively.

2. Theorem 5.1 follows Ising model arguments of Lebowitz and Penrose (1974). The main difficulties are technical ones involving boundary conditions.

3. By even state, we mean one obtained by applying an even number of fields to the vacuum. In more physical terms, Theorem 5.2 asserts that in a $(\phi^4)_2$ theory there are no even "G-parity" bound states (at least, below the two-particle threshold).

4. Spencer's proof of Theorem 5.2 uses Lebowitz inequalities and a very clever trick. Newman (1975a) has a simple proof using his inequality (*) quoted in 4. For, if n and m are even, the sum in (*) has exp(-2mt) falloff if the prime and unprimed indices are sepa-rated by Euclidean time, t.

§6. "Coupling Constants" Variation

Another subject of considerable interest involves monotonicity and smoothness information about physical parameters (mass, vertex functions at special points, etc.) as functions of bare "coupling constants" (bare mass, bare coupling constant, etc.). The earliest applications of correlation inequalities by Guerra, Rosen, and Simon (1973) include a statement of this type, which in one form says that in a $P(\phi)_2$ theory with $P(X) = a_{2m}X^{2m} + \ldots + a_2X^2$ (even P), the physical mass is monotone increasing as a_2 increases (with $a_{2m},\ldots,a_4$, m_0 fixed). Recently, Glimm and Jaffe, in a series of papers [Glimm, Jaffe (1974a,c,d,e); see Guerra, Rosen, Simon (1974) for an additional critical index result] have proven a large number of such bounds and bounds also on critical exponents. We quote an example:

<u>Theorem 6.1</u> (Glimm-Jaffe (1974a)) For fixed m_0 and λ, let $m(\sigma)$ denote the physical mass of the $(\lambda\phi^4 + \frac{1}{2}\sigma\phi^2)_2$ theory. Then for $\sigma > \sigma_c$ (the critical value where the even theory stops possessing a mass gap) $m(\sigma)$ is Lipschitz continuous and for $\sigma > \sigma' > \sigma_c$:

$$m(\sigma)^2 - m(\sigma')^2 \leq (\sigma - \sigma')$$

<u>Remarks</u> 1. If $m(\sigma)$ were differentiable, the GRS result quoted above says that $dm^2/d\sigma \geq 0$ and the Glimm-Jaffe results say that $dm^2/d\sigma \leq 1$.
    2. If $m(\sigma) \to 0$ as $\sigma \to \sigma_c$ [Baker (1974b) has announced a closed related result], then one has that $m(\sigma) \leq (\sigma - \sigma_c)^{1/2}$.

<u>§7. The :cos ϕ:$_2$ Theory</u>

I would like to briefly describe some recent work of Fröhlich (1975) which doesn't fit into either the general topic of my talk or of Glimm's talk but which I feel should be mentioned at this conference. Fröhlich has constructed a :cos ϕ:$_2$ theory (quantized (massive) sine-Gordon equation) or more accurately the $\int_0^\alpha d\nu(\alpha)[:\cos(\alpha\phi+\eta(\alpha)):]$ theory where η is a function, ν a signed measure and one needs $\alpha < \sqrt{4\pi}$ for the finite volume and small coupling theories and $\alpha < 4/\sqrt{\pi}$ for the strongly coupled theories. The introduction of a new beast to the zoo of two-dimensional field theories (and a beast which is non-renormalizable in three or more dimensions at that!) does not, in itself, seem cause for excitement. But this theory has one extremely striking property: <u>for sufficiently small coupling constant, the Feynman series for all the</u>

<u>Schwinger functions are convergent</u>! (This is definitely false for $(\phi^4)_2$ - Jaffe (1965)).

<u>Remarks</u> 1. The intuitive reason the series are convergent <u>once the theory</u> <u>is proven to exist</u> is quite simple. $:\cos \phi:_2$ is equivalent to $:\sin \phi:_2$ by translation of the field. But $\lambda:\sin \phi:_2$ and $-\lambda:\sin \phi:_2$ are equally good theories via $\phi \rightarrow -\phi$ covariance.

  2. The key requirement is the control of $<\exp(-\lambda U_\Lambda)>$ for finite Λ and $\lambda \in R$, for then the cluster expansion machine can be turned to prove analyticity of the Schwinger functions.

  3. Fröhlich uses the idea of Albeverio-Hoegh Krohn (1973) that these theories are equivalent to certain statistical mechanical theories. By some clever use of Guerra, Rosen, Simon (1973), he reduces control of $<\exp(-\lambda U_\Lambda)>$ to control of a purely Coulomb system in two dimensions (in finite volume and with image charges) and then appeals to methods of Deutsch-Lavand (1974).

REFERENCES

S. Albeverio and R. Hoegh Krohn (1973), Commun. Math. Phys. <u>30</u>, 171-200.

G. Baker (1974a) Brookhaven preprint on field theories.

G. Baker (1974b) Brookhaven preprint on correlation lengths in Ising
  models.

P. Cartier (1974) in preparation.

C. Deutsch and M. Lavand (1974) Phys. Rev. <u>A9</u> (1974) 2598.

F. Dunlop and C. Newman (1975), Indiana-IHES preprint, in preparation.

J. Feldman (1974) Canad. J. Phys. <u>52</u> (1974) 1583.

J. Fröhlich (1974a) Helv. Phys. Acta, to appear.

J. Fröhlich (1974b) Ann. Inst. H. Poincaré, to appear.

J. Fröhlich (1975) Manuscript in preparation.

J. Glimm and A.Jaffe (1972) J. Math. Phys. <u>13</u> 1568.

J. Glimm and A. Jaffe (1974a) Phys.Rev. <u>D10</u> 536.

J. Glimm and A. Jaffe (1974b) Phys. Rev. Lett. <u>33</u> 440.

J. Glimm and A. Jaffe (1974c) Preprint: "On the approach to the critical
  point".

J. Glimm and A. Jaffe (1974d) Preprint: "Absolute bounds on vertices and couplings".

J. Glimm and A. Jaffe (1974e) Preprint: "Two and three body equations in quantum field models".

F. Guerra, L. Rosen, and B. Simon (1973) Ann. Math., to appear.

F. Guerra, L. Rosen, and B. Simon (1974) Commun. Math. Phys., to appear.

F. Guerra, L. Rosen, and B. Simon (1975) in preparation.

A. Jaffe (1965) Commun. Math. Phys. $\underline{1}$ 127.

J. Lebowitz (1974) Commun. Math. Phys. $\underline{35}$ 87.

J. Lebowitz and S. Penrose (1974) Yeshiva preprint.

C. Newman (1975a) Preprint in preparation on new inequalities.

C. Newman (1975b) Preprint in preparation on random variables of type $\mathcal{L}$.

K. Osterwalder and R. Schrader (1975) Preprint.

Y. Park (1974), IBM preprint.

J. Percus (1974) N.Y.U. preprint.

B. Simon (1973) Commun. Math. Phys. $\underline{31}$, 127.

B. Simon (1974a) Ann. Math., to appear.

B. Simon (1974b) "The $P(\phi)_2$ Euclidean (Quantum) Field Theory", Princeton Press.

B. Simon (1974c), Proc. Int. Cong. Math.

B. Simon and R. Griffiths (1973) Commun. Math. Phys. $\underline{33}$ 145.

T. Spencer (1974) Commun. Math. Phys., to appear.

G. Sylvester (1974) Harvard preprint.

G. Velo and A.S. Wightman (1973) "Constructive Quantum Field Theory", Springer Lecture Notes in Physics No. 25.

Arafune, J. 240
Arens, R. 312
Borchers, H. J. 283
Brüning, E. 72
Chikashige, Y. 187
Combes, J. M. 467
DeWitt-Morette, C. 535
Di Castro, C. 342
Doplicher, S. 264
Emch, G. G. 315
Feldman, J. S. 151
Freund, P. G. O. 240
Fujii, Y. 261
Fukushima, Masahisa 528
Fukushima, Masatoshi 224
Glimm, J. 118
Goebel, C. J. 240
Gürsey, F. 189
Hasegawa, H. 433
Hepp, K. 138
Hosoya, A. 238
Iagolnitzer, D. 1
Ikebe, T. 458
Ikeda, K. 520
Ishida, J. 238
Itô, K, ...218
Izuyama, T. 353
Jackiw, R. 319,394
Jaffe, A. 118
Jona-Lasinio, G. 326
Kamefuchi, S. 107
Kametaka, Y. 401
Kamimura, K. 187
Karaki, Y. 524
Kashiwara, M. 30
Kato, Y. 170
Kawai, T. 38
Kikkawa, K. 128
Kinoshita, K. 116
Kinoshita, T. 55
Klauder, J. R. 160
Konisi, G. 184
Kubo, I. 230
Kubo, Reijiro 199
Kubo, Ryogo 274
Kuramoto, Y. 420
Kuroda, S. T. 472
Lassner, G. 297
Lebowitz, J. L. 370,432
Lewis, J. T. 516
Lions, J.-L. 356
Maskawa, T. 242
Matsuura, S. 380
Mayer, M. E. 532
Mimura, M. 408

Minami, M. 175
Miyamoto, M. 228
Mochizuki, K. 486
Mori, H. 413
Morimoto, M. 49
Mühlschlegel, B. 437
Nakajima, H. 242
Nakanishi, N. 245
Nakano, H. 336
Nakazawa, N. 202
Ninomiya, M. 111
Nishida, T. 408
Nishijima, K. 205
Niwa, T. 236
Ogawa, T. 339
Ohnuki, Y. 107
Osterwalder, K. 151
Pohlmeyer, K. 59
Pulè, J. V. 516
Reeh, H. 249
Reents, G. 293
Robinson, D. W. 303
Ruelle, D. 449
Saito, T. 184
Saitō, Y. 476
Sasakawa, T. 491
Sato, M. 13
Schlieder, S. 85
Schroer, B. 92
Sekine, K. 196
Sewell, G. L. 510
Shelest, V. P. 114
Simon, B. 543
Slavnov, A. A. 214
Sommer, G. 66
Souillard, B. 497
Stapp, H. P. 38
Stichel, P. 72
Suzuki, M. 423
Symanzik, K. 102
Takahashi, M. 446
Takahashi, Y. 261
Thomas, L. E. 483
Toda, M. 387
Ukawa, A. 55
Verbeure, A. 504
Vladimirov, V. S. 281
Watanabe, K. 111
Yamada, M. 202
Yamazaki, Y. 349
Yokoyama, K. 199
Yoneya, T. 180
Zavialov, O. I. 256
Zittartz, J. 330

Subject Index

(A) <u>Quantum field theory and theory of elementary particles</u>
  (1) Constructive field theory.

     J. S. Feldman and K. Osterwalder: The Wightman axioms
      and the mass gap for weakly coupled $(\Phi^4)_3$
      quantum field theories 151

     J. Glimm and A. Jaffe: Particles and bound states
      and progress toward unitarity and scaling 118

     Y. Kato: Bound states and asymptotic fields in
      model field theories 170

     J. R. Klauder: Soluble models and the meaning of
      nonrenormalizability 160

     B. Simon: Bose quantum field theory as an Ising
      ferromagnet: recent developments 543

  (2) Algebraic approach.

     R. Arens: A method for making relativistic non-
      quantum systems by constructing the Hamiltonian
      and the other nine generating functions 312

     H. J. Borchers: Algebraic aspects of Wightman
      quantum field theory 283

     S. Doplicher: The statistics of particles in local
      quantum theories 264

     G. G. Emch: An algebraic approach to the theory of
      K-flows and K-entropy 315

     G. Lassner: Continuous representations of the test
      function algebra and the existence problem for
      quantum fields 297

     Y. Ohnuki and S. Kamefuchi: The locality condition
      in parafermi field theory 107

     G. Reents: The infrared problem and non-Fock-
      representations of the canonical commutation
      relations 293

     D. W. Robinson: Unbounded derivations of C* algebras 303

  (3) Spin, statistics and symmetry.

     S. Doplicher: The statistics of particles in local
      quantum theories 264

F. Gürsey: Algebraic methods and quark structure 189

Y. Ohnuki and S. Kamefuchi: The locality condition
    in parafermi field theory 107

(4) Structure of scattering amplitudes and related topics.

D. Iagolnitzer: Analyticity properties of scattering
    amplitudes: a review of some recent developments 1

T. Kawai and H. P. Stapp: Microlocal study of
    S-matrix singularity structure 38

T. Kinoshita and A. Ukawa: Mass singularities of
    Feynman amplitudes 55

K. Pohlmeyer: Large momentum behaviour of Feynman
    amplitudes 59

G. Sommer: Spectral dependence of the analyticity
    domain of local vertex functions 66

(5) Asymptotic behavior of scattering amplitude and light-
    cone singularity.

E. Brüning and P. Stichel: Asymptotics and light-cone
    singularities in quantum field theory 72

M. Ninomiya and K. Watanabe: Bound state nature and
    deep inelastic structure functions 111

K. Pohlmeyer: Large momentum behaviour of Feynman
    amplitudes 59

S. Schlieder: Structure of singularities of Wightman-
    distributions and the Wilson-Zimmermann-expansion
    respectively lightcone-expansion 85

B. Schroer: Conformal invariance in Minkowskian quantum
    field theory and global operator expansions 92

(6) Renormalization group approach.

Y. Fujii and Y. Takahashi: On the regularization in
    Callan-Symanzik equation 261

G. Jona-Lasinio: Critical behaviour of stationary
    random fields 326

K. Symanzik: On some massless superrenormalizable
    and non-renormalizable theories 102

(7) Renormalization theory.

A. A. Slavnov: Renormalization of supersymmetric gauge
    theories 214

O. I. Zavialov: Structure of renormalized Feynman
    amplitudes 256

(8) Gauge theory and broken symmetry.

J. Arafune, P. G. O. Freund and C. J. Goebel: Topology

of Higgs fields 240

A. Hosoya and J. Ishida: New exact solutions of the
classical Yang-Mills field equation 238

R. Jackiw: Quantization of non-linear waves 394

T. Maskawa and H. Nakajima: Spontaneous breaking of
chiral symmetry in a vector-gluon model 242

N. Nakanishi: Quantum field-theoretical approach to
spontaneously broken gauge invariance 245

N. Nakazawa and M. Yamada: Space-time approach to
anomalies in the Ward-Takahashi identities and
implications in physical processes 202

K. Nishijima: Dispersion approach to Ward-Takahashi
identities 205

H. Reeh: Reviews on axiomatic study of symmetry
breaking 249

A. A. Slavnov: Renormalization of supersymmetric
gauge theories 214

K. Yokoyama and Reijiro Kubo: Invariant gauge families
inherent in Abelian-gauge field theory 199

(9) Dual-resonance model.

Y. Chikashige and K. Kamimura: The canonical quantization of a relativistic string 187

K. Kikkawa: Field theory of relativistic strings 128

G. Konisi and T. Saito: Physical states satisfying
supergauge conditions in dual resonance models 184

M. Minami: Feynman propagators associated with the
Veneziano model 175

T. Yoneya: Dual string models and quantum gravity 180

(10) Phenomenological approaches.

K. Kinoshita: Field theoretical approach to composite
particle reactions 116

M. Ninomiya and K. Watanabe: Bound state nature and
deep inelastic structure functions 111

V. P. Shelest: The physical content of the statistical
bootstrap and high energy hadron interaction 114

(11) Miscellaneous topics.

C. DeWitt-Morette: Path integrals in Riemannian
manifolds 535

K. Sekine: Unitarity and asymptotic condition in a
model with dipole ghost 196

(B) <u>Statistical Mechanics</u>

  (1) Exact results (equilibrium).

     Masahisa Fukushima: Operator-valued-measure approach to spectra of two-dimensional classical harmonic lattices.. 528

     K. Ikeda: Distribution of zeros and the equation of state ... 520

     Y. Karaki: New theorems on algebraic equation and its application to statistical physics ... 524

     J. L. Lebowitz: Uniqueness, analyticity and decay properties of correlations in equilibrium systems ... 370

     D. Ruelle: Equilibrium statistical mechanics of one-dimensional classical lattice systems ... 449

     B. Souillard: Links between decay properties of correlations and analyticity of the pressure and correlation functions ... 497

  (2) Exact results (non-equilibrium).

     K. Hepp: Results and problems in irreversible statistical mechanics of open systems ... 138

     J. L. Lebowitz: Equilibrium states and ergodic properties of infinite systems ... 432

     T. Nishida and M. Mimura: Global solutions to the Broadwell's model of Boltzmann equation for a simple discrete velocity gas ... 408

     M. Suzuki: Thermodynamic limit of non-equilibrium systems —— extensive property, fluctuation and nonlinear relaxation ... 423

  (3) Renormalization group approach.

     C. Di Castro: Generalized Gell-Mann and Low group transformations: scaling variables,and cross-over effects ... 342

     G. Jona-Lasinio: Critical behaviour of stationary random fields ... 326

     Y. Yamazaki: Generalizations and applications of Callan-Symanzik equations to statistical mechanics in critical phenomena ... 349

  (4) Mathematical foundation (equilibrium).

     Masatoshi Fukushima: Asymptotic properties of the spectral distributions of disordered systems ... 224

     T. Izuyama: On the rigorous definition of superfluidity and superconductivity ... 353

     J. L. Lebowitz: Uniqueness, analyticity and decay

properties of correlations in equilibrium
systems 370

D. Ruelle: Equilibrium statistical mechanics of one-
dimensional classical lattice systems 449

G. L. Sewell: Equilibrium and metastable states of
classical systems 510

M. Takahashi: On the validity of collective variable
description of Bose systems 446

(5) Mathematical foundation (non-equilibrium).

G. G. Emch: An algebraic approach to the theory of
K-flows and K-entropy 315

H. Hasegawa: Variational principles for Markov processes
and Onsager principle 433

K. Hepp: Results and problems in irreversible
statistical mechanics of open systems 138

I. Kubo: The ergodicity of the motion of a particle
in a potential field 230

Ryogo Kubo: Relaxation and fluctuation of
macrovariables 274

J. L. Lebowitz: Equilibrium states and ergodic
properties of infinite systems 432

J. T. Lewis and J. V. Pulè: Dynamical theories of
Brownian motion 516

H. Mori: Scaling method for asymptotic evaluation
in nonequilibrium statistical mechanics 413

T. Niwa: Ergodicity of some simple model systems
of infinitely many particles 236

M. Suzuki: Thermodynamic limit of non-equilibrium
systems —— extensive property, fluctuation and
nonlinear relaxation 423

A. Verbeure: Linear response, stability, cluster
properties 504

(6) Exactly solvable models.

K. Hepp: Results and problems in irreversible
statistical mechanics of open systems 138

Y. Kuramoto: Self-entrainment of a population of
coupled non-linear oscillators 420

D. Ruelle: Equilibrium statistical mechanics of one-
dimensional classical lattice systems 449

M. Toda: Wave propagation in a non-linear lattice 387

J. Zittartz: Phase transitions of continuous order 330

(7) Phase transitions.
    C. Di Castro: Generalized Gell-Mann and Low group
       transformations: scaling variables and cross-
       over effects 342
    T. Izuyama: On the rigorous definition of superfluidity
       and superconductivity 353
    R. Jackiw: Symmetry restoration at finite temperature 319
    G. Jona-Lasinio: Critical behaviour of stationary
       random fields 326
    H. Nakano: Order of the phase transition related
       to the degeneracy and interaction with proposal
       for an exact problem 336
    T. Ogawa: The Heisenberg model on the Cayley tree 339
    G. L. Sewell: Equilibrium and metastable states of
       classical systems 510
    Y. Yamazaki: Generalizations and applications of
       Callan-Symanzik equations to statistical mechanics
       in critical phenomena 349
    J. Zittartz: Phase transitions of continuous order 330
(8) Methods in statistical physics.
    J. M. Combes: On the Born-Oppenheimer approximation 467
    I. Kubo: The ergodicity of the motion of a particle
       in a potential field 230
    M. E. Mayer: Models for relativistic statistical
       mechanics 532
    H. Mori: Scaling method for asymptotic evaluation in
       nonequilibrium statistical mechanics 413
    B. Mühlschlegel: Functional averages in statistical
       physics 437
(C) <u>Mathematics</u>
  (1) Probability theory and ergodic theory.
    C. DeWitt-Morette: Path integrals in Riemannian
       manifolds 535
    G. G. Emch: An algebraic approach to the theory of
       K-flows and K-entropy 315
    Masatoshi Fukushima: Asymptotic properties of the
       spectral distributions of disordered systems 224
    K. Itô: Stochastic calculus 218
    G. Jona-Lasinio: Critical behaviour of stationary
       random fields 326

I. Kubo: The ergodicity of the motion of a particle
in a potential field 230

J. T. Lewis and J. V. Pulè: Dynamical theories of
Brownian motion 516

M. Miyamoto: A remark to Harris's theorem on
percolation 228

T. Niwa: Ergodicity of some simple model systems of
infinitely many particles 236

(2) Analytic functions.

D. Iagolnitzer: Analyticity properties of scattering
amplitudes: a review of some recent developments 1

G. Sommer: Spectral dependence of the analyticity
domain of local vertex functions 66

V. S. Vladimirov: Holomorphic functions with a
nonnegative imaginary part in the future tube 281

(3) Hyperfunction theory.

M. Kashiwara: Micro-local calculus 30

T. Kawai and H. P. Stapp: Microlocal study of
S-matrix singularity structure 38

M. Morimoto: Convolutors for ultrahyperfunctions 49

M. Sato: Recent development in hyperfunction theory
and its application to physics. (Microlocal
analysis of S-matrices and related quantities.) 13

(4) Linear and non-linear equations.

R. Jackiw: Quantization of non-linear waves 394

Y. Kametaka: On the non-linear diffusion equation
of Kolmogorov-Petrovskii-Piskunov type 401

J.-L. Lions: Variational problems and free boundary
problems 356

S. Matsuura: On the propagation of support of solutions
to general systems of partial differential
equations 380

T. Nishida and M. Mimura: Global solutions to the
Broadwell's model of Boltzmann equation for a
simple discrete velocity gas 408

M. Toda: Wave propagation in a non-linear lattice 387

(5) Scattering theory.

T. Ikebe: A look at the development of spectral and
scattering theory in Japan 458

Y. Kato: Bound states and asymptotic fields in model
field theories 170

S. T. Kuroda: Abstract approaches to spectral and
scattering theory, contributions from Japan 472

K. Mochizuki: Decay and asymptotics for wave
equations with dissipative term 486

Y. Saitō: Eigenfunction expansions for differential
operators with operator-valued coefficients and
their applications to the Schrödinger operators
with long-range potentials 476

T. Sasakawa: Treatment of three-body problems in
coordinate space 491

L. E. Thomas: Asymptotic completeness in three-
particle quantum mechanical scattering 483

(6) Theory of operator algebras.

H. J. Borchers: Algebraic aspects of Wightman quantum
field theory 283

G. G. Emch: An algebraic approach to the theory of
K-flows and K-entropy 315

G. Lassner: Continuous representations/of the test
function algebra and the existence problem for
quantum fields 297

G. Reents: The infrared problem and non-Fock-
representations of the canonical commutation
relations 293

D. W. Robinson: Unbounded derivations of C* algebras 303

Lecture Notes in Physics

Bisher erschienen/Already published

Vol. 1: J. C. Erdmann, Wärmeleitung in Kristallen, theoretische Grundlagen und fortgeschrittenene experimentelle Methoden. 1969. DM 22,–

Vol. 2: K. Hepp, Théorie de la renormalisation. 1969. DM 20,–

Vol. 3: A. Martin, Scattering Theory: Unitarity, Analyticity and Crossing. 1969. DM 18,–

Vol. 4: G. Ludwig, Deutung des Begriffs physikalische Theorie und axiomatische Grundlegung der Hilbertraumstruktur der Quantenmechanik durch Hauptsätze des Messens. 1970. Vergriffen.

Vol. 5: M. Schaaf, The Reduction of the Product of Two Irreducible Unitary Representations of the Proper Orthochronous Quantummechanical Poincaré Group. 1970. DM 18,–

Vol. 6: Group Representations in Mathematics and Physics. Edited by V. Bargmann. 1970. DM 27,–

Vol. 7: R. Balescu, J. L. Lebowitz, I. Prigogine, P. Résibois, Z. W. Salsburg, Lectures in Statistical Physics. 1971. DM 20,–

Vol. 8: Proceedings of the Second International Conference on Numerical Methods in Fluid Dynamics. Edited by M. Holt. 1971. Out of print.

Vol. 9: D. W. Robinson, The Thermodynamic Pressure in Quantum Statistical Mechanics. 1971. DM 18,–

Vol. 10: J. M. Stewart, Non-Equilibrium Relativistic Kinetic Theory. 1971. DM 18,–

Vol. 11: O. Steinmann, Perturbation Expansions in Axiomatic Field Theory. 1971. DM 18,–

Vol. 12: Statistical Models and Turbulence. Edited by C. Van Atta and M. Rosenblatt. Reprint of the First Edition 1975. DM 28,–

Vol. 13: M. Ryan, Hamiltonian Cosmology. 1972. DM 20,–

Vol. 14: Methods of Local and Global Differential Geometry in General Relativity. Edited by D. Farnsworth, J. Fink, J. Porter and A. Thompson. 1972. DM 20,–

Vol. 15: M. Fierz, Vorlesungen zur Entwicklungsgeschichte der Mechanik. 1972. DM 18,–

Vol. 16: H.-O. Georgii, Phasenübergang 1. Art bei Gittergasmodellen. 1972. DM 20,–

Vol. 17: Strong Interaction Physics. Edited by W. Rühl and A. Vancura. 1973. DM 32,–

Vol. 18: Proceedings of the Third International Conference on Numerical Methods in Fluid Mechanics, Vol. I. Edited by H. Cabannes and R. Temam. 1973. DM 20,–

Vol. 19: Proceedings of the Third International Conference on Numerical Methods in Fluid Mechanics, Vol. II. Edited by H. Cabannes and R. Temam. 1973. DM 29,–

Vol. 20: Statistical Mechanics and Mathematical Problems. Edited by A. Lenard. 1973. DM 24,–

Vol. 21: Optimization and Stability Problems in Continuum Mechanics. Edited by P. K. C. Wang. 1973. DM 18,–

Vol. 22: Proceedings of the Europhysics Study Conference on Intermediate Processes in Nuclear Reactions. Edited by N. Cindro, P. Kulišić and Th. Mayer-Kuckuk. 1973. DM 29,–

Vol. 23: Nuclear Structure Physics. Proceedings of the Minerva Symposium on Physics. Edited by U. Smilansky, I. Talmi, and H. A. Weidenmüller. 1973. DM 29,–

Vol. 24: R. F. Snipes, Statistical Mechanical Theory of the Electrolytic Transport of Non-electrolytes. 1973. DM 22,–

Vol. 25: Constructive Quantum Field Theory. The 1973 "Ettore Majorana" International School of Mathematical Physics. Edited by G. Velo and A. Wightman. 1973. DM 29,–

Vol. 26: A. Hubert, Theorie der Domänenwände in geordneten Medien. 1974. DM 28,–

Vol. 27: R. Kh. Zeytounian, Notes sur les Ecoulements Rotationnels de Fluides Parfaits. 1974. DM 28,–

Vol. 28: Lectures in Statistical Physics. Edited by W. C. Schieve and J. S. Turner. 1974. DM 24,–

Vol. 29: Foundations of Quantum Mechanics and Ordered Linear Spaces. Advanced Study Institute Held in Marburg 1973. Edited by A. Hartkämper and H. Neumann. 1974. DM 26,–

Vol. 30: Polarization Nuclear Physics. Proceedings of a Meeting held at Ebermannstadt October 1–5, 1973. Edited by D. Fick. 1974. DM 24,–

Vol. 31: Transport Phenomena. Sitges International School of Statistical Mechanics, June 1974. Edited by G. Kirczenow and J. Marro. DM 39,–

Vol. 32: Particles, Quantum Fields and Statistical Mechanics. Proceedings of the 1973 Summer Institute in Theoretical Physics held at the Centro de Investigacion y de Estudios Avanzados del IPN – Mexico City. Edited by M. Alexanian and A. Zepeda. 1975. DM 18,–

Vol. 33: Classical and Quantum Mechanical Aspects of Heavy Ion Collisions. Symposium held at the Max-Planck-Institut für Kernphysik, Heidelberg, Germany, October 2–5, 1974. Edited by H. L. Harney, P. Braun-Munzinger and C. K. Gelbke. 1975. DM 28,–

Vol. 34: One-Dimensional Conductors, GPS Summer School Proceedings, 1974. Edited by H. G. Schuster. 1975. DM 32,–

Vol. 35: Proceedings of the Fourth International Conference on Numerical Methods in Fluid Dynamics. June 24–28, 1974, University of Colorado. Edited by R. D. Richtmyer. 1975. DM 37,–

Vol. 36: R. Gatignol, Théorie Cinétique des Gaz à Répartition Discrète de Vitesses. 1975. DM 23,–

Vol. 37: Trends in Elementary Particle Theory. Proceedings 1974. Edited by H. Rollnik and K. Dietz. 1975. DM 37,–

Vol. 38: Dynamical Systems, Theory and Applications. Proceedings 1974. Edited by J. Moser. 1975. DM 46,–

Vol. 39: International Symposium on Mathematical Problems in Theoretical Physics. Proceedings 1975. Edited by H. Araki. 1975. DM 44,–